IFIP Advances in Information and Communication Technology 312

IFIP – The International Federation for Information Processing

IFIP was founded in 1960 under the auspices of UNESCO, following the First World Computer Congress held in Paris the previous year. An umbrella organization for societies working in information processing, IFIP's aim is two-fold: to support information processing within its member countries and to encourage technology transfer to developing nations. As its mission statement clearly states,

> IFIP's mission is to be the leading, truly international, apolitical organization which encourages and assists in the development, exploitation and application of information technology for the benefit of all people.

IFIP is a non-profitmaking organization, run almost solely by 2500 volunteers. It operates through a number of technical committees, which organize events and publications. IFIP's events range from an international congress to local seminars, but the most important are:

- The IFIP World Computer Congress, held every second year;
- Open conferences;
- Working conferences.

The flagship event is the IFIP World Computer Congress, at which both invited and contributed papers are presented. Contributed papers are rigorously refereed and the rejection rate is high.

As with the Congress, participation in the open conferences is open to all and papers may be invited or submitted. Again, submitted papers are stringently refereed.

The working conferences are structured differently. They are usually run by a working group and attendance is small and by invitation only. Their purpose is to create an atmosphere conducive to innovation and development. Refereeing is less rigorous and papers are subjected to extensive group discussion.

Publications arising from IFIP events vary. The papers presented at the IFIP World Computer Congress and at open conferences are published as conference proceedings, while the results of the working conferences are often published as collections of selected and edited papers.

Any national society whose primary activity is in information may apply to become a full member of IFIP, although full membership is restricted to one society per country. Full members are entitled to vote at the annual General Assembly, National societies preferring a less committed involvement may apply for associate or corresponding membership. Associate members enjoy the same benefits as full members, but without voting rights. Corresponding members are not represented in IFIP bodies. Affiliated membership is open to non-national societies, and individual and honorary membership schemes are also offered.

Adam Korytowski   Kazimierz Malanowski
Wojciech Mitkowski   Maciej Szymkat (Eds.)

System Modeling and Optimization

23rd IFIP TC 7 Conference
Cracow, Poland, July 23-27, 2007
Revised Selected Papers

 Springer

Volume Editors

Adam Korytowski
Wojciech Mitkowski
Maciej Szymkat
AGH University of Science and Technology
Chair of Automatics, Al. Mickiewicza 30, 30-059 Kraków, Poland
E-mail: {akor, wmi, msz}@ia.agh.edu.pl

Kazimierz Malanowski
Polish Academy of Sciences
ul. Newelska 6, 01-447 Warszawa, Poland
E-mail: kmalan@ibspan.waw.pl

CR Subject Classification (1998): C.0, C.4, C.2, G.1.6, D.4.8, I.6

ISSN   1868-4238
ISBN-13   978-3-642-26093-3 Springer Berlin Heidelberg New York

springer.com

© IFIP International Federation for Information Processing 2010
Softcover reprint of the hardcover 1st edition 2010

Typesetting: Camera-ready by author, data conversion by Scientific Publishing Services, Chennai, India
Printed on acid-free paper   SPIN: 12768561   06/3180   5 4 3 2 1 0

Preface

This book constitutes a collection of extended versions of papers presented at the 23rd IFIP TC7 Conference on System Modeling and Optimization, which was held in Cracow, Poland, on July 23–27, 2007. It contains 7 plenary and 22 contributed articles, the latter selected via a peer reviewing process. Most of the papers are concerned with optimization and optimal control. Some of them deal with practical issues, e.g., performance-based design for seismic risk reduction, or evolutionary optimization in structural engineering. Many contributions concern optimization of infinite-dimensional systems, ranging from a general overview of the variational analysis, through optimization and sensitivity analysis of PDE systems, to optimal control of neutral systems. A significant group of papers is devoted to shape analysis and optimization. Sufficient optimality conditions for ODE problems, and stochastic control methods applied to mathematical finance, are also investigated. The remaining papers are on mathematical programming, modeling, and information technology.

The conference was the 23rd event in the series of such meetings biennially organized under the auspices of the Seventh Technical Committee "Systems Modeling and Optimization" of the International Federation for Information Processing (IFIP TC7). It was attended by over 200 participants from 33 countries and 5 continents, who presented talks on a wide spectrum of topics covering all areas of interest of the IFIP TC7, including optimization theory, algorithms of nonlinear optimization, discrete optimization, stochastic optimization, distributed parameter systems: modeling and optimization, optimal control, stochastic control and financial mathematics, structural design, stability and sensitivity analysis in optimal control, and complex systems.

The conference was organized by the AGH University of Science and Technology in Cracow, one of the greatest technical universities in Poland. The organization was mainly carried out at the Faculty of Electrical Engineering, Automatics, Computer Science and Electronics.

We would like to thank the members of the Program Committee of the conference for their help and valuable advice. In particular we are grateful to Irena Lasiecka, the chairperson of the IFIP TC7, who also chaired the PC of the conference. Thanks are due to G. Augusti, L.T. Biegler, H. Furuta, M. Grötschel, A. Lewis, H. Maurer, B.S. Mordukhovich, S. Scholtes, V. Schulz, Ł. Stettner, and J.-P. Zolésio for delivering plenary talks, and to N.U. Ahmed, G. Avalos, L.T. Biegler, T. Burczyński, J. Cagnol, M. Delfour, D. Dentcheva, D. Dolk, J. Granat, J. Henry, I. Lasiecka, U. Ledzewicz, T. Lewiński, M. Makowski, S. Migórski, M. Pietrzyk, H. Schättler, V. Schulz, H.-J. Sebastian, J. Sokołowski, Ł. Stettner, A. Świerniak, R. Tichatschke, F. Tröltsch, S. Volkwein, J. Zabczyk, J.-P. Zolésio, and A. Żochowski, who organized the invited sessions, thus greatly contributing to the success of the conference. We would also like to thank all the reviewers, who refereed the abstracts of talks delivered at the conference as well as the articles published in this book.

The organization of the conference was largely dependent on the financial support of the AGH UST, which is gratefully acknowledged. The success of the conference and the publication of this book would not have been possible without the help and support of many individuals. We owe sincere thanks to them all, and in particular to Antoni Tajduś, Rector of the AGH UST, Tomasz Szmuc, the then Dean of the Faculty of Electrical Engineering of AGH, and Karol Musioł, Rector of the Jagiellonian University. Our special thanks go to Anna Sury, whose help in all organizational matters was invaluable. We also thank Grzegorz J. Nalepa and his co-workers for their help in the preparation of the camera ready form of this volume.

June 2009

Adam Korytowski
Kazimierz Malanowski
Wojciech Mitkowski
Maciej Szymkat

Table of Contents

Part I: Plenary Papers

Part II: Regular Papers

Part I

Plenary Papers

Performance-Based Design as a Strategy for Risk Reduction: Application to Seismic Risk Assessment of Composite Steel-Concrete Frames

Giuliano Augusti and Marcello Ciampoli

Department of Structural and Geotechnical Engineering, Sapienza Università di Roma
giuliano.augusti@uniroma1.it, marcello.ciampoli@uniroma1.it

Abstract. Performance-based design is an efficient strategy for assessing and reducing the risk that a construction violates some performance requirement. In this paper, a procedure for performance-based assessment of seismic risk is illustrated with reference to a composite steel-concrete frame structure. Such risk is conventionally evaluated in a simplified formulation, i.e. as the mean annual frequency of exceeding a threshold level of damage in any significant structural element. The procedure is applied to evaluate the site seismic hazard, the structural damage, the corresponding capacity, and finally the seismic risk of a plane frame, extrapolated from a 3-D structure that was subjected to experimental tests at the ELSA-JRC Laboratory in Ispra, Italy. Specific attention is given to the choice of the intensity and damage measures for use in performance-based seismic risk assessment of composite steel-concrete frames.

1 Introduction

By definition, "Performance-Based Design" (usually referred to by the acronym PBD) requires the satisfaction of the relevant performance requirements with a sufficiently high probability throughout the lifetime of an engineering system.

This definition might appear a truism, since design is always addressed to fulfill one or more performance objectives. Indeed, PBD is a new concept not because it refers to performance objectives, but in the way it aims at fulfilling these objectives.

As a matter of fact, up to a few years ago, the fulfillment of performance objectives was based on engineering experience and practice; PBD is instead a design philosophy specifically constructed in order to reach rationally and with a given reliability the chosen objectives [14,15,2,27,28,26].

The basic document for all Structural Eurocodes, i.e. EN 1990: "Eurocode - Basis for Structural Design" (CEN, 2002), begins with the statement: "A structure shall be designed and executed in such a way that it will, during its intended life, with appropriate degrees of reliability and in an economical way: (i) sustain all actions and influences likely to occur during execution and use, and (ii) remain fit for the use for which it is required". Thus, in principle the Code recognizes correctly that the fulfillment of the chosen objectives cannot be guaranteed in deterministic terms, but rather tackled in a probabilistic context ("appropriate degrees of reliability"); recognizes also that such fulfillment is conditioned by cost/benefit problems ("in an economical way"). However,

A. Korytowski et al. (Eds.): System Modeling and Optimization, IFIP AICT 312, pp. 3–20, 2009.

the Structural Eurocodes prescribe practical design rules that have little "performance-based" character, and in actual fact make them prescriptive codes; the same happens with the 2005 Italian Building Code [24].

A general definition of PBD, derived from the specific definition given by SEAOC for Performance Based Seismic Design [29], is the following: "Performance Based Engineering consists of those actions, including site selection, development of conceptual, preliminary and final design, construction and maintenance of a structure over its life, to ensure that it is capable of predictable and reliable performances. The life of a structure includes also its decommissioning and/or demolition. Each of the above actions can have significant impact on the ability of the structure to reliably reach desired performances".

At present, PBD is still far away from providing a consistent set of design rules, and will probably remain for some time only a philosophy on the verge of utopia. However, this should not be regarded as a negative aspect: as a matter of fact, the very scope of engineering is to conceive, design and build facilities that are better, more effective and economical than present ones. This most probably will never be fully achieved and only partial steps will be possible, but such a high objective cannot be pursued at without an ideal vision: and PBD appears to be at the moment the best [21].

The difficulties encountered in applying the PBD philosophy are put in evidence by the everyday experience of the engineers and their limited capacity of governing the choices in the several stages of the facility lifetime: choice of the location, conceptual design, preliminary and final design, construction, maintenance, decommissioning and/or demolition. Already the choice of the site of a facility is often not the result of a rational strategy but of many casual chances. Conceptual design includes aesthetic, functional and structural aspects: these aspects are interrelated and should be tackled in a systematic way, aimed at fulfilling the performance objectives; however, much too often the engineer's role is limited to the choice of the load-carrying system. Preliminary and final design are the stages in which the influence of the engineer is highest: however, he often cannot control choices essential for the performance objectives, such as construction details, heating and electrical systems, etc. Similar considerations apply to other stages, from construction to demolition.

According to the International Code Council [17], the performance capacities of a structure can be verified in three ways: by design computations; by experimental tests; by the respect of appropriate code guidelines (the last being the actual way in which most regulations, and specifically the Eurocodes, are implemented).

Thus, rather than being an alternative to prescriptive codes, PBD is in reality an approach that does not exclude guidelines, but is aimed directly at the rational achievement of well specified performance objectives and/or their optimization: once the performance objectives are defined, the designer can limit himself to demonstrate that he has respected the code guidelines aimed at fulfilling those objectives [6].

PBD has been originally formulated in a systematic way for structures subject to seismic hazard: but applications to other hazards have also been implemented.

An example is Performance-Based Blast Engineering that had a significant development in the US after the terrorist attacks to World Trade Center in 1993 and 2001, to Murrah Building in 1995 and to Pentagon in 2001 [33]. To take into account in the design the possible effects of explosions is now considered an added value by building owners and insurance companies.

A field of great potential for developments of PBD procedure is Wind Engineering: the first steps in this direction go back to an Italian research project (PERBACCO: 2003/2005), in which the expression "Performance-based Wind Engineering" (PBWE) was coined [4]. A recent example can be found in [30].

Also this paper will be limited to seismic design. In what follows, after a brief presentation of the general framework, the application of PBD to the evaluation of the seismic risk (although conventionally defined) of a composite steel-concrete frame will be illustrated in detail.

2 Formulation of the Problem

Concerning the demonstration of the capacity of a structure to fulfill the stated performance objectives under seismic hazard, one can refer to the equation proposed by the Pacific Earthquake Research Center (PEER) for PB seismic design [11].

In quantitative terms, the performance requirement is identified with an acceptable value of the mean annual frequency $\lambda(LS)$ of exceeding a limit state LS of the structure, assessed combining the site seismic hazard and the structural vulnerability by the Total Probability Theorem, through the relation:

$$\lambda(LS) = \iint G(LS|DM) \cdot dG(DM|IM) \cdot d\lambda(IM) \tag{1}$$

where:

1. IM is an appropriate "intensity measure" of the earthquake;
2. $\lambda(IM)$ is the mean annual occurrence rate in the considered site of a quake of intensity equal or larger than IM (i.e. the measure of the seismic hazard in the relevant site);
3. DM is a parameter obtained from the analysis of the structural response and measuring the structural damage;
4. G(DM|IM) is the conditional probability of overcoming DM for a given IM;
5. LS is the considered "limit state";
6. G(LS|DM) is the conditional probability of overcoming LS for a given DM.

Each of the quantities in Eq. 1 is highly uncertain: thus, their choice and probabilistic characterization is essential for the reliability of the results and the efficiency of the approach in practical application.

2.1 The Site-Seismic Hazard

In Eq. 1, the site seismic hazard is referred to a single parameter, usually an intensity measure IM, and is measured by the average frequency $\lambda(IM)$ with which each IM value is exceeded every year, while the seismic vulnerability of the construction is identified with the probability of overcoming each value of a structural damage parameter DM for any given IM.

The ground motion intensity measure, IM, serves as an interface between the characterization of the seismic hazard and the assessment of the structural behavior.

In turn, the limit condition is described by an appropriate limit state, LS, corresponding to an appropriate performance requirement, whose achievement is probabilistically related to DM: the choice of the limit state and its relation with DM depend on several factors, such as the construction type, the structural model, the method of analysis.

As concerns the characterization of the seismic hazard, Eq. 1 is significant only insofar as:

(i) the seismic hazard, hence the damage potential of the earthquake, is consistently defined by the probabilistic characterization of the chosen IM;

(ii) in the relevant site, the frequency curve λ(IM) can be evaluated by a probabilistic seismic hazard analysis;

(iii) the number of calculations that are needed to evaluate the structural response, taking into account the uncertainties of both the seismic motion and the structural properties, is not excessive.

These conditions are met if the parameter IM satisfies, respectively, the requirements of sufficiency, hazard valuableness and efficiency, as defined in [32] and discussed in [19], [13].

According to these definitions, an intensity measure IM is sufficient if DM for a given IM can be evaluated independently on any other property of the seismic motion, like magnitude of the generating event, source-to-site distance, fault type, soil type, and directivity effects. The adoption of a sufficient IM (i) permits an unbiased evaluation of λ(LS) by Eq. 1, (ii) makes it not essential a careful selection of records to be used in non-linear dynamic analyses to take into account the record-to-record variability, (iii) legitimizes scaling the accelerograms [32], (iv) allows decoupling seismic hazard and structural analysis, and (iv) gives an exhaustive description (at least in a statistical sense) of the damaging power of the seismic event, thus overcoming the need of adopting a vector of parameters.

The hazard valuableness depends on the possibility of describing the source activity and the source-to-site attenuation law in terms of the chosen IM.

Finally, an IM is efficient if the uncertainties in the structural response, evaluated by considering a set of ground motion records scaled to different values of IM, are not excessive. This last condition can be quantified by limiting the value of the dispersion $\beta_{DM|IM}$ of DM for each value of IM; since in Eq. 1 the structural response, thence the function G(DM|IM) are evaluated by nonlinear dynamic analyses, the use of a more efficient IM implies a reduction in the number of analyses necessary to estimate λ(LS) with the same confidence level.

From the definitions given above, it appears that sufficiency is an essential property of an IM. Once the sufficiency of a candidate IM is established, efficiency and hazard valuableness are two relative criteria that can be used to select a candidate IM among the alternative ones. However, the choice of IM depends strongly on the structural typology and the damage measure DM.

2.2 The Structural Vulnerability

Also the choice of the proper DM, that completely characterize the structural response, should be carried out by checking both sufficiency and efficiency.

The DM sufficiency is related to the possibility of describing the damage state of the structure by the candidate DM, and of relating the seismic risk of the construction to the mean occurrence rate of overcoming each given value of DM. However, it may be really difficult to justify this choice, even only in a statistical sense.

The choice of DM is then related (i) to the relative efficiency of the candidate parameters, that can be measured (as for IM) by the dispersion $\beta_{DM|IM}$ of the values of DM resulting from nonlinear dynamic analyses for each value of IM, and (ii) to the possibility of estimating, in a consistent way, the type and parameters of the conditional distribution G(DM|IM).

In the present context, the structural response is evaluated by nonlinear incremental dynamic analysis (IDA). IDA [32] is a parametric nonlinear analysis that involves subjecting the structural model to a certain number of ground motion time histories, each of them scaled to multiple levels of intensity, that is, to a seismic action characterized by different occurrence rate, intensity and frequency content.

The use of IDA has many advantages: indeed, it describes the evolution of the structural response in the whole investigated range of seismic intensities and gives a visual description of the effects of the record-to-record variability.

The results of IDAs are given as plots of the structural response, that is, of the assumed damage measure, DM, as a function of the ground motion intensity measure, IM. Each IDA plot depends on the adopted DM and on the specific record, but not on the considered IM: IDA plots in terms of different IMs are easily obtained by scaling the ordinate of each point on the curve according to the relation between the originally adopted IM and the candidate IM.

2.3 Seismic Risk Assessment

Once the expressions of the relevant probability functions are defined, Eq. 1 can be evaluated by a numerical procedure.

However, if some simplifying assumptions hold, it is possible to attribute an analytic expression to λ(LS), as described in detail in [19]. These assumptions are crucial and may introduce considerable epistemic uncertainties in the solution to Eq. 1, that cannot be quantified; therefore their contribution is disregarded in what follows.

As suggested by [23], let us assume that the site-seismic hazard λ(IM) is described by a curve that is approximated, in the region of interest, by the power-law relationship:

$$\lambda\,(\mathrm{IM}) = P\,[\mathrm{IM} \geq x] = k_0 \cdot (x)^{-k} \tag{2}$$

where k_0 and k are the parameters that define the shape of this hazard curve.

Observations of demand values are obtained from the results of incremental dynamic analyses, performed for various levels of IM. Considering any set of ground motion records applied to the structure, there will be a certain variability in the values of DM for each given level of IM. The record-to-record variability is a consequence of the variability of the generating events; however, also the uncertainties characterizing the structural parameters can be taken into account.

In order to build a simplified probabilistic model of the demand (the damage to the structure), DM can be written as:

$$DM = \eta_{DM|IM} \cdot \varepsilon \tag{3}$$

that is, as the product of the conditional median $\eta_{DM|IM}$ of DM given IM and a log-normal random variable ε; the median and the standard deviation of ln ε are set equal, respectively, to 1 and $\beta_{DM|IM}$, that in turn depends on the level of IM.

Let us introduce an approximate functional relationship between each IM value and the median $\eta_{DM|IM}$ of the demand parameter DM, based on a regression of the results of time-history analyses. If the conditional median demand is approximated as a power-law function of IM, the relationship is given by:

$$\eta_{DM|IM} = a \cdot IM^b \tag{4}$$

The conditional demand is thus the random variable:

$$DM = a \cdot IM^b \cdot \varepsilon \tag{5}$$

The standard deviation of the natural logarithm of DM given IM is equal to:

$$\sigma_{\ln DM|IM}(IM) = \beta_{DM|IM} \tag{6}$$

According to this formulation, it is possible to obtain the demand hazard, that is, the mean annual frequency (i.e. the annual probability if some simplifying assumptions hold) of exceeding a certain level of DM, dm, for each given level of IM, im, by the relationship:

$$\lambda(DM) = \int_0^\infty P[DM > dm | IM = im] \cdot |d\lambda(IM)| = k_0 \left(\frac{DM}{a}\right)^{\frac{-k}{b}} \cdot e^{\frac{1}{2} \cdot \frac{k^2}{b^2} \cdot \beta_{DM|IM}^2} \tag{7}$$

According to Eq. 7 the demand hazard is given by the seismic hazard curve 2 evaluated at the IM value corresponding to the DM value times a factor related to the dispersion in demand for a given level of IM.

In order to derive the limit state annual frequency $\lambda(LS)$, that is a measure of the probability that the damage to the structure exceeds a threshold corresponding to the relevant limit state, which is an uncertain quantity representing the structural capacity C, let us finally assume that the demand and the capacity are independent random variables and that the capacity is a lognormal random variable itself, with the following parameters: median$(C) = \eta_C$, $\sigma_{\ln(C)} = \beta_C$. As:

$$\lambda(IM(\eta_C)) = k_0 \left(\frac{\eta_C}{a}\right)^{\frac{-k}{b}} \tag{8}$$

gives the value of the seismic hazard for the IM value corresponding to the median capacity η_C, after some rearrangements [19], it is straightforward to obtain:

$$\lambda(LS) = k_0 \left(\frac{\eta_C}{a}\right)^{\frac{-k}{b}} \cdot e^{\frac{1}{2} \cdot \frac{k^2}{b^2} \cdot \beta_{DM|IM}^2} \cdot e^{\frac{1}{2} \cdot \frac{k^2}{b^2} \cdot \beta_C^2} \tag{9}$$

In Eq. 9, the limit state probability is expressed as the seismic hazard curve evaluated at the IM value corresponding to η_C (the median capacity) multiplied by two coefficients that take into account respectively the randomness in DM for a given IM (the demand factor) and the randomness in the capacity C itself (the capacity factor).

3   An Application

In this paper, the procedure summarized in Sect. 2 is applied to assess the seismic risk of the composite steel-concrete planar frame of Fig. 1, that is part of a spatial frame that has been subjected to pseudo-dynamic seismic tests [7].

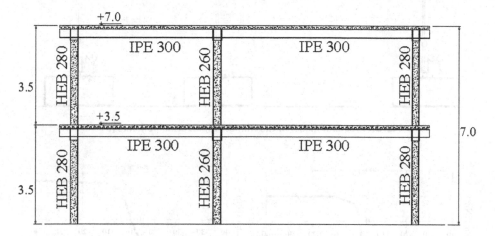

Fig. 1. Plane frame from a structure subject to pseudo-dynamic seismic tests at ELSA Laboratory, JRC, Ispra (Research project: ECOLEADER HPR-CT-1999-00059; ECSC 7210-PR-250) [7]

The structural model is shown in Fig. 2. The beam-column joints have been simulated as assemblages of translational and rotational nonlinear springs; their parameters have been calibrated by the results of experiments carried out on the whole spatial frame [5]. The moment-rotation relationships of the rotational springs for external and internal joints are plotted in Fig. 3. The nonlinear dynamic incremental analyses have been carried out by IDARC2D [18].

3.1   Choice of IM

As stated in Sect. 2.1, sufficiency and efficiency are the measures of the IM capability to eliminate the bias in formulating Eq. 1 and to reduce the variability of the structural response evaluated by IDA.

The most appropriate IM has been chosen by evaluating the sufficiency of various IMs, with reference also to other example cases of composite steel-concrete plane frames similar to the one represented in Fig. 1, and by choosing the most efficient IM among the sufficient ones.

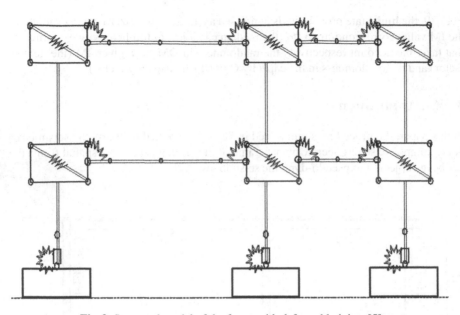

Fig. 2. Structural model of the frame with deformable joints [5]

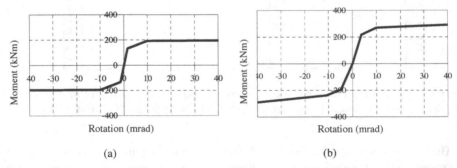

(a)   (b)

Fig. 3. Moment–rotation relationships for the external (a) and internal (b) joints derived from the experimental tests

The results illustrated in the following are specifically referred to the frame in Fig. 1; similar conclusions have been drawn by considering the other example cases [8].

After a series of preliminary analyses, presented in detail in [13], the selected candidate IMs were:

a) The peak ground acceleration, PGA.
b) The peak ground velocity, PGV.
c) The Housner Intensity, I_H, given by [16]:

$$I_H = \int_{0.1}^{2.5} S_V(T, \xi)\, dT \qquad (10)$$

where: $S_V(\cdot)$ is the pseudo-spectral velocity; T is the natural period; ξ is the damping ratio.

d) The spectral intensity, $Sa(T_1)$, that is, the ξ damped spectral pseudo-acceleration at the fundamental period T_1 of the structure.

e) The two-parameter scalar IM proposed by [10], SaC, and defined as:

$$SaC = Sa(T_1)\left(\frac{Sa(c \cdot T_1)}{Sa(T_1)}\right)^{\alpha} \tag{11}$$

where: $Sa(\cdot)$ is the ordinate of the pseudo-acceleration response spectrum; c and α are parameters that are usually derived from an ad-hoc calibration. In numerical calculations, α has been taken equal to 0.5; c has been taken equal either to 2, as suggested in [10], to take into account the effect of period lengthening due to nonlinear behavior, or to T_2/T_1 (where T_2 is the period of the second mode of vibration of the considered structure). Both alternatives for c have been considered because the latter, as suggested in [13], led to a more efficient IM in case of slender structures whose behavior is influenced by higher modes of vibration.

f) The spectral pseudo-acceleration averaged over a period interval [20], defined as:

$$Sa_{av} = \frac{1}{b' - a'}\int_{a'}^{b'} Sa(T)\,dT. \tag{12}$$

As suggested in [20], in numerical calculations the following alternative pairs have been introduced: $a' = T_1$, $b' = 1.5T_1$, and: $a' = T_2$, $b' = T_1$.

As one of the objectives of this study is the selection of appropriate IM and DM for composite steel-concrete frames, the structural response has been evaluated by using as input 30 accelerograms obtained from the PEER data base [http://peer.berkeley.edu/smcat], but not selected on the basis of the Probabilistic Seismic Hazard Analysis of any specific site; the accelerograms have been scaled to different intensity levels, until the collapse of the frame was reached, it being identified by the lack of convergence of the numerical solution (that is, by global dynamic instability). Records with potential near-source effects, like directivity, have been discarded; no other specific criteria have been adopted in selecting the records.

The selected records had moment magnitude M between 5.5 and 7.0 and source-to-site distance R between 13 km and 60 km.

The sufficiency of each candidate IM has been evaluated by a slightly refined version of the procedure based on the analysis of a set of linear regressions together with tests of hypothesis, proposed in [28], [19].

The significant ground motion characteristics are those included in the assumed attenuation law (in the example case, the law by [1]), namely, the moment magnitude M, the source-to-site distance R, and the error term ε describing the model uncertainty.

As the sufficiency of IM depends on the structural typology and the damage measure, four different DMs have been considered:

I. The maximum interstory drift ratio, MIDR.
II. The global interstory drift ratio or roof drift angle, RDA.
III. The peak floor (absolute) acceleration, PFA.
IV. The peak floor (relative) velocity, PFV.

MIDR is commonly adopted to describe the level of seismic damage, as it is relatively easy to establish a correlation between its values and the level of damage to both structural and non structural components. RDA is a global indicator, strictly related to the first one. PFA and PFV can be considered as consistent indicators of the structural response and of the damage to the building content [9].

To evaluate the sufficiency of the candidate IMs, residual-residual plots have been obtained for each IM, considering the alternative DMs. These plots indicate the significance of each additional regression variable (M, R and ε) by indicating whether a second variable introduced in the model of the site-seismic hazard significantly improves the prediction of the structural response.

The plots, illustrated in detail in [8] are constructed by performing a regression of the dependent variable (the selected DM) versus the first independent variable (the candidate IM); then by performing the regressions of each secondary independent variable (M, R or ε) on the first variable; finally, by plotting the residuals of the primary regression against the residuals of each secondary regression.

By observing in these residual-residual plots a statistically significant trend between the two sets of residuals, the potential dependence of the dependent variable on the secondary variable can be investigated.

The trend is quantified by the slope b* of the regression line in each residual-residual plot; if b* is "different" from 0, the residuals of the primary regression are correlated to the residuals of the secondary regression, and the candidate IM is not sufficient with respect to the considered secondary variable. A test of hypothesis based on the Student-t distribution has been applied to verify if the null hypothesis (b* = 0) can be rejected; this happens if the p^- value is smaller than 0.05.

As a case example, consider MIDR as the damage measure. The results of the tests of hypothesis referred to M and lnR are reported in Tables 1 and 2; comparable results have been obtained for the other DMs [8].

The a* values reported in Tables 1 and 2 are the intercepts of the regression lines in the residual-residual plots. These values are very small (they should be equal to 0), as the regression prediction always passes through the average values of the two sets of horizontal and vertical components, and these average values must be equal to 0 as the two sets of components are themselves residuals of linear regression.

Table 1. Check of sufficiency with respect to M

Magnitude	σ_0	σ	a*	b*	σ_b	p-value	Check of sufficiency
PGA	0.4367	0.3138	-1.33E-15	0.2124	0.1189	0.04250	No
PGV	0.4312	0.4146	-2.25E-15	-0.2228	0.1470	0.0704	Yes
I_H	0.4372	0.4285	1.82E-15	-0.1581	0.1474	0.1464	Yes
Sa(T$_1$)	0.2511	0.2309	-3.02E-15	0.1511	0.0667	0.0157	No
SaC(c =2, α = 0.5)	0.3162	0.3161	-2.31E-15	-0.0160	0.0980	0.4358	Yes
SaC(c = T$_2$/T$_1$, α = 0.5)	0.3793	0.3505	-3.60E-16	0.2191	0.2000	0.0685	Yes
Saav(T$_1$÷1.5T$_1$)	0.2702	0.2662	-5.98E-16	0.0725	0.0792	0.1838	Yes
Saav(T$_2$÷1.5T$_1$)	0.2519	0.2503	-8.39E-16	0.0462	0.0751	0.2715	Yes

Table 2. Check of sufficiency with respect to ln R

Source-to-site distance R	σ_0	σ	a*	b*	σ_b	p-value	Check of sufficiency
PGA	0.4367	0.4190	-1.01E-15	0.2661	0.1713	0.06582	Yes
PGV	0.4312	0.4306	-1.86E-15	-0.0522	0.1781	0.3858	Yes
I_H	0.4372	0.4367	1.77E-15	-0.0473	0.1805	0.3977	Yes
$Sa(T_1)$	0.2511	0.2268	-2.83E-15	0.2328	0.0926	0.0090	No
SaC(c =2, α = 0.5)	0.3162	0.3090	-2.58E-15	0.1458	0.1261	0.1288	Yes
SaC(c = T_2/T_1, α = 0.5)	0.3793	0.3508	-1.70E-16	0.3131	0.2437	0.1900	Yes
$Saav(T_1 \div 1.5T_1)$	0.2702	0.2608	-6.38E-16	0.1522	0.1065	0.0820	Yes
$Saav(T_2 \div 1.5T_1)$	0.2519	0.2467	-7.89E-16	0.1107	0.1008	0.1407	Yes

To refine the statistical estimate, the test of hypothesis has been repeated by characterizing the regression in the residual-residual plot by the Pearson's or Spearman's coefficients, and by adopting the procedure of non parametric regression proposed by [31]. The obtained results have confirmed those reported in Tables 1 and 2, as illustrated in [8].

In Tables 1 and 2, also the square root of the mean of squares (rms) σ_0 of the primary regression of the dependent variable (the considered DM) on the first variable (the candidate IM) and the rms σ of the residual-residual regression are reported. These two quantities can be used to measure the significance of each additional variable in reducing the variability of data, as the smaller the ratio of these two quantities, the more significant is the role of the secondary variable.

Moreover, if the standard deviation σ_b (also shown in Tables 1 and 2) is very large relative to the slope b* itself, the slope may not be statistically different from zero.

It can be concluded that PGA is not sufficient with respect to M: indeed, σ_0 is quite different from σ, and the p-value is significantly lower than 0.05. The same result applies to $Sa(T_1)$ if both M and R are considered.

In order to evaluate the efficiency of the three left IMs, that satisfy sufficiency, their values corresponding to each level of DM have been calculated, and the cross-sectional counted percentiles (e.g., $im^{16\%}$, $im^{50\%}$, $im^{84\%}$) estimated. If the distribution of IM given DM is assumed to be lognormal, the estimator of the dispersion of the values of IM is given by [19]:

$$\beta_{IM|DM} (dm) = \frac{1}{2} \ln \left[\frac{im^{84\%}}{im^{16\%}} \right] \tag{13}$$

The value of $\beta_{IM|DM}(dm)$ has thus been assumed as the measure of the efficiency of each candidate IM for the considered value of DM: indeed, the number of nonlinear dynamic analyses to be run in order to evaluate $\lambda(LS)$ in Eq. 1 with a given confidence level, increase if $\beta_{IM|DM}(dm)$ is higher.

It should also be noted that each limit state condition is usually associated to a threshold value of DM: the values of $\beta_{IM|DM}(dm)$ could also be used as the estimators of the dispersion of the structural capacity at the corresponding successive damage states [13].

In Table 3, the lowest values of $\beta_{IM|DM}$ are evidenced by *italic*, for some discrete values of the considered DMs, representative of low, moderate or significant damage

Table 3. Dispersion of the results for two different values of the various DMs and the sufficient IMs

DM	MIDR		RDA		PFA [m/s2]		PFV [m/s]	
IM	0.01	0.05	0.005	0.02	3.00	9.00	0.40	1.10
PGV	0.399	0.451	0.393	0.462	0.406	0.447	0.333	0.476
I_H	0.269	0.382	0.256	0.389	0.366	0.407	0.277	0.409
SaC(c=2; α = 0.5)	0.100	0.347	0.142	0.349	0.371	0.458	0.233	0.366
SaC(c=T_2/T_1; α = 0.5)	0.251	0.446	0.267	0.436	0.240	0.386	0.257	0.455
SaC(c = T_2/T_1, α = 0.5)	0.364	0.465	0.379	0.468	0.252	0.445	0.275	0.473
Saav($T_1 \div 1.5T_1$)	0.269	0.301	0.379	0.305	0.393	0.304	0.228	0.300
Saav($T_2 \div 1.5T_1$)	0.224	0.282	0.304	0.312	0.417	0.305	0.136	0.313

states. As demonstrated in [13] with reference to reinforced concrete frames, SaC (with parameters: c = 2; α = 0.5) gives acceptable results also when the damage is limited.

In the end, SaC has been chosen as the "best" intensity parameter, because: it is sufficient; it is efficient; it is easy to derive the hazard in the relevant site in terms of SaC, as it is a weighted average of two values of the spectral intensity Sa(T) and the attenuation laws in terms of Sa(T) are well known and reliable.

3.2   Choice of DM

In order to define the damage to the structure, various DMs have been compared in terms of their efficiency. [34] provide a detailed summary of the available DMs.

In addition to those considered in evaluating the efficiency of the candidate IMs (cf. Sect. 3.1), specific attention has been given to four indices representing a measure of the local damage to the structure. They are all able to take into account the nonlinear response and the dissipation of energy due to plastic deformations. Their choice is thus justified by the model adopted for evaluating the structural response, and specifically for the beam-column joints, that, as demonstrated by the experiments, are the most vulnerable components of the structural system. The additional candidate DMs were:

V. The ratio μ between the maximum required and the available curvature ductility in a set of control points.
VI. The maximum normalized dissipated energy, E, in the joints.
VII. The Park-Ang index, P&A, evaluated as a linear combination of the maximum ductility response and the total hysteretic energy dissipation [25] as:

$$P\&A = \frac{\mu_{max}}{\mu_{mon}} + \beta \frac{E_H}{M_y \cdot \theta_{mon}} \tag{14}$$

where: μ_{max} and μ_{mon} represent the maximum ductility response for a given earthquake history and the available ductility in the corresponding plastic zone; β is an empirical factor, that came out equal to 0.30 according to the calibration described below; E_H is the total hysteretic energy dissipated in the plastic zone for the given earthquake history; M_y is the yielding moment; θ_{mon} is the maximum admissible rotation. This damage measure, while referring to a local behavior, gives some information about the seismic damage, as it takes into account both the maximum required ductility and the dissipated energy, that is, the maximum value of the structural response and the degradation of strength and stiffness due to cyclic behavior. The factor β in Eq. 14 has been calibrated by submitting several examples of composite steel-concrete frames to IDA, and putting P&A = 1 when the collapse of each examined structure occurred in each nonlinear dynamic analysis. The final value $\beta = 0.30$ has been the average value of all obtained results.

VIII. The Banon-Veneziano index, B&V [3], given by:

$$B\&V = \sqrt{(D_1^*)^2 + (D_2^*)^2} \qquad (15)$$

where, indicating by x_{max} and x_y the maximum and the yield displacements respectively, and by F_y the yielding strength in the corresponding plastic zone,

$$D_1^* = D_1 - 1 = \mu_s - 1; \quad D_2^* = a' \cdot D_2^{b'} = a'[2(\mu_e - 1)]^{b'} \qquad (16)$$

$$D_1 = \frac{x_{max}}{x_y} = \mu_s; \quad D_2 = \frac{E_H}{0.5 F_y x_y} \qquad (17)$$

The parameters a' and b' have been set: $a' = 1.1$; $b' = 0.38$, as suggested by Banon and Veneziano themselves.

The efficiency of all DMs has been quantified by the dispersion $\beta_{DM|IM}$ of the IDA plots, evaluated by Eq. 13. It has been found that, due to the assumed structural model, the most significant DMs are the latter four [8].

Table 4. Mean value of $\beta_{DM|IM}$ obtained by assuming SaC as IM and averaging on three values of the corresponding DM

$\beta_{DM|IM}$	MIDR	P&A	B&V	μ	E
SaC	0.64	0.51	0.57	0.60	0.55

The values of $\beta_{DM|IM}$ for these four DMs and three values of SaC are compared in Fig. 4. In Table 4 the values of $\beta_{DM|IM}$ averaged over three different levels of damage (light, moderate, severe) are shown. They have been evaluated by considering the IDA plots reported in Fig. 5.

3.3 Site-Seismic Hazard

The seismic hazard curve of Fig. 6 has been elaborated from data available for Reggio Calabria (South Italy) in the Seismic Catalogue elaborated by the Italian National Group for Seismic Risk Reduction [22]; the parameters k_0 and k in Eq. 2 have been obtained by a regression in the logarithmic plane.

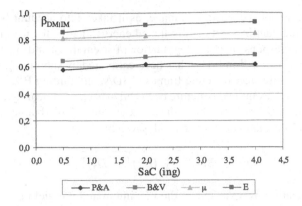

Fig. 4. Values of $\beta_{DM|IM}$ evaluated for the four local DMs in correspondence to three values of SaC

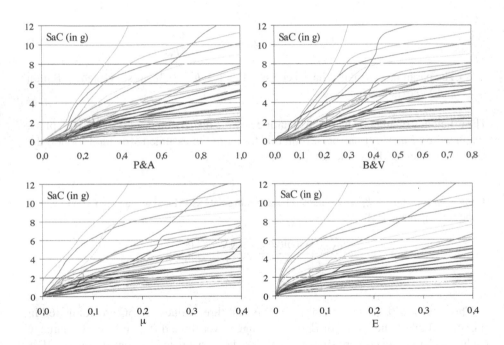

Fig. 5. IDA plots: the four local damage indices are assumed as DM; SaC is assumed as IM

3.4   Seismic Risk Assessment

The median values of P&A derived by IDAs are plotted in Fig. 7 as a function of SaC; the parameters a and b in Eq. 4 are obtained by the plotted regression, while the values of $\beta_{DM|IM}$ in Eq. 7 have been obtained by the plots in Fig. 5.

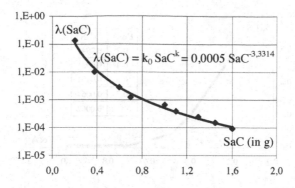

Fig. 6. Assumed seismic hazard curve

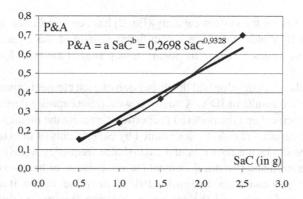

Fig. 7. Median values of P&A as a function of SaC

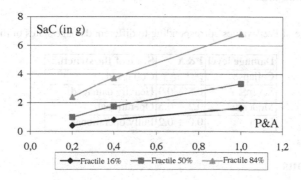

Fig. 8. Median and fractile values of βIM/DM (IM = SaC; DM=P&A)

In Fig. 8 the 50%, 16% and 84% percentiles of $\beta_{IM|DM}$ are reported for three values of P&A; by considering the median values, the median value η_C of the system capacity (Eq. 9) has been obtained.

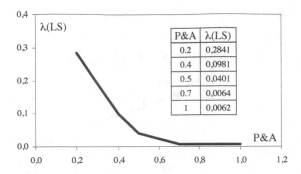

Fig. 9. Seismic risk of the frame: mean yearly frequency of exceeding the corresponding value of Park & Ang index in a significant structural element (beam, column or joint)

The dispersion β_C of the system capacity (Eq. 9) has been evaluated by considering all points of the IDA plots in which a significant change of the slope occurs. By a regression of these values (linear in the logarithmic plane), it has been found that $\beta_C = 0.395$.

In this paper, the median value and the dispersion of each element capacity have been determined from the results of IDA. Actually however, the capacity depends on the dimensions of the element and the material properties: therefore the median value and the dispersion of the capacity should be determined by experiments or field observations.

Finally, Fig. 9 shows the mean annual exceedance frequency $\lambda(LS)$ of the largest P&A index in a significant structural element (beam, column, or beam-column joint).

The curve in Fig. 9 quantifies the seismic risk of the frame: in fact, it allows to evaluate the seismic performance of the system considering appropriate threshold values of P&A, e.g. those reported in Table 4, and checking if the correspondent exceeding probability is admissible.

Table 5. Range of P&A values corresponding to different damage states of the structure

Damage level	P&A	State of the structure
Collapse	≥ 1	Collapsed
Severe	$0.5 \div 0.9$	Heavily damaged
Moderate	$0.2 \div 0.5$	Operational
Light	$0.0 \div 0.2$	Fully operational

4 Conclusions

As noted in the Introduction, PBD should cover all stages of the facility design, but at present is still more a philosophy than a consistent framework, and has been mostly developed for seismic design of structures. This is also the limitation of this paper, that intends only to show critically the application of PBD to a concrete example. In particular the need has been underlined of a careful selection of the parameters that characterize the seismic hazard and the structural response.

Acknowledgements. The numerical results presented in this paper were obtained in the framework of a "Research Project of National Interest" (PRIN) financially supported by the (Italian) Ministry for Education, University and Research MIUR (2003/2005). A preliminary version of this paper was presented orally at the "Henderson Colloquium" of the IABSE British Group (Cambridge, July 2005).

References

1. Abrahamson, N.A., Silva, W.J.: Empirical response spectral attenuation relations for shallow crustal earthquakes. Seismological Research Letters 68(1), 94–127 (1997)
2. Applied Technology Council (ATC): NEHRP guidelines for seismic rehabilitation of buildings: Report No. FEMA - 273, Federal Emergency Management Agency, WA, D.C. (1997)
3. Banon, H., Veneziano, D.: Seismic safety of reinforced concrete members and structures. Earthquake Engineering and Structural Dynamics 10, 179–193 (1982)
4. Bartoli, G., Ricciardelli, F., Saetta, A., Sepe, V. (eds.): Performance of Wind-Exposed Structures: Results of the PERBACCO Project. Firenze University Press (2006)
5. Braconi, A., Bursi, O., Ferrario, F., Salvatore, W.: Seismic design of beam-to-column connections for steel-concrete composite moment resisting frames. In: Fourth International Conference STESSA 2003: Behaviour of Steel Structures in Seismic Areas, Napoli, Italy (2003)
6. Burns, S., Bostrom, A.: Barriers to implementation of performance based engineering: lessons from the Tennessee 2003 international building code hearings. In: 2004 ANCER Annual Meeting: Networking of Young Earthquake Engineering Researchers and Professionals, Honolulu, Hawaii (2004)
7. Bursi, O.S., Molina, J., Salvatore, W., Taucer, F.: Dynamic characterization of a 3-d full scale steel-concrete composite building at elsa. Technical report, European Commission, Joint Research Centre, ELSA, EUR 21206 EN (2004)
8. Ciampoli, M.: Performance reliability of earthquake-resistant steel-concrete composite structures. In: Scientific research programmes of relevant national interest (2003), http://www2.ing.unipi.it/dis/sismoresistenti/national.html (in Italian)
9. Ciampoli, M., Augusti, G.: Vulnerabilità sismica degli oggetti esibiti nei musei: interventi per la sua riduzione. In: Vulnerabilità dei beni archeologici e degli oggetti esibiti nei musei, pp. 85–104. CNR-Gruppo Nazionale per la Difesa dai Terremoti (2000)
10. Cordova, P.P., Deierlein, G.G., Mehanny, S.S.F., Cornell, C.A.: Development of a two-parameter seismic intensity measure and probabilistic assessment procedure. In: 2nd U.S.-Japan Workshop on Performance-Based Earthquake Engineering for Reinforced Concrete Building Structures, Sapporo, Japan (2000)
11. Cornell, C.A., Krawinkler, H.: Progress and challenges in seismic performance assessment (2000), http://peer.berkeley.edu/news/2000spring/performance/html. febwetg4wt4
12. European Committee for Standardization (CEN): EN 1990: Eurocode - Basis for Structural Design, Technical Report, Brussels, Belgium (2002)
13. Giovenale, P., Ciampoli, M., Jalayer, F.: Comparison of ground motion intensity measure using the incremental dynamic analysis. In: ICASP'9, San Francisco, USA, pp. 1483–1491. Millpress, Rotterdam (2003)
14. Hadjian, A.H.: A general framework for risk-consistent seismic design. Earthquake Engineering and Structural Dynamics 31, 601–626 (2002)

15. Hamburger, R.O., Moehle, J.P.: State of performance based-engineering in the United States. In: Second US-Japan Workshop on Performance-Based Design Methodology for Reinforced Concrete Building Structures, Sapporo, Japan (2000)
16. Housner, G.: Spectrum intensities of strong motion earthquakes. In: Symposium on Earthquake and Blast Effect on Structures, Los Angeles, California, pp. 20–36 (1952)
17. ICC: International performance code for buildings and facilities. Technical report, International Code Council, Falls Curch, VA (2001)
18. IDARC2D: A computer program for seismic inelastic structural analysis - version 5.5. University at Buffalo, Department of Civil, Structural and Environmental Engineering (1987)
19. Jalayer, F.: Direct probabilistic seismic analysis. Implementing non-linear dynamic assessments. PhD thesis, Stanford University (2003)
20. Kennedy, R.P., Short, S.A., Merz, K.L., Tokarz, F.J., Idriss, I.M., Power, M.S., Sadigh, K.: Engineering characterization of ground motion – task I: Effects of characteristic of free field motion on structural response. Technical report, NUREG/CR-3805, U.S. NRC, Washington D.C (1984)
21. Krawinkler, H.: Challenges and progress in performance-based earthquake engineering. In: International Seminar on Seismic Engineering for Tomorrow, Tokyo, Japan (1999)
22. Lucantoni, A., Bosi, V., Bramerini, F., De Marco, R., Lo Presti, T., Naso, G., Sabetta, F.: Il rischio sismico in Italia. Ingegneria Sismica 1, 5–36 (2001)
23. Luco, N., Cornell, C.A.: Structure-specific scalar intensity measures for near-source and ordinary earthquake ground motions. Earthquake Spectra 23(2), 357–392 (2007)
24. Ministero delle Infrastrutture e dei Trasporti: Norme tecniche per le costruzioni, Technical Report, Decreto 14 Settembre 2005, Roma (2005)
25. Park, Y.J., Ang, A.-H.S.: Mechanistic seismic damage model for reinforced concrete. Journal of Structural Engineering 111, 722–739 (1985)
26. SAC Joint Venture: Recommended post-earthquake evaluation and repair criteria for welded steel moment-frame buildings. Technical Report FEMA - 352, Federal Emergency Management Agency, WA, D.C. (2000)
27. SAC Joint Venture: Recommended seismic design criteria for new steel moment-frame buildings. Technical Report FEMA - 350, 350, Federal Emergency Management Agency, WA, D.C. (2000)
28. SAC Joint Venture: Recommended seismic evaluation and upgrade criteria for existing welded steel moment-frame buildings. Technical Report FEMA - 351, Federal Emergency Management Agency, WA, D.C. (2000)
29. SEAOC: Vision 2000 - a framework for performance based design (I-III). Technical report, Structural Engineers Association of California, Sacramento, CA (1995)
30. Sibilio, E., Ciampoli, M.: Performance-based wind design for footbridges: evaluation of pedestrian comfort. In: Proceedings, ICASP10, Japan (2007) (in press)
31. Theil, H.: Economics and Information Theory. Rand McNally and Company, Chicago (1967)
32. Vamvatsikos, D., Cornell, C.A.: Incremental dynamic analysis. Earthquake Engineering and Structural Dynamics 31(3), 491–514 (2002)
33. Whittaker, A.S., Hamburger, R.O., Mahoney, M.: Performance-based engineering of buildings and infrastructure for extreme loadings. In: Proceedings, AISC-SINY Symposium on Resisting Blast and Progressive Collapse. American Institute of Steel Construction, New York (2003)
34. Williams, M.S., Sexsmith, R.G.: Seismic damage indices for concrete structures: a state-of-the-art review. Earthquake Spectra 11(2), 319–349 (1995)

Efficient Nonlinear Programming Algorithms for Chemical Process Control and Operations

Lorenz T. Biegler

Chemical Engineering Department, Carnegie Mellon University, Pittsburgh, PA 15213 USA
biegler@cmu.edu

Abstract. Optimization is applied in numerous areas of chemical engineering including the development of process models from experimental data, design of process flowsheets and equipment, planning and scheduling of chemical process operations, and the analysis of chemical processes under uncertainty and adverse conditions. These *off-line* tasks require the solution of nonlinear programs (NLPs) with detailed, large-scale process models. Recently, these tasks have been complemented by *time-critical, on-line* optimization problems with differential-algebraic equation (DAE) process models that describe process behavior over a wide range of operating conditions, and must be solved sufficiently quickly. This paper describes recent advances in this area especially with dynamic models. We outline large-scale NLP formulations and algorithms as well as NLP sensitivity for on-line applications, and illustrate these advances on a commercial-scale low density polyethylene (LDPE) process.

1 Introduction

Manufacturing processes for petroleum products, basic chemicals, pharmaceuticals, specialty chemicals, consumer products, agricultural chemicals and fertilizers form essential and irreplaceable components of our day-to-day existence. In the US alone, these products lead to revenues of over 10^{12}/yr. Their manufacture is dominated by raw material and energy costs and a strong competitive market, which drives down operating margins. These factors emphasize the need for systematic, model-based process optimization strategies, both in the original design of the process and in day-to-day operations.

Mathematical models for process optimization reflect processing tasks such as mixing, reaction and separation at appropriate conditions, through calculation of state variables, e.g., stream flowrates, temperature, pressure and composition. Modeling equations include conservation laws for mass, energy, and momentum along with constitutive relations and equilibrium conditions (such as physical properties, hydraulics, rate laws and interface behavior). Moreover, with advances in computing hardware and numerical algorithms there has been a steady evolution of model sophistication from steady state to dynamic behavior and from lumped to spatially distributed systems.

Nonlinear programming strategies have been used for process optimization for almost 50 years. These have been essential for plant and equipment design, retrofitting and operations planning. Over the past 25 years *real-time optimization* has also evolved as a standard practice in the chemical and petroleum industry. In particular, the ability to optimize predictive models provides a major step towards linking on-line performance

A. Korytowski et al. (Eds.): System Modeling and Optimization, IFIP AICT 312, pp. 21–35, 2009.
© IFIP International Federation for Information Processing 2009

to higher-level corporate planning decisions. As described in [11,15], these tasks form a well-known pyramidal hierarchy with levels of decision-making including planning at the top, followed by scheduling, site-wide and real-time optimization, model predictive control and regulatory control at the bottom. In this pyramid, the frequency of decision-making increases from top to bottom, while the impact and importance of decision-making increases from bottom to top. Moreover, while planning and scheduling decision models are often characterized by linear models with many discrete decisions, site-wide and real-time optimization require detailed nonlinear process models which usually reflect steady-state performance of the plant. On the other hand, model predictive control (MPC) is often formulated with linear dynamic models.

Interaction among decision-making levels requires that higher-level actions be feasible at lower levels. Moreover, the performance described by lower level models must be reflected accurately in decisions made at higher levels. A particularly close integration is needed for real-time optimization and control, especially for nonlinear processes that may never really be in steady state. Examples of these include batch processes, processes with load changes and grade transitions, such as power plants and polymerization processes, and production units that operate in a periodic manner, such as Simulated Moving Beds (SMBs) [17] and Pressure Swing Adsorption (PSA) [14]. Treating these nonlinear processes requires on-line optimization with nonlinear dynamic models, including strategies such as nonlinear model predictive control (NMPC) [2]. Research in this direction includes development and application of detailed and accurate first-principle differential-algebraic equation (DAE) models for off-line dynamic optimization [5,15,23]. A comprehensive research effort on real-time dynamic optimization is described in [12] and, more recently, large-scale industrial NMPC applications have been reported at ExxonMobil [2], BASF [21] and ABB [10]. Moreover, in addition to enabling NLP solvers, there is a much better understanding of NMPC stability properties and associated dynamic optimization problem formulations that provide them (see [20]). Along with these theoretical developments, NMPC robustness properties have also been developed and analyzed [19]. From the comprehensive treatment of dynamic real-time optimization in [12], it is clear that with improved optimization formulations and algorithms, the role of on-line dynamic optimization can be greatly expanded to consider economic objectives directly, allow longer time horizons with additional constraints and degrees of freedom to improve the objective, and incorporate multiple operating stages over the predictive horizon, including transitions in the predictive horizon due to product change-overs, nonstandard cyclic operations, or anticipated shutdowns [24,12].

The next section provides a background of dynamic optimization strategies and their application to process optimization. An LDPE (low density polyethylene) process case study is also introduced to illustrate the application of these strategies. Sect. 3 then considers simultaneous collocation methods for *off-line* dynamic optimization. A parameter estimation for the LDPE reactor is presented to demonstrate the effectiveness of this approach. Sect. 4 then discusses methods for dynamic optimization for *on-line, time-critical* applications and introduces an NLP sensitivity-based nonlinear model predictive controller, which relies on a "background solution" of the NLP optimization. This is illustrated on a grade transition optimization for the LDPE process. Finally, Sect. 5 concludes the paper and outlines areas for future work.

2   Background

To develop the NLP formulation and solution strategy we consider the dynamic optimization problem in the following form:

$$\min \qquad \varphi(z(t_f)) \tag{1}$$

$$\text{s.t.} \quad \frac{dz(t)}{dt} = f(z(t), y(t), u(t), p), \ \ z(0) = z_0 \tag{2}$$

$$g(z(t), y(t), u(t), p) = 0, \ \ t \in [0, t_f] \tag{3}$$

$$g_f(z(t_f)) = 0 \tag{4}$$

$$u_L \le u(t) \le u_U, \ y_L \le y(t) \le y_U, \ z_L \le z(t) \le z_U \tag{5}$$

where $t \in [t_0, t_f]$ (e.g., time) is the independent variable, $z(t) \in \mathfrak{R}^{n_z}$ is the vector of differential state variables, $u(t) \in \mathfrak{R}^{n_u}$ is the vector of control variables, $y(t) \in \mathfrak{R}^{n_y}$ is a vector of algebraic state variables, and p is a set of optimization variables independent of time. The process model is described by semi-explicit differential and algebraic equations (DAEs) (2),(3) which we assume without loss of generality, are index one.

A number of approaches can be taken to solve (1)-(5). Until the 1970s, these problems were solved using an *indirect* or *variational approach*, based on the first order necessary conditions for optimality obtained from Pontryagin's Maximum Principle; a review of these approaches can be found in [12]. However, if the problem requires the handling of active inequality constraints, finding the correct switching structure as well as suitable initial guesses for state and adjoint variables may be difficult. This limitation has made the *indirect* approach less popular for NMPC applications and can be overcome by *direct* methods that apply NLP solvers.

Sequential methods with NLP solvers, also known as *control vector parameterization*, represent the control variables as piecewise polynomials [26] and perform the optimization with respect to the coefficients of these polynomials. Given initial conditions and a set of control parameters, the DAE model is solved over time within an inner loop of the NLP iteration; the control variables are then updated by the NLP solver itself. Gradients of the objective function with respect to the control coefficients and parameters are calculated either from direct DAE sensitivity equations or by integration of the adjoint equations. Sequential strategies are relatively easy to construct and to apply as they contain the components of reliable DAE solvers (e.g., DASSL, DASOLV, DAEPACK) and NLP solvers (e.g., NPSOL, SNOPT). On the other hand, repeated numerical integration of the DAE model is required, which may become time consuming for large problems. Moreover, sequential approaches may fail with unstable dynamics [1]. Instead, for unstable systems *Multiple Shooting*, which inherits many of the advantages of sequential approaches should be applied. Here, the time domain is partitioned into N time elements, i. e., $t \in [t_{k-1}, t_k], k = 1, \ldots N$, and the DAE models are integrated separately in each element [4,8]. Control variables are parameterized as in the sequential approach and gradient information is obtained for both control variables as well as the initial conditions of the state variables in each element. Finally, equality constraints are added in the NLP to link the elements and ensure that the states are continuous across each element. As with the sequential approach, bound constraints for states and controls are normally imposed only at the grid points t_k.

In the *simultaneous collocation approach*, also known as *direct transcription*, we represent both the state and control profiles as piecewise polynomials in time using collocation on finite elements $t \in [t_{k-1}, t_k], k = 1, \dots N$. This approach corresponds to a fully implicit Runge-Kutta method with high order accuracy and excellent stability properties. It is also a desirable way to obtain accurate solutions for boundary value problems and related optimal control problems. On the other hand, simultaneous approaches also require efficient, large-scale optimization strategies [7,3] because they directly couple the solution of the DAE system with the optimization problem. The DAE system is solved only once, at the optimal point, and therefore can avoid intermediate solutions that require excessive computational effort or may not even exist. Moreover, in the *simultaneous approach* the control variables can be discretized at the same level as the state variables and, under mild conditions, (see [13,16]) the Karush-Kuhn-Tucker (KKT) conditions of the simultaneous NLP are consistent with the optimality conditions of the discretized variational problem, and fast convergence rates to the solution of the variational problem have been shown. Moreover, simultaneous approaches can deal with unstable systems and allow the direct enforcement of state and control variable constraints, at the same level of discretization as the state variables of the DAE system.

2.1   LDPE Case Study

Low density polyethylene (LDPE) is currently the most widely produced polymer. Its uses span many packaging applications including plastic bags, food wrap, squeeze bottles, and plastic films in construction; different polymer grades are produced to ensure

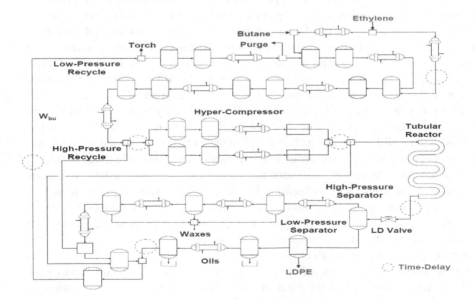

Fig. 1. High-pressure LDPE Process Flowsheet

the best material properties for each of these applications. The high-pressure process for LDPE manufacture is described in [6,30] and serves as a case study for dynamic optimization. As seen in Fig. 1, ethylene is polymerized in a long tubular reactor at high pressures (1500-3000 atm) and temperatures (130-300 °C) through a free-radical mechanism. Accordingly, many compression stages are required to obtain these extreme operating conditions. The LDPE product is recovered after several stages of vapor-liquid separation. These flexible processes obtain several different polymer grades by adjusting the reactor operating conditions. The process model contains a number of challenges for optimization. In the next section we will focus on parameter estimation of a detailed LDPE reactor model, while in Sect. 4, we deal with the important on-line problem of grade changes.

3   Simultaneous Collocation Approach

The DAE optimization problem can be converted into an NLP by approximating state and control profiles by piecewise polynomials on finite elements ($t_0 < t_1 < \ldots < t_N = t_f$). Using a monomial basis representation for the differential profiles, which is popular for Runge-Kutta discretizations, leads to:

$$z(t) = z_{i-1} + h_i \sum_{q=1}^{K} \Omega_q \left(\frac{t - t_{i-1}}{h_i} \right) \frac{dz}{dt}_{i,q} \tag{6}$$

where z_{i-1} is the value of the differential variable at the beginning of element i, h_i is the length of element i, $dz/dt_{i,q}$ is the value of its first derivative in element i at the collocation point q, and Ω_q is a polynomial of order K, satisfying

$$\Omega_q(0) = 0, \ \Omega_q'(\rho_r) = \delta_{q,r}, \ q,r = 1,\ldots,K, \tag{7}$$

where $\rho_r \in [0,1]$ is the normalized location of the r-th collocation point within each element. Continuity of the differential profiles is enforced by

$$z_i = z_{i-1} + h_i \sum_{q=1}^{K} \Omega_q(1) \frac{dz}{dt}_{i,q}. \tag{8}$$

From a number of studies (see [1,16]), we prefer Radau collocation points (with $\rho_K = 1$) as it has a stronger stability property. In addition, the control and algebraic profiles are approximated using a Lagrange basis representation of the form:

$$y(t) = \sum_{q=1}^{K} \psi_q \left(\frac{t - t_{i-1}}{h_i} \right) y_{i,q}, \quad u(t) = \sum_{q=1}^{K} \psi_q \left(\frac{t - t_{i-1}}{h_i} \right) u_{i,q}, \tag{9}$$

where $y_{i,q}$ and $u_{i,q}$ represent the values of the algebraic and control variables, respectively, in element i at collocation point q. ψ_q is the Lagrange polynomial of degree $K - 1$ satisfying $\psi_q(\rho_r) = \delta_{q,r}$ for $q,r = 1,\ldots,K$. From (6), the differential variables are required to be continuous throughout the time horizon, while the control and algebraic variables are allowed to have discontinuities at the boundaries of the elements. Substitution of (6) and (9) into (1)-(5) leads to the following NLP:

$$\min_{\frac{dz}{dt}_{i,q}, u_{i,q}, y_{i,q}, p} \varphi(z_N) \tag{10}$$

s.t. $$\frac{dz}{dt}_{i,q} = f(z_{i,q}, y_{i,q}, u_{i,q}, p), \quad g(z_{i,q}, y_{i,q}, u_{i,q}, p) = 0$$

$$u_{i,q} \in [u_L, u_U], \; y_{i,q} \in [y_L, y_U], \; z_{i,q} \in [z_L, z_U], \quad i = 1, \ldots N, \; q = 1, \ldots K$$

$$\text{and } (8), \; g_f(z_N) = 0$$

This NLP can be rewritten as:

$$\min_{x \in \Re^n} \varphi(\mathbf{x}), \quad \text{s.t. } c(\mathbf{x}) = 0, \; \mathbf{x}_L \leq \mathbf{x} \leq \mathbf{x}_U \tag{11}$$

where $\mathbf{x} = \left(\frac{dz}{dt}_{i,q}, z_i, y_{i,q}, u_{i,q}, p \right)^T$, $f : \Re^n \longrightarrow \Re$ and $c : \Re^n \longrightarrow \Re^m$. To address the resulting large-scale NLP, we apply a full space, interior point (or barrier) solver, embodied in a code called IPOPT. IPOPT applies a Newton strategy to the optimality conditions that result from the primal-dual barrier subproblem,

$$\min \; \varphi(\mathbf{x}) - \mu \sum_{i=1}^{n} [\ln(\mathbf{x}^{(i)} - \mathbf{x}_L^{(i)}) + \ln(\mathbf{x}_U^{(i)} - \mathbf{x}^{(i)})], \quad \text{s.t. } c(\mathbf{x}) = 0. \tag{12}$$

Problem (12) is solved for a sequence of decreasing values of the barrier parameter μ; under typical regularity conditions this sequence of solutions $\mathbf{x}(\mu)$ converges to the solution of (11) [9].

The IPOPT code [27] includes a novel filter based line-search strategy and also allows the use of exact second derivatives. Under mild assumptions, the filter-based barrier algorithm has global and superlinear convergence properties; correspondingly the IPOPT code performs very well when compared to state-of-the-art NLP solvers. Originally developed in FORTRAN, the IPOPT code was recently redesigned to allow for structure dependent specialization of all linear algebra operations. Implemented in C++ and freely available through the COIN-OR foundation, IPOPT can be obtained from the following website:

http://projects.coin-or.org/Ipopt.

A key step in the IPOPT algorithm is the solution of linear systems derived from the linearization of the first order optimality conditions (in primal-dual form) of the barrier subproblem. The linear KKT system can be solved with any direct linear solver configured with IPOPT. However, as the problem size grows, the time and memory requirements can make this approach expensive. Instead, specialized decompositions such as Schur complements lead to efficient (and often parallizable) solution strategies. This allows the efficient solution of very large NLPs on the order of several *million* variables, constraints and degrees of freedom [22]. A detailed description of IPOPT's internal decomposition features and their implementation in the IPOPT software environment is given in [18,25].

Because of these features, the simultaneous collocation approach has lower complexity bounds than competing dynamic optimization strategies, especially since exact second derivatives can be obtained very cheaply and expensive DAE integration and direct sensitivity steps are avoided. This comparison and complexity analysis can be found in [31].

3.1   Parameter Estimation for LDPE Reactor

An important off-line optimization problem is the estimation of reactor parameters from experimental data. The LDPE tubular reactor seen in Fig. 1 can be described as a jacketed, multi-zone device with a predefined sequence of reaction and cooling zones. Different configurations of monomer and initiator mixtures enter in feed and multiple sidestreams, and are selected to maximize the reactor productivity and obtain desired polymer properties. The total reactor length ranges between 0.5 to 2 km, while its internal diameter does not exceed 70-80 mm. Models of this reactor typically comprise detailed polymerization kinetic mechanisms and reacting mixture thermodynamic and transport properties at extreme conditions. A first-principles model describing the gas-phase free-radical homopolymerization of ethylene in the presence of several different initiators and chain-transfer agents at supercritical conditions is considered in [30]. The reaction mechanism consists of 35 reactions with 100 kinetic parameters for each polymer chain of a given length. Here, the method of moments is used to describe macromolecular properties of the copolymer including number- and weight-average molecular weights and polydispersity as described in [25,30]. The steady-state evolution of the reacting mixture along the multiple reactor zones can be formulated as a multi-stage DAE system of the form,

$$\mathbf{F}_{k,j}\left[\frac{dz_{k,j}(t)}{dt}, z_{k,j}(t), y_{k,j}(t), p_{k,j}, \Pi\right] = 0$$

$$z_{k,j}(0) = \phi(z_{k,j-1}(t_{L_{k,j-1}}), w_{k,j-1}), \quad k = 1, \ldots, NS, \ j = 1, \ldots, NZ$$

(13)

where the stage index j denotes a particular reactor zone and index k pertains to a product grade or operating scenario; this formulation allows estimation over different reactor configurations. At zone boundaries, these DAE models are coupled through material and energy balances $\phi(\cdot)$ while additional inputs, $w_{k,j}$, are introduced for monomer, initiator, and cooling water. Also, $t_{L_{k,j}}$ denotes the total length of zone j in scenario k, $p_{k,j}$ denotes local parameters (such as heat transfer coefficients and initiator efficiencies) in each zone j and scenario k and Π corresponds to the kinetic rate constants which apply to all stages. The reactor model contains around 130 ordinary differential equations and 500 algebraic equations for each instance k. Because of significant coupling among the state variables and parametric sensitivity, the reactor DAE model is also highly nonlinear and stiff.

Using (13), we estimate kinetic parameters, Π, to match the plant reactor operating conditions and polymer properties. However, due to the uncertainty associated to the fouling and initiator decomposition mechanisms, it is also necessary to include the local parameters as well. To capture the interaction of $p_{k,j}$ and Π and to account for the measurement errors in the multiple of flow rates, concentrations, temperatures and pressures around the reactor, we consider a multi-scenario estimation problem of the form:

$$\min_{\Pi, p_{k,j}, w_{k,j}} \sum_{k=1}^{NS} \sum_{j=1}^{NZ_k} \sum_{i=1}^{NM_{k,j}} \left(y_{k,j}(t_i) - \bar{y}_{k,j,i} \right)^T \mathbf{V_y}^{-1} \left(y_{k,j}(t_i) - \bar{y}_{k,j,i} \right)$$
$$+ \sum_{k=1}^{NS} \sum_{j=1}^{NZ_k} \left(w_{k,j} - w_{k,j}^M \right)^T \mathbf{V_w}^{-1} \left(w_{k,j} - w_{k,j}^M \right) \tag{14}$$
$$\text{s.t. (13), } \mathbf{H}_{k,j} \left[z_{k,j}(t), y_{k,j}(t), p_{k,j}, \Pi \right] \leq 0$$

where the output variables are matched to the corresponding available plant measurements for each operating scenario or data set k. The vector of outputs contains the reactor temperature profile, jacket inlet and outlet temperatures in each zone, as well as macromolecular properties and product quality at the reactor outlet.

To formulate the estimation problem (14) as a multi-scenario NLP, we perform a full discretization of the differential and algebraic variables and group the resulting set of variables by data sets or scenarios k. For each data set, we use a total of 16 finite elements for the reaction zones, 2 finite elements for the cooling zones and 3 collocation points for the discretization in (14), so that each scenario has around 12,000 constraints and 92 degrees of freedom (corresponding to 32 local parameters $p_{k,j}$, 25 global parameters Π and 35 input variables $w_{k,j}$). In order to obtain exact first and second derivative information, the NLP instances are implemented as NS separate AMPL models that internally *indicate* the set of variables corresponding to the global parameters Π.

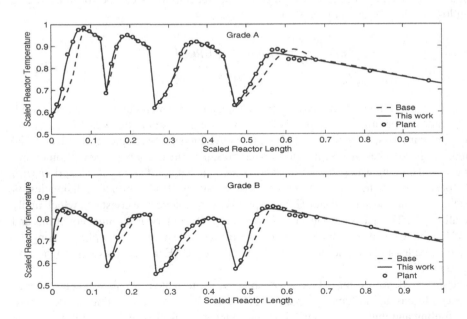

Fig. 2. Comparison of reactor temperature profiles for using simultaneous and "zone-by-zone" parameter estimation

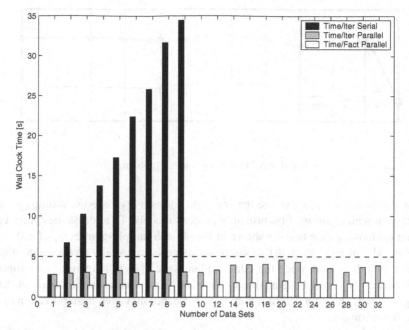

Fig. 3. Wall clock time per iteration and per KKT matrix factorization for multi-scenario parameter estimation with IPOPT. Serial and parallel implementations.

Using the internal decomposition strategy in optimization strategy we consider the solution of multi-scenario NLPs with $NS \leq 32$ data sets. A result of the model fit to two typical product grade data sets can be seen in Fig. 2, where the dashed line depicts a suboptimal "zone-by-zone" estimation with global parameters fixed. The results were obtained in a Beowulf type cluster using standard Intel Pentium IV Xeon 2.4GHz, 2GB RAM processors running under Linux. These are compared against serial solutions of the multi-scenario problems on a single processor with similar characteristics. Fig. 3 presents both computational results. The serial solution of the multi-scenario NLPs exhausts the available memory when the number of data sets exceeds nine, while the parallel implementation overcomes this memory bottleneck and solves problems with up to 32 data sets. For the parallel approach, notice that the effect of parallelism is reflected less in the time required per iteration than in the time per factorization of the KKT matrix. Nevertheless, we see that the time per iteration can be consistently kept below 5 seconds, while the factorization in the serial approach can take as much as 35 seconds before running out of memory. More information on this application and details of the optimization strategy can be found in [25,30].

4   Fast NMPC Based on IPOPT Sensitivity

As described in the previous section, efficient dynamic optimization solvers enable fast solution times even for large-scale models. However, on-line optimization demands time-limited, robust calculations that may exceed the capabilities of current solvers. To

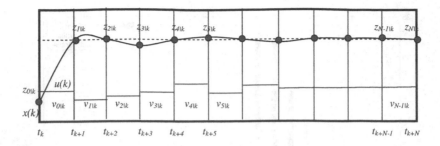

Fig. 4. NMPC moving horizon problem

address this issue, we now explore the concept of sensitivity-based real-time dynamic optimization with rigorous, first principle process models. To address these concepts, consider the moving time horizon shown in Fig. 4, with sampling times $t_{k+l}, l = 0, \dots N$. For chemical processes we note that sampling intervals are usually on the order of minutes. On the other hand, once the current plant state $x(k)$ is known, the appropriate control action $u(k)$ must be available to the plant. Any *computational delay* in determining $u(k)$ will lead to a deterioration of performance and even destabilization of the on-line optimization.

To satisfy these restrictions, we partition the optimization calculations into *background* and *on-line* steps. We assume the NLP can be solved within only a few sampling intervals in "background" for an initial condition "close" to the measured (or estimated) state. Once this state is obtained, a perturbed problem is solved quickly to update the NLP solution, using a particular NLP sensitivity formulation.

To describe this approach, we consider the dynamic optimization problem written over the moving time horizon shown in Fig. 4. After temporal discretization, the dynamic optimization problem can be written as the following simplified NLP,

$$\mathscr{P}_N(x(k)): \quad \min_{z_{l|k}, v_{l|k}} \Phi(z_{N|k}) + \sum_{l=0}^{N-1} \psi(z_{l|k}, v_{l|k}) \tag{15}$$

$$\text{s. t. } z_{l+1|k} = f(z_{l|k}, v_{l|k}), \ l = 0, \dots N-1$$
$$z_{0|k} = x(k), \ z_{l|k} \in \mathbb{X}, \ z_{N|k} \in \mathbb{X}_f, \ v_{l|k} \in \mathbb{U}, \tag{16}$$

where $z_{l|k}$ and $v_{l|k}$ are the states and controls, respectively, over the prediction horizon. From the solution of this problem for current time t_k, we obtain $u(k) = v_{0|k}$ and inject it into the plant. In the nominal case, this drives the state of the plant towards $x(k+1) = z_{1|k} = f(x(k), u(k))$. Once $x(k+1)$ is known, the prediction horizon is shifted forward by one sampling interval and problem $\mathscr{P}_N(x(k+1))$ is solved to find $u(k+1)$. This recursive strategy gives rise to the ideal NMPC controller (neglecting computational delay).

Now consider the state of the plant at the previous sampling time, $x(k-1)$, where we already have the control $u(k-1)$. In the nominal case the system evolves according to the dynamic model (16), and we can predict the future state by solving $\mathscr{P}_N(f(x(k-1), u(k-1)))$ in advance. For instance, if this problem can be solved between t_{k-1} and

t_k, then $u(k)$ will already be available at t_k. For this, we define the equivalent NLP of the form,

$$\mathscr{P}_{N+1}(x(k-1),u(k-1)): \min_{z_{l|k-1},v_{l|k-1}} \Phi(z_{N|k-1}) + \psi(x(k-1),u(k-1))$$

$$+ \sum_{l=0}^{N-1} \psi(z_{l|k-1},v_{l|k-1})$$

$$\text{s.t. } z_{l+1|k-1} = f(z_{l|k-1},v_{l|k-1}), \quad l = 0,\ldots N-1$$

$$z_{0|k-1} = f(x(k-1),u(k-1)),$$

$$z_{l|k-1} \in \mathbb{X}, \; z_{N|k-1} \in \mathbb{X}_f, \; v_{l|k-1} \in \mathbb{U}$$

(17)

In the nominal case, it is clear that the solution of this problem is *equivalent to* $\mathscr{P}_N(x(k))$ and that $\mathscr{P}_{N+1}(x(k-1),u(k-1))$ can be solved in advance to obtain $u(k) = v_{0|k-1}^*$ *without computational delay*. Moreover, under the NMPC assumptions posed in [20], it is easy to see that such a controller has the *same nominal stability properties* as the ideal NMPC controller [28].

On the other hand, a realistic controller must also be robust to model mismatch, unmeasured disturbances and measurement noise. As noted in [19], ideal NMPC provides a mechanism to react to these features along with some inherent robustness. In particular, tolerance to mismatch and disturbances can be characterized by input-to-state stability [19,28]. In [28] we focus on sensitivity-based NMPC schemes and show their inherent robustness properties through input-to-state stability concepts. The key to this extension comes by noting that problem $\mathscr{P}_{N+1}(x(k),u(k))$ is parametric in its initial conditions so we can define the dummy parameter vector $p_0 = x(k)$. Here we rewrite $\mathscr{P}_{N+1}(x(k),u(k))$ as the following NLP,

$$\min \varphi(\mathbf{x},p_0), \text{ s.t. } c(\mathbf{x},p_0) = 0, \; \mathbf{x}_L \leq \mathbf{x} \leq \mathbf{x}_U$$

(18)

and we define $\mathbf{x}$ as the vector of all variables in $\mathscr{P}_{N+1}(x(k),u(k))$. From the optimality conditions of (18), and under mild regularity conditions of the NLP [9], we obtain a first order estimate of the perturbed solution of (18), i.e., $\Delta\mathbf{x} = \mathbf{x}_*(p) - \mathbf{x}_*(p_0)$. This can be calculated very cheaply in IPOPT from the factorization of the KKT matrix in the final NLP iteration. Therefore in the presence of uncertainty, we apply the sensitivity equations of $\mathscr{P}_{N+1}(x(k),u(k))$ to find the approximate solution of $\mathscr{P}_N(x(k+1))$.

Moreover, to maintain a consistent active set for the solution of $\mathscr{P}_N(x(k+1))$, we modify the sensitivity calculation to determine the value of p that enforces the relation $z_{0|k} = x(k+1)$ in the perturbed $\mathscr{P}_{N+1}(x(k),u(k))$, instead of a direct change ($\Delta p = x(k+1) - x(k)$) in the initial conditions. Coupled to the linearized optimality conditions, the added constraint, $\Delta z_{0|k} = x(k+1) - z_{0|k}^*$, gives rise to an extended set of linear sensitivity equations, which can be solved efficiently through a Schur complement approach. This approach takes advantage of the already factorized KKT matrix at the solution of $\mathscr{P}_{N+1}(x(k),u(k))$ with IPOPT. Therefore, once the next state is known, the desired approximate solution can be obtained from the background Schur decomposition and a single on-line backsolve [31]. As described in [31,28], the on-line step requires less than 1% of the (already fast) dynamic optimization calculation. We denote this sensitivity-based approach the Advanced Step NMPC (as-NMPC) controller.

This controller can be viewed as a fast linear model predictive controller *linearized about the optimal nonlinear model at the previous measurement.* Moreover, it inherits the stability and robustness properties of ideal NMPC while avoiding the difficulties of computational delay.

4.1 NMPC for LDPE Process

To demonstrate the advantages of the sensitivity-based NMPC strategy, we return to the LDPE case study. The process represents a difficult dynamic system; reactor dynamics are much faster than responses in the recycle loops and long time delays are present throughout the compression and separation systems. Due to the complex, exothermic nature of the polymerization, the reactor temperature and pressure are enforced strictly along the operating horizon following fixed recipes. The main operational problem in these processes consists of providing fast adjustments to the butane feed and purge stream to keep the melt index at a desired reference value. This is especially important during transitions (switching between two different operating points). As shown in [6], dynamic optimization can lead to significant reduction in the grade transition time; in one case, it was reduced from about 5 h to no more than 2.8 h, leading to reduction of at least 23 tons of off-spec product.

The resulting DAE model of the LDPE process (with a simplified reactor model) contains 289 differential and 64 algebraic state variables. We now consider an appropriate optimal feedback policy that minimizes the switching time between steady states corresponding to the production of different polymer grades. This poses a severe test of the NMPC algorithm as it needs to optimize over a large dynamic transition. The following moving horizon problem is solved on-line at every sampling time t_k:

$$\min \int_{t_k}^{t_{k+N}} \left(w_{C_4}(t) - w_{C_4}^r\right)^2 + \left(F_{C_4}(t) - F_{C_4}^r\right)^2 + \left(F_{Pu}(t) - F_{Pu}^r\right)^2 dt \tag{19}$$

s.t. DAEs for LDPE Model

where the inputs are the flowrates of butane and purge streams, F_{C_4} and F_{Pu}, respectively, the output is the butane weight fraction in the recycle stream, w_{C_4}, and superscript r denotes a reference value. Using the simultaneous collocation approach, problem (19) is converted into a large-scale NLP with 15 finite elements with 3 collocation points in each element. The resulting NLP contains 27,135 constraints, and 30 degrees of freedom. For the dynamic optimization, we set $N = 15$ and sampling interval to 6 min.

To compare ideal and as-NMPC strategies, we ignore the effect of computational delay in the closed-loop response. To assess robust performance, the plant response is also subjected to strong, random disturbances in the transportation delays in the recycle loops. Performance of both NMPC approaches is presented in Fig. 5. Note that the optimal feedback policy involves the saturation of both control valves for the first 2500 seconds of operation, with the final flowrates set to values corresponding to the new operating point. It is interesting to observe that the output profile for as-NMPC is indistinguishable from the full optimal solution, with only small differences in the input profiles.

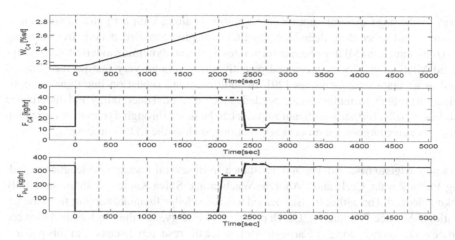

Fig. 5. Closed-loop performance of the ideal NMPC (solid) and as-NMPC (dashed) approaches with output w_{C_4} and inputs F_{C_4} and F_{Pu}

The on-line and background computational times are especially worth comparing. Ideal NMPC requires around 351 CPU seconds and about 10 IPOPT iterations of on-line computation while as-NMPC requires a negligible on-line time (1.04 CPU seconds) for the solution of the Schur complement system and a final backsolve to obtain the updated solution vector. As a result, as-NMPC reduces the on-line computation time (and associated computational delay) by over two orders of magnitude with virtually no loss in performance. Moreover, as-NMPC also serves as an excellent basis for effective initialization of the *next* NLP problem solved in background. From the perturbed solution provided by the sensitivity calculation, as-NMPC provides very accurate NLP initializations at all sampling times. Leading to only 2-3 IPOPT iterations, as-NMPC also reduces the *background* NLP computation by up to a factor of five.

5 Conclusions

This paper addresses the increasing value of dynamic optimization for chemical process operations. Both off-line and on-line optimization tasks demand fast and robust optimization strategies, often for challenging large-scale applications. Current dynamic optimization formulations and algorithms are reviewed with an emphasis on the simultaneous collocation approach. This strategy has advantages for unstable systems, and with the suitable application of large-scale NLP solvers (such as IPOPT), it is especially effective for time-critical applications. Moreover, for on-line applications, NLP sensitivity can be calculated very cheaply from IPOPT; this leads to a nonlinear model predictive control strategy with fast on-line performance and minimal computational delay. All of these aspects are demonstrated on a case study for a large-scale polymerization process.

Nevertheless, this summary represents only a beginning in addressing dynamic real-time optimization. Future challenges include effective off-line solution strategies for large, multi-stage dynamic optimization problems along with a tighter integration of planning and scheduling decisions. On-line strategies can also benefit from moving horizon estimation (MHE) which incorporates nonlinear dynamic models. A sensitivity-based MHE strategy was developed recently and exhibited very fast performance as well as accurate state estimates [29]. In addition, more robust on-line dynamic optimization problem formulations are needed to include model uncertainty and disturbance models. Finally, further significant impacts can be made through dynamic optimization on challenging large-scale process applications, such as the LDPE process.

Acknowledgements. Advice and assistance from several research colleagues including Victor Zavala, Carl Laird, Andreas Wächter and Shiva Kameswaran are gratefully acknowledged. The author is also pleased to acknowledge financial support from the US National Science Foundation, US Department of Energy, and the Center for Advanced Process Decision-making at Carnegie Mellon for the research discussed in this paper.

References

1. Ascher, U.M., Petzold, L.R.: Computer Methods for Ordinary Differential Equations and Differential-Algebraic Equations. SIAM, Philadelphia (1998)
2. Bartusiak, R.D.: NLMPC: A platform for optimal control of feed- or product-flexible manufacturing. In: Assessment and Future Directions of NMPC, p. 338. Springer, Berlin (2007)
3. Betts, J.: Practical Methods for Optimal Control Using Nonlinear Programming, SIAM Series on Advances in Design and Control, Philadelphia, PA (2001)
4. Bock, H.G.: Numerical treatment of inverse problem in differential and integral equations. In: Recent Advances in Parameter Identification Techniques for ODE. Federal Republic of Germany, Heidelberg (1983)
5. Busch, J., Oldenburg, M., Santos, J., Cruse, A., Marquardt, W.: Dynamic predictive scheduling of operational strategies for continuous processes using mixed-logic dynamic optimization. Comput. Chem. Eng. 31, 574–587 (2007)
6. Cervantes, A.M., Tonelli, S., Brandolin, A., Bandoni, J.A., Biegler, L.T.: Large-scale dynamic optimization for grade transitions in a low density polyethylene plant. Computers & Chemical Engineering 26, 227–237 (2002)
7. Cervantes, A.M., Wächter, A., Tutuncu, R., Biegler, L.T.: A reduced space interior point strategy for optimization of differential algebraic systems. Comp. Chem. Engr. 24, 39–51 (2000)
8. Diehl, M., Bock, H.G., Schlöder, J.P., Allgöwer, F., Findeisen, R., Nagy, Z.: Real-time optimization and nonlinear MPC of processes governed by differential-algebraic equations. Journal of Process Control 12(4), 577–585 (2002)
9. Fiacco, A.V.: Introduction to Sensitivity and Stability Analysis in Nonlinear Programming. Academic Press, New York (1983)
10. Franke, R., Doppelhamer, J.: Integration of advanced model based control with industrial it. In: Assessment and Future Directions of NMPC, p. 368. Springer, Berlin (2007)
11. Grossmann, I.E.: Enterprise-wide optimization: A new frontier in process systems engineering. AIChE J. 51(7), 1846–1857 (2005)
12. Grötschel, M., Krumke, S., Rambau, J. (eds.): Online Optimization of Large Systems. Springer, Berlin (2001)
13. Hager, W.W.: Runge-Kutta methods in optimal control and the transformed adjoint system. Numer. Math. 87, 247–282 (2000)

14. Jiang, L., Biegler, L.T., Fox, V.G.: Simulation and optimization of pressure swing adsorption systems for air separation. AIChE J. 49(5), 1140–1157 (2003)
15. Kadam, J., Marquardt, W.: Integration of economical optimization and control for intentionally transient process operation. In: Findeisen, R., Allgoewer, F., Biegler, L. (eds.) Assessment and Future Directions of Nonlinear Model Predictive Control. LNCIS, vol. 358, pp. 419–434. Springer, Heidelberg (2007); Sensitivity-based solution updates in closed-loop dynamic optimization. In: Proceedings of the DYCOPS 7 Conference. Elsevier, Amsterdam (2004)
16. Kameswaran, S., Biegler, L.T.: Convergence rates for direct transcription of optimal control problems using collocation at Radau points. Computational Optimization and Applications 41, 81–126 (2008)
17. Kawajiri, Y., Biegler, L.T.: Optimization strategies for simulated moving bed and powerfeed processes. AIChE Journal 52(4), 1343–1350 (2006)
18. Laird, C.D., Biegler, L.T.: Large-scale nonlinear programming for multi-scenario optimization. In: Modeling, Simulation and Optimization of Complex Processes, pp. 323–336. Springer, Heidelberg (2008)
19. Magni, L., Scattolini, R.: Robustness and robust design of MPC for nonlinear discrete-time systems. In: Assessment and Future Directions of NMPC. LNCIS, vol. 358, pp. 239–254. Springer, Heidelberg (2007)
20. Mayne, D.Q.: Nonlinear model predictive control: challenges and opportunities. In: Zheng, A., Allgöwer, F. (eds.) Nonlinear Model Predictive Control, pp. 3–22. Birkhaüser-Verlag, Basel (2000)
21. Nagy, Z.K., Franke, R., Mahn, B., Allgöwer, F.: Real-time implementation of nonlinear model predictive control of batch processes in an industrial framework. In: Assessment and Future Directions of NMPC. LNCIS, vol. 358, pp. 465–472. Springer, Heidelberg (2007)
22. Schenk, O., Waechter, A., Hagemann, M.: Matching-based preprocessing algorithms to the solution of saddle-point problems in large-scale nonconvex interior-point optimization. Computational Optimization and Applications 36(2-3), 321–341 (2007)
23. Oldenburg, J., Marquardt, W., Heinz, D., Leineweber, D.B.: Mixed-logic dynamic optimization applied to batch distillation process design. AIChE J. 49(11), 2900–2917 (2003)
24. Toumi, A., Diehl, M., Engell, S., Bock, H.G., Schlöder, J.P.: Finite horizon optimizing control of advanced SMB chromatographic processes. In: 16th IFAC World Congress, Prague, Czech Republic (2005)
25. Laird, C.D., Zavala, V.M., Biegler, L.T.: Interior-point decomposition approaches for parallel solution of large-scale nonlinear parameter estimation problems. Chemical Engineering Science 63, 4834–4845 (2008)
26. Vassiliadis, V.S., Sargent, R.W.H., Pantelides, C.C.: Solution of a class of multistage dynamic optimization problems. part i - algorithmic framework. Ind. Eng. Chem. Res. 33, 2115–2123 (1994)
27. Wächter, A., Biegler, L.T.: On the implementation of a primal-dual interior point filter line search algorithm for large-scale nonlinear programming. Math. Program. 106(1), 25–57 (2006)
28. Zavala, V.M., Biegler, L.T.: The advanced-step NMPC controller: optimality, stability and robustness. Automatica 45, 86–93 (2009)
29. Zavala, V.M., Laird, C.D., Biegler, L.T.: A fast computational framework for large-scale moving horizon estimation. Journal of Process Control 18(9), 876–884 (2008)
30. Zavala, V.M., Biegler, L.T.: Large-scale parameter estimation in low-density polyethylene tubular reactors. I &EC Research 45(23), 7867–7881 (2006)
31. Zavala, V.M., Laird, C.D., Biegler, L.T.: Fast solvers and rigorous models: can both be accommodated in NMPC? J. Robust and Nonlinear Control 18(8), 800–815 (2008)

Application of Evolutionary Optimization in Structural Engineering

Hitoshi Furuta[1], Koichiro Nakatsu[2], Takahiro Kameda[1],
and Dan M. Frangopol[4]

[1] Department of Informatics, Kansai University, Takatsuki, Osaka 569-1095, Japan
furuta@res.kutc.kansai-u.ac.jp,
kamechan@sc.kutc.kansai-u.ac.jp
[2] Graduate School of Informatics, Kansai University, Takatsuki, Osaka 569-1095, Japan
inside2@sc.kutc.kansai-u.ac.jp
[3] Department of Civil and Environmental Engineering, Lehigh University, Bethlehem,
PA 18015-4729, USA
dan.frangopol@lehigh.edu

Abstract. Practical optimization methods including *genetic algorithms* are introduced, based on *evolutionary computing* or *soft computing*. Several application examples are presented to demonstrate and discuss the efficiency and applicability of the described methods.

1 Introduction

Due to the recent advance and development of computer and information technologies, it becomes possible to obtain useful information for decision making with ease. To resolve some problems facing in real life, it is necessary to find out an appropriate solution among possible candidates under several constrained conditions. Therefore, most of these problems are belonging to a kind of optimization problems. However, most of optimization problems being studied are solved under ideal circumstances. Most of real life problems are very large and complicated ones with vague or uncertain objective functions and constrained conditions, different from the ideal circumstances.

Under such circumstances, evolutionary computing has been paid great attention and recognized as a powerful tool for optimization of various practical problems. Evolutionary computing uses iterative procedures, such as growth or development in a population. This population is then selected in a guided random search using parallel processing to achieve the desired end. Such processes are often inspired by biological mechanisms of evolution. Evolutionary computing includes evolutionary programming, genetic algorithm, genetic programming, immune algorithm, learning classifier system, particle swarm optimization, ant colony optimization, etc. Among them, genetic algorithm has been widely used in various fields, because it is a representative method of the evolutionary computing and has a good ability to find out quasi-optimum solutions with ease.

In this paper, several practical optimization methods including GA are introduced, which are based on "evolutionary computing" or "soft computing". Soft computing covers fuzzy decision making, neural networks, and so forth. Several application examples

A. Korytowski et al. (Eds.): System Modeling and Optimization, IFIP AICT 312, pp. 36–81, 2009.
© IFIP International Federation for Information Processing 2009

are presented to demonstrate and discuss the efficiency and applicability of the methods described here.

2 Structural Vibration Control Using Soft Computing Techniques

In Japan, many high-rise structures have been constructed due to the recent advance of structural material and construction technology. Since the high-rise structures are generally very flexible, vibration control is essential to maintain safety and reliability of the structures. Especially, natural disasters which cause the strong vibration, like typhoon and earthquake should be considered in their design and construction.

Under such situations, a lot of researches on the vibration control have been done in the past [1], most of which do not consider the structural and environmental changes in time. Those systems may lose the performance when the environment has a perceptible change. In actual cases, the characteristics of structures may change due to the structural degradation and the addition of other facilities to the structure. This implies that it is necessary to deal with the structural and environmental changes when designing the vibration control system. There are some robust and adaptive systems that can consider the change of environment. However, in those systems, the structural performance in the static environment may be reduced. In the case, it is essential to adapt to the changing environment quickly while maintaining the performance in the static environment.

In this paper, a new structural vibration control system is described, which can adapt to the change of structural systems and environments, by introducing the learning ability. In this system, it is necessary to prevent the reduction of the performance in the static state, while improving the effectiveness of adaptation and the performance after learning. This system has two different controllers: a robust controller used for the static state and an adaptive controller following the change of environment. By using these two controllers properly, it is possible to achieve a good control performance under any situation. Fuzzy controller is employed for the adaptive controller that can adapt to the change of structural systems, in which the steepest descent method is employed for the learning method. In addition to the two controllers, the system has a judgment ability to recognize the change of environment based on structural response, external force, and control power. Through numerical and model experiments, it is concluded that the system can provide a good control for the unforeseen and incidental changes of external loads and conditions.

2.1 Vibration Control System

The present system has two controllers and one judgment machine. While both controllers calculate control force from structural response and external force all the time, the system uses one of the outputs provided by the two controllers. The structure of the system is shown in Fig. 1.

2.1.1 Robust Controller

In the robust controller, the control force is calculated by fuzzy-neural network system [2, 3]. Although fuzzy-neural network is one of neural networks [4], it can provide the

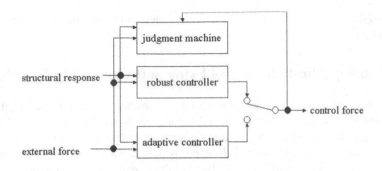

Fig. 1. Structure of the system

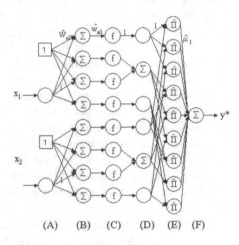

(A)   (B)   (C)   (D)   (E)   (F)

Fig. 2. Structure of fuzzy-neural network

calculating results equal to the fuzzy reasoning method. Therefore, it has characteristics of both neural network and fuzzy reasoning, namely, learning ability of neural network and robustness of fuzzy reasoning. The structure of fuzzy-neural network system is shown in Fig. 2.

This system uses the simplified fuzzy reasoning in which the consequent part is expressed in terms of crisp numbers. In Fig. 2, layers A to D correspond to antecedent parts, and layers E to F correspond to consequent parts. Membership functions are expressed by sigmoid function, and their central values are depicted by the weights between A and B and gradients are depicted by the weights between B and C. Consequent parts are weighted from E to F. Using the sigmoid function, it is possible to define the shape of membership functions in anti-symmetrical forms. Thus, it is possible to realize a better fitting and a better control than with symmetrical membership functions. Those three kinds of weights are learned using the back propagation algorithm.

The robust controller has a learning ability, which is not performed under control, and can avoid over-learning and over-reaction to the change of state. In the system, the neural network has already learned good control patterns by the optimal control theory with the same structure. This means that the neural network may be able to have a control performance equivalent to the optimal control theory. The fuzzy-neural network used here can calculate faster and has robustness, because its structure includes the fuzzy reasoning process. In other words, it can calculate the control force providing good performance equivalent to the optimal control theory in shorter time, and the deterioration of performance at the change of environment is smaller than with the optimal control theory.

2.1.2 Adaptive Controller

For the adaptive controller, the control force is calculated by fuzzy control with a learning mechanism using the steepest decent method. Since the learning method is the same as for the neural network, this fuzzy control system is called neuro-fuzzy system. The neuro-fuzzy system has also the characteristics of both neural network and fuzzy reasoning that are the learning ability of the neural network and the robustness of the fuzzy reasoning. However, neuro-fuzzy structure is simpler than fuzzy-neural network, and therefore the learning is faster and convergence is generally quick. This implies that neuro-fuzzy system is well suited to adaptive controller. In the present system, Gauss function is employed for the membership function. The neuro-fuzzy system can tune the membership functions whose consequent parts are tuned with the steepest decent method by fitting the teaching data. Then, it is necessary to prepare teaching data to learn, however it is impossible to collect the complete teaching data, because the state is so quickly changing that the optimal solution cannot be identified. In other words, if the state of structure is changing, it is not able to know the exact vibration characteristics under the control. In this research, structure response after control is given to the neuro-fuzzy system, in which the system learns the teaching data that are created when the structural response velocity is smaller than that under the previous control. Namely, the teaching data are supposed to be a vector with the inverse direction of structural response. The learning is done at each step. In this way, good rules of reducing the vibration of structure are obtained. Moreover, the calculation is implemented in a faster manner, and the controller has robustness, because the calculation method is based on fuzzy reasoning.

2.1.3 Judgment Machine

Judgment machine is always checking the performance of robust controller to find out the reduction of the performance based on the structural response, external force, and control power. When the vibration characteristics change, some difference appears on those three data. If the structural response differs even for the same external force and the same control power is outputted, judgment machine can recognize some change of environment.

2.1.4 Model Experiment

To validate the performance of each controller, model experiment using rigid frame model is conducted. In the experiment, the structure and environment are not changed. As external force, actual earthquake data observed at Kobe in 1995 are used. The results

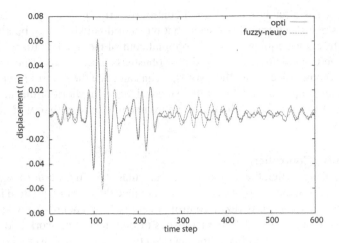

Fig. 3. Experiment result of robust controller

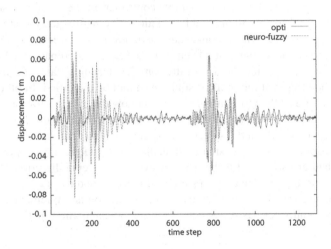

Fig. 4. Experiment result of adaptive controller

of both controllers are shown in Figs. 3 and 4, together with the results of optimal control theory. The robust controller has seven membership functions, which are allocated at even intervals at first. The robust controller has already learned a part of results given by the optimal control theory. However, the input patterns to the system are unknown at the experiment. From Fig. 3, it can be confirmed that the robust controller has as good performance as that by the optimal control theory. In other words, the robust controller can obtain good control rules that are equally efficient as those of the optimal control theory. On the other hand, the adaptive controller has no information regarding the structure and environment at first. The controller has also seven membership functions, in which all consequent parts are initialized to be zero. At the initial stage, the adaptive controller cannot control well and there are big differences from that of the optimal control theory, because it has no information about the environment. However, the performance of control is gradually improved step by step as the learning proceeds. Finally, the performance of adaptive controller becomes equal to the optimal control theory, as shown in Fig. 4. From this result, it is confirmed that this controller can adapt quickly and rightly even though it has no knowledge about the structure and environment. This means that the present system can adapt to the changing environment, therefore it is useful for the control under the dynamic environment.

2.1.5 Numerical Example

To demonstrate the effectiveness of the present system, numerical experiment is done. In this experiment, the frequency characteristic of the structure is changed. The weight of the structure is increased from time to time. In the experiment, the external force is calculated from the actual wind velocity of the 9th typhoon in Osaka on July 26, 1997. This data are observed at the interval of 0.05 second and the total number of data is 30,000. The numerical results by the present system are shown in Fig. 5, comparing with the results by the optimal control theory. In the experiment, the structural weight becomes double after 500 steps. At first, both the systems show almost the same performance at the static state. When the structural characteristics begin to change, some difference appears. While the difference is small at first, the performance of the present system provides better results than the optimal control theory, because of the robustness of the present system. As time proceeds, the difference of performance becomes larger gradually. This shows that the proposed system can adapt to the change of structural characteristic. Since the present system could identify the change of structure and environment, it switched the controller from the robust one to the adaptive one efficiently. The adaptive controller could adapt to the new environment quickly.

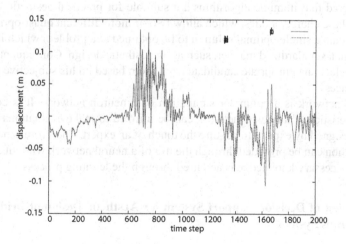

Fig. 5. The numerical results by the present system

3 Aesthetic Design of Bridge Structures

In recent years it is becoming important to consider the aesthetic design factors in the design of bridges. Various researches (e.g. [6]) on the decision-support systems for aesthetic design of a bridge have been made in the past. Here, a practical decision support system for aesthetic design of bridge handrails is introduced. In the design of a bridge handrail, outsourcing the design works to a specialized designer to ensure an aesthetically satisfactory design can cause not only budgetary problems, but also a gap between what the designer imagines and what the engineers think on the structural design.

This paper takes the view that it is possible to obtain new designs by combining components for handrails which were designed in the past, because it would not be easy to create an original new design. Moreover, such use of handrail components designed in the past will be considered to allow candidate aesthetic designs smoothly because there will be no structural problems with components designed in the past.

In this paper, several attempts are presented to develop a decision support system for aesthetic design of bridge handrails. The decision support system consists of the evaluation system using neural network and the optimization system based upon immune algorithm. Thus, it is confirmed that the present system is effective for the aesthetic design of bridge handrails by means of several numerical examples. Furthermore, some of the results obtained are visualized through the use of computer graphics (CGs) and compared.

3.1 Immune Algorithm and Neural Network

Immune algorithms [7] are a kind of optimal solution search algorithms allowing the diversity of solutions to be retained and multiple quasi-optimal solutions to be obtained. It is considered that immune algorithms are suitable for practical aesthetic designing because of these characteristics, which allow two or more different quasi-optimal solutions rather than a single optimal solution to be obtained to a problem which is difficult to evaluate in a standardized manner, such as an aesthetic design. Consequently, an engineer can select an appropriate candidate from them based on his subjective judgment and preferences.

A neural network is a computer simulation of a neuron network. It is considered that characteristics of a neuron network can be utilized to make evaluations of bridge handrails designed by experts to acquire the touch of an expert. Therefore, a near-expert level evaluation can be provided through the use of a neural network without an expert, if once the necessary knowledge is acquired through the learning process.

3.2 Overview of Decision Support System for Aesthetic Design of Bridge Handrails System

The user (decision maker) who wishes to create aesthetic designs of a bridge handrail inputs the following data for the bridge handrail to be designed:

1. The design concepts.
2. The surrounding environment.
3. The configuration of the bridge.
4. The color of the bridge components other than the handrail.

An interactive input system is employed by selecting from items stored in it. Based on the input data two or more quasi-optimal solutions are searched and found using an immune algorithm and a neural network, and candidate aesthetic designs are presented. The flow of the processing by the system is shown in Fig. 6.

One hundred and five photographs of existing bridge handrails are evaluated individually and the results are learned using the neural network. The touch of an expert can be acquired by making evaluation in this manner of bridge handrails that were designed by experts. Therefore, a near-expert level evaluation can be provided through the use of the neural network without an expert once the necessary knowledge is acquired through the learning process.

Then a search for two or more quasi-optimal solutions is made using the neural network that learned the necessary knowledge as an evaluation function for the immune algorithm. As mentioned above, immune algorithms are suitable for aesthetic designing because they allow two or more different quasi-optimal solutions rather than a single optimal solution to be obtained so that the human decision makers can select an appropriate candidate from them based on their subjective judgment and preferences.

3.2.1 Aesthetic Design Items
The surrounding environment, bridge configuration, colors (handrails and except handrails) and handrail components are employed as the aesthetic design items. The photographs of existing bridges are used as the data for the learning with the neural

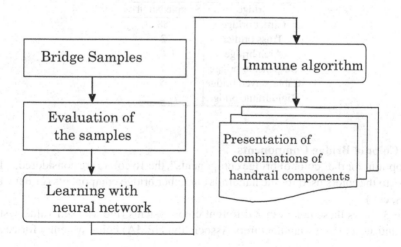

Fig. 6. Flow of the processing by the system

network. Each of these photographs is treated as a sample (i.e. one learning data piece). Explanation of each of aesthetic parameters follows.

3.2.2   Surrounding Environment

The 14 kinds of surrounding environments assumed from the photographs were set up. These are shown in Table 1. In the genetic representation used in this system, 1 or 0 is used to indicate that each of the environmental components is present or not, respectively.

Table 1. Environmental factors

Blue sky	Rice paddies
Cloudy	River
White clouds	Sea
Mountains (green leaves)	Urban area (buildings)
Mountains (brown leaves)	Residences (houses)
Mountains (red leaves)	Snow
Rock or soil	Pavement

3.2.3   Bridge Configuration

The classification of bridge configurations shown in Table 2 is used. The 105 samples are also broken down in Table 2.

Table 2. Bridge configuration classification and sample number

Bridge	Sample Number
Girder bridge	88
Truss bridge	2
Arch bridge	6
Suspension bridge	2
Cable-stayed bridge	5
Rigid frame bridge	2

3.2.4   Color of Bridge Components

As the options for the "color of bridge components", the 16 colors are considered, which are close to the colors used for the handrails and other bridge components (girders, arch sections, etc.).

Table 3 shows these colors in 3 different color systems (the Mansell value system, the JIS and Japan Paint Manufacturers Association (JPMA) color systems) for ease of use.

Table 3. Color options

JIS	JPMA	Mansell
Red	T07-40X	7.5R4/14
Brown	T15-30F	5YR3/3
Cream	T25-85F	5Y8.5/3
Celadon	T35-70H	5GY7/4
Jasper green	T37-50D	7.5GY5/2
Light greenish blue	T55-50P	5BG5/8
Light blue	T65-80D	5B8/2
Baby blue	T72-70D	2.5PB7/2
Saxon gray	TN-50	N-5
Dayflower	T69-50T	10B5/10
Rose pink	T12-70L	2.5YR7/6
Snow white	TN-95	N-9.5
Yellowish brown	T19-60F	10YR6/3
Sky gray	TN-80	N-8
India ink	TN-10	N-1
Silver		

3.2.5 Handrail Components

In this system, bridge handrails are made up of upper, intermediate and lower bridge handrail components and bridge handrail columns. Each part was classified as shown in Fig. 7.

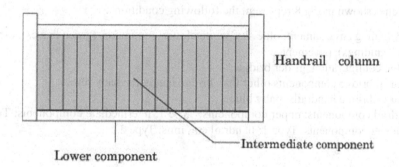

Fig. 7. Handrail components

3.2.6 Design Concepts

To realize the aesthetic design of bridges in this system, several concepts are prepared, which are summarized in Table 4. The user can select the desired design concepts. Two or more concepts can be selected.

Table 4. Design concepts

Symbolic value
Uniqueness
Reliability (Peaceful)
Friendly
Nobleness
Internationality
Harmony with the surrounding environment

3.2.7 Evaluation

105 photographs of the bridges that really exist are used for evaluation. The aesthetic parameters of each sample (photograph) were coupled with the corresponding data that were input into the neural network and were represented as a gene for the immune algorithm. Genes that represent aesthetic parameters in binary figures, i.e., 0 and 1 were used so that genetic codes are represented as one-dimensional bit rows.

1. Surrounding environment: 14 types, 14 bits.
2. Bridge configuration: 6 types, 3 bits.
3. Colors (handrails and bridge components other than handrails): 16 types, 4 bits.
4. Upper handrail components: 4 types, 2 bits.
5. Intermediate handrail components: 64 types, 6 bits.
6. Lower handrail components: 4 types, 2 bits.
7. Handrail columns: 16 types, 4 bits.

As shown in Fig 8, these are represented as 39-bit genes.

The genes shown in Fig 8 represent the following conditions:

1. Surrounding environment: blue sky, white clouds, mountains (green leaves), urban area (buildings), pavement.
2. Bridge configuration: girder bridge.
3. Color of bridge components other than bridge handrails: baby blue.
4. Color of bridge handrails: baby blue.
5. Handrail components: upper components: Type 1; intermediate components: Type 54; lower components: Type 1; handrail columns: Type 2.

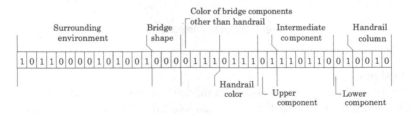

Fig. 8. Gene row of aesthetic parameters

In the application of immune algorithm to a gene, the "surrounding environment", "bridge configuration" and "color of bridge components other than handrails" parameters are fixed to the values (options) selected by the system user.

Each bridge sample (photograph) is evaluated with respect to the degree to which it matches the design concepts selected out of the seven design concepts shown in Table 4, using an integer scale of zero to ten. The results are used as the training data for the neural network, which are used as the evaluation function for the immune algorithm after it has learned the necessary knowledge. The result of an evaluation of a sample may differ from another evaluation of the same sample made at a different time or on a different day or may become a different one when the order of the examination of the samples (photographs) is changed. Some fluctuations are inevitable because it is humans that make evaluations, but it is possible to make the range of fluctuation narrow by utilizing the knowledge of experts, who are more consistent than non-experts in making technical evaluations. It is also considered that looking through all samples before starting evaluations will help reduce fluctuations.

By having the neural network learn the results of samples evaluations (learning data), it becomes possible to evaluate the data about which learning has not been done. Thus by using the neural network that learned the necessary knowledge as the evaluation function for the immune algorithm, new data can be evaluated to obtain two or more quasi-optimal solutions.

The neural network outputs the evaluation result for the selected design concepts, and the total of the evaluation values for the target design concepts is used as the evaluation value for the candidate of aesthetic design. In the neural computing the following parameters are used:

1. Number of layers of the network: 3
2. Number of patterns learned: 105
3. Number of first layer units: 39
4. Number of second layer units: 46
5. Number of third layer units: 7
6. Number of learning runs (integer value): 1000000
7. Allowable error range (real number value): 0.0000001
8. Learning coefficient (real number value): 0.9
9. Inertia coefficient (real number value): 0.6
10. Gradient of sigmoid function: 1.0

3.2.8 Application Example

The immune algorithm parameters were set as follows:

1. Initial number of antibodies: 20 (gene of the initial antibodies-generation of random number as binary figures 0, 1).
2. Upper limit for the number of memory cells: 5.
3. Number of generations: 300.
4. Manipulation of crossover: uniform crossover, crossover rate: 70.
5. Manipulation of mutation: reversing of a selected bit, mutation rate: 0.3.

6. Thresholds: Tc = 0.8, Tac1 = 0.8, Tac2 = 0.7, Tac3 = 0.8, Tac4 = 0.88.
7. Suppress power: 1.

Examples of actual application of the present system are described as follows.

3.2.9 Design Case 1

1. Surrounding environment: blue sky, white clouds, mountains (green leaves), mountains (brown leaves), rock or soil, river, residence (houses), pavement.
2. Bridge configuration: girder bridge.
3. Color of bridge components other than handrails: cream ((T25-85F)/(5Y8.5/3)).
4. Design concepts: friendly, harmony with the surrounding environment.

Under the above conditions, five design candidates are obtained as shown in Table 5. Table 5 shows that the present system provided five design plans with different characteristics. It is seen that the color selected for the handrail is calm and unremarkable and the configuration of the handrail is simple so that all the plans are satisfactory for such design concepts as "friendly" and "harmony with the surrounding environment".

Table 5. Design candidates for Case 1

	Plan 1	Plan 2	Plan 3	Plan 4	Plan 5
Color of handrail	Yellowish brown	Sky gray	Brown	Brown gray	Brown gray
Upper component	Type 2	Type 1	Type 1	Type 1	Type 3
Middle component	Type 7	Type 3	Type 3	Type 1	Type 35
Lower component	Type 2	Type 2	Type 2	Type 2	Type 0
Column	Type 2	Type 0	Type 10	Type 10	Type 10
Friendly	0.8	0.9	0.8	0.9	0.8
Harmony	0.9	0.9	0.9	0.8	0.8

3.2.10 Design Case 2

1. Surrounding environment: blue sky, white clouds, mountains (green leaves), mountains (brown leaves), rock or soil, river, residence, pavement.
2. Bridge configuration: girder bridge.
3. Color of bridge components other than handrails: cream ((T25-85F)/(5Y8.5/3)).
4. Design concepts: unique, harmony with the surrounding environment.

Design Case 2 is the same as Design Case 1 except that the design concept "unique" is used in place of the "friendly". Table 6 presents the design plans obtained by the present system. As shown in Table 6, the configuration of the handrail becomes complicated compared with that of Case 1. The larger the type number of handrail elements is, the more complicated the configuration of handrail becomes. In addition, the color obtained for Plan 3 is red that is not chosen for Case 1. Regarding the design concepts, the second concept of "harmony" is difficult to get high score, when the plan shows the good match to the first concept of "unique", because these two concepts have contradictory characteristics. Thus, this result is considered to be reasonably satisfactory.

Table 6. Design candidates for Case 2

	Plan 1	Plan 2	Plan 3	Plan 4	Plan 5
Color of handrail	Saxon gray	Sky gray	Red	Brown	Brown
Upper component	Type 1	Type 1	Type 2	Type 0	Type 2
Middle component	Type 45	Type 47	Type 61	Type 63	Type 37
Lower component	Type 0	Type 2	Type 0	Type 0	Type 0
Column	Type 2	Type 2	Type 2	Type 6	Type 3
Unique	1.0	0.9	1.0	1.0	1.0
Harmony	0.6	0.8	0.7	0.6	0.6

3.2.11 Design Case 3

1. Surrounding environment: blue sky, white clouds, mountains (green leaves), mountains (brown leaves), rock or soil, river, residence (houses), pavement.
2. Bridge configuration: girder bridge.
3. Color of bridge components other than handrails: cream ((T25-85F)/(5Y8.5/3)).
4. Design concepts: internationality, harmony with the surrounding environment.

Design Case 3 has the same design requirements except the design concept: "internationality", whereas "friendly" in Case 1 and "unique" in Case 2. Table 7 presents the design plans obtained for Case 3. By changing the design concept from "friendly" to "internationality", five design plans different from those of Case 1 and Case 2 are obtained. As shown in Table 7, there are more color variations than in Case 1 and the components are different from those used in Case 1, but it is rather debatable whether these candidates of aesthetic designs have internationality. It had been anticipated that intermediate handrail components with European looks would be selected, but in reality relatively simple ones were selected. Also, the colors return to calm ones.

Table 7. Design candidates for Case 3

	Plan 1	Plan 2	Plan 3	Plan 4	Plan 5
Color of handrail	Sky gray	Light-greenish blue	Silver	Baby blue	Silver
Upper component	Type 3	Type 3	Type 1	Type 2	Type 0
Middle component	Type 3	Type 19	Type 39	Type 7	Type 1
Lower component	Type 2	Type 0	Type 3	Type 2	Type 2
Column	Type 2	Type 10	Type 10	Type 8	Type 2
Internationality	0.7	0.6	0.8	0.8	0.9
Harmony	0.9	0.8	0.7	0.8	0.6

3.2.12 Design Case 4

1. Surrounding environment: blue sky, white clouds, river, urban area (buildings).
2. Bridge configuration: girder bridge.

3. Color of bridge components other than handrails: cream ((T25-85F)/(5Y8.5/3)).
4. Design concepts: internationality, harmony with the surrounding environment.

Design Case 4 is almost the same as Design Case 3 except for the "surrounding environment", which is changed from the "residence" to "business area". Table 8 presents the candidates of aesthetic design obtained for Case 4. The evaluation values are higher than those given in Case 3 for both "internationality" and "harmony with the surrounding environment". This shows that different results can be obtained only by changing the surrounding environment parameter setting. The handrail components used are little different from those used in Case 3, but the colors are almost the same as those used in Case 3.

Table 8. Design candidates for Case 4

	Plan 1	Plan 2	Plan 3	Plan 4	Plan 5
Color of handrail	Baby blue	Silver	Baby blue	Light-greenish blue	Silver
Upper component	Type 0	Type 0	Type 2	Type 0	Type 2
Middle component	Type 21	Type 3	Type 8	Type 1	Type 33
Lower component	Type 1	Type 3	Type 0	Type 2	Type 1
Column	Type 2	Type 8	Type 10	Type 2	Type 10
Internationality	1	1	0.9	1	1
Harmony	0.8	0.9	1	0.9	0.9

3.2.13 Design Case 5

1. Surrounding environment: blue sky, white clouds, mountains (green leaves), mountains (brown leaves), rock or soil, river, residence (houses), pavement.
2. Bridge configuration: girder bridge.
3. Color of bridge components other than handrails: jasper green ((T37-50D)/(7.5GY5/2)).
4. Design concepts: friendly, harmony with the surrounding environment.

This application example is the same as Design Case 1 except that "jasper green" is used in place of the "cream" as the color of bridge components other than the handrails.

Table 9. Design candidates for Case 5

	Plan 1	Plan 2	Plan 3	Plan 4	Plan 5
Color of handrail	Sky gray	Brown gray	Jasper green	Brown gray	Sky gray
Upper component	Type 1	Type 1	Type 3	Type 3	Type 1
Middle component	Type 3	Type 3	Type 38	Type 7	Type 23
Lower component	Type 3	Type 2	Type 2	Type 2	Type 2
Column	Type 2	Type 0	Type 10	Type 0	Type 10
Friendly	0.8	0.9	0.8	0.8	0.8
Harmony	0.9	0.9	0.9	0.8	0.9

The colors obtained are calm and unremarkable, and the configuration of the handrail is rather simple. It can be said that the candidates of aesthetic design provide similar atmospheres to those provided by the candidates of aesthetic design in Case 1 and thus match the design concepts selected.

Fig. 9. Plan 1 of Design Case 1

Fig. 10. Plan 3 of Design Case 1

Fig. 11. Plan 1 of Design Case 2

Fig. 12. Plan 1 of Design Case 3

Fig. 13. Plan 1 of Design Case 4

Fig. 14. Plan 1 of Design Case 5

3.2.14   Visualization of Application Example Using Computer Graphics

Some of the candidates of aesthetic design presented by the system in the application example were compared mutually using figures visualized by computer graphics. Plans 1 of Design Cases 1 to 5 and Plan 3 of Design Case 1 were selected to be visualized (a total of 6 plans). To facilitate comparison, the same background was used for all of the plans selected. The visualized candidates of aesthetic design are shown in Figs. 9 to 14.

The plans shown in Figs. 9 and 10 are ones from the same application example (Design Case 1). Although these plans have different features, it is considered that they are ones with similar atmospheres because both of them seem to match the design concepts selected.

The candidate of aesthetic design shown in Fig. 11 is a plan from Case 2, which is the same as Case 1 except that the concept "unique" is used in place of the "friendly". It is considered that the plan matches the design concept "unique" because the intermediate handrail components have a more complex shape than those of the intermediate handrail components shown in Figs. 9 and 10 and the evaluation value is high.

The plan shown in Fig. 12 is a plan from Case 3, which is the same as Case 1 except that the concept "internationality" is used in place of the "friendly". This plan may appear to have an international appearance at a first glance in the sense that the handrail color and the shape of the upper handrail components are more uncommon than those used in Figs. 9 and 10, but it is debatable whether this plan has an international appearance because the shape of the intermediate handrail components is rather simple.

The plan (candidate of aesthetic design) shown in Fig. 13 is a plan from Case 4, which is the same as Case 3 except that the surrounding environment is different. It is not possible to compare Fig. 12 with Fig. 13 in terms of the surrounding environment because the same background is used. However, it is considered that this plan matches the design concept "internationality" because the evaluation point is high.

The plan shown in Fig. 14 is a plan from Case 5, which is the same as Case 1 except that "jasper green" is used in place of the "cream" as the color of the bridge components other than the bridge handrails. Although the plans shown in Figs. 9, 10 and 14 have different features, it is considered that they are ones with similar atmospheres because they match the design concepts selected.

4   Optimal Restoration Scheduling for Earthquake Disaster

Japan has been suffering from many natural disasters such as typhoons, tsunami and earthquakes. However, road networks have not been designed to protect against all such natural hazards. Moreover, even the newest design theory cannot guarantee the absolute safety due to the economic constraints. Therefore, it is necessary to develop a synthetic disaster prevention program based on the recognition that road networks may be unavoidably damaged when big earthquakes occur.

In this paper, the early restoration of road networks after the earthquake disasters is focused on. Three issues are dealt with, the first of which is an allocation problem: which groups restore which disaster places, the second is a scheduling problem: what order is the best for the restoration, and the third is an allocation problem: which restoring method is suitable for which disaster places. In order to solve the three problems simultaneously, Genetic Algorithm (GA) is applied, because it has been proven to be very powerful in solving combinatorial problems. In this paper, the relationships among early restoration, minimization of Life-Cycle Cost (LCC), and target safety level of road network are discussed by using Multi-Objective Genetic Algorithm (MOGA). Namely, the following three objective functions are considered:

1. Restoring days are minimized.
2. LCC is minimized.
3. Performance level of road network is maximized.

Then, it is possible to solve the multi-objective optimization problem by making two of the three objective functions as the constrained conditions. For example, the minimization of restoring days is required according to the prescribed performance level and LCC constraints. The predetermination of these requirements may be detrimental. Namely, taking the same restoring days, it may be possible to find solutions that can largely reduce the LCC if the performance level can be slightly decreased. Alternatively, taking the same restoring days, it may be possible to find solutions that can substantially increase the performance level if the LCC is slightly increased.

By introducing the concept of multi-objective optimization into the restoration scheduling for earthquake disasters, it is intended to find several near-optimal restoration scheduling plans. Although single-objective optimization can provide various solutions by changing the constraints, it requires enormous computation time. When selecting a practical restoration schedule, it is desirable to compare feasible optional solutions obtained under various conditions. Thus, a decision support system that can provide several alternative restoration schedules was developed by applying MOGA. Several numerical examples are presented to demonstrate the applicability and efficiency of the present method.

4.1 Road Network Models

Here, it is assumed that a road network is damaged, in which multiple portions are suffered from damage so that it cannot function well. The objective is the realization of quick restoration of the lifeline system. It is intended to determine the optimal allocation of restoring teams and optimal scheduling of restoring process, and optimal allocation of restoring methods. Then, the following conditions should be taken into account:

1. The optimal allocation of restoring teams, optimal scheduling of restoring process, and optimal allocation of restoring methods must be determined simultaneously.
2. A portion of the road network is suffered from several kinds of damage that have a hierarchical relation in time.

As an example of restoration, a road network is considered, which has 164 nodes as shown in Fig. 15. This model corresponds to an area damaged by the 1995 Kobe

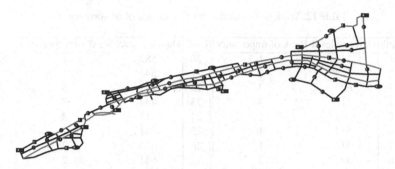

Fig. 15. Road network model

Table 10. Data of work (*A*)

Team of work (*A*)	Ability	Previous works before scheduling
1	15	0
2	30	0
3	12	0
4	17	37
5	18	0
6	23	36
7	25	0
8	35	38

Table 11. Data of work (*B*)

Team of work (*B*)	Ability	Previous works before scheduling
1	15	43
2	20	46
3	25	48
4	10	0
5	17	0
6	30	49
7	23	0
8	27	0

earthquake. For this road network, the following restoration works are necessary to recover the function:

1. Work (*A*): work to clear the interrupted things, 38 sites (1 - 38)
2. Work (*B*) : work to restore the roads, 50 sites (1 - 50)

Table 12. Work (A)'s damage level and rank of importance

Link	Damage level	Rank of importance	Link	Damage level	Rank of importance
1	242	1	20	582	1
2	223	2	21	542	1
3	625	3	22	451	3
4	312	3	23	434	3
5	554	1	24	311	3
6	514	1	25	441	2
7	311	1	26	412	2
8	473	3	27	531	2
9	300	3	28	156	2
10	321	3	29	556	1
11	656	1	30	520	2
12	380	1	31	551	3
13	501	3	32	166	1
14	302	1	33	513	1
15	312	3	34	531	3
16	321	3	35	495	1
17	231	1	36	424	1
18	534	2	37	337	3
19	171	1	38	564	2

Then, the limitations and restrictions of each work should be considered, for instance, work (B) should be done after work (A). Work (B) consists of the following three works: work to repair the roads, work to reinforce the roads and work to rebuild the roads. The waiting places of restoring teams for work (A) and work (B) are shown by the numbers A (1-8) and B (1-8), respectively. Tables 10 and 11 show the ability to restore and the previous works before scheduling of each team. Tables 12 and 13 show the damage level and the rank of importance.

4.2   Restoration Scheduling

Weighting factors are prescribed for the links with damage, which are denoted by W_i ($i = 1,\ldots,n_L$). n_L is the total number of damaged links. Then, the restoring rate after q days, $R^{(q)}$, is expressed as follows:

$$R^{(q)} = \frac{\sum\limits_{i \in J^{(q)}} W_i \times l_i}{\sum\limits_{i \in J^{(0)}} W_i \times l_i} \tag{1}$$

where l_i is the distance of the i-th link, $J^{(0)}$ is the set of damaged links, $J^{(q)}$ is the set of restored links until q days after the disaster, and W_i is the weighting factor of the

Table 13. Work (*B*)'s damage level and rank of importance

Link	Damage level	Rank of importance	Link	Damage level	Rank of importance
1	153	1	26	146	2
2	453	2	27	366	2
3	496	3	28	311	2
4	464	3	29	145	1
5	133	1	30	425	2
6	415	1	31	413	3
7	355	1	32	231	1
8	531	3	33	245	1
9	246	3	34	353	3
10	623	3	35	461	1
11	445	1	36	131	1
12	154	1	37	455	3
13	613	3	38	564	2
14	444	1	39	631	1
15	366	3	40	322	2
16	615	3	41	464	3
17	641	1	42	114	1
18	151	2	43	415	1
19	254	1	44	700	1
20	654	1	45	311	3
21	561	1	46	211	3
22	125	3	47	344	3
23	345	3	48	407	3
24	462	3	49	512	2
25	456	2	50	423	2

i-th link. Then, the objective function can be calculated by using the restoring days and the restoring rate. The relation between restoring days and restoring rate is shown in Fig. 16. The area of the uncolored portion should be minimized to obtain the optimal solution, because this enables not only to shorten the restoring days but also to restore the important links faster.

Restoring days are calculated for each restoring work, and the minimum days necessary for each work is given as

$$d = \frac{h}{t_1} \tag{2}$$

where h is the restoration time required to complete the restoration work.

In this paper, the restoration time is calculated by using the restoration rate for each work and the capability value. The relation between the restoration rate for each work and the capability of the teams are shown in Fig. 17. The restoration rate is given as follows:

a) Small damage: In the small damage, there is no difference in capability of each team. The restoration will be completed during a fixed time. Here, 4 hours are assumed.

$$h = h_t \tag{3}$$

b) Moderate damage: In the moderate damage, there are some differences in capability between teams. However, every team can restore the damage.

$$h = \frac{D}{A} \tag{4}$$

where D is the amount of damage and A is the capability of the team, that is, the restoring amount per hour.

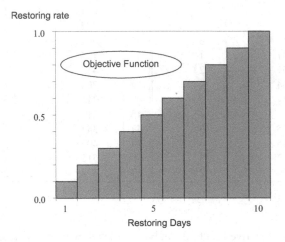

Fig. 16. Objective function

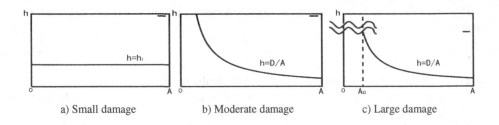

Fig. 17. Relations between restoration rate for each work and capability of teams

c) Large damage: In the large damage, only some teams can restore, because other teams have no restoring equipment and facility necessary for the large damage.

$$h = \begin{cases} \infty, & \text{if } A < A_c, \\ D/A, & \text{if } A \geq A_c. \end{cases} \tag{5}$$

where A_c is the minimum capability which the team can work.

The working hours per day of a restoration team are calculated by Equation (6), where t_m is the moving time to a site given by Equation (7). The shortest distance from the waiting place of the restoration team to the site is considered as L (km), and the moving speed of the team is set to be v (km/h). h_c is the preparation time that is necessary for every work.

$$t_1 = t_0 - 2t_m - h_c \tag{6}$$

$$t_m = \frac{L}{v} \tag{7}$$

4.3 Optimal Restoration Scheduling for Earthquake Disaster Using Life-Cycle Cost

Road networks have not been designed to sustain all natural hazards. Moreover, even the newest design theory cannot guarantee the absolute safety due to the economic constraints. Therefore, it is necessary to develop a synthetic disaster prevention program based on the recognition that road networks may be unavoidably damaged when big earthquakes occur. The purpose of this research is the early restoration of road networks after the earthquake disasters. Three issues are focused on, the first of which is an allocation problem: which groups restore which disaster places, the second is a scheduling problem: what order is the best for the restoration, and the third is an allocation problem: which restoring method is suitable for which disaster places. In order to solve the three problems simultaneously, Genetic Algorithm (GA) is applied. In this paper, an attempt is made to discuss the relationships among early restoration, minimization of LCC, and target safety level of road network by using Multi-Objective Genetic Algorithm (MOGA). Namely, the following three objective functions are considered:

1. Restoring days are minimized.
2. LCC is minimized.
3. Target safety level of road network is maximized.

Then, it is possible to solve the multi-objective optimization problem by making two of the three objective functions as the constrained conditions. For example, the minimization of restoring days is required according to the prescribed target safety level and LCC constraints. The predetermination of these requirements may be detrimental. Namely, taking the same restoring days, it may be possible to find solutions that can largely reduce the LCC if the safety level can be slightly decreased. Alternatively, taking the same restoring days, it may be possible to find solutions that can substantially increase the safety level if the LCC is slightly increased.

By introducing the concept of multi-objective optimization into the restoration scheduling for earthquake disasters, it is intended to find several near-optimal restoration schedules. Although single-objective optimization can provide various solutions by changing the constraints, it requires enormous computation time. When selecting a practical restoration schedule, it is desirable to compare feasible optional solutions obtained under various conditions. Thus, an attempt is made in this study to develop a decision support system that can provide several alternative restoration schedules by applying MOGA.

4.4   Objective Functions

In this study, restoring days, LCC and safety level are used as objective functions. Restoring days are minimized, LCC is minimized, and safety level is maximized. There are trade-off relations among the three objective functions. For example, safety level decreases when LCC decreases, and safety level is extended due to the extension of restoring days. Then, multi-objective optimization can provide a set of Pareto solutions that cannot improve an objective function without making other objective functions worse. In this study, DNA structure is constituted as shown in Fig. 18, in which DNA of each individual consists of three parts such as restoring method, allocation of restoring teams, and schedule of restoring process. Using the coding, it is possible to determine the optimal allocation of restoring teams, optimal scheduling of restoring process, and optimal allocation of restoring methods simultaneously.

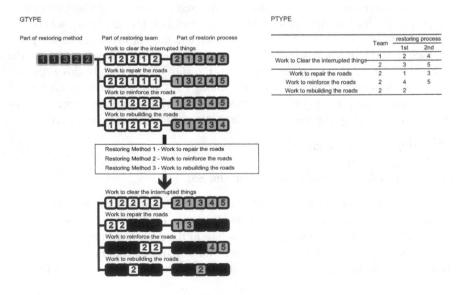

Fig. 18. DNA structure

Then, the three objective functions are expressed as follows.

4.4.1 Restoring Days

The relation between restoring days and restoring rate is shown in Fig. 16. The area of the uncolored portion should be minimized to obtain the optimal solution, because this enables not only to shorten the restoring days but also to restore the important links faster.

4.4.2 Life-Cycle Cost

Life-Cycle Cost (LCC) is defined as the total maintenance cost in terms of road network and all the entire bridges during their lives. In this paper, restoring method is defined by three kinds of methods: work to repair the roads, work to reinforce the roads, and work to rebuild the roads. Then, restoring cost of each work is defined by

$$RC = C_b \times D_{degree} \tag{8}$$

where C_b is the basic restoring cost and D_{degree} is the level of damage defined in Table 14. Table 15 presents the basic costs and safety levels by the restoring methods. Fig. 19 shows the performance levels of restoring methods. Maintenance cost of each work after restoring is defined by

$$MC = M_b \times D_r \tag{9}$$

where M_b is the basic maintenance cost presented in Table 15 and D_r is the level of deterioration defined in Table 14. Here, the accumulated maintenance cost is considered for 50 years.

Then, the objective function is defined by

$$LCC = \sum_{i \in J^{(0)}} (RC_i + MC_i) \tag{10}$$

where RC_i is the restoring cost of the i-th link, MC_i is the maintenance cost of the i-th link, and $J^{(0)}$ is the set of damaged links.

4.4.3 Safety Level

Safety level depends on the traffic volume and the condition of links. In this research, safety level (SL) of the road network is maximized, which is defined by

$$SL = \sum_{i \in J^{(0)}} (I_i + SL_i) \tag{11}$$

where I_i is the importance of the i-th link, S_i is the safety level of the i-th link, and $J^{(0)}$ is the set of damaged links.

Table 14. Levels of damage and levels of deterioration

Link	Work (A)	Work (B)	Level of deterioration	Link	Work (A)	Work (B)	Level of deterioration
1	1.70	0.47	0.8	26	1.36	0.45	1.2
2	0.73	0.96	1.0	27	1.33	1.12	1.2
3	1.91	1.61	1.5	28	0.95	1.87	1.2
4	0.94	1.42	1.5	29	1.33	0.44	0.8
5	1.96	0.41	1.0	30	1.26	1.30	1.0
6	1.53	1.27	0.8	31	1.65	1.26	1.5
7	0.63	1.09	0.8	32	0.50	0.71	1.0
8	1.38	1.62	1.5	33	1.54	0.75	0.8
9	1.97	0.75	1.5	34	1.59	1.08	1.5
10	1.02	1.91	1.5	35	0.74	1.10	1.0
11	2.00	1.36	1.0	36	1.27	0.40	1.0
12	1.27	0.47	1.0	37	1.01	0.78	1.5
13	1.56	1.87	1.5	38	1.69	1.72	1.8
14	0.94	1.36	1.0	39		1.93	0.8
15	0.97	1.12	1.2	40		0.98	1.9
16	0.71	1.88	1.2	41		1.42	1.5
17	0.74	1.96	1.0	42		0.35	0.8
18	2.00	0.46	1.0	43		1.27	0.8
19	1.87	0.78	0.8	44		2.14	0.8
20	0.94	2.00	0.8	45		1.56	1.8
21	0.67	1.72	0.8	46		0.65	1.5
22	1.56	0.38	1.2	47		1.05	1.5
23	0.49	1.06	1.2	48		1.24	1.5
24	1.67	0.65	1.5	49		1.57	1.0
25	1.63	1.64	1.0	50		1.29	1.0

Work (A): The level of damage for work to clear the interrupted things
Work (B): The level of damage for work to restoring the roads

Table 15. The basic costs and safety levels by the restoring methods

Restoring method	Basic restoring cost	Basic maintenance cost	Safety level
Repair	700	3500	0.6
Reinforcement	1200	2000	0.8
Rebuilding	5000	1500	0.9

Basic restoring cost (Ten thousand yen)
Basic maintenance cost (Ten thousand yen)

4.5 Multi-objective Optimization

A multi-objective optimization problem has two or more objective functions that cannot be integrated into a single objective function. In general, the objective functions cannot be simultaneously minimized (or maximized). It is the essence of the problem that trade-off relations exist among the objective functions. The concept of "Pareto optimum" becomes important in order to balance the trade-off relations. The Pareto

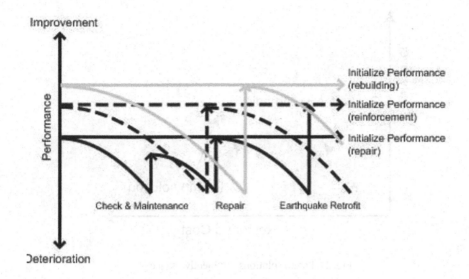

Fig. 19. The performance levels of restoring methods

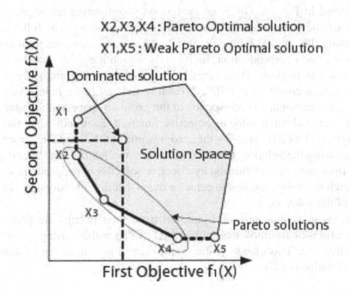

Fig. 20. Cost-effective domain including Pareto solutions

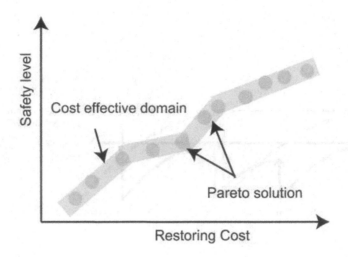

Fig. 21. Pareto solutions in objective space

optimum solution is a solution that cannot improve an objective function without sacrificing other functions (Figs. 20 and 21). A dominated, also called non-dominant, solution is indicated in Fig. 20. GA is an evolutionary computing technique, in which candidates of solutions are mapped into GA space by encoding. The following steps are used to obtain the optimal solutions: a) initialization, b) crossover, c) mutation, d) natural selection, and e) reproduction. Individuals, which are solution candidates, are initially generated at random. Then, steps b, c, d, and e are repeatedly implemented until the termination condition is fulfilled. Each individual has a fitness value to the environment. The environment corresponds to the problem space and the fitness value corresponds to the evaluation value of objective function. Each individual has two aspects: Gene Type (GTYPE) expressing the chromosome or DNA and Phenomenon Type (PTYPE) expressing the solution. GA operations are applied to GTYPE and generate new children from parents (individuals) by effective searches in the problem space, and extend the search space by mutation to enhance the possibility of individuals other than the neighbor of the solution.

GA operations that generate useful children from their parents are performed by crossover operations of chromosomes or genes (GTYPE) without using special knowledge and intelligence. This characteristic is considered as one of the reasons of the successful applications of GA.

4.6   Application of MOGA to Restoration Scheduling

Table 16 presents the parameters of MOGA used here. Fig. 22 present the results obtained by MOGA. Table 17 shows the evaluation values of each solution. Then, the efficiency of MOGA is expressed as follows; for example, comparing solution C with

Table 16. Parameters of MOGA

Population	Probability of crossover	Probability of mutation	Generation
1000	0.6	0.005	1000

Table 17. Evaluation values of each solution

Solution	Restoring days	LCC	Safety level
A	14.234	241343	61.224
B	15.069	242300	62.123
C	16.309	254316	63.496
D	17.421	260416	63.756
E	17.565	265017	74.116
F	17.576	264866	80.011
G	17.546	284954	80.023
H	17.779	289676	80.229
I	17.898	292191	82.054
J	18.325	291942	82.234
K	18.649	301471	82.268
L	18.623	303837	82.302

solution D in safety level, there is no significant difference between the two solutions. However, in restoring days and LCC, solution D is worse than solution C. On the other hand, comparing solution E with solution F, there is no significant difference in restoring days between the two solutions. However, solution F is better than solution E in LCC. Moreover, in safety level, solution F is substantially better than solution E. In Table 17, comparing solution E with solution G, there is no significant difference in restoring days between the two solutions. However, in safety level and LCC, there are significant differences between the two solutions.

In Fig. 23, the vertical axis represents LCC, whereas the horizontal axis represents restoring days. Fig. 24 presents the relation between restoring days and safety level. In Figs. 23 and 24, since restoring days and the other two objective functions have a rather perfect positive linear correlation, it can be said that the other two can have a positive effect if restoring days can be increased. Fig. 25 presents the relation between LCC and safety level. In Figs. 24 and 25, it should be noted that it can be said that the other two objective function can have a positive effect when there is no significant difference in safety level.

Figs. 26 to 29 show the detailed restoration schedule associated with the solution K shown in Fig. 22. In Figs. 26 to 29, compared to the number of work to repair the roads,

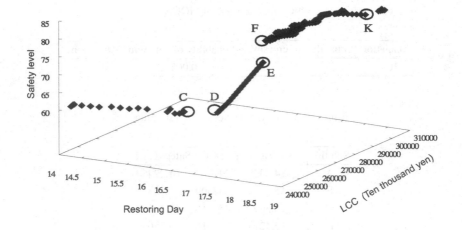

Fig. 22. Pareto solutions obtained by MOGA

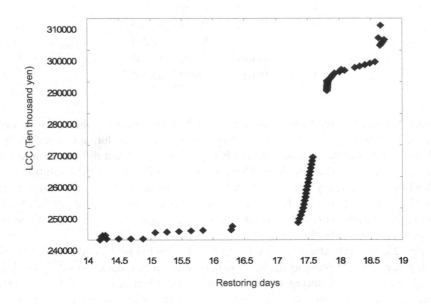

Fig. 23. Relation between restoring days and LCC

there are many works to reinforce the roads and to rebuild the roads, which can increase safety level, in some important links. As will be appreciated from Table 17, comparing solution K with the other solutions, solution K is worse than the other solutions in restoring days. However, solution K is better than the other solutions in safety level.

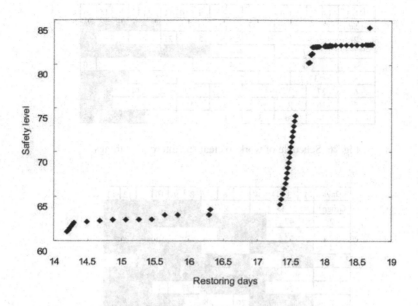

Fig. 24. Relation between restoring days and safety level

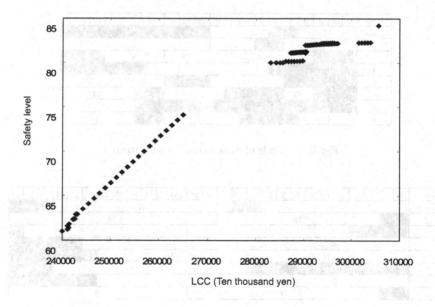

Fig. 25. Relation between LCC and safety level

Days	1	2	3	4	5	6	7	8	9	10	11	12	13	14	15	16
Group1	20			32	7				27							
2	25				13		23		5		21	6		8		
3	15				10											
4	38				24							9				30
5	33			28		34					22					
6	37			19			17		29			14				
7	16		2		18				4			3				
8	36		26		31		1		11		35	12				

Fig. 26. Schedule of work to clear the interrupted things

Days	1	2	3	4	5	6	7	8	9	10	11	12	13
Group1	43				32								
2	47												
3	49							13					
4					2								
5			37										29
6	50												
7	39							7	1				
8					26		38						

Fig. 27. Schedule of work to repair the roads

Days	1	2	3	4	5	6	7	8	9	10	11	12	13	14	15	16	17	18	19	20
Group1	42							31												
2	45							23					17					3		
3						25					35				6					
4						19				4										
5					28								22			30				
6			33					24							14					
7															9					
8	41							10							8					

Fig. 28. Schedule of work to reinforce the roads

Days	1	2	3	4	5	6	7	8	9	10	11	12	13	14	15	16	17	18	19	20	21	22	23	24	25	26	27
Group1	46											12															
2				16										34													
3			15										11														
4	40								18																		
5	48								21																		
6											27																
7	44								20																		
8			36										5														

Fig. 29. Schedule of work to rebuild the roads

From the above results, it is confirmed that various kinds of solutions can be obtained by the proposed method. Namely, when selecting a practical restoration schedule, the proposed method enables to compare feasible optional solutions obtained under various conditions.

5 Optimal Maintenance Planning of Bridge Structures Using MOGA

It has been widely recognized that maintenance work is important, because the number of existing bridges requiring repair or replacement increases in the future. In order to establish a rational maintenance program, it is necessary to develop a cost-effective decision-support system that can provide us with a practical and economical plan. Although low-cost maintenance plans are desirable for bridge owner, it is necessary to consider various constraints when choosing an appropriate actual maintenance program. For example, the minimization of maintenance cost requires to prescribe the target safety level and the expected service life time. The predetermination of requirements may loose the variety of possible maintenance plans. Namely, it may be possible to find out a better solution that can largely extend the service life if the safety level can be sensitively decreased even with the same maintenance cost.

It is desirable to discover many alternative maintenance plans with different characteristics. Although a single-objective optimization can provide various solutions by changing the constraints, it requires enormous computation time. When selecting a practical maintenance plan, it is useful to compare feasible solutions obtained under the various conditions. This process is inevitable and effective for the accountability by the disclosure of information. Then, an attempt was made to develop a decision support system for the bridge maintenance that can provide us with several alternative plans by applying Multi-Objective Genetic Algorithm (MOGA). However, it is not easy for the decision maker to choose an appropriate solution from many Pareto solutions. In order to help the decision maker, a 3D graphical system was developed using JAVA techniques. It is important to find out the appropriate repair methods and the branching points of cost effectiveness. Several numerical examples are presented to demonstrate the applicability and efficiency of the present system.

5.1 Concrete Bridge Model

A group of ten concrete highway bridges are considered. The locations of all these bridges along the coast of Japan are indicated in Fig. 30. Maintenance management planning for ten consecutive piers and floor slabs (composite structure of steel girders and reinforced concrete (RC) slabs) is considered. Each bridge has the same structure and is composed of six main structural components: upper part of pier, lower part of pier, shoe, girder, bearing section of floor slab, and central section of floor slab (Fig. 31). In this study, an attempt was made to develop a new searching method for optimization problem. Environmental conditions can significantly affect the degree of deterioration of the structures and may vary from location to location according to geographical characteristics such as wind direction, amount of splash, etc. To take the environmental conditions into account, the deterioration type and year from completion of each bridge are summarized in Table 18.

Table 18. Years from completion and type of deterioration caused by environmental conditions

Bridge number	Years from completion	Deterioration type
B01	2	neutralization of concrete
B02	2	neutralization of concrete
B03	0	chloride attack (slight)
B04	0	chloride attack (medium)
B05	0	chloride attack (severe)
B06	0	chloride attack (medium)
B07	0	chloride attack (severe)
B08	1	chloride attack (medium)
B09	1	chloride attack (slight)
B10	1	chloride attack (slight)

Environmental corrosion due to neutralization of concrete, chloride attack, frost damage, chemical corrosion, or alkali-aggregate reaction are considered as major deteriorations. The structural performance of each bridge component i is evaluated by the associated safety level (also called durability level) P_i which is defined as the ratio of current

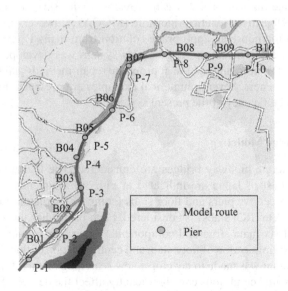

Fig. 30. Location of ten bridges in Japan

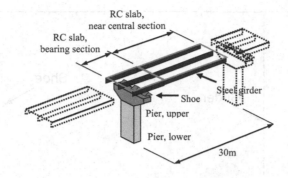

Fig. 31. Main components of a bridge

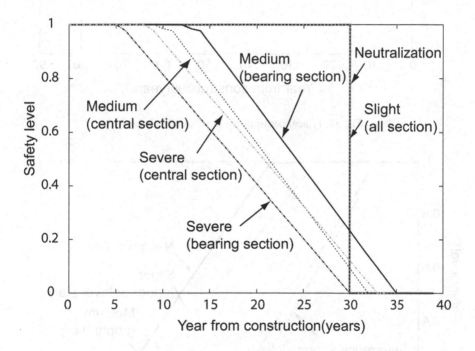

Fig. 32. Typical performance of RC slabs

safety level to initial safety level. Deterioration of a bridge due to corrosion depends on the concrete cover of its components and environmental conditions, among other factors. For each component, the major degradation mechanism and its rate of deterioration are assumed corresponding to associated environmental conditions. Figs. 32, 33,

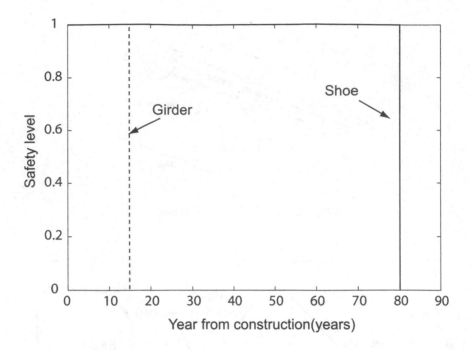

Fig. 33. Typical performance of shoes and girders

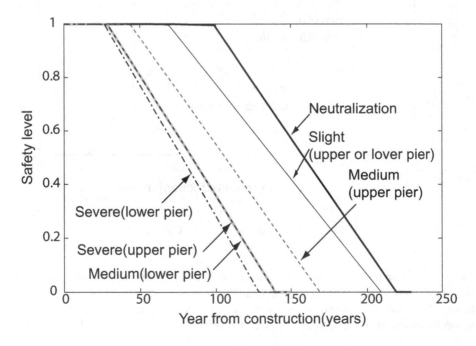

Fig. 34. Typical performance of piers

and 34 show the decreasing patterns of safety levels for RC slabs, shoes and girders, and piers, respectively. Average values are employed here as representative values for each level of chloride attack because the deteriorating rates can vary even in the same environment. The decrease of RC slab performance is assumed to depend on corrosion. Hence, the safety level depends on the remaining cross-section of reinforcement bars. For shoe and girder, the major deterioration mechanism is considered to be fatigue due to repeated loadings. The decrease in performances occurs as the rubber bearing of shoe or paint coating of girder deteriorates. For pier, the major mechanism for deterioration is assumed to be only corrosion. Thus the reduced performance of pier is expressed by the remaining section of reinforcement bars. The development of reinforcement corrosion is determined in accordance with Standard Specification for Design and Construction of Concrete in Japan.

5.2 Maintenance Methods

In order to prevent deterioration in structural performance, several options such as repair, restoring, and reconstruction are considered. Their applicability and effects on each component are shown in Table 19. Since the effects may differ even under the same conditions, average results are adopted here. Maintenance methods applicable to RC slab may vary according to the environmental conditions and are determined considering several assumptions.

Table 19. Effects of repair, restoring, and reconstruction

Structural component	Maintenance type	Average effect
Pier or Slab	Surface painting	Delays P_i decrease for 7 years
	Surface covering	Delays P_i decrease for 10 years
	Section restoring	Restores P_i to 1.0, and then allows it to deteriorate with the same slope as the initial deterioration curve
	Desalting (Re-alkalization)	P_i deteriorates with the same slope as the initial deterioration curve
	Cathodic protection	Delays P_i decrease for 40 years
	Section restoring with surface covering	Restores P_i to 1.0, delays P_i decrease for 10 years, and then P_i deteriorates with the same slope as the initial deterioration curve
Girder	Painting	Maintains initial performance until the end of the specified lifetime
Shoe	Replacement of bearing	Maintains initial performance until the end of the specified lifetime
Slab	Recasting	Maintains initial performance until the end of the specified lifetime
All	Reconstruction	Restore P_i to 1.0, delays P_i decrease for 10 years, and then P_i deteriorates with the same slope as the initial deterioration curve

5.3   LCC

LCC is defined as the total maintenance cost for the entire bridge group during its life. This is obtained by the summation of the annual maintenance costs through the service life of the bridges. The future costs are discounted to their present values. Other costs, such as indirect construction costs, general costs, administrative costs, etc., are calculated in accordance with Cost Estimation Standards for Civil Construction. The direct construction costs consist of material and labor costs and the cost of scaffold. The breakdown of the material and labor costs and the cost of the scaffold are shown in Tables 20 and 21. The construction costs are based upon the market prices.

Table 20. Material and labor costs

Maintenance action	Upper pier (yen/m^2)	Lower pier (yen/m^2)	Shoe (yen/part)	Girder (yen/m^2)	Slab -central section- (yen/m^2)	Slab -bearing section- (yen/m^2)
Surface painting	780,000	1,920,000	–	–	1,640,000	3,280,000
Surface covering	2,730,000	6,720,000	–	–	4,100,000	8,200,000
Section restoring	20,670,000	50,880,000	–	–	22,140,000	44,280,000
Desalting (Re-alkalization)	3,510,000	8,640,000	–	–	7,380,000	14,760,000
Cathodic protection	3,900,000	9,600,000	–	–	8,200,000	16,400,000
Section restoring with surface covering	22,620,000	55,680,000	–	–	26,240,000	52,480,000
Reconstruction	–	–	4,200,000	5,400,000	12,300,000	24,600,000

Table 21. Scaffold costs

Upper pier (yen/m^2)	Lower pier (yen/m^2)	Shoe (yen/part)	Girder (yen/m^2)	Slab -central section- (yen/m^2)	Slab -bearing section- (yen/m^2)
360,000	190,000	360,000	4,830,000	690,000	510,000

5.4   Application of MOGA to Maintenance Planning

It is desirable to determine an appropriate life-cycle maintenance plan by comparing several solutions for various conditions. A new decision support system is described here from the viewpoint of multi-objective optimization, in order to provide various solutions needed for the decision-making.

LCC, safety level and service life are used as objective functions. LCC is minimized, safety level is maximized, and service life is maximized. There are trade-off relations among the three objective functions. For example, LCC increases when service life is extended, and safety level and service life decrease due to the reduction of LCC. Then, multi-objective optimization can provide a set of Pareto solutions that cannot improve an objective function without making other objective functions worse.

In the present system, DNA structure is constituted as shown in Fig. 35, in which DNA of each individual consists of three parts such as repair method, interval of repair, and shared service life (Fig. 36). In this figure, service life is calculated as the sum of repairing years and their interval years. In Fig. 36, service life is obtained as 67 years which is expressed as the sum of 30 years and 37 years. The repair part and the interval part have the same length. Gene of repair part has ID number of repair method.

The interval part has enough length to consider service life. In this system, ID 1 means surface painting, ID 2 surface coating, ID 3 section restoring, ID 4 desalting (re-alkalization) or cathodic protection, and ID 5 section restoring with surface covering. DNA of service life part has a binary expression with six bits and its value is changed

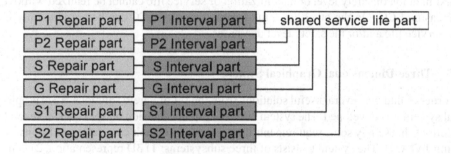

Fig. 35. Structure of DNA

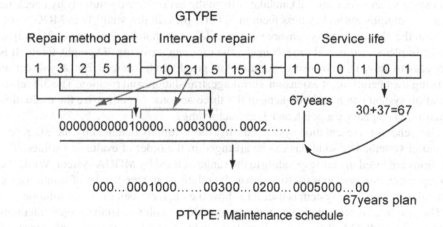

Fig. 36. Coding rule

to decimal number. For mutation, the system shown in Fig. 36 is used. Then, objective functions are defined as follows:

$$\text{Objective function 1:} \quad C_{total} = \sum LCC_i \rightarrow min \tag{12}$$

where LCC_i is LCC for bridge i.

$$\text{Objective function 2:} \quad Y_{total} = \sum Y_i \rightarrow max \tag{13}$$

subject to $Y_i > Y_{required}$, where Y_i is service life of bridge i, $Y_{required}$ is required service life.

$$\text{Objective function 3:} \quad P_{total} = \sum P_i \rightarrow max \tag{14}$$

subject to $P_i > P_{target}$, where P_{target} is target safety level.

The above objective functions have trade-off relations to each other. Namely, the maximization of safety level or maximization of service life cannot be realized without increasing LCC. On the other hand, the minimization of LCC can be possible only if the service life and/or the safety level decreases.

5.5 Three-Dimensional Graphical Systems

In order to find out several useful solutions from the set of Pareto solutions, a 3D graphical system was developed. The system aims to help the decision maker to select several solutions that satisfy some requirements through checking their constraint conditions by using JAVA3D. The system consists of three subsystems: 1) 3D representation, 2) general representation, and 3) graphical representation. Each representation is implemented using JAVA language.

The 3D representation system is the most important among the three subsystems. This system can select several candidates from the set of Pareto solutions by checking various requirements and express them in 3D graphics. In this study, both MOGA system and the 3D graphical system are written in JAVA so that the rendering of 3D graphs can be implemented in real time. Namely, the user can move the 3D graphs freely. It is very easy to move them by using a mouse. Any graph can be viewed from any direction by using the operations of extension, shrinkage, translation, and rotation. The 3D representation system can mainly implement the three actions: 1) emphasize the evaluation function, 2) emphasize a point, and 3) extract a range.

The general representation system can list the solutions obtained by the 3D representation system. The solutions can be arranged in the order of evaluation values. The solutions are listed up, corresponding to the range defined by MOGA system. While the 3D representation system is useful to grasp the relations and tendencies of solutions, the general representation system is useful to show the characteristics of each solution.

The graphical representation system can provide us with the detail of repair methods calculated by MOGA. The system can play a role in checking the appropriateness of the obtained repair methods and in finding out the tendency or pattern of repair program.

Observing and comparing the patterns obtained, it is possible to discover the branching points of short, medium and long term repair plans.

5.6   Application Example

Fig. 37 shows the Pareto solutions calculated by MOGA system. This graph is given in the conventional way of representation.

Fig. 38 shows the representation of the same graph by the proposed 3D representation system. Fig. 39 presents the same graph that is rotated and shrunken. As seen form Fig. 39, it is possible for the user to check the Pareto solutions from any desired direction by using the proposed representation system based on JAVA3D.

For example, it is possible to find out a gap among the solutions, which may be caused by the cost reduction by the common usage of scaffold. Apparently, it is possible to obtain more useful information by using the 3D representation instead of the usual 2D representation. Comparing the long term repair plans and short term repair plans, it is made clear that the long term plan is superior in monetary term to the plan with the repetition of short term repair (Table 22). The branching point between the short term plan and long term plan can be found out. Moreover, it is also possible to obtain the solutions for the cases in which the safety of the structure is emphasized and the LCC is emphasized, respectively. This means that the proposed system can provide the user with an appropriate solution for any case.

By considering LCC, safety level, and service life as objective functions, it is possible to obtain the relationships among these three performance indicators and provide

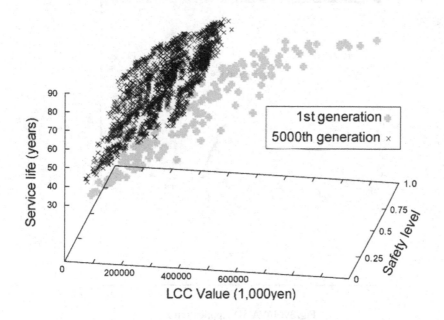

Fig. 37. Pareto solutions obtained by MOGA

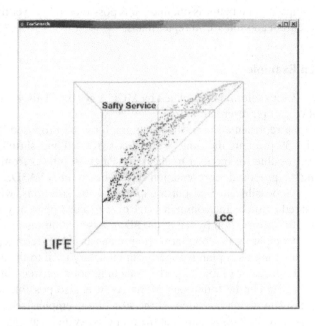

Fig. 38. JAVA 3D Application 1

Fig. 39. JAVA 3D Application 2

Table 22. Maintenance terms

Maintenance term	LCC	Safety level	Service life
long term	359501	5.18887	84
middle term	308605	5.18132	66
short term	255104	5.18154	40

bridge maintenance management engineers with various maintenance plans with appropriate allocations of resources. Since the optimal maintenance problem is a very complex combinatorial problem, it is difficult to obtain reasonable solutions by the current optimization techniques. Although GA is applicable to solve multi-objective problems, it is difficult to apply it to large and very complex bridge maintenance problems. By introducing the technique of Non-Dominated Sorting GA-2 (NSGA2), it is possible to obtain efficient near-optimal solutions for the maintenance planning of a group of bridge structures. However, it is not easy for the decision maker to choose an appropriate solution from many Pareto solutions. In order to help the decision maker, a 3D graphical system is developed using JAVA techniques. It is important to find the appropriate repair methods and the branching points of cost effectiveness.

6 Conclusions

In this paper, several practical optimization methods including GA were introduced, which are based on "evolutionary computing" or "soft computing". Several application examples in structural engineering are presented to discuss the efficiency and applicability of the methods described here. Through the numerical computations, the following conclusions were derived:

1. The optimization problems in real life are very difficult to solve, because they have objective functions and constraint conditions which have uncertainty and vagueness.
2. The evolutionary computing including GA is useful in solving real life problems, because of their superior ability such as understandable thinking way, high searching performance, easiness of programming, and robustness to peculiar characteristics of problems.
3. The structural vibration control system presented in this paper has an advantage that it can adapt to the change of structural systems and environments. Through the model and numerical experiments, it was validated that the systems can follow the change of vibration characteristics of structure. Using the descent method for the learning in fuzzy reasoning, quick and right adaptation can be achieved.
4. A decision support system for the aesthetic design of bridge handrails can be applied for practical use. In order to obtain several satisfactory design alternatives, the immune algorithm was applied to the optimization procedure and the neural network was used for the learning of the necessary knowledge. The effectiveness of the system was confirmed through numerical calculations.

5. The optimal restoration scheduling was formulated as a multi-objective optimization problem. By considering restoring days, LCC and performance level as objective functions, it is possible to obtain the relationships among these three performance indicators and to compare feasible optional solutions obtained under various conditions.
6. An optimal maintenance planning problem was also formulated as a multi-objective optimization. Furthermore, a 3D graphical representation system was introduced to find out several useful solutions from the set of Pareto solutions obtained by the optimal maintenance planning system using Multi-Objective Genetic Algorithm (MOGA).
7. By comparing the method with the current methods, it was proven that the present method can reduce the computation time, improve the convergence of searching procedure, regardless of vague or uncertain objective functions and constraint functions.

References

1. Soong, T.T.: Active Structural Control: Theory and Practice. Addison-Wesley, Reading (1990)
2. Furuhashi, T., Hayashi, I.: Fuzzy Neural Network. Asakura Publishing (1996) (in Japanese)
3. Uchikawa, Y.: Fuzzy Neural System. Nikkan Kogyo (1995) (in Japanese)
4. Dayhoff, J.E.: Neural Network Architectures: An Introduction. Van Nostrand Reinhold (1990)
5. Wang, L.X., Mendel, J.M.: Back-propagation fuzzy systems as nonlinear dynamic system identifiers. In: Proceedings of the IEEE International Congress on Fuzzy System (1992)
6. Furuta, H., et al.: Development of a decision support system for aesthetic design of existing and new girder bridges. Journal of Structural Engineering, JSCE 46A, 321–331 (2000) (in Japanese)
7. Mori, K., et al.: An immunity-based algorithm that offers diversity and its application to load allocation problems. Proceedings of Institute of Electrical Engineers of Japan, Part C 113(10), 872–878 (1993) (in Japanese)
8. Furuta, H., Nakatsu, K.: Optimal restoration scheduling for earthquake disaster by emergent computing. In: Proceedings of the IFIP WG 7.5 on Reliability and Optimization of Structural Systems, Banff, Alberta, Canada, November 2-5 (2003); Maes, M.A., Huyse, L. (eds.): Reliability and Optimization of Structural Systems, pp. 267–274. A.A. Balkema Publ., Leiden (2004)
9. Furuta, H., Nakatsu, K., Frangopol, D.M.: Optimal restoration scheduling for earthquake disaster using life-cycle cost. In: Proceedings of the Fourth International Workshop on Life-Cycle Cost Analysis and Design of Civil Infrastructure Systems, Cocoa Beach, Florida, May 8-11 (2005); Nowak, A.S., Frangopol, D.M. (eds.): Advances in Life-Cycle Analysis and Design of Civil Infrastructure Systems, pp. 47–54. University of Nebraska (2005)
10. Furuta, H., Sugimoto, H.: Applications of Genetic Algorithm to Structural Engineering. Morikita Publishing, Tokyo (1997) (in Japanese)
11. Frangopol, D.M., Furuta, H. (eds.): Life-Cycle Cost Analysis and Design of Civil Infrastructure Systems. ASCE, Reston (2001)
12. Goldberg, D.E.: Genetic Algorithms in Search, Optimization and Machine Learning. Addison Wesley Publishing Company, Inc., Reading (1989)

13. Furuta, H., Kameda, T., Fukuda, Y., Frangopol, D.M.: Life-Cycle Cost Analysis for Infrastructure Systems: Life Cycle Cost vs. Safety Level vs. Service Life. In: Proceedings of Joint International Workshops LCC03/IABMAS and fip/JCSS, EPFL, Lausanne, March 24-26 (2003) (keynote lecture); Frangopol, D.M., Brühwiler, E., Faber, M.H., Adey, B. (eds.): Life-Cycle Performance of Deteriorating Structures: Assessment, Design and Management, pp. 19–25. ASCE, Reston (2004)
14. Liu, M., Frangopol, D.M.: Probabilistic maintenance prioritization for deteriorating bridges using a multi-objective genetic algorithm. In: Proceedings of the Ninth ASCE Joint Specialty Conference on Probabilistic Mechanics and Structural Reliability, Omnipress, Albuquerque, July 26-28. Hosted by Sandia National Laboratories, New Mexico (2004), 6 pages on CD-ROM
15. Liu, M., Frangopol, D.M.: Optimal bridge maintenance planning based on probabilistic performance prediction. Engineering Structures 26(7), 991–1002 (2004)
16. Furuta, H., Kameda, T., Nakahara, K., Takahashi, Y.: Genetic algorithm for optimal maintenance planning of bridge structures. In: Proc. of GECCO, Chicago, USA (2003)
17. Kitano, H. (ed.): Genetic Algorithm 3. Sangyo-tosho, Tokyo (1995) (in Japanese)

Sufficient Conditions and Sensitivity Analysis for Optimal Bang-Bang Control Problems with State Constraints

Helmut Maurer[1] and Georg Vossen[2]

[1] Institut für Numerische und Angewandte Mathematik, Westfälische Wilhelms-Universität Münster, Einsteinstrasse 62, D-48149 Münster, Germany
maurer@math.uni-muenster.de
[2] Fraunhofer-Institut für Lasertechnik, Steinbachstr. 15, 52074 Aachen, Germany
georg.vossen@ilt.fraunhofer.de

Abstract. Bang-bang control problems subject to a state inequality constraint are considered. It is shown that the control problem induces an optimization problem, where the optimization vector assembles the switching and junction times for bang-bang and boundary arcs. Second order sufficient conditions (SSC) for the state-constrained control problem are given which require that SSC for the induced optimization problem are satisfied and a generalized strict bang-bang property holds at switching and junction times. This type of SSC ensures solution differentiability of optimal solutions under parameter perturbations and allows to compute parametric sensitivity derivatives. A numerical algorithm is presented that simultaneously determines a solution candidate, performs the second-order test and computes parametric sensitivity derivatives. We illustrate the algorithm with two state-constrained optimal control problems in biomedicine.

1 Introduction

Second-order sufficient optimality conditions (SSC) for bang-bang controls without state constraints have been derived in Agrachev, Stefani and Zezza [1] on the basis of an induced optimization problem where the control process is optimized with respect to the unknown switching times of the bang-bang control. The equivalence of this type of SSC with a different form of SSC obtained earlier in the literature has recently been shown in [20,21,23]. Numerical methods for the verification of optimization based SSC have been developed in Maurer, Büskens, Kim and Kaya [18] using the so-called arc-parameterization method where the arc-lengths of the bang-bang arcs are optimized. Basic ideas for a sensitivity analysis of bang-bang controls may be found in Kim, Maurer [11].

The purpose of this paper is to extend the results and techniques in [18,11] to bang-bang control problems with a state constraint. In Sect. 2, we review the necessary conditions for an optimal control problem with a state constraint of order one. The regularity conditions in assumptions (A1), (A2) allow to determine the multiplier associated with the state constraint. In Sect. 3, we formulate an *induced optimization problem* where the

A. Korytowski et al. (Eds.): System Modeling and Optimization, IFIP AICT 312, pp. 82–99, 2009.

optimization vector assembles the switching times of bang-bang arcs, resp., the entry- and exit-times (junction times) of boundary arcs, and the free final time. The crucial point in this optimization approach is the fact that the optimal control is given in feedback form along boundary arcs.

Based on second-order sufficient conditions (SSC) for the induced optimization problem, we give SSC for the control problem which require the strict bang-bang property in assumption (A4) and the strict complementarity in (A5). In Sect. 4, the arc-parameterization method from [18] is extended to incorporate boundary arcs. The main result on sensitivity analysis is given in Sect. 5. We include a formula for the sensitivity derivatives of the switching and junction times which can be implemented into the routine NUDOCCCS developed in [4]; cf. also [5,6]. In Sect. 6, we discuss the drug displacement problem in [22] in the light of the SSC presented in Theorem 2. In Sect. 7, we determine the optimal control in a two-compartment model for cancer therapy [12], when a state constraint on the number of tumor cells is imposed.

2   Optimal Bang-Bang Control Problems with a State Constraint

Let $x(t) \in \mathbb{R}^n$ denote the state variable and $u(t) \in \mathbb{R}$ the control variable at time $t \in [0, t_f]$, where the final time $t_f > 0$ is either fixed or free. For simplicity, the control is assumed to be scalar. The following autonomous optimal control problem with control variable appearing linearly will be denoted by (OC): determine a measurable control function $u : [0, t_f] \to \mathbb{R}$ and a terminal time $t_f > 0$ such that the pair of functions $(x(\cdot), u(\cdot))$ minimizes the cost functional of Mayer type

$$J(x, u, t_f) := g(x(t_f), t_f) \tag{1}$$

subject to the constraints in the interval $[0, t_f]$,

$$\dot{x}(t) = f(x(t), u(t)) = f_0(x(t)) + f_1(x(t))u(t), \tag{2}$$

$$x(0) = x_0, \quad \varphi(x(t_f), t_f) = 0, \tag{3}$$

$$u_{\min} \leq u(t) \leq u_{\max}, \tag{4}$$

and the scalar state inequality constraint

$$S(x(t)) \leq 0 \text{ for } 0 \leq t \leq t_f. \tag{5}$$

The functions $g : \mathbb{R}^n \times \mathbb{R} \to \mathbb{R}$, $f_0, f_1 : \mathbb{R}^n \to \mathbb{R}^n$, $\varphi : \mathbb{R}^n \times \mathbb{R} \to \mathbb{R}^r$, $0 \leq r \leq n$, and $S : \mathbb{R}^n \to \mathbb{R}$ are assumed to be twice continuously differentiable. The state constraint is assumed to be of *order one* [9,16], i.e., the total time derivative of the function $S(x(t))$ contains the control explicitly,

$$S^1(x, u) := S_x(x)(f_0(x) + f_1(x)u) = a(x) + b(x)u, \tag{6}$$

where $b(x) = S_x(x)f_1(x) \not\equiv 0$. Here and in the sequel, partial derivatives are denoted by subscripts. A subinterval $[\tau_1, \tau_2] \subset [0, t_f]$ is called an *interior arc* if $S(x(t)) < 0$ holds on (τ_1, τ_2). The interval $[\tau_1, \tau_2]$ is called a *boundary arc* if $S(x(t)) \equiv 0$ holds for all

$t \in [\tau_1, \tau_2]$. If $[\tau_1, \tau_2]$ is maximal with this property, then τ_1 is called *entry-time* and τ_2 is called *exit-time* of the boundary arc; τ_1, τ_2 are also called *junction times*. The following assumption is a standard regularity condition for a boundary arc [9,15,17].

(A1) $$b(x(t)) \neq 0 \, \forall \, t \in [\tau_1, \tau_2].$$

Under this assumption, the *boundary control* on a boundary arc is determined by the equation $S^1(x, u) = a(x) + b(x)u = 0$ as the feedback expression

$$u_b(x) = -a(x)/b(x), \quad u(t) = u_b(x(t)). \tag{7}$$

The following assumption will be needed to determine the multiplier associated with the state constraint explicitly.

(A2) The boundary control lies in the interior of the control region:

$$u_{\min} < u(t) = u_b(x(t)) < u_{\max} \, \forall \, t \in [\tau_1, \tau_2]. \tag{8}$$

Assumptions (A1) and (A2) allow us to formulate first order necessary conditions of Pontryagin's minimum principle in a computationally convenient form. We recall from [9,17] that the Lagrange multiplier associated with the state constraint (5) is a measure that is represented by a function μ of bounded variation. Using (A1) and (A2), it has been shown in [16,17,15,14] that the measure has a Radon-Nikodym derivative η which allows to write the following adjoint equation (11) in differential form. In the *direct adjoining approach* [9,17], the augmented Pontryagin or Hamiltonian function is defined by

$$H(x, u, \lambda, \mu) = \lambda f(x, u) + \eta S(x) = \lambda f_0(x) + \lambda f_1(x)u + \eta S(x), \tag{9}$$

where the adjoint variable $\lambda \in \mathbb{R}^n$ is a row vector and η is the multiplier associated with the state constraint.

Suppose now that $\bar{u} : [0, \bar{t}_f] \rightarrow [u_{\min}, u_{\max}]$ is an optimal control with optimal final time $\bar{t}_f$ and corresponding trajectory $\bar{x} : [0, \bar{t}_f] \rightarrow \mathbb{R}^n$. Assume that (A1), (A2) hold and the state constraint (5) is not active at $t = 0$ and $t = \bar{t}_f$,

$$S(\bar{x}(0)) < 0 \text{ and } S(\bar{x}(\bar{t}_f)) < 0. \tag{10}$$

In the sequel, we will use the junction theorem in [16], Corollary 5.2 (ii), where it was shown that the adjoint variables are *continuous* at junction times provided that the state constraint is of *first order* and the control is *discontinuous* at junctions. Note that the latter property follows from (A2). Then there exist an absolutely continuous (a.c.) adjoint function $\lambda : [0, \bar{t}_f] \rightarrow \mathbb{R}^n$, a piecewise a.c. multiplier function $\eta : [0, \bar{t}_f] \rightarrow \mathbb{R}$ and a multiplier $\rho \in \mathbb{R}^r$ (row vector) such that the following conditions hold a.e. on $[0, \bar{t}_f]$:

$$\dot{\lambda}(t) = -H_x(\bar{x}(t), \bar{u}(t), \lambda(t), \eta(t)), \tag{11}$$

$$\lambda(\bar{t}_f) = l_x(\bar{x}(\bar{t}_f), \bar{t}_f, \rho), \tag{12}$$

$$H(\bar{x}(t), \bar{u}(t), \lambda(t), \eta(t))|_{t=\bar{t}_f} + l_{t_f}(\bar{x}(\bar{t}_f), \bar{t}_f, \rho) = 0, \text{ if } t_f \text{ is free}, \tag{13}$$

$$H(\bar{x}(t), \bar{u}(t), \lambda(t), \eta(t)) = \min\{H(\bar{x}(t), u, \lambda(t), \eta(t)) \,|\, u \in [u_{\min}, u_{\max}]\}, \tag{14}$$

$$\eta(t) \geq 0, \ \eta(t) = 0, \text{ if } S(x(t)) < 0, \tag{15}$$

where $l(x,t_f,\rho) := (g + \rho\varphi)(x,t_f)$ is the endpoint Lagrangian function. The *switching function* is defined by

$$\sigma(x,\lambda) := H_u = \lambda f_1(x), \quad \sigma(t) = \sigma(x(t),\lambda(t)). \tag{16}$$

On *interior arcs* with $S(x(t)) < 0$ the minimum condition (14) yields the control law

$$u(t) = \begin{cases} u_{min}, & \text{if } \sigma(t) > 0, \\ u_{max}, & \text{if } \sigma(t) < 0. \end{cases} \tag{17}$$

The switching times of the control are zeroes of the switching function. A *singular arc* occurs if the switching function $\sigma(t)$ vanishes identically on an interval $I_{sing} \subset [0,\bar{t}_f]$. In this paper, we assume that the optimal control does not contain singular arcs.

Along a *boundary arc* $[\tau_1, \tau_2]$, assumption (A2) requires that the control takes values in the interior of the control set. Hence, the minimum condition (14) implies

$$\sigma(t) = \lambda(t) f_1(x(t)) = 0 \ \forall t \in [\tau_1, \tau_2]. \tag{18}$$

This relation can be interpreted as the property that a boundary control behaves formally like a singular control, a fact that was exploited in [16] to obtain junction theorems. By differentiating (18) and using the adjoint equation (11) we find the following explicit representation of the multiplier $\eta = \eta(x,\lambda)$ in (11) (cf. [17,14]),

$$\eta(x,\lambda) = \lambda[(f_1)_x(x)f(x,u_b(x)) - f_x(x,u_b(x))f_1(x)]/b(x), \tag{19}$$

where $u_b(x(t))$ is the boundary control (7). In short, the multiplier is given by the Lie bracket $\eta(x,\lambda) = \lambda[f_1,f]$; we then set $\eta(t) = \eta(x(t),\lambda(t))$.

3 Induced Optimization Problem and Second Order Sufficient Conditions

In order to transcribe the control problem into a finite-dimensional optimization problem, we make the following assumption:

(A3) The optimal control has finitely many bang-bang and boundary arcs.

Under assumptions (A1)-(A3), the optimal control problem can be transcribed into an optimization problem in the following way. We assume that the structure of the optimal control, i.e., the sequence of finitely many bang-bang and boundary arcs, is known. Let $t_j, j = 1, \ldots, s$, be the switching and junction times which are ordered as

$$0 =: t_0 < t_1 < \ldots < t_j < \ldots < t_s < t_{s+1} := t_f. \tag{20}$$

For simplicity, assume that there exists only a single boundary arc $[\tau_1, \tau_2] = [t_k, t_{k+1}]$ with an index $1 \leq k \leq s$. Then $[0, t_k]$ and $[t_{k+1}, t_f]$ are the interior arcs. By assumption, in every interval $I_j := [t_{j-1}, t_j]$ there exists a function $u^j(x)$ with the property that the optimal control is given by

$$u(t) = u^j(x(t)), \quad t_{j-1} \leq t \leq t_j, \quad (j = 1, \ldots, s, s+1). \tag{21}$$

The interval I_{k+1} then represents the boundary arc. The function $u^j(x)$ is either the constant value of the bang-bang control on interior arcs or the boundary control $u^{k+1}(x) = u_b(x) = -a(x)/b(x)$.

Consider now the *optimization variable* $z := (t_1, \ldots, t_{s+1})^* \in \mathbb{R}^{s+1}$ with $t_{s+1} := t_f$ in case of a free final time, resp., $z := (t_1, \ldots, t_s)^* \in \mathbb{R}^s$ for fixed final time t_f, where the asterisk denotes the transpose. Denote by $x(t; z)$ the absolutely continuous solution of the ODE system

$$\dot{x}(t) = f(x(t), u^j(x(t))), \quad t_{j-1} \leq t \leq t_j, \tag{22}$$

with initial condition $x(0) = x_0$. Then the control problem (OC) can be reformulated as the following finite-dimensional optimization problem (OP) with equality constraints:

$$\text{Min} \quad G(z) := g(x(t_{s+1}; z), t_{s+1})$$

$$\text{s.t.} \quad \Phi(z) := \varphi(x(t_{s+1}; z), t_{s+1}) = 0, \tag{23}$$

$$\mathcal{S}(z) := S(x(t_k; z)) = 0.$$

The last equation arises from the entry-condition for the boundary arc. We consider the Lagrangian for the induced optimization problem (OP) in normal form,

$$\mathcal{L}(z, \rho, \beta) = G(z) + \rho \Phi(z) + \beta \mathcal{S}(z), \tag{24}$$

with multipliers $\rho \in \mathbb{R}^r$ (row vector) and $\beta \in \mathbb{R}$. First order necessary and second order sufficient conditions (SSC) for (23) are well known in the literature. In the following theorem, we consider control problems with free final time which involve the optimization vector $z \in \mathbb{R}^{s+1}$.

Theorem 1. SSC FOR THE INDUCED OPTIMIZATION PROBLEM.
Let $\bar{z}$ be feasible for the optimization problem (23). Suppose there exist multipliers $\rho \in \mathbb{R}^r$ and $\beta \in \mathbb{R}$ such that the following three conditions hold:

(a) $\text{rank} \, [\Phi_z(\bar{z}) \,|\, \mathcal{S}_z(\bar{z})] = r + 1$,
(b) $L_z(\bar{z}, \rho, \beta) = 0$,
(c) $v^* L_{zz}(\bar{z}, \rho, \beta) v > 0$ *for all* $v \in \mathbb{R}^{s+1} \setminus \{0\}$ *with* $\Phi_z(\bar{z}) v = 0$, $\mathcal{S}_z(\bar{z}) v = 0$.

Then $\bar{z}$ is a strict local minimizer of the optimization problem (OP).

Arguments similar to those in [18,23] reveal that the first order conditions in part (a) and (b) of Theorem 1 are closely related to those in (11)-(13) involving the adjoint function $\lambda(t)$. Namely, using the multiplier ρ in the Lagrangian (24), we define the adjoint function $\lambda(t)$ through the transversality condition $\lambda(\bar{t}_f) = (g + \rho \varphi)_x(\bar{x}(\bar{t}_f), \bar{t}_f)$ in (12) and the adjoint equation (11) where the multiplier η is given in (19). However, on the boundary arc $[t_k, t_{k+1}]$ there exists another multiplier η^1 and an adjoint function $\lambda^1(t)$ which correspond to the so-called *indirect adjoining approach* [9,17]. Here, the Hamiltonian is defined by $H^1 = \lambda^1 f(x, u) + \eta^1 S^1(x, u)$ with the function $S^1(x, u)$ given in (6). Then the adjoint equation is $\dot{\lambda}^1 = -H_x^1$ and the multiplier η^1 is determined via the equation $H_u^1 = 0$ as $\eta^1(x, \lambda^1) = -\lambda^1 f_u(x, u_b(x))/b(x)$. The multiplier β in the Lagrangian (24) yields the jump condition $\lambda^1(t_k+) = \lambda^1(t_k-) - \beta S_x(x(t_k))$ at the entry-time t_k. Moreover, one can show the relation $\beta = \int_{t_k}^{t_{k+1}} \eta(t) dt \geq 0$.

For bang-bang control problems without state constraints, it was shown in [1,20,21,23] that one further needs the so-called *strict bang-bang property* to obtain SSC for the bang-bang control problem. The following assumption gives an extension of the strict bang-bang property to control problems with state constraints.

(A4) **(a)** on *interior arcs* for $j = 1, ..., k-1, k+2, ..., s$:
$$\sigma(t_j) = 0, \ \dot{\sigma}(t_j)(u(t_j^-) - u(t_j^+)) > 0, \ \sigma(t) \neq 0 \text{ for } t \neq t_j .$$
   (b) at the entry-time t_k and exit-time t_{k+1} of the *boundary arc*:
$$\dot{\sigma}(t_k-)(u(t_k^-) - u(t_k^+)) > 0, \ \dot{\sigma}(t_{k+1}^+)(u(t_{k+1}^-) - u(t_{k+1}^+)) > 0.$$

Finally, we need the following *strict complementarity* condition:

(A5) The multiplier $\eta(t)$ satisfies $\eta(t) > 0 \, \forall \, t \in [t_k, t_{k+1}]$.

Note that assumptions (A4) and (A5) have also been used in [13,14] to construct a local field of extremals near boundary arcs. This has enabled the authors to prove sufficient conditions (SSC) which correspond to the following form of SSC. Detailed proofs will be given elsewhere.

Theorem 2. SSC FOR THE STATE-CONSTRAINED CONTROL PROBLEM.
Let $\bar{u}$ be a feasible control for the control problem (1)-(5) which has finitely many switching and junction times $\bar{t}_j, j = 1, ..., s$, and let $\bar{x}$ be the corresponding trajectory. Suppose there exists an adjoint function $\lambda : [0, t_f] \rightarrow \mathbb{R}^n$ and a multiplier $\rho \in \mathbb{R}^r$ such that assumptions (A1)-(A5) hold with multiplier function $\eta : [0, t_f] \rightarrow \mathbb{R}$ defined by (19). Suppose further that the vector $\bar{z} = (\bar{t}_1, ..., \bar{t}_s, \bar{t}_{s+1})^* \in \mathbb{R}^{s+1}, \bar{t}_{s+1} = \bar{t}_f$, satisfies the SSC in Theorem 1. Then the control $\bar{u}$ provides a strict strong minimum for the control problem (OC).

4 Numerical Methods for Solving the Induced Optimization Problem

In this section, we shall extend the *arc-parameterization* method in [10,18] to solve state-constrained control problems. Instead of directly optimizing the switching and junction times $t_j, j = 1, ..., s$, one determines the arc-lengths (arc durations)
$$\xi_j := t_j - t_{j-1}, \ j = 1, ..., s, s+1, \tag{25}$$
of bang-bang and boundary arcs. Therefore, the optimization variable $z = (t_1, ..., t_s, t_{s+1})^*$ is replaced by the optimization variable
$$\xi := (\xi_1, ..., \xi_s, \xi_{s+1})^* \in \mathbb{R}^{s+1}, \ \xi_j := t_j - t_{j-1}. \tag{26}$$
The variables z and ξ are related by a linear transformation involving the regular $(s+1) \times (s+1)$-matrix R,

$$\xi = Rz, \quad z = R^{-1}\xi, \quad R = \begin{pmatrix} 1 & 0 & ... & 0 \\ -1 & 1 & \ddots & \vdots \\ & \ddots & \ddots & 0 \\ 0 & & -1 & 1 \end{pmatrix}. \tag{27}$$

In the arc-parameterization method, the time interval $[t_{j-1}, t_j]$ is mapped to the fixed interval $I_j := \left[\frac{j-1}{s+1}, \frac{j}{s+1}\right]$ by the linear transformation

$$t = a_j + b_j \tau, \quad \tau \in I_j = \left[\frac{j-1}{s+1}, \frac{j}{s+1}\right], \tag{28}$$

where $a_j = t_{j-1} - (j-1)\xi_j$, $b_j = (s+1)\xi_j$. Identifying $x(\tau) \cong x(a_j + b_j \tau) = x(t)$ in the relevant intervals, we obtain the ODE system

$$\dot{x}(\tau) = (s+1)\,\xi_j\, f(x(\tau), u^j(x(\tau))) \text{ for } \tau \in I_j. \tag{29}$$

By concatenating the solutions in the intervals I_j we get the *continuous* solution $x(t) = x(t; \xi)$ in the normalized interval $[0, 1]$. When expressed via the new optimization variable ξ, the optimization problem (OP) in (23) is equivalent to the following optimization problem $(\widetilde{OP})$ with $t_f = \sum\limits_{j=1}^{s+1} \xi_j$:

$$\text{Min } \tilde{G}(\xi) := g(x(1; \xi), t_f),$$

$$\text{s.t. } \widetilde{\Phi}(\xi) := \varphi(x(1; \xi), t_f) = 0, \tag{30}$$

$$\widetilde{\mathscr{S}}(\xi) := S(x(\tfrac{k}{s+1}; \xi)) = 0.$$

The Lagrangian function is given by

$$\mathscr{L}(\xi, \rho, \beta) = \widetilde{G}(\xi) + \rho \widetilde{\Phi}(\xi) + \beta \widetilde{\mathscr{S}}(\xi). \tag{31}$$

Using the linear transformation (27), it can easily be seen that the SSC for the optimization problems (OP) and $(\widetilde{OP})$ are equivalent; cf. similar arguments in [18]. To solve this optimization problem, we use a suitable adaptation of the control package NUDOCCCS in Büskens [4,6]. Then we can take advantage of the fact that this routine also provides the Jacobian of the equality constraints and the Hessian of the Lagrangian which are needed to check the second order condition in Theorem 1.

5   Sensitivity Analysis for Bang-Bang Control Problems with a State Constraint

The SSC given in Theorem 2 pave the way to stability and sensitivity analysis for *parametric* bang-bang control problems with a state constraint. Suppose that the control problem (OC) in (1)-(5) depends as well on a parameter $p \in P \subset \mathbb{R}^q$ in the following way: Minimize

$$J(x, u, t_f, p) := g(x(t_f), t_f, p) \tag{32}$$

subject to the constraints on the interval $[0, t_f]$,

$$\dot{x}(t) = f(x(t), u(t), p) = f_0(x(t), p) + f_1(x(t), p)u(t), \tag{33}$$

$$x(0) = x_0, \quad \varphi(x(t_f), t_f, p) = 0, \tag{34}$$

$$u_{\min} \leq u(t) \leq u_{\max}, \tag{35}$$

$$S(x(t), p) \leq 0. \tag{36}$$

All functions are supposed to be sufficiently smooth. This parametric control problem will be denoted by $(OC(p))$. For a fixed parameter $p_0 \in P$, the problem $(OC(p_0))$ is considered as the *nominal control problem*.

We shall assume that the state constraint is of order one uniformly in the parameter p, i.e., for

$$S^1(x,u,p) = S_x(x,p)(f_0(x,p) + f_1(x,p)u) = a(x,p) + b(x,p)u \qquad (37)$$

we have $b(x,p) \not\equiv 0$ for all $p \in P$. Then the boundary control is given by

$$u_b(x,p) = -a(x,p)/b(x,p). \qquad (38)$$

Assuming (A1)-(A5) for the nominal control problem $(OC(p_0))$, we arrive at an *induced optimization problem* (23) *in parametric form* upon inserting the parametric boundary control (38). The Lagrangian function (24) becomes

$$\mathscr{L}(z,\rho,\beta,p) = G(z,p) + \rho\,\Phi(z,p) + \beta\,\mathscr{S}(z,p). \qquad (39)$$

Using a well-known sensitivity result for finite-dimensional parametric optimization problems (cf. [7,5,6]) we arrive at the following sensitivity result for the parametric control problem.

Theorem 3. SENSITIVITY ANALYSIS OF THE CONTROL PROBLEM $(OC(p))$.
Suppose that $\bar{u}(t) =: u(t,p_0)$ is a feasible control for the nominal control problem $(OC(p_0))$. Assume that $u(t,p_0)$ has switching and junction times $\bar{t}_j, j = 1, ..., s$, and a final time $\bar{t}_f = \bar{t}_{s+1}$ such that the SSC in Theorem 2 are satisfied. Then there exists a neighborhood $P_0 \subset P$ of the nominal parameter p_0 and functions $t_j : P_0 \to \mathbb{R}$ $(j = 1, ..., s, s+1)$, $\rho : P_0 \to \mathbb{R}^r$ and $\beta : P_0 \to \mathbb{R}$ with the following properties:

(1) $t_j(p_0) = \bar{t}_j, \ j = 1, ..., s, s+1,$
(2) for every $p \in P_0$, the control $u(t,p)$ with switching and junction times $t_j(p), j = 1, ..., s$, and final time $t_f(p) = t_{s+1}(p)$ is a strict strong local minimum for $(OC(p))$. The values of $u(t,p)$ agree with those of the nominal control $\bar{u}(t)$ on every interior bang-bang interval and are determined on boundary arcs by $u(t,p) = u_b(x(t),p)$ in view of (38).

The parametric sensitivity derivatives of $z(p) := (t_1(p), ..., t_s(p), t_{s+1}(p))^* \in \mathbb{R}^{s+1}$ and of $\rho(p), \beta(p)$ are given by the formula

$$\begin{pmatrix} dz/dp \\ d\rho^*/dp \\ d\beta/dp \end{pmatrix} = -\begin{pmatrix} \mathscr{L}_{zz} & (\Psi_z)^* \\ \Psi_z & 0 \end{pmatrix}^{-1} \begin{pmatrix} \mathscr{L}_{zp} \\ \Psi_{zp} \end{pmatrix}, \qquad (40)$$

where we have put

$$\Psi(z,p) = \begin{pmatrix} \Phi(z,p) \\ \mathscr{S}(z,p) \end{pmatrix}. \qquad (41)$$

The right hand side in (40) is evaluated for the argument $(z(p), \rho(p), \beta(p))$.

Proof. We sketch the main ideas of the proof. Since SSC hold for the nominal induced optimization problem $(OC(p_0))$, the sensitivity theorem in [7,5,6] tells us that there exist a neighborhood $P_0 \subset P$ of p_0 and, for every $p \in P_0$, an optimal solution and multipliers

$$z(p) = (t_1(p), ..., t_s(p), t_f(p)), \ \rho(p), \ \beta(p), \tag{42}$$

for the parametric problem $(OC(p))$. The triple $(z(p), \rho(p), \beta(p))$ satisfies the assumptions of Theorem 1 and is a C^1-function with respect to $p \in P_0$. We define the parametric control $u(t, p)$ in the following way. On bang-bang arcs $[t_j(p), t_{j+1}(p)]$, $j \neq k$, the control $u(t, p)$ takes the values of the nominal control on the corresponding nominal arcs $[\bar{t}_j, \bar{t}_{j+1}]$. On the boundary arc $[t_k(p), t_{k+1}(p)]$, the parametric control is defined by $u(t, p) = u_b(x(t), p)$ using the boundary control (38). Then corresponding trajectory $x(t, p)$ is determined by

$$\dot{x} = f(x, u(t, p)), \ x(0, p) = x_0. \tag{43}$$

Then the adjoint function $\lambda(t, p)$ is defined by the transversality condition

$$\lambda(t_f(p), p) := (g + \rho(p)\varphi)_x(x(t_f(p), p), t_f(p)) \tag{44}$$

and the solution to the adjoint equation

$$\dot{\lambda}(t, p) = -H_x(x(t, p), \lambda(t, p), u(t, p), \eta(t, p)), \tag{45}$$

where the multiplier $\eta(t, p)$ is given by the Lie-bracket

$$\eta(t, p) = \lambda(t, p)[f_1, f](x(t, p), u(t, p), p). \tag{46}$$

By shrinking the set P_0 if necessary, it is easy to verify that assumptions (A1)-(A5) hold for all $p \in P_0$. This proves the optimality of the triple $(x(t, p), u(t, p), \lambda(t, p))$. The sensitivity formula (40) follows from standard results in [7,5,6]. Note that Theorem 3 represents an extension of a sensitivity result for bang-bang control problems without state constraints; cf. Kim, Maurer [11].   □

We point out that the code NUDOCCCS [4] provides also the numerical evaluation of sensitivity formula (40).

6   Numerical Example: Time-Optimal Drug Displacement Problem

We consider the time-optimal control problem discussed in Bell, Katusiime [3] and Maurer, Wiegand [22]. The model simulates the interaction of the two drugs, warfarin and phenylbutazone in the human blood stream. The state variables x_1 and x_2 represent the concentration of warfarin and phenylbutazone, respectively. The problem is to control the rate of infusion (control u) of the pain-killing drug phenylbutazone such that both drugs reach a stable steady-state in minimal time while the concentration of warfarin is bounded by a given toxic level:

$$\text{Minimize} \quad t_f \tag{47}$$

$$\text{subject to} \quad \dot{x}_1 = D^2(C_2(0.02 - x_1) + 46.4x_1(u - 2x_2))/C_3, \tag{48}$$

$$\dot{x}_2 = D^2(C_1(u - 2x_2) + 46.4x_2(0.02 - x_1))/C_3, \tag{49}$$

$$x_1(0) = 0.02, \quad x_2(0) = 0, \tag{50}$$

$$x_1(t_f) = 0.02, \quad x_2(t_f) = 2, \tag{51}$$

$$0 = u_{\min} \leq u(t) \leq u_{\max} = 8 \quad \forall t \in [0, t_f], \tag{52}$$

and the state constraint

$$S(x(t)) = x_1(t) - \alpha \leq 0 \qquad \forall t \in [0, t_f], \tag{53}$$

where

$$D = D(x) = 1 + 0.2x_1 + 0.2x_2, \tag{54}$$

$$C_1 = C_1(x) = D^2 + 232 + 46.4x_2, \tag{55}$$

$$C_2 = C_2(x) = D^2 + 232 + 46.4x_1, \tag{56}$$

$$C_3 = C_3(x) = C_1 C_2 - (46.4)^2 x_1 x_2. \tag{57}$$

The augmented Hamiltonian associated with the state constraint (53) is given by

$$H(x, u, \lambda, \eta) = \lambda f(x, u) + \eta S(x) = \frac{D^2}{C_3} \cdot K + \eta(x_1 - \alpha), \tag{58}$$

where

$$K := \lambda_1(C_2(0.02 - x_1) + 46.4x_1(u - 2x_2))$$
$$+ \lambda_2(C_1(u - 2x_2) + 46.4x_2(0.02 - x_1)). \tag{59}$$

Then the adjoint equations (11) are

$$\dot{\lambda}_1 = -H_{x_1} = -\frac{(D^2)_{x_1} C_3 - (C_3)_{x_1} D^2}{C_3^2} \cdot K - \frac{D^2}{C_3} \cdot K_{x_1} - \eta, \tag{60}$$

$$\dot{\lambda}_2 = -H_{x_2} = -\frac{(D^2)_{x_2} C_3 - (C_3)_{x_2} D^2}{C_3^2} \cdot K - \frac{D^2}{C_3} \cdot K_{x_2}, \tag{61}$$

with the following partial derivatives of K:

$$K_{x_1} = 0.02\lambda_1(C_2)_{x_1} - \lambda_1(C_2 + (C_2)_{x_1} x_1) + 46.4\lambda_1 u - 92.8x_2\lambda_1 \tag{62}$$
$$+ \lambda_2 u(C_1)_{x_1} - 2x_2\lambda_2(C_1)_{x_1} - 46.4x_2\lambda_2, \tag{63}$$

$$K_{x_2} = 0.02\lambda_1(C_2)_{x_2} - x_1\lambda_1(C_2)_{x_2} - 92.8x_1\lambda_1 + \lambda_2 u(C_1)_{x_2} \tag{64}$$
$$- 2\lambda_2(C_1 + (C_1)_{x_2} x_2) + 0.928\lambda_2 - 46.4x_1\lambda_2. \tag{65}$$

The switching function (16) is given by

$$\sigma(x,\lambda) = H_u = \frac{D^2}{C_3}(46.4x_1\lambda_1 + C_1\lambda_2). \tag{66}$$

The state constraint (53) is of order one, as the function $S^1(x,u)$ in (6) contains the control explicitly:

$$S^1(x,u) = S_x(x)f(x,u) = \frac{D^2}{C_3}(C_2(0.02 - x_1) + 46.4x_1(u - 2x_2)). \tag{67}$$

The boundary control $u_b(x)$ in (7) is determined by the feedback expression

$$u_b(x) = \frac{\alpha - 0.02}{46.4\alpha}C_2(x) + 2x_2. \tag{68}$$

Using boundary value methods it was shown in [22] that for toxic levels $0 < \alpha \leq 0.0246$ the control has the following structure with two bang-bang arcs encompassing a boundary arc:

$$u(t) = \begin{cases} 8, & \text{for } t \in [0,t_1), \\ u_b(x(t)) = \frac{\alpha-0.02}{46.4\alpha}C_2(x(t)) + 2x_2(t), & \text{for } t \in [t_1,t_2], \\ 0, & \text{for } t \in (t_2,t_f]. \end{cases} \tag{69}$$

We choose the toxic level $\alpha = 0.024$ and compute the numerical solution using the routine IPOPT by Wächter and Biegler [27]. Using 1000 discretization points and the method of Heun for the approximation of the differential equations, IPOPT provides the solution depicted in Figs. 1-3. The final time is computed as $t_f = 358.731$ and the entry- and exit-time are $t_1 = 29.7747$ and $t_2 = 333.261$.

Now we solve the induced optimization problem (23) with optimization variables $z = (t_1,t_2,t_3)$, resp., problem (30) with variables $\xi = (\xi_1,\xi_2,\xi_3)$, where $\xi_1 = t_1, \xi_2 = t_2 - t_1, \xi_3 = t_f - t_2$. The routine NUDOCCCS [4] yields the entry-time $t_1 = 29.90806$, the exit-time $t_2 = 333.1561$ and the final time $t_f = 358.7085$. NUDOCCCS also provides the Hessian $\mathcal{L}_{\xi\xi}$ and Jacobian $\tilde{\Phi}_\xi$,

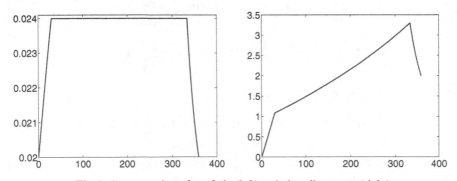

Fig. 1. Concentration of warfarin (left) and phenylbutazone (right)

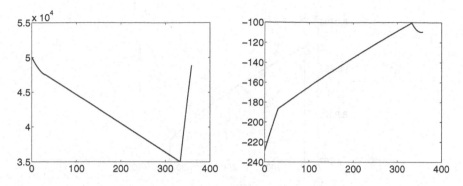

Fig. 2. Adjoint variables $\lambda_1(t)$ and $\lambda_2(t)$

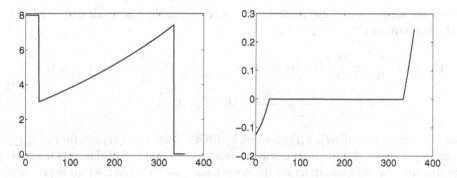

Fig. 3. Optimal control and switching function

$$\mathcal{L}_{\xi\xi} = \begin{pmatrix} -0.046251 & -0.014521 & 0.026232 \\ -0.014521 & -0.0011993 & 0.0052637 \\ 0.026232 & 0.0052637 & 0.054240 \end{pmatrix}, \tag{70}$$

$$\tilde{\Phi}_\xi = \begin{pmatrix} 0.0000000 & 0.0000000 & -0.0000954 \\ 0.0298122 & 0.0043075 & -0.0336047 \\ 0.0001245 & 0.0000000 & 0.0000000 \end{pmatrix}. \tag{71}$$

Since $\text{rank}(\tilde{\Phi}_\xi) = 3$ holds, condition (a) of Theorem 1 is fulfilled with $s = r = 2$. Furthermore, it is clear that the first order necessary conditions hold in the induced problem, i.e., condition (b) in Theorem 1, is fulfilled. Finally, the second order condition (c) is trivially satisfied, since the matrix $\tilde{\Phi}_\xi$ is regular. Next, we check the regularity assumption (A1). The data provided by IPOPT give the following estimate:

$$b(x(t)) = 46.4 \frac{D(x(t))^2}{C_3(x(t))} x_1(t) \geq 0.23 \cdot 10^{-3}. \tag{72}$$

Hence, (A1) is satisfied. It can be seen in Fig. 3 that also the assumptions (A2) and (A4) hold.

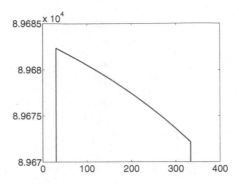

Fig. 4. Multiplier $\eta(t)$ from (73)

For the verification of assumption (A5), we compute the multiplier $\eta(t) = \eta(x(t), \lambda(t))$ via formula (19),

$$
\begin{aligned}
\eta(x,\lambda) = {} & \frac{2\dot{D}C_3 - D\dot{C}_3}{46.4DC_3 x_1}(46.4 x_1 \lambda_1 + C_1 \lambda_2) + \frac{1}{46.4 x_1}(\dot{C}_1 \lambda_2 + C_1 \dot{\lambda}_2) + \frac{1}{x_1}\dot{x}_1 \lambda_1 \\
& - x_3 \cdot \left(\frac{(D^2)_{x_1} C_3 - (C_3)_{x_1} D^2}{C_3^2} \cdot K + \frac{D^2}{C_3} \cdot K_{x_1} \right),
\end{aligned}
\tag{73}
$$

and insert the values of $x(t)$, $\lambda(t)$ provided by IPOPT. Note that $\eta(t) \equiv 0$ for $t \notin [t_1, t_2]$. Then Fig. 4 shows that the strict complementarity condition (A5) holds. It is noteworthy that the Lagrange multiplier for the discretized state constraint, which is provided directly by IPOPT, has the same values as the multiplier $\eta(t)$. Finally, we may conclude from Theorem 2 that the control (69) provides a strict strong minimum for the problem (47)-(53).

7   Optimal Control for a Two-Compartment Model in Cancer Chemotherapy

Ledzewicz, Schättler [12] considered a two-compartment model in cancer chemotherapy and established optimality using extremal field theory [13]. The state constraint (77) below has been studied in de Pinho, Ferreira, Ledzewicz, Schättler [24] using the methods developed in [13,14]. Here, we prove optimality on the basis of the SSC in Theorem 2 which allows us to apply the sensitivity result in Theorem 3. The description of the control model is taken from [12]: "The cell cycle is broken into two compartments of which the first combines the first growth phase G_1 and the synthesis phase S while the second contains the second growth phase G_2 and mitosis M. Let $x_i(t), i = 1,2$, denote the number of cancer cells in the i-th compartment at time t." The control u is the drug treatment which is measured by its cell-killing effect. The control problem is to *minimize* the cost functional with *fixed* final time t_f,

$$
J(x,u) = r_1 x_1(t_f) + r_2 x_2(t_f) + \int_0^{t_f} u(t)dt,
\tag{74}
$$

subject to

$$\dot{x}_1 = -a_1x_1 + 2(1-u)a_2x_2, \quad x_1(0) = x_{10},$$
$$\dot{x}_2 = a_1x_1 - a_2x_2, \quad\quad\quad x_2(0) = x_{20}, \quad\quad (75)$$
$$0 \le u(t) \le 1 \quad \forall t \in [0, t_f].$$

The cost functional (74) can be transformed to a functional (1) of Mayer type by introducing the equation $\dot{x}_3 = u$, $x_3(0) = 0$, which yields

$$J(x, u) = g(x(t_f)) = r_1x_1(t_f) + r_2x_2(t_f) + x_3(t_f). \quad\quad (76)$$

In addition, we consider the state constraint of order one:

$$S(x(t)) := x_1(t) + x_2(t) - \alpha \le 0 \quad \forall t \in [0, t_f], \quad\quad (77)$$

which imposes an upper bound on the total number of tumor cells in both compartments. The first total time derivative (6) of $S(x)$ is given by

$$S^1(x, u) = a_2x_2 - 2a_2x_2u. \quad\quad (78)$$

Obviously, assumption (A1) is satisfied since $b(x(t)) = -2a_2x_2(t) \ne 0$ holds on $[0, t_f]$. Data for (74) and (75) are taken from [12]:

$$r_1 = 6.94, \ r_2 = 3.94, \ a_1 = 0.197, \ a_2 = 0.356, \ t_f = 10. \quad\quad (79)$$

We choose the initial conditions $x_1(0) = x_{10} = 0.86$, $x_2(0) = x_{20} = 0.55$, which match approximately the solution in [12], where the terminal state $x(t_f)$, $i = 1, 2$, was fixed and the initial state was free. The parameter α in the state constraint (77) will be assigned the value $\alpha = 1.7$ for which the state constraint becomes active. The augmented Hamiltonian (9) is given by

$$H = \lambda_1(-a_1x_1 + 2a_2x_2) + \lambda_2(a_1x_1 - a_2x_2) + \sigma u + \eta(x_1 + x_2 - \alpha), \quad\quad (80)$$

with *switching function*

$$\sigma = \sigma(x, \lambda) = 1 - 2a_2x_2\lambda_1. \quad\quad (81)$$

The adjoint equation (11) and the transversality condition (12) yield

$$\dot{\lambda}_1 = a_1(\lambda_1 - \lambda_2) - \eta, \quad\quad \lambda_1(t_f) = r_1,$$
$$\dot{\lambda}_2 = a_2(2(u-1)\lambda_1 + \lambda_2) - \eta, \ \lambda_2(t_f) = r_2. \quad\quad (82)$$

The *boundary control* $u_b(x)$ satisfies the equation $S^1(x, u_b(x)) \equiv 0$ which gives

$$u_b(x) \equiv \frac{1}{2}. \quad\quad (83)$$

Hence, the boundary control lies in the interior of the control set and satisfies assumption (A2). The multiplier η for the state constraint (77) is determined by equation (19):

$$\eta(t) = a_1\lambda_1(t)\left(\frac{x_1(t)}{x_2(t)} + 1\right) - a_2\lambda_1(t) - a_1\lambda_2(t). \tag{84}$$

To determine the structure of the optimal control we first discretize the control problem with 500 gridpoints and apply the program NUDOCCCS of Büskens [4]. We find that the control has two bang-bang arcs and one *boundary arc*:

$$u(t) = \begin{cases} 0, & t \in [0,t_1), \\ u_b(x(t)) = \frac{1}{2}, & t \in [t_1,t_2], \\ 1, & t \in (t_2,t_f]. \end{cases} \tag{85}$$

Fig. 5 (left) displays the optimal control and the switching function. It clearly shows that the optimal control satisfies assumptions (A3) and (A4) since, in particular, for $k = 1$ in (A4) we have $\dot{\sigma}(t_1-) < 0$ and $\dot{\sigma}(t_2+) < 0$. Fig. 5 (right) depicts the state constrained function $x_1(t) + x_2(t)$ and the multiplier $\eta(t)$, which is seen to satisfy the *strict* complementarity condition (A5). State and adjoint variables are shown in Fig. 6, resp., Fig. 7.

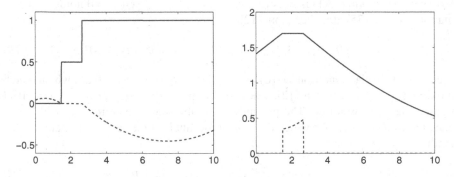

Fig. 5. Left: control with switching function (dashed); right: state constrained function $x_1(t) + x_2(t)$ with multiplier $\eta(t)$ (dashed)

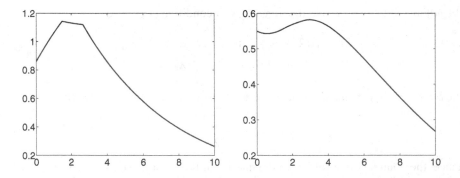

Fig. 6. State variables x_1 and x_2

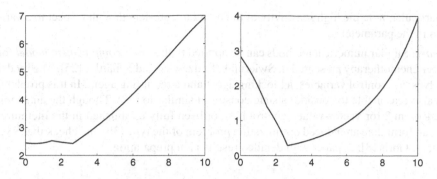

Fig. 7. Adjoint variables λ_1 and λ_2

It remains to verify the SSC in Theorem 1 for the optimization problem (30). The optimization variable is defined by (25) as

$$\xi = (\xi_1, \xi_2) = (t_1, t_2 - t_1). \tag{86}$$

Then the arc-length of the terminal time interval is given by $t_f - \xi_1 - \xi_2$ with $t_f = 10$. Since no terminal state boundary conditions are prescribed, the only equality constraint is the entry-condition of the boundary arc,

$$x_1(1/3; \xi) + x_2(1/3; \xi) = \alpha = 1.7. \tag{87}$$

The code NUDOCCCS gives the following results:

$$\begin{aligned}
t_1 &= 1.490713, \quad t_2 = 2.653005, \\
\lambda_1(0) &= 2.44417, \quad \lambda_2(0) = 2.82883, \\
x_1(t_f) &= 0.2635156, \, x_2(t_f) = 0.2673589, \\
J(x, u) &= 10.81033.
\end{aligned} \tag{88}$$

The Hessian of the Lagrangian for (30) is computed as

$$\mathscr{L}_{\xi\xi} = \begin{pmatrix} 0.225319 & 0.128060 \\ 0.128060 & 0.099212 \end{pmatrix} \tag{89}$$

while the Jacobian of the equality constraint is given by $\mathscr{S}_\xi = (0.1979670, 0)$. Obviously, the Hessian $\mathscr{L}_{\xi\xi}$ is positive definite and we have rank $(\mathscr{S}_\xi) = 1$. Hence, we may conclude that the control (85) with data (88) satisfies the SSC in Theorem 1 and provides a strict local minimum of the optimal control problem.

Using the sensitivity result in Theorem 3 and the sensitivity formula (38), we obtain the following sensitivity derivatives for the arc-length of the first bang-bang arc and the boundary arc:

$$\begin{aligned}
d\xi_1/da_1 &= -1.513, \, d\xi_2/da_1 = 11.99, \\
d\xi_1/da_2 &= -3.350, \, d\xi_2/da_2 = 0.5165, \\
d\xi_1/dx_{10} &= -5.359, \, d\xi_1/dx_{10} = 4.421, \\
d\xi_1/dx_{20} &= -7.233, \, d\xi_1/dx_{20} = 4.077.
\end{aligned} \tag{90}$$

In particular, note the high sensitivity of the boundary arc-length with respect to a variation in the parameter a_1.

Remark. Similar numerical methods can be applied to the *three-compartment model* for cancer chemotherapy presented in Swierniak, Ledzewicz and Schättler [25]. The model involves two control variables, a blocking agent and a recruiting agent. In this problem, it is also reasonable to consider a state constraint similar to (77). Though the analogon to Theorem 2 for vector-valued control has not been fully established in the literature, one can formulate an induced optimization problem of the type (30) and check that SSC hold; cf. Goris [8]. A paper with detailed results is in preparation.

Acknowledgements. We are indebted to Inga Altrogge [2] and Nadine Goris [8] for their numerical assistance.

References

1. Agrachev, A.A., Stefani, G., Zezza, P.L.: Strong optimality for a bang-bang trajectory. SIAM J. Control and Optimization 41, 991–1014 (2002)
2. Altrogge, I.: Hinreichende Optimalitätsbedingungen für optimale Steuerprozesse mit bang-bang Steuerungen und Zustandsbeschränkungen mit verschiedenen Anwendungsbeispielen, Diploma thesis, Institut für Numerische und Angewandte Mathematik, Universität Münster (2005)
3. Bell, D.J., Katusiime, F.: A time-optimal drug displacement probem. Optimal Control Applications and Methods 1, 217–225 (1980)
4. Büskens, C.: Optimierungsmethoden und Sensitivitätsanalyse für optimale Steuerprozesse mit Steuer- und Zustands-Beschränkungen, Dissertation, Institut für Numerische Mathematik, Universität Münster (1998)
5. Büskens, C., Maurer, H.: Sensitivity analysis and real-time optimization of parametric nonlinear programming problems. In: Grötschel, M., et al. (eds.) Online Optimization of Large Scale Systems, pp. 3–16. Springer, Heidelberg (2001)
6. Büskens, C., Maurer, H.: SQP-methods for solving optimal control problems with control and state constraints: adjoint variables, sensitivity analysis and real-time control. J. of Computational and Applied Mathematics 120, 85–108 (2000)
7. Fiacco, A.V.: Introduction to sensitivity and stability analysis in nonlinear programming. In: Mathematics in Science and Engineering, vol. 165. Academic Press, New York (1983)
8. Goris, N.: Hinreichende Optimalitätsbedingungen für optimale Steuerprozesse mit Zustandsbeschränkungen und linear auftretender Steuerung: Beispiele aus der Medizin und Physik, Diploma thesis, Institut für Numerische und Angewandte Mathematik, Universität Münster (2005)
9. Hartl, R.F., Sethi, S.P., Vickson, R.G.: A survey of the maximum principles for optimal control problems with state constraints. SIAM Review 17, 181–218 (1995)
10. Kaya, C.Y., Noakes, J.L.: Computational method for time-optimal switching control. J. of Optimization Theory and Applications 117, 69–92 (2003)
11. Kim, J.-H.R., Maurer, H.: Sensitivity analysis of optimal control problems with bang-bang controls. In: Proc. of the 42nd IEEE Conf. on Decision and Control, Maui, USA, pp. 3281–3286 (2003)
12. Ledzewicz, U., Schättler, H.: Optimal bang-bang controls for a 2-compartment model in cancer chemotherapy. J. Optimization Theory and Applications 114, 609–637 (2002)

13. Ledzewicz, U., Schättler, H.: A local field of extremals for single-input systems with state space constraints. In: Proc. of the 43nd IEEE Conf. on Decision and Control, Nassau, The Bahamas, USA, pp. 923–928 (2004)
14. Ledzewicz, U., Schättler, H.: A local field of extremals for optimal control problems with state constraints of relative degree 1. J. of Dynamical and Control Systems 12, 563–599 (2006)
15. Malanowski, K.: On normality of Lagrange multipliers for state constrained optimal control problems. Optimization 52, 75–91 (2003)
16. Maurer, H.: On optimal control problems with bounded state variables and control appearing linearly. SIAM J. Control and Optimization 15, 345–362 (1977)
17. Maurer, H.: On the minimum principle for optimal control problems with state constraints. Schriftenreihe des Rechenzentrums der Universität Münster (1977)
18. Maurer, H., Büskens, C., Kim, J.-H.R., Kaya, C.Y.: Optimization methods for the verification of second order sufficient conditions for bang-bang controls. Optimal Control Applications and Methods 26, 129–156 (2005)
19. Maurer, H., Kim, J.-H.R., Vossen, G.: On a state-constrained control problem in optimal production and maintenance. In: Deissenberg, C., Hartl, R.F. (eds.) Optimal Control and Dynamic Games, Applications in Finance, Management Science and Economics, pp. 289–308. Springer, Heidelberg (2005)
20. Maurer, H., Osmolovskii, N.P.: Quadratic sufficient optimality conditions for bang-bang control problems. Control and Cybernetics 33, 555–584 (2003)
21. Maurer, H., Osmolovskii, N.P.: Second order sufficient conditions for time-optimal bang-bang control problems. SIAM J. Control and Optimization 42, 2239–2263 (2004)
22. Maurer, H., Wiegand, M.: Numerical solution of a drug displacement problem with bounded state variables. Optimal Control Applications and Methods 13, 43–55 (1992)
23. Osmolovskii, N.P., Maurer, H.: Equivalence of second order optimality conditions for bang-bang control problems. Part 1: Main results. Control and Cybernetics 34, 927–950 (2005); Part 2: Proofs, variational derivatives and representations. Control and Cybernetics 36, 5–45 (2007)
24. de Pinho, M.R., Ferreira, M.M., Ledzewicz, U., Schättler, H.: A model for cancer chemotherapy with state space constraints. Nonlinear Analysis 63, 2591–2602 (2005)
25. Swierniak, A., Ledzewicz, U., Schättler, H.: Optimal control for a class of compartment models in cancer chemotherapy. Intern. J. of Applied Mathematics and Computer Science 13, 357–368 (2003)
26. Vossen, G.: Numerische Lösungsmethoden, hinreichende Optimalitätsbedingungen und Sensitivitätsanalyse für optimale bang-bang und singuläre Steuerungen, Dissertation, Institut für Numerische und Angewandte Mathematik, Universität Münster (2005)
27. Wächter, A., Biegler, L.T.: On the implementation of a primal-dual interior point filter line search algorithm for large-scale nonlinear programming. Mathematical Programming 106, 25–57 (2006)

New Applications of Variational Analysis to Optimization and Control*

Boris S. Mordukhovich

Department of Mathematics, Wayne State University, Detroit, Michigan 48202, USA
boris@math.wayne.edu

Abstract. We discuss new applications of advanced tools of variational analysis and generalized differentiation to a number of important problems in optimization theory, equilibria, optimal control, and feedback control design. The presented results are largely based on the recent work by the author and his collaborators. Among the main topics considered and briefly surveyed in this paper are new calculus rules for generalized differentiation of nonsmooth and set-valued mappings; necessary and sufficient conditions for new notions of linear subextremality and suboptimality in constrained problems; optimality conditions for mathematical problems with equilibrium constraints; necessary optimality conditions for optimistic bilevel programming with smooth and nonsmooth data; existence theorems and optimality conditions for various notions of Pareto-type optimality in problems of multiobjective optimization with vector-valued and set-valued cost mappings; Lipschitzian stability and metric regularity aspects for constrained and variational systems.

1 Introduction

Variational analysis has been recognized as a rapidly growing and fruitful area in mathematics and its applications concerning mainly the study of optimization and equilibrium problems, while also applying perturbation ideas and *variational principles* to a broad class of problems and situations that may be not of a variational nature. It can be viewed as a modern outgrowth of the classical calculus of variations, optimal control theory, and mathematical programming with the focus on *perturbation/approximation* techniques, sensitivity issues, and applications. We refer the reader to the now classical monograph by Rockafellar and Wets [58] for the key issues of variational analysis in finite-dimensional spaces and to the recent books by Attouch, Buttazzo and Michaelle [1], Borwein and Zhu [7], and Mordukhovich [31,32] devoted to new aspects of variational analysis in finite-dimensional and infinite-dimensional spaces with numerous applications to different areas of mathematics, engineering, economics, mechanics, computer science, ecology, biology, etc.

One of the most characteristic features of modern variational analysis is the intrinsic presence of *nonsmoothness*, i.e., the necessity to deal with nondifferentiable functions,

---

* Research was partially supported by the US National Science Foundation under grants DMS-0304989 and DMS-0603846 and also by the Australian Research Council under grant DP-04511668.

A. Korytowski et al. (Eds.): System Modeling and Optimization, IFIP AICT 312, pp. 100–128, 2009.

sets with nonsmooth boundaries, and set-valued mappings. Nonsmoothness naturally enters not only through initial data of optimization-related problems (particularly those with inequality and geometric constraints) but largely via *variational principles* and other optimization, approximation, and perturbation techniques applied to problems with even smooth data. In fact, many fundamental objects frequently appearing in the framework of variational analysis (e.g., the distance function, value functions in optimization and control problems, maximum and minimum functions, solution maps to perturbed constraint and variational systems, etc.) are inevitably of nonsmooth and/or set-valued structures requiring the development of new forms of analysis that involve *generalized differentiation*. Besides the aforementioned books, we refer the reader to the very recent texts by Jeyakumar and Luc [22] and Schirotzek [59], which present new developments on generalized differentiation and their applications to a variety of optimization-related as well as nonvariational problems.

It is important to emphasize that even the simplest and historically earliest problems of *optimal control* are *intrinsically nonsmooth*, in contrast to the classical calculus of variations. This is mainly due to *pointwise constraints* on control functions that often take only discrete values as in typical problems of automatic control, a primary motivation for developing optimal control theory. Optimal control has always been a major source of inspiration as well as a fruitful territory for applications of advanced methods of variational analysis and generalized differentiation; see, e.g., the books by Clarke [9], Mordukhovich [31,32], and Vinter [60] with the references therein.

In this paper we discuss some new trends and developments in variational analysis and its applications that are based on the 2-volume book by the author [31,32] and mostly survey more recent and/or brand new results obtained by the author and his collaborators. As mentioned, generalized differentiation lies at the heart of variational analysis and its applications. We systematically develop a *geometric dual-space approach* to generalized differentiation theory revolving around the *extremal principle*, which can be viewed as a local *variational* counterpart of the classical convex separation in nonconvex settings. This principle allows us to deal with *nonconvex* derivative-like constructions for sets (normal cones), set-valued mappings (coderivatives), and extended-real-valued functions (subdifferentials). These constructions are defined directly in dual spaces and, being nonconvex-valued, cannot be generated by any derivative-like constructions in primal spaces (like tangent cones and directional derivatives). Nevertheless, our basic nonconvex constructions enjoy *comprehensive/full calculus*, which happens to be significantly better than those available for their primal and/or convex-valued counterparts. The developed generalized differential calculus based on variational principles provides the *key tools* for various applications.

Observe to this end that *dual objects* (multipliers, adjoint arcs, shadow prices, etc.) have always been at the center of variational theory and applications used, in particular, for formulating the main optimality conditions in the calculus of variations, mathematical programming, optimal control, and economic modeling. The usage of variations of optimal solutions in primal spaces can be considered just as a convenient tool for deriving necessary optimality conditions. There are no essential restrictions in such a "primal" approach in smooth and convex frameworks, since primal and dual derivative-like constructions are equivalent for these classical settings. It is not the case any more in the framework of modern variational analysis, where even *nonconvex primal space* local approximations (e.g., tangent cones) inevitably yield, *under duality, convex sets* of normals and subgradients. This convexity of dual objects leads to significant restrictions for the theory and applications. Moreover, there are many situations particularly identified in [31,32], where primal space approximations simply cannot be used for variational analysis, while the employment of dual space constructions provides comprehensive treatments and results.

The main attention of this paper is paid to the description of certain basic constructions of generalized differentiation in variational analysis and their applications to important and also new classes of problems in *constrained optimization* and *optimal control* that happen to be intrinsically *nonsmooth*, even in the case of smooth initial data. In Sect. 2 we define these *dual-space generalized differential constructions* and discuss new *calculus results* for them. Sect. 3 is devoted to recent applications of the generalized differential calculus to studying the notion of *linear suboptimality* in constrained optimization, where the usage of these generalized differential constructions allows us to *fully characterize* linearly suboptimal solutions, in the sense of deriving verifiable *necessary and sufficient* conditions for them.

In Sect. 4 we discuss new results for a broad class of optimization problem known as *mathematical programs with equilibrium constraints* (MPECs) significant in optimization theory and its applications. Besides characterizations of the aforementioned notion of linear suboptimality for MPECs, we present new necessary optimality conditions for the conventional notion of optimal solutions to MPECs whose equilibrium constraints are governed by parameterized *quasivariational inequalities* that are challenging in the MPEC theory and highly important for applications.

Sect. 5 is devoted to new results on the so-called *bilevel programming*, which is a remarkable class of *hierarchical optimization* problems somehow related to MPECs while generally independent. Concentrating on the *optimistic version* of bilevel programs and using our basic tools of generalized differentiation, we present advanced necessary optimality conditions in finite-dimensional bilevel programming that are new even for problems with *smooth* data on both lower level and upper levels.

Sect. 6 concerns various problems of *multiobjective optimization* and *equilibria*, which are among the most challenging theoretically and the most important for numerous applications (to economics, mechanics, and other areas). We pay the main attention to new existence theorems and necessary optimality conditions for Pareto-type solutions to constrained multiobjective problems with vector-valued and set-valued objectives. Our approach is based on developing and implementing advanced variational principles for multifunctions with values in partially ordered spaces.

In Sect. 7 we consider several important issues revolving around *Lipschitzian stability* and *metric regularity* properties for set-valued mappings and their applications to structural systems arising in numerous aspects of variational analysis, optimization, and control. Our approach is based on the *dual coderivative criteria* for such properties established earlier by the author; they can be applied to a variety of structural systems due to well-developed *coderivative calculus* in finite-dimensional and infinite-dimensional spaces. In this way, along with deriving positive results in this direction, we come up to a rather surprising conclusion that major classes of variational/optimality systems, which are the most interesting from the both viewpoints of the theory and applications, *do not exhibit metric regularity.*

Sect. 8 presents new results on *optimal control* dealing mainly with evolution systems governed by constrained *difference, differential,* and *delay-differential inclusions* in infinite-dimensional spaces. We develop the method of *discrete approximations* for continuous-time evolution systems and investigate both *qualitative* and *quantitative* aspects of this approach. Our results include *stability/convergence* of discrete approximations, deriving necessary optimality conditions for discrete-time systems and then for the original continuous-time control problems by passing to the limit from discrete approximations and employing advanced tools of variational analysis and generalized differentiation.

The concluding Sect. 9 is devoted to problems of *feedback control design* of constrained parabolic systems in *uncertainty conditions.* Control problems of these type are undoubtedly among the most important for various (in particular, engineering and ecological) applications; at the same time they are among the most challenging in control theory. Especially serious difficulties arise in studying and solving such problems in the presence of *hard/pointwise constraints* on control and state variables, which is the case considered in the concluding section motivated by some practical applications to environmental systems. The approach discussed in Sect. 9 and the results presented therein are based on certain specific features of the parabolic dynamics related to *monotonicity* and *turnpike* behavior on the *infinite horizon,* as well as on approximation techniques typical in variational analysis. In this way we justify implementable *suboptimal* structures of *feedback control regulators* acting through boundary conditions and compute their optimal parametric ensuring the *best* behavior of the systems under *worst* perturbations and *robust stability* of the closed-loop systems for arbitrary perturbations from the feasible area.

Throughout the paper we use the standard notation of variational analysis; see, e.g., [31,58]. Recall that $\mathbb{B}$ stands for the closed unit ball of the space in question and that $\mathbb{N} := \{1, 2, \ldots\}$. Given a set-valued mapping $F: X \rightrightarrows X^*$ between a Banach space X and its topological dual X^*, the symbol

$$\operatorname*{Lim\,sup}_{x \to \bar{x}} F(x) := \left\{ x^* \in X^* \middle| \exists \text{ sequences } x_k \to \bar{x} \text{ and } x_k^* \xrightarrow{w^*} x^* \right.$$

$$\left. \text{with } x_k^* \in F(x_k) \text{ for all } k \in \mathbb{N} \right\} \tag{1}$$

signifies the *sequential Painlevé-Kuratowski upper/outer limit* of F at $\bar{x}$ in the norm topology of X and weak* topology of X^*.

2   Generalized Differentiation

In this section we define, for the reader's convenience, some basic constructions and properties from variational analysis and generalized differentiation needed in what follows. All these are taken from the book by Mordukhovich [31], where the reader can find more details, discussions, and references. The reader may also consult with the books by Borwein and Zhu [7], Rockafellar and Wets [58], and Schirotzek [59] for related and additional material.

Most results presented in this paper are obtained in the framework of Asplund spaces; so our *standing assumption* is that all the spaces under consideration are *Asplund* unless otherwise stated. One of the equivalent descriptions of an Asplund space is that it is a Banach space for which every separable subspace has a separable dual. It is well known that any reflexive Banach space is Asplund as well as any space with a separable dual; see [31, Sect. 2.2] for more discussions and references. The generalized differential constructions and properties presented below generally rely on the Asplund structure; see [7,20,31] for the corresponding modifications in other (including arbitrary) Banach space settings.

Given a nonempty set $\Omega \subset X$, define the *Fréchet normal cone* to Ω at $\bar{x} \in \Omega$ by

$$\hat{N}(\bar{x};\Omega) := \left\{ x^* \in X^* \,\Big|\, \limsup_{x \xrightarrow{\Omega} \bar{x}} \frac{\langle x^*, x - \bar{x}\rangle}{\|x - \bar{x}\|} \leq 0 \right\}, \tag{2}$$

where the symbol $x \xrightarrow{\Omega} \bar{x}$ signifies that $x \to \bar{x}$ with $x \in \Omega$. Construction (2) looks as an adaptation of the idea of Fréchet derivative to the case of sets; that's where the name comes from. However, this construction does not have a number of natural properties expected for an appropriate notion of normals. In particular, we may have $\hat{N}(\bar{x};\Omega) = \{0\}$ for boundary points of Ω even in simple finite-dimensional nonconvex settings; furthermore, inevitable required calculus rules often fail for (2). The situation is dramatically improved while applying the regularization procedure

$$N(\bar{x};\Omega) := \operatorname{Lim\,sup}_{x \xrightarrow{\Omega} \bar{x}} \hat{N}(x;\Omega) \tag{3}$$

via the sequential outer limit (1) in the norm topology of X and the weak* topology of X^*. The construction (3) is known as the *(basic, limiting, Mordukhovich) normal cone* to Ω at $\bar{x} \in \Omega$; it was introduced in [27] in an equivalent form in finite dimensions. Both constructions (2) and (3) reduce to the classical normal cone of convex analysis for convex sets Ω. In contrast to (2), the basic normal cone (3) is often *nonconvex* while satisfying the required properties and *calculus rules* in the Asplund space setting, together with the corresponding coderivative constructions for set-valued mappings and subdifferential constructions for extended-real-valued functions generated by it; see below. All this calculus and the required properties are mainly due to the *extremal/variational principles* of variational analysis; see [31] for more discussions.

Given a set-valued mapping/multifunction $F : X \rightrightarrows Y$ with the *graph*

$$\operatorname{gph} F := \left\{ (x,y) \in X \times Y \,\big|\, y \in F(x) \right\}, \tag{4}$$

and following the pattern introduced in [28], define the *coderivative* constructions for F used in this paper. The *Fréchet coderivative* of F at $(\bar{x},\bar{y}) \in \operatorname{gph} F$ is given by

$$\hat{D}^* F(\bar{x},\bar{y})(y^*) := \left\{ x^* \in X^* \,\big|\, (x^*, -y^*) \in \hat{N}\big((\bar{x},\bar{y});\operatorname{gph} F\big) \right\}, \quad y^* \in Y^*, \tag{5}$$

and the *normal coderivative* of F at the reference point is given by

$$D_N^* F(\bar{x},\bar{y})(y^*) := \left\{ x^* \in X^* \,\big|\, (x^*, -y^*) \in N\big((\bar{x},\bar{y});\operatorname{gph} F\big) \right\}, \quad y^* \in Y^*. \tag{6}$$

We also need the following modification of the normal coderivative (5) called the *mixed coderivative* of F at $(\bar{x}, \bar{y})$ and defined by

$$D_M^* F(\bar{x}, \bar{y})(y^*) := \left\{ x^* \in X^* \middle| \exists \, (x_k, y_k) \xrightarrow{\mathrm{gph}\, F} (\bar{x}, \bar{y}), \, x_k^* \xrightarrow{w^*} x^*, \, y_k^* \xrightarrow{\|\cdot\|} y^* \right.$$

$$\left. \text{with } (x_k^*, -y_k^*) \in \hat{N}\big((x_k, y_k); \mathrm{gph}\, F\big), \, k \in \mathbb{N} \right\}, \tag{7}$$

where $\xrightarrow{\|\cdot\|}$ stands for the norm convergence in the dual space; we usually omit the symbol $\|\cdot\|$ indicating the norm convergence simply by "$\to$" and also skip $\bar{y} = f(\bar{x})$ in the coderivative notation if $F = f \colon X \to Y$ is a single-valued mapping. Clearly $D_M^* F(\bar{x}, \bar{y}) = D_N^* F(\bar{x}, \bar{y})$ if $\dim Y < \infty$, and then we use the same notation D^* for both coderivatives. The above equality also holds in various infinite-dimensional settings, while not in general; see [31, Subsect. 1.2.1 and 4.2.1]. If $F = f$ is single-valued and *smooth* around $\bar{x}$ (or merely *strictly differentiable* at this point), then we have the representations

$$\hat{D}^* f(\bar{x})(y^*) = D_M^* f(\bar{x})(y^*) = D_N^* f(\bar{x})(y^*) = \left\{ \nabla f(\bar{x})^* y^* \right\}, \quad y^* \in Y^*, \tag{8}$$

which show that the *coderivative* notion is a natural extension of the *adjoint derivative* operator to nonsmooth and set-valued mappings.

Given an extended-real-valued function $\varphi \colon X \to \bar{\mathbb{R}} := (-\infty, \infty]$, consider the associated *epigraphical multifunction* $\mathcal{E}_\varphi \colon X \rightrightarrows \mathbb{R}$ and define the *Fréchet/regular subdifferential* of φ at $\bar{x} \in \mathrm{dom}\, \varphi$ in the two equivalent (geometric and analytic) ways

$$\hat{\partial}\varphi(\bar{x}) := \hat{D}^* \mathcal{E}_\varphi\big(\bar{x}, \varphi(\bar{x})\big)(1) = \left\{ x^* \in X^* \middle| \liminf_{x \to \bar{x}} \frac{\varphi(x) - \varphi(\bar{x}) - \langle x^*, x - \bar{x} \rangle}{\|x - \bar{x}\|} \geq 0 \right\}. \tag{9}$$

The *basic/limiting/Mordukhovich subdifferential* of φ at $\bar{x}$ is defined by

$$\partial\varphi(\bar{x}) = D^* \mathcal{E}_\varphi\big(\bar{x}, \varphi(\bar{x})\big)(1) = \operatorname*{Lim\,sup}_{x \xrightarrow{\varphi} \bar{x}} \hat{\partial}\varphi(x), \tag{10}$$

where the symbol $x \xrightarrow{\varphi} \bar{x}$ stands for $x \to \bar{x}$ with $\varphi(x) \to \varphi(\bar{x})$. Note that the Fréchet subdifferential agrees with the Crandall-Lions subdifferential in the sense of *viscosity solutions* to partial differential equations independently introduced in [10], while the limiting construction (10) reduces to that introduced in [27] motivated by applications to optimal control. The *convexification* of (10) for locally Lipschitzian functions agrees with the generalized gradient introduced by Clarke via different relationships; see [9]. For non-Lipschitzian functions φ it makes sense to consider the *singular* counterpart of φ given by

$$\partial^\infty \varphi(\bar{x}) = D^* \mathcal{E}_\varphi\big(\bar{x}, \varphi(\bar{x})\big)(0) = \operatorname*{Lim\,sup}_{\substack{x \xrightarrow{\varphi} \bar{x} \\ \lambda \downarrow 0}} \lambda \hat{\partial}\varphi(x), \tag{11}$$

which reduces to $\{0\}$ if φ is locally Lipschitzian around $\bar{x}$.

Among the main advantages of the robust limiting constructions (3), (6), (7), (10), and (11), we particularly mention *full pointwise calculi* available for them, the possibility to *characterize* in their terms *Lipschitzian, metric regularity*, and *openness* properties of set-valued and single-valued mappings that play a fundamental role in nonlinear analysis and its applications, and to derive in their terms refined conditions for *optimality* and *sensitivity* in various problems of optimization, equilibria, control, etc. Besides variational principles, extended *calculus is the key* for major theoretical advances and applications.

Referring the reader to [31] for a variety of calculus rules for the basic normals, subgradients, and coderivatives under consideration, let us mention several recent ones (in addition to [31]) motivated by the required applications presented in the corresponding papers.

In [18], we develop certain calculus rules for the so-called *reversed mixed coderivative* of $F: X \rightrightarrows Y$ at $(\bar{x}, \bar{y}) \in \mathrm{gph}\, F$ defined by

$$\tilde{D}_M^* F(\bar{x}, \bar{y})(y^*) := \{x^* \in X^* | -y^* \in D_M^* F^{-1}(\bar{y}, \bar{x})(-x^*)\}, \tag{12}$$

which is different from both coderivative constructions (6) and (7) in infinite dimensions while playing a crucial role in characterizing *metric regularity*. In contrast to (6) and (7), the reversed construction (12) does *not* generally enjoy satisfactory calculus rules, since taking the inverse in (12) dramatically complicates some major operations (e.g., sums) for single-valued and set-valued mappings. The calculus rules derived in [18] for the reversed coderivative (12) mainly address a special class of set-valued mappings known as solution maps to *generalized equations* (in the sense of Robinson [57]):

$$S(x) = \{y \in Y \mid 0 \in f(x,y) + Q(y)\} \quad \text{with} \quad f: X \times Y \to Z \text{ and } Q: Y \rightrightarrows Z, \tag{13}$$

which are highly important in many aspects of variational analysis and optimization; see, e.g., [17,31,56,58] and the references therein. The calculus results obtained in [18] and related developments allow us to make a principal conclusion on the *failure of metric regularity* for major classes of parametric *variational systems*; see Sect. 7 below.

Another important setting that requires new coderivative calculus rules is described by set-valued mappings in the form

$$Q(x,y) = N(y; \Lambda(x,y)) \quad \text{with} \quad \Lambda: X \times Y \rightrightarrows Y, \tag{14}$$

which corresponds to the so-called *quasivariational inequalities* in the generalized equation framework (13) with $Q = Q(x,y)$ of type (14). Advanced results in this direction are obtained in [50] in finite-dimensional spaces and are applied there to *sensitivity analysis* of quasivariational inequalities and *necessary optimality conditions* for the corresponding MPECs; see Sect. 4 and 7 for more details.

Let us also mention new *intersection rules* for coderivatives obtained in [46] in general infinite-dimensional settings and applied therein to sensitivity analysis of extended parametric models of type (13) arising in various applications, particularly to *bilevel programs*; see Sect. 5 for more discussions.

Several new calculus rules for the (basic and singular) *limiting subdifferentials* (10) and (11) of the important classes of *marginal/value functions* are derived in [49] with applications to sensitivity analysis and optimality conditions in problems of mathematical programming in finite-dimensional and infinite-dimensional spaces. In [48], rather surprising *exact* (versus "fuzzy") calculus rules are obtained for the *Fréchet subdifferential* (9) of various compositions and marginal functions with applications to some classes of optimization problems; see Sect. 3. Among them the following striking *difference rule*:

$$\hat{\partial}(\varphi_1 - \varphi_2)(\bar{x}) \subset \bigcap_{x^* \in \hat{\partial}\varphi_2(\bar{x})} \left[\hat{\partial}\varphi_1(\bar{x}) - x^* \right] \subset \hat{\partial}\varphi_1(\bar{x}) - \hat{\partial}\varphi_2(\bar{x}) \qquad (15)$$

is derived in general Banach spaces provided that $\hat{\partial}\varphi_2(\bar{x}) \neq \emptyset$. Counterparts of such exact calculus results for the so-called *proximal subgradients* can be found in [45].

3 Constrained Optimization

It has been well recognized that, except convex programming and related problems with a convex structure, *necessary* conditions are usually *not sufficient* for conventional notions of optimality. Observe also that major necessary optimality conditions in all the branches of the classical and modern optimization theory (e.g., Lagrange multipliers and Karush-Kuhn-Tucker conditions in nonlinear programming, the Euler-Lagrange equation in the calculus of variations, the Pontryagin maximum principle in optimal control, etc.) are expressed in *dual* forms involving *adjoint* variables. At the same time, the very *notions of optimality*, in both scalar and vector frameworks, are formulated of course in *primal* terms.

A challenging question is to find certain modified notions of local optimality so that *first-order* necessary conditions known for the previously recognized notions become *necessary and sufficient* in the new framework. Such a study has been initiated by Kruger (see [25] and the references therein), where the corresponding notions are called "weak stationarity". It seems that the main difference between the conventional notions and those of the type [25] is that the latter relate to a certain *suboptimality* not at the point in question but in a *neighborhood* of it, and that they involve a *linear rate* similar to that in *Lipschitz continuity* (in contrast merely to continuity) as well as in modern concepts of *metric regularity* and *linear openness*, which distinguishes them from the classical regularity and openness notions of nonlinear analysis. On this basis we suggested in [32] to use the names of *linear subextremality* for set systems and of *linear suboptimality* for the corresponding notions in optimization problems.

As has been fully recognized just in the framework of modern variational analysis (even regarding the classical settings), the *linear rate nature* of the fundamental properties involving Lipschitz continuity, metric regularity, and openness for single-valued and set-valued mappings is the *key issue* allowing us to derive *complete characterizations* of these properties via appropriate tools of generalized differentiation; see the books [31,58] and their references. Precisely the same linear rate essence of the (sub)extremality and (sub)optimality concepts considered below is the driving force ensuring the possibility to justify the validity of known necessary extremality and optimality conditions for the conventional notions as *necessary and sufficient* conditions for the new notions under consideration.

In contrast to [25], where dual criteria for "weak stationarity" are obtained in "fuzzy" forms involving Fréchet-like constructions at points *nearby* the reference ones, in [32, Chapter 5] and in the more recent developments [36,38] we pay the main attention to *pointwise* conditions expressed via the basic *robust* generalized differential constructions discussed in Sect. 2, which are defined *exactly at* the points in question. Besides the latter being more convenient for applications, we can significantly gain from such pointwise characterizations due to the well-developed *full calculus* enjoyed by the robust constructions, which particularly allows us to cover problems with various *constrained* structures important for both the optimization theory and its applications.

A major role in our approach to variational analysis and optimization systematized and developed in [31,32] is played by the so-called *extremal principle*; see [31, Chapter 2] with the references and comprehensive discussions therein. Recall that a point $\bar{x} \in \Omega_1 \cap \Omega_2 \subset X$ is *locally extremal* for the set system $\{\Omega_1, \Omega_2\}$ if there exists a neighborhood U of $\bar{x}$ such that for any $\varepsilon > 0$ there is $a \in \varepsilon \mathbb{B}$ with

$$(\Omega_1 + a) \cap \Omega_2 \cap U = \emptyset. \tag{16}$$

Loosely speaking, the local extremality of sets at a common point means that they can be locally "pushed apart" by a small perturbation/translation of one of them. It has been well recognized that set extremality encompasses various notions of optimal solutions to problems of scalar and vector/multiobjective optimization, equilibria, etc.

It is easy to observe that $\bar{x} \in \Omega_1 \cap \Omega_2$ is locally extremal for $\{\Omega_1, \Omega_2\}$ *if and only if*

$$\vartheta\left(\Omega_1 \cap B_r(\bar{x}), \Omega_2 \cap B_r(\bar{x})\right) = 0 \text{ with some } r > 0, \tag{17}$$

where $B_r(\bar{x}) := \bar{x} + r\mathbb{B}$, and where the *measure of overlapping* $\vartheta(\Omega_1, \Omega_2)$ for the sets Ω_1, Ω_2 is defined by

$$\vartheta(\Omega_1, \Omega_2) := \sup\left\{v \geq 0 \mid v\mathbb{B} \subset \Omega_1 - \Omega_2\right\}. \tag{18}$$

Following [25] and the terminology in [32, Sect. 5.4], we say that the set system $\{\Omega_1, \Omega_2\}$ is *linearly subextremal* around $\bar{x} \in \Omega_1 \cap \Omega_2$ if

$$\vartheta_{\text{lin}}(\Omega_1, \Omega_2, \bar{x}) := \liminf_{\substack{x_i \xrightarrow{\Omega_i} \bar{x} \\ r\downarrow 0}} \frac{\vartheta\left([\Omega_1 - x_1] \cap r\mathbb{B}, [\Omega_2 - x_2] \cap r\mathbb{B}\right)}{r} = 0 \tag{19}$$

with $i = 1, 2$ under the "$\liminf$" sign in (19); see [25,32,36] for more discussions.

To formulate the following results about the extremal principle also for the subsequent use in the paper, recall that a set $\Omega \subset X$ is *sequentially normally compact* (SNC) at $\bar{x} \in \Omega$ if for any sequences $x_k \xrightarrow{\Omega} \bar{x}$ and $x_k^* \xrightarrow{w^*} 0$ we have

$$\|x_k^*\| \to 0 \text{ provided that } x_k^* \in \hat{N}(x_k; \Omega) \text{ as } k \to \infty. \tag{20}$$

In finite dimensions, every subset is obviously SNC. For arbitrary Banach space, Ω is SNC at $\bar{x}$ if it is "compactly epi-Lipschitzian" in the sense of Borwein and Strójwas; see [31, Subsect. 1.1.4] for this and other sufficient conditions. If Ω is *convex* in infinite dimensions, then its SNC property is closely related to Ω being of *finite codimension*.

The *extremal principle* from [31, Theorem 2.20] says that for any local extremal point $\bar{x} \in \Omega_1 \cap \Omega_2$ of the system $\{\Omega_1, \Omega_2\}$ of closed subsets of an Asplund space X there is $x^* \in X^*$ satisfying the relationship

$$0 \neq x^* \in N(\bar{x}; \Omega_1) \cap \left(-N(\bar{x}; \Omega_2)\right) \tag{21}$$

provided that either Ω_1 or Ω_2 is SNC at $\bar{x}$. This result can be treated as a *variational counterpart* of the classical convex separation theorem in *nonconvex* settings. In fact, its role in variational analysis is similar to that of convex separation in convex analysis and its "convexified" versions; see [31,32] for more details and discussions.

An appropriate "necessary and sufficient" modification of the extremal principle for *linear subextremality* reads as follows; cf. [32, Theorem 5.89] and [36, Theorem 1].

Theorem 1. (Necessary and sufficient conditions for linear subextremality via the extremal principle). *Let Ω_1 and Ω_2 be subsets of an Asplund space X that are locally closed around $\bar{x} \in \Omega_1 \cap \Omega_2$. If the system $\{\Omega_1, \Omega_2\}$ is linearly suboptimal around $\bar{x}$, then there is $x^* \in X^*$ satisfying the extremal principle (21). Furthermore, the extremal principle (21) is necessary and sufficient for the linear suboptimality of $\{\Omega_1, \Omega_2\}$ around $\bar{x}$ if $\dim X < \infty$.*

Based on this theorem and on well-developed *robust calculus* rules for our *limiting* generalized differential constructions, we derive in [32, Sect. 5.4], [36,38] a number of *necessary* as well as *necessary and sufficient* conditions for the notions of *linear suboptimality* generated by the set subextremality (19) for various optimization and equilibrium problems involving constraints of geometric, operator, functional, and equilibrium types. It should be emphasized that to derive in this way necessary and sufficient conditions for constraint problems, we need to use generalized differential results ensuring *equalities* in the corresponding calculus rules. Such results are largely available in [31] and are employed in [32,36,38].

Among other recent applications to optimization, let us mention new necessary optimality conditions for *sharp minimizers* and also to *DC (difference of convex) programs* derived in [45,48,49] on the basis of the subdifferential calculus rules developed therein in both finite-dimensional and infinite-dimensional settings.

A series of new results on necessary conditions for nonsmooth *infinite-dimensional* optimization problems are established in [35] based on advanced methods of variational analysis, on extended calculus rules of *generalized differentiation* as well as on efficient *calculus/preservation* rules for the *sequential normal compactness property* (20) and its *partial* counterparts. These results include several new versions of the *Lagrange principle* for nonsmooth optimization problems with functional and geometric constraints and also refined necessary conditions for problems with *operator constrains* given by nonsmooth *Fredholm-type* mappings with values in infinite dimensions. The latter result is applied to constrained *optimal control* problems governed by *discrete-time inclusions*; see Sect. 8 for more details.

4 Mathematical Programs with Equilibrium Constraints

The modern terminology of *mathematical programs with equilibrium constraints* (MPECs) generally concerns optimization problems given in the following form:

$$\text{minimize } \varphi_0(x,y) \text{ subject to } y \in S(x), \ (x,y) \in \Omega, \tag{22}$$

which contain, among other constraints, the so-called *equilibrium constraints* defined by solution maps to the parameterized *generalized equations/variational conditions*

$$S(x) := \{y \in Y \mid 0 \in f(x,y) + Q(x,y)\} \tag{23}$$

that are described by single-valued *base* mappings $f\colon X \times Y \to Z$ and set-valued *field* mappings $Q\colon X \times Y \rightrightarrows Z$; see, e.g., [17,31,56] for more discussions. Variational systems of type (23) are introduced in the seminal work by Robinson [57] in the setting when $Q(y) = N(y;\Lambda)$ is the *normal cone mapping* to a *convex* set Λ, in which case the generalized equation (23) reduces to the parametric *variational inequality*:

$$\text{find } y \in \Lambda \text{ such that } \langle f(x,y), v-y \rangle \geq 0 \text{ for all } v \in \Lambda. \tag{24}$$

The classical parametric *complementarity system* corresponds to (24) when Λ is the nonnegative orthant in $\mathbb{R}^n$. It is well known that the latter model covers sets of *optimal solutions* with the associated *Lagrange multipliers* and sets of *Karush-Kuhn-Tucker* (KKT) vectors satisfying first-order necessary optimality conditions in parametric problems of *nonlinear programming* with *smooth* data. General models with parameter-dependent field mappings $Q = Q(x,y)$ in (23) have been also, but to much lesser extent, considered in the literature. They are related, in particular, to the *quasivariational inequalities*

$$\text{find } y \in \Lambda \text{ such that } \langle f(x,y), v-y \rangle \geq 0 \text{ for all } v \in \Lambda(x,y) \tag{25}$$

in the extended framework of (24); see [50] for more discussions and references. Note that in *infinite-dimensional* spaces models of these types are closely associated with variational problems arising in *partial differential equations*.

Variational systems most important for optimization/equilibrium theory and applications mainly relate to generalized equations (23) with *subdifferential fields* when Q is given by a *subdifferential/normal cone operator* $\partial\varphi$ generated by an extended-real-valued lower semicontinuous (l.s.c.) function φ, which is often labeled as *potential*. As mentioned above, this is the case of the classical variational inequalities (24) and complementarity problems generated by convex *indicator* functions $\varphi(\cdot) = \delta(\cdot;\Lambda)$ as well as of their quasivariational counterparts in (25). Formalism (23) with $Q = \partial\varphi$ encompasses also other types of variational and extended variational inequalities generated by *nonconvex* potentials, e.g., the so-called *hemivariational inequalities* with Lipschitzian potentials.

In this vein, two remarkable classes of equilibrium constraints are of particular interest for optimization/equilibrium theory and applications. The first one is given in the form

$$S(x) := \big\{ y \in Y \,\big|\, 0 \in f(x,y) + \partial(\psi \circ g)(x,y) \big\}, \tag{26}$$

where $g\colon X \times Y \to W$ and $f\colon X \times Y \to X^* \times Y^*$ are single-valued mappings between Banach spaces, and where $\partial\varphi\colon X \times Y \rightrightarrows X^* \times Y^*$ is the basic subdifferential mapping (10) generated by the *composite potential* $\varphi = \psi \circ g$ with $\psi\colon W \to \bar{\mathbb{R}}$. The aforementioned variational systems are special cases of the composite formalism (26).

The second class of remarkable equilibrium constraints is described by the generalized equations with *composite subdifferential fields*

$$S(x) := \big\{ y \in Y \,\big|\, 0 \in f(x,y) + (\partial\psi \circ g)(x,y) \big\}, \tag{27}$$

where $g\colon X \times Y \to W$, $\psi\colon W \to \bar{\mathbb{R}}$, and $f\colon X \times Y \to W^*$. Formalism (27) encompasses, in particular, perturbed *implicit complementarity problems* of the type: find $y \in Y$ satisfying

$$f(x,y) \geq 0, \quad y - g(x,y) \geq 0, \quad \langle f(x,y), y - g(x,y) \rangle = 0, \tag{28}$$

where the inequalities are understood in the sense of some order on Y.

It occurs nevertheless that generalized equation and variational inequality models of the types discussed above with *single-valued base* mappings $f(x, y)$ do not cover a number of variational systems important in optimization theory and applications. Consider, e.g., the *parametric optimization* problem

$$\text{minimize } \phi(x, y) + \vartheta(x, y) \text{ over } y \in Y \tag{29}$$

described by a *cost* function ϕ and a *constraint* function ϑ that generally take their values in the extended real line $\mathbb{R}$. The *stationary point multifunction* associated with (29) is

$$S(x) := \left\{ y \in Y \mid 0 \in \partial_y \phi(x, y) + \partial_y \vartheta(x, y) \right\} \tag{30}$$

via collections of partial subgradients of the cost and constraint functions with respect to the decision variable. If the cost function ϕ in (29) is *smooth*, then $\partial_y \phi(x, y) = \{\nabla_y \phi(x, y)\}$ and thus (30) can be written as the solution map to a generalized equation of type (23) with the base $f(x, y) = \nabla_y \phi(x, y)$ and the field mapping $Q(x, y) = \partial_y \vartheta(x, y)$. However, in the case of *nonsmooth optimization* in (29) corresponding, e.g., to *nonsmooth bilevel programs* (see Sect. 5), the stationary point multifunction (30) cannot be written as the standard generalized equation (23) while requiring the *extended formalism*

$$0 \in F(x, y) + Q(x, y), \tag{31}$$

where both the base mapping F and the field mapping Q are *set-valued*.

Another interesting and important class of variational systems that can be written in the extended generalized equation form (31) but not in the conventional one (23) is described by the so-called *set-valued/generalized variational inequalities*:

$$\text{find } y \in \Omega \text{ such that } y^* \in F(x, y) \text{ with } \langle y^*, v - y \rangle \geq 0 \text{ for } v \in \Lambda, \tag{32}$$

which provide a set-valued extension of (24); see, e.g., the handbook [61] for the theory and applications of (32) and related models.

In the recent papers [2,4,37,38,46,50,51] we derive *necessary optimality conditions* for various MPECs (22) as well as for related multiobjective optimization and equilibrium problems with equilibrium constraints governed by generalized equations/variational conditions (23)–(27), (30)–(32) and their specifications. A major role in these conditions is played by the *Fredholm constraint qualification*, which reads, in the particular case of the generalized equation in (22) with a smooth base, as that the *adjoint generalized equation*

$$0 \in \nabla f(\bar{x}, \bar{y})^* z^* + D^* Q(\bar{x}, \bar{y}, \bar{z})(z^*) \text{ with } \bar{z} := -f(\bar{x}, \bar{y}) \tag{33}$$

has only the *trivial solution* $z^* = 0$. Furthermore, in [38] we derive *necessary and sufficient* conditions for *linear suboptimality* in some of such problems.

Following the pattern developed in [32, Sect. 5.2], the results obtained in the aforementioned papers are generally expressed via *coderivatives* of the base and/or field mappings, while for *subdifferential systems* of types (26) and (27) we employ the *second-order subdifferentials* of extended-real-valued functions defined by the scheme

$$\partial^2 \varphi(\bar{x}, \bar{y}) := (D^* \partial \varphi)(\bar{x}, \bar{y}) \text{ for } \bar{y} \in \partial \varphi(\bar{x}) \tag{34}$$

via the corresponding coderivatives of the first-order subdifferential mappings; see [31] and the references therein for more details, calculus rules, explicit computations, and a number of applications of the second-order subdifferential constructions.

5 Bilevel Programming

Bilevel programming deals with a broad class of problems in *hierarchical optimization* that consist of minimizing *upper-level* objective functions subject to upper-level constraints given by set-valued mappings whose values are sets of *optimal solutions* to some *lower-level* problems of parametric optimization. There are several frameworks of bilevel programs and a number of approaches to their study and applications; see the book [11] and the extended introduction to [12] for more discussions and references. The so-called *optimistic version* in bilevel programming reads as follows:

$$\text{minimize } \varphi_0(x) \text{ subject to } x \in \Omega \text{ with } \varphi_0(x) := \inf\{\varphi(x,y) \mid y \in \Psi(x)\}, \qquad (35)$$

where the sets $\Psi(x)$ of *feasible solutions* to the upper-level problem in (35) consist of *optimal solutions* to the parametric lower-level optimization problem

$$\Psi(x) := \operatorname{argmin}\{\psi(x,y) \mid f_i(x,y) \leq 0, \ i = 1,\ldots,m\}, \qquad (36)$$

which may also contain constraints of other types (e.g., given by equalities).

Note that problems of this type are *intrinsically nonsmooth*, even for smooth initial data, and can be treated by using appropriate tools of modern variational analysis and generalized differentiation. In [12], we develop the so-called *value function approach* to bilevel programs in (35) and (36) that reduces them to the single-level framework of nondifferentiable programming formulated via (nonsmooth) optimal value functions of parametric lower-level problems in the original model.

It is important to observe that standard *constraint qualifications* in mathematical programming (e.g., the classical Mangasarian-Fromovitz one and the like) are *violated* for single-level programs obtained in this way. An appropriate qualification condition for bilevel programs related to a certain exact penalization was introduced in [62] under the name of "partial calmness". Using the latter constraint qualification and advanced formulas for computing and estimating *limiting subgradients* of *value/marginal functions* in parametric optimization obtained in [31,49], we derive new necessary optimality conditions for bilevel programs reflecting significant phenomena that have never been observed earlier. In particular, the necessary optimality conditions for bilevel programs established in [12] do *not depend* on the *partial derivatives* with respect to *parameters* of smooth objective functions in parametric lower-level problems. Efficient implementations of this approach are developed in [12] for bilevel programs with differentiable, convex, linear, and locally Lipschitzian functions describing the initial data of lower-level and upper-level problems.

The results obtained in [12] have been recently improved in [47] by deriving and applying new formulas for value functions in parametric optimization, which allow us to fully *avoid convexification* in the necessary optimality conditions established in [12].

In particular, under the same assumptions as in [12, Theorem 3.1] with the upper-level constraint set Ω in (35) described by the inequalities

$$\Omega := \left\{ x \in \mathbb{R}^n \,\middle|\, g_j(x) \leq 0, \; j = 1, \dots, p \right\} \tag{37}$$

involving the smooth initial data φ, ψ, f_i, and g_j in (35)–(37), we get the following necessary conditions for a local optimal solution $(\bar{x}, \bar{y})$ to the bilevel program under consideration: there are $\gamma > 0$ and nonnegative multipliers $\lambda_1, \dots, \lambda_m, \alpha_1, \dots, \alpha_m$, and $\beta_1, \dots, \beta_p$ such that

$$
\begin{aligned}
& \nabla_x \varphi(\bar{x}, \bar{y}) + \sum_{i=1}^{m} (\alpha_i - \gamma \lambda_i) \nabla_x f_i(\bar{x}, \bar{y}) + \sum_{j=1}^{p} \beta_j \nabla g_x(\bar{x}) = 0, \\
& \nabla_y \varphi(\bar{x}, \bar{y}) + \gamma \nabla_y \psi(\bar{x}, \bar{y}) + \sum_{i=1}^{m} \alpha_i \nabla f_y(\bar{x}, \bar{y}) = 0, \\
& \nabla_y \psi(\bar{x}, \bar{y}) + \sum_{i=1}^{m} \lambda_i \nabla_y f_i(\bar{x}, \bar{y}) = 0, \\
& \lambda_i f_i(\bar{x}, \bar{y}) = 0, \; \alpha_i f_i(\bar{x}, \bar{y}) = 0 \; \text{for} \; i = 1, \dots, m, \; \beta_j g_j(\bar{x}) = 0 \; \text{for} \; j = 1, \dots, p.
\end{aligned}
\tag{38}
$$

In [12,47], the reader can find more results and discussions on bilevel programs with nonsmooth data, and also with fully convex and linear structures.

6 Multiobjective Optimization and Equilibria

It is difficult to overstate the importance of multiobjective optimization and related equilibrium problems for both optimization/equilibrium theory and practical applications; see, e.g., [6,7,8,17,19,21,22,32,56,61,63] with the discussions and references therein. It has been well recognized that the advanced methods of variational analysis and generalized differentiation provide useful tools for the study of such problems and lead to significant progress in the theory and applications. In this section we discuss some latest advances in this direction based mostly on the recent research by the author and his collaborators.

A large class of constrained *multiobjective optimization* problems is described as:

$$\text{minimize } F(x) \text{ subject to } x \in \Omega \subset X, \tag{39}$$

where the *cost mapping* $F \colon X \rightrightarrows Z$ is generally *set-valued*, and where "minimization" is understood with respect to some *partial ordering* on Z. Thus (39) is a problem of *set-valued optimization*, while the term of *vector optimization* is usually used when $F = f \colon X \to Z$ is a single-valued mapping. We prefer to unify both set-valued and vector optimization problems under the name of multiobjective optimization. It is well known that various notions of *equilibrium* can be written in (or reduce to) form (39).

In [32, Sect. 5.3] and in the subsequent papers [2,37,40] we paid the main attention to the study of *generalized order optimality* defined as follows: given an ordering set $\Theta \subset Z$ with $0 \in \Theta$, we say tat $\bar{x} \in \Omega$ is a locally (f, Θ, Ω)-*optimal* if there are a neighborhood U of $\bar{x}$ and a sequence $\{z_k\} \subset Z$ with $\|z_k\| \to 0$ as $k \to \infty$ such that

$$f(x) - f(\bar{x}) \notin \Theta - z_k \text{ for all } x \in \Omega \cap U, \; k \in \mathbb{N}. \tag{40}$$

The (generally nonconvex and nonconical) set Θ in (40) can be viewed as a generator of an extended *order/preference relation* on Z and encompasses standard notions of multiobjective optimization and equilibria. In fact the above notion of generalized order optimality is induced by the notion of *local extremal points* of sets discussed in Sect. 3; see [32, Subsect. 5.3.1] for more details and examples.

The main results of [2,37,40] provide *necessary optimality conditions* for multiobjective problems with respect to the above generalized order optimality under various constraints (geometric, functional, operator, equilibrium, and their specifications) in finite and infinite dimensions. The results obtained are expressed via the robust/limiting generalized differential constructions discussed in Sect. 2. In [32, Subsect. 5.4.2] and [36], pointbased *necessary and sufficient* conditions are derived for *linearly suboptimal solutions* to multiobjective problems generated by linear subextremality of sets considered in Sect. 3.

Paper [51] is devoted to the study and applications of a remarkable and rather new class of *equilibrium problems with equilibrium constraints* (EPECs), which can be treated as *hierarchical games* defined by some equilibrium notions on both lower and upper levels of hierarchy. In [51], we pay a particular attention to the case of *weak Pareto optimality/equilibrium* on the upper level and *mixed complementarity constraints* on the lower level. Such problems can be modeled in the above framework of *multiobjective optimization with equilibrium constraints*. The necessary optimality conditions derived in [51] are based on the robust generalized differentiation constructions of Sect. 2, while they are finally presented fully in terms of the initial data and used in developing and implementing *numerical techniques*. The applications given in [51] concern *oligopolistic market* models that primarily motivate the research.

Paper [39] concerns a thorough study of multiobjective optimization problems with equilibrium constraints, where the notion of optimality is generated by *closed preference relations*. Given a subset $\Xi \subset Z \times Z$, we define the *preference* $\prec$ on Z by

$$z_1 \prec z_2 \text{ if and only if } (z_1, z_2) \in \Xi \qquad (41)$$

and say that $\prec$ is *locally closed* around $\bar{z}$ if there is a neighborhood U of $\bar{z}$ such that:

(a) preference $\prec$ is *nonreflexive*, i.e., $(z,z) \notin \Xi$;
(b) preference $\prec$ is *locally satiated* around $\bar{z}$, i.e., $z \in \mathrm{cl}\,\mathscr{L}(z)$ for all $z \in U$, where the *level set* $\mathscr{L}(z)$ corresponding to $\prec$ is defined by

$$\mathscr{L}(z) := \{ u \in Z \mid u \prec z \}; \qquad (42)$$

(c) preference $\prec$ is *almost transitive* on Z, i.e.

$$v \prec z \text{ whenever } v \in \mathrm{cl}\,\mathscr{L}(u), \ u \prec z, \ \text{and} \ v, z, u \in U. \qquad (43)$$

Observe that ordering relations on Z given by the generalized order optimality as in (40) and by closed preferences in (43) are generally independent. In particular, the almost transitivity of a *Pareto-type* preference given by

$$z_1 \prec z_2 \text{ if and only if } z_2 - z_1 \in \Theta \qquad (44)$$

via a closed cone $\Theta \subset Z$ is *equivalent* to the *convexity* and *pointedness* of the cone Θ, which means that $\Theta \cap (-\Theta) = \{0\}$. The latter is not required in (40) and does not hold in fact for a number of useful preferences important in the theory and applications, e.g., for the *lexicographical ordering* on $\mathbb{R}^n$; see [32, Subsect. 5.3.1] for more details and discussions.

Note that the necessary optimality conditions obtained in [39] for multiobjective problems described via closed preferences employ the notion of the *extended normal cone* to parameterized/moving sets $\Omega(\cdot)$ defined by

$$N_+\big(\bar{x}; \Omega(\bar{z})\big) := \underset{(z,x) \overset{\mathrm{gph}\,\Omega}{\to} (\bar{z},\bar{x})}{\mathrm{Lim\,sup}}\ \hat{N}\big(x, \Omega(z)\big) \text{ at } \bar{x} \in \Omega(\bar{z}). \qquad (45)$$

We refer the reader to the recent paper [54] for a comprehensive study of the extended normal cone (45) and associated coderivative and subdifferential constructions for moving objects (calculus rules, various relationships, normal compactness properties, etc.). In [39], the extended normal cone construction (45) is applied to express a part of necessary optimality conditions related to the moving level sets (42).

The main focus of [3] is on the study of the constrained multiobjective optimization problems (39) with general set-valued costs. We consider there two classical notions of minimizers/equilibria: *Pareto* and *weak Pareto*. The first notion corresponds to the preference on Z given by a closed and convex cone $\Theta \subset Z$ (which is assumed to be pointed in [3]), while the weak one assumes in addition that $\mathrm{int}\,\Theta \neq \emptyset$. Although the latter is a serious restriction, the vast majority of publications on multiobjective optimization, even in the simplest frameworks, concern *weak* Pareto minimizers, which are much more convenient to deal with in the vein of the conventional *scalarization* techniques.

In [3], we derive necessary conditions for both Pareto and weak Pareto minimizers in terms of our coderivatives discussed in Sect. 2 and also using new *subdifferential* constructions for *set-valued mappings* with values in *partially ordered spaces* that are extensions of those in (9)–(11) to the case of vector-valued and set-valued mappings. The basic techniques of [3] involves new versions of *variational principles* that are set/vector-valued counterparts of the classical Ekeland variational principle [16] and the subdifferential variational principle given in [31, Subsect. 2.3.2]. Furthermore, paper [3] contains new *existence theorems* for optimal solutions to (39) that employ, in particular, the following *subdifferential Palais-Smale condition* expressed in terms of the aforementioned analog of the basic subdifferential (10) for set/vector-valued mappings with values in partially ordered spaces: every sequence $\{x_k\} \subset X$ such that

$$\text{there are } z_k \in F(x_k) \text{ and } x_k^* \in \partial F(x_k, z_k) \text{ with } \|x_k^*\| \to 0 \text{ as } k \to \infty \qquad (46)$$

contains a convergent subsequence, provided that $\{z_k\}$ is (quasi)bounded from below.

In [4], we obtain a number of extensions of the existence theorems and necessary optimality conditions from [3] to multiobjective problems with various *constraints*, including those of the *equilibrium* type. This becomes possible due to the availability of coderivative/subdifferential *calculus* for the generalized differential constructions used in [3,4] (including the aforementioned new subdifferentials for set/vector-valued mappings), which particularly allows us to deal with various constraint structures.

Paper [6] addresses the study of the new notions of *relative Pareto minimizers* to constrained multiobjective problems that are defined via several kinds of *relative interiors* of ordering cones and occupy intermediate positions between the classical notions of Pareto and weak Pareto efficiency/optimality in finite-dimensional and infinite-dimensional spaces. Using advanced tools of variational analysis and generalized differentiation, we establish the *existence* of *relative Pareto minimizers* to general multiobjective problems under a *refined version of the subdifferential Palais-Smale condition* for set-valued mappings with values in partially ordered spaces and then derive *necessary optimality conditions* for these minimizers (as well as for conventional efficient and weak efficient counterparts) that are new in both finite-dimensional and infinite-dimensional settings. The proofs in [6] are mainly based on *variational and extremal principles* of variational analysis including certain new versions of them derived in the paper.

Finally in this section, we mention the recent developments in [5] devoted to so-called *super minimizers* to multiobjective optimization problems (39) with generally set-valued cost mappings. This notion is induced by the concept of *super efficiency* introduced in [8], which refines and/or unifies various modifications of *proper efficiency* and reflects crucial features of solutions to vector optimization problems important from the viewpoints of both the theory and applications. We derive necessary conditions for super minimizers using advanced tools of variational analysis and generalized differentiation that are new in both finite-dimensional and infinite-dimensional settings for problems with single-valued and set-valued objectives. The results obtained are expressed in generally independent *coderivative* and *subdifferential* forms. Then a part of [5] concerns establishing relationships between these notions for *set/vector-valued* mappings with values in *partially ordered* spaces, which are also important for further developments and applications.

7   Metric Regularity and Lipschitzian Stability of Parametric Variational Systems

It has been well recognized that the property of set-valued mappings known as *metric regularity*, as well as the *linear openness/covering* property equivalent to it, play an important role in many aspects of nonlinear and variational analysis and their applications; see, e.g., [7,14,18,20,23,24,31,32,58] with the extensive bibliographies therein. In the aforementioned references, the reader can find verifiable conditions ensuring these properties and their implementations in specific situations mainly related to the *implicit functions* and *multifunctions* frameworks and to the so-called *parametric constraint systems* in nonlinear analysis and optimization. The latter class of systems incorporates, in particular, sets of *feasible* solutions to various constrained optimization and equilibrium problems.

Recall that $F \colon X \rightrightarrows Y$ is *metrically regular* around $(\bar{x}, \bar{y}) \in \operatorname{gph} F$ if there are neighborhoods U of $\bar{x}$ and V of $\bar{y}$ and a number $\mu > 0$ such that

$$\operatorname{dist}\big(x; F^{-1}(y)\big) \leq \mu \operatorname{dist}\big(y; F(x)\big) \quad \text{whenever } x \in U \text{ and } y \in V. \tag{47}$$

Further, we say that $F: X \rightrightarrows Y$ is *Lipschitz-like* around $(\bar{x}, \bar{y}) \in \mathrm{gph}\, F$ is there are neighborhoods U of $\bar{x}$ and V of $\bar{y}$ and a number $\ell \geq 0$ such that

$$F(x) \cap V \subset F(u) + \ell \|x - u\| \mathbb{B} \quad \text{for all } x, u \in U. \tag{48}$$

The latter property is also known as the Aubin "pseudo-Lipschitz" property of set-valued mappings; see [31,58]. When $V = Y$ in (48), it reduces to the classical (Hausdorff) *local Lipschitzian* property of F around $\bar{x} \in \mathrm{dom}\, F$. Note that the Lipschitzian properties under consideration are *robust*, i.e., stable with respect to small perturbations of the initial data.

It is well known and utilized in nonlinear and variational analysis that the metric regularity property of F around $(\bar{x}, \bar{y})$ is *equivalent* to the Lipschitz-like property of its *inverse* around $(\bar{y}, \bar{x})$ with the same modulus in (47) and (48). Similar relationships hold true for certain *semilocal* and *global* modifications of the above *local* metric regularity and Lipschitzian properties and their *linear openness/covering* counterparts; see, [31, Sect. 1.2] for more details and discussions.

Observe that both metric regularity and Lipschitzian properties are defined in *primal* spaces and are *derivative-free*, i.e., they do not depend on any derivative-like construction. It turns out nevertheless that, due to *variational/extremal principles*, they admit *complete dual-space characterizations* in both finite-dimensional and infinite-dimensional spaces via appropriate *coderivatives* of set-valued mappings; see [29], [31, Chapter 4], and [58, Chapter 9] with comprehensive references and commentaries.

We have discussed in Sect. 4 a significant role of *parametric variational systems* of the types considered therein in variational analysis, optimization/equilibrium theory, and their numerous applications. It is shown in many publications that *robust Lipschitzian properties* are *intrinsic* for such systems being fulfilled under natural assumptions; see, e.g., the recent developments in [31, Sect. 4.4] and [33] based on *coderivative analysis* that largely revolves around the *Fredholm qualification condition* (33). It surprisingly happens, however, that it is *not the case for metric regularity* and the equivalent properties of linear openness/covering, which fail to be fulfilled for major classes of parametric variational systems.

In what follows, we present some results in this direction recently obtained in [43]. They are largely based on the *equivalence* [18] between metric regularity of the solution maps in systems (23), (26), and (27) and the Lipschitz-like property of the field/subdifferential mappings in these systems under the assumptions made. The latter property does not hold in the major cases under considerations; see [15,26,43] for more details.

Theorem 2. (Failure of metric regularity for generalized equations with monotone fields). *Let $f: X \times Y \to Y^*$ be a mapping between Asplund spaces that is strictly differentiable at $(\bar{x}, \bar{y})$ with the surjective partial derivative $\nabla_x f(\bar{x}, \bar{y})$, and let $Q: Y \rightrightarrows Y^*$ be locally closed-graph around $(\bar{y}, \bar{y}^*)$ with $\bar{y}^* := -f(\bar{x}, \bar{y}) \in Q(\bar{y})$. Assume in addition that Q is monotone and that there is no neighborhood of $\bar{y}$ on which Q is single-valued. Then the solution map $S: X \rightrightarrows Y$ in (23) with $Q = Q(y)$ is not metrically regular around $(\bar{x}, \bar{y})$.*

Since the *set-valuedness* of field mappings is a *characteristic* feature of *generalized* equations as a satisfactory model to describe *variational systems* (otherwise they reduce just to standard equations, which are not of particular interest in the variational framework under consideration), the conclusion of Theorem 2 reads that parametric variational systems with *monotone fields* are *not metrically regular* under the strict differentiability and surjectivity assumptions on base mappings, which do not seem to be restrictive. A major consequence of Theorem 2 is the following corollary concerning *subdifferential* systems with *convex* potentials, which encompass the classical cases of variational inequalities and complementarity problems in (24) that correspond to the *highly nonsmooth* (extended-real-valued) case of the convex *indicator functions* $\varphi(y) = \delta(y;\Omega)$ in (23) with $Q(y) = \partial\varphi(y)$.

Corollary 1. (Failure of metric regularity for subdifferential variational systems with convex potentials). *Let $Q(y) = \partial\varphi(y)$ in (23), where $f: X \times Y \to Y^*$ is a mapping between Asplund spaces that is strictly differentiable at $(\bar{x},\bar{y})$ with the surjective partial derivative $\nabla_x f(\bar{x},\bar{y})$, and where $\varphi: Y \to \bar{\mathbb{R}}$ is a l.s.c. convex function finite at $\bar{y}$ and such that there is no neighborhood of $\bar{y}$ on which φ is Gâteaux differentiable. Then the solution map S in (23) is not metrically regular around $(\bar{x},\bar{y})$.*

In fact, essentially more general *composite* subdifferential structures of parametric variational systems prevent the fulfillment of metric regularity for solutions maps with *no* reduction to the field monotonicity. In particular, it is proved in [43, Theorem 5.3 and 5.4] that *metric regularity fails* for the composite subdifferential systems (26) and (27) in Asplund spaces with $g = g(y)$ provided that f satisfies the assumptions of Theorem 2, that g is continuously differentiable around $\bar{y}$ in (27) while twice continuously differentiable around $\bar{y}$ in (26) with the surjective derivative $\nabla g(\bar{y})$ in both cases, and that ψ is l.s.c., convex, and *not* Gâteaux differentiable around the point $g(\bar{y})$.

In the case of *Hilbert spaces*, the results of Corollary 1 and the aforementioned ones for the composite structures (26) and (27) can be extended to subdifferential variational systems generated by essentially larger (than convex) classes of extended-real-valued functions. Recall [58] that $\varphi: X \to \bar{\mathbb{R}}$ is *subdifferentially continuous* at $\bar{x}$ for some subgradient $\bar{x}^* \in \partial\varphi(\bar{x})$ if $\varphi(x_k) \to \varphi(\bar{x})$ whenever $x_k \to \bar{x}$, $x_k^* \xrightarrow{w^*} \bar{x}^*$ as $k \to \infty$ with $x_k^* \in \partial\varphi(x_k)$ for all $k \in \mathbb{N}$. Further, φ is *prox-regular* at $\bar{x} \in \text{dom}\,\varphi$ for some $\bar{x}^* \in \partial\varphi(\bar{x})$ if it is l.s.c. around $\bar{x}$ and there are $\gamma > 0$ and $\eta \geq 0$ such that

$$\varphi(u) \geq \varphi(x) + \langle \bar{x}^*, u - x \rangle - \frac{\eta}{2}\|u - x\|^2 \text{ for all } x^* \in \partial\varphi(x)$$
$$\text{with } \|x^* - \bar{x}^*\| \leq \gamma, \|u - \bar{x}\| \leq \gamma, \|x - \bar{x}\| \leq \gamma, \text{ and } \varphi(x) \leq \varphi(\bar{x}) + \gamma. \tag{49}$$

Both properties above hold for broad classes of functions important in variational analysis and optimization. This is the case, in particular, for the so-called *strongly amenable* functions; see [58] and also [31,32] for more details, references, and applications.

Theorem 3. (Failure of metric regularity for composite subdifferential variational systems with prox-regular potentials). *Let $(\bar{x},\bar{y}) \in \text{gph}\,S$ for S given in (26), where $g: Y \to W$ is twice continuously differentiable around $\bar{y}$ with the surjective derivative $\nabla g(\bar{y})$, where $f: X \times Y \to Y^*$ is strictly differentiable at $(\bar{x},\bar{y})$ with the surjective partial derivative $\nabla_x f(\bar{x},\bar{y})$, where the spaces $X, Y,$ and Y^* are Asplund while W is Hilbert. Set $\bar{w} := g(\bar{y})$ and assume in addition that:*

(i) either ψ is locally Lipschitzian around $\bar{w}$;

(ii) or ψ is prox-regular and subdifferential continuous at $\bar{w}$ for the basic subgradient $\bar{v} \in \partial \psi(\bar{w})$, which is uniquely determined by $\nabla g(\bar{y})^* \bar{v} = -f(\bar{x}, \bar{y})$.

Then S is not metrically regular around $(\bar{x}, \bar{y})$ provided that there is no neighborhood of $\bar{w}$ on which ψ is Gâteaux differentiable.

Theorem 4. (Failure of metric regularity for subdifferential variational systems with composite fields and prox-regular potentials). *Let $(\bar{x}, \bar{y}) \in$ gph S for S defined by (27), where $g: Y \to W$ is strictly differentiable at $\bar{y}$ with the surjective derivative $\nabla g(\bar{y})$, where $f: X \times Y \to W$ is strictly differentiable at $(\bar{x}, \bar{y})$ with the surjective partial derivative $\nabla_x f(\bar{x}, \bar{y})$, where the spaces X and Y are Asplund while W is Hilbert. Set $\bar{w} := g(\bar{y})$ and assume in addition that either (i) or (ii) of Theorem 3 is satisfied, and that there is no neighborhood of $\bar{w}$ on which ψ is Gâteaux differentiable. Then the solution map S is not metrically regular around $(\bar{x}, \bar{y})$.*

In [43], the reader can find the proofs of these theorems and more discussions on them and related results for metric regularity and Lipschitzian stability of variational systems.

8 Optimal Control of Constrained Evolution Inclusions with Discrete and Continuous Time

As discussed in Sect. 1, problems of *optimal control* and related problems of *dynamic optimization* have always been among the strongest motivations and most important areas for applications of advanced methods and constructions of modern variational analysis and generalized differentiation. In this section we briefly review recent results on optimal control and related problems obtained by the author and his collaborators in [13,34,35,52,55].

In [35], we study the following problem of dynamic optimization governed by *discrete-time inclusions* with endpoint constraints of inequality, equality, and geometric types:

$$\begin{cases} \text{minimize } \varphi_0(x_0, x_K) \text{ subject to } (x_0, x_K) \in \Omega, \\ x_{j+1} \in F_j(x_j), \ j = 0, \ldots, K-1, \\ \varphi_i(x_0, x_K) \leq 0, \ i = 1, \ldots, m, \ \varphi_i(x_0, x_K) = 0, \ i = m+1, \ldots, m+r, \end{cases} \tag{50}$$

where $F_j: X \rightrightarrows X$, $\varphi_i: X^2 \to \mathbb{R}$, $\Omega \subset X^*$ and $K \in \mathbb{N}$. Observe that the inclusion model in (50) encompasses more conventional *discrete control systems* of the parameterized type

$$x_{j+1} = f_j(x_j, u_j), \ u_j \in U_j \text{ as } j = 0, \ldots, K-1 \tag{51}$$

with *explicit* control variables u_j taking values in some admissible control regions U_j.

The following major result is established in [35] based on the reduction to the *Lagrange principle* for non-dynamic constrained optimization problems discussed at the end of Sect. 3 and then on employing appropriate rules of *generalized differential and SNC calculi.*

Theorem 5. (Extended Euler-Lagrange conditions for discrete optimal control).
Let $\{\bar{x}_j| \ j=0,\dots,K\}$ be a local optimal solution to the discrete optimal control problem
(50). *Assume that X is Asplund, that φ_i are locally Lipschitzian around $(\bar{x}_0,\bar{x}_K)$ for all*
$i = 0,\dots,m+r$ *while Ω is locally closed around this point, and that the graphs of F_j are*
locally closed around $(\bar{x}_j,\bar{x}_{j+1})$ for every $j = 0,\dots,K-1$. Suppose also that all but one
of the sets Ω and $\mathrm{gph}\, F_j$, $j = 0,\dots,K-1$, are SNC at the points $(\bar{x}_0,\bar{x}_K)$ and $(\bar{x}_j,\bar{x}_{j+1})$,
respectively. Then there are multipliers $(\lambda_0,\dots,\lambda_{m+r})$ and an adjoint discrete trajectory
$\{p_j \in X^*| \ j=0,\dots,K\}$, *not all zero, satisfying the relationships:*

- *the Euler-Lagrange inclusion*

$$-p_j \in D_N^* F_j(\bar{x}_j,\bar{x}_{j+1})(-p_{j+1}) \ \text{for} \ j = 0,\dots,K-1, \tag{52}$$

- *the transversality inclusion*

$$(p_0, -p_K) \in \partial \Big(\sum_{i=0}^{m+r} \lambda_i \varphi_i \Big)(\bar{x}_0,\bar{x}_K) + N\big((\bar{x}_0,\bar{x}_K);\Omega\big), \tag{53}$$

- *the sign and complementary slackness conditions*

$$\lambda_i \geq 0 \ \text{for} \ i = 0,\dots,m, \ \lambda_i \varphi_i(\bar{x}_0,\bar{x}_K) = 0 \ \text{for} \ i = 1,\dots,m. \tag{54}$$

Note that if F_j is inner/lower semicontinuous at $(\bar{x}_j,\bar{x}_{j+1})$ and *convex-valued* around
these points for all $j = 0,\dots,K-1$, then the Euler-Lagrange inclusion (52) implies the
relationships of the *discrete maximum principle*:

$$\langle p_{j+1},\bar{x}_{j+1}\rangle = \max_{v \in F(\bar{x}_j)} \langle p_{j+1},v\rangle \ \text{for all} \ j = 0,\dots,K-1, \tag{55}$$

which provide necessary optimality conditions for problem (50) along with (52)–(54).

Observe that the results of Theorem 5 allow us to establish necessary optimality
conditions (52)–(54) and the maximum principle (55) with *no SNC* (or *finite codimen-
sionality*, or *interiority*) assumptions imposed on the endpoint constraint/target set Ω
and to cover, e.g., the classical two-point constraint case in (50) that has always been an
obstacle in infinite-dimensional optimal control, including that for smooth systems (51).

By using generalized differential and SNC calculus rules, Theorem 5 induces the corresponding necessary optimality conditions for optimal control problems of constrained
parametric discrete-time evolution inclusions of the type

$$x_{j+1} \in x_j + hF_j(x_j), \ j = 0,\dots,K-1. \tag{56}$$

It is worth mentioning that explicit control counterparts as in (51) of the parametric
discrete-time systems (56), considered as a *process* with $h \downarrow 0$, possess a number of important specific features that are *not* inherent in general parametric discrete systems with
fixed parameters h. An especially remarkable fact for optimal control of such systems
with *smooth* velocity mappings f_j is the validity of necessary optimality conditions in
the form of the *approximate maximum principle* with *no convexity* requirements. The
approximate maximum condition means that the exact one as in (55) is replaced by its

$\varepsilon(h)$-*perturbation* with $\varepsilon(h) \to 0$ as $h \downarrow 0$; see [32, Sect. 6.4] for more details, references, and commentaries.

Systems of type (56) arise, in particular, from *discrete/finite-difference approximations* of *continuous-time* evolution systems governed by *differential inclusions*

$$\dot{x}(t) \in F(x(t),t), \ x \in X, \ \text{a.e.} \ t \in [a,b]. \tag{57}$$

In fact, the approach to the study of continuous-time systems of type (57) and optimization problems for them via *well-posed discrete approximations* has been among the author's main interests and developments for a long time; see, e.g., [30], [32, Chapter 5] with the references and commentaries therein. The *major steps* of this approach to derive necessary optimality conditions for various constrained optimal control problems governed by continuous-time systems are as follows:

(a) To construct a *well-posed* sequence of discrete-time problems that *approximate* in an *appropriate sense* the original continuous-time problem of dynamic optimization.
(b) To derive *necessary optimality* conditions for the approximating *discrete-time* problems by reducing them to non-dynamic problems of mathematical programming and employing then *generalized differential calculus*.
(c) By *passing to the limit* in the obtained results for discrete approximations to establish necessary conditions for the *given optimal solution* to the original problem.

Note that each of the above steps in the study of relationships between continuous-time systems and their discrete approximations is certainly of its own interests regardless of deriving necessary optimality conditions for the continuous-time dynamics. In particular, step (a) and its modifications are important for *numerical analysis* of continuous-time systems.

In this vein, paper [52] deals with establishing the *epi-convergence* of discrete approximations to the so-called *generalized Bolza problem* of dynamic optimization, which encompasses a number of the most interesting optimal control problems governed by differential inclusions of type (57) with *finite-dimensional* state spaces $X = \mathbb{R}^n$. The methods developed in this study and the results obtained seem to be suitable for extensions to *higher dimensions* (versus $t \in \mathbb{R}$) in the framework of *finite element methods*.

Paper [13] also goes in the direction of the aforementioned step (a) and is devoted to the study of well-posedness of discrete approximations to *nonconvex* differential inclusions of type (57) with *Hilbert* state spaces X. The underlying feature of the problems under consideration in [13] is a *one-sided Lipschitz* condition imposed on $F(\cdot,t)$, which is a significant improvement of the conventional Lipschitz continuity studied in prior publications. Among the main results of [13] we mention establishing efficient conditions that ensure the *strong approximation* (in the $W^{1,p}$-norm as $p \geq 1$) of feasible trajectories for one-sided Lipschitzian differential inclusions by those for their discrete approximations and also the *strong convergence* of optimal solutions to the corresponding dynamic optimization problems under discrete approximations. To proceed with the latter issue, we derive a new extension of the Bogolyubov-type *relaxation/density* theorem to the case of differential inclusions satisfying the modified one-sided Lipschitzian condition. All the results obtained are new not only in the infinite-dimensional Hilbert space framework but also in finite-dimensional spaces.

Paper [34] develops *all the three* of the aforementioned steps (a)–(c) in the implementation of the *method of discrete approximations* to derive new necessary optimality conditions for *nonconvex evolution/differential inclusions* of type (57) in the case of *Asplund* state spaces X. Dynamic optimization problems (of the Bolza and Mayer types) are considered in [34] subject to *finitely many* of the Lipschitzian *endpoint constraints*

$$\varphi_i\big(x(b)\big) \leq 0, \; i = 1,\ldots,m, \;\; \varphi_i\big(x(b)\big) = 0, \; i = m+1,\ldots,m+r, \tag{58}$$

on the trajectories for the evolution inclusion (57) with $x(a) = x_0$. The optimality conditions derived in [34] do *not impose* any *SNC/finite codimension* requirements on the target sets in (58) in contrast to geometric endpoint constraints of the type $x(b) \in \Omega$ studied previously in the author's book [32, Sect. 6.1 and 6.2]. The continuous-time counterpart of the extended *Euler-Lagrange inclusion* obtained in [34] is given by

$$\dot{p}(t) \in \mathrm{clco}\, D_N^* F\big(\bar{x}(t),\dot{\bar{x}}(t)\big)\big(-p(t)\big) \quad \text{a.e. } t \in [a,b] \tag{59}$$

together with the corresponding transversality, sign, complementary slackness, and maximum conditions as in (53)–(55). Note that, in contrast to the discrete case of (52), the Euler-Lagrange inclusion (59) involves the *convexification* of the coderivative values, while the *maximum condition*

$$\langle p(t),\dot{\bar{x}}(t)\rangle = \max_{v\in F(\bar{x}(t))} \langle p(t),v\rangle \quad \text{a.e. } t \in [a,b] \tag{60}$$

does *not* require *any convexification*. The latter is due the "hidden convexity" property (of the Lyapunov-Aumann type), which is automatically generated by the continuous-time dynamics; see [32,34] for more results and discussions in this direction.

Finally in this section, we mention new results on the well-posedness of discrete approximations and necessary optimality conditions obtained in [55] for dynamic optimization problems governed by constrained *delay-differential inclusions* of the type

$$\begin{cases} \dot{x}(t) \in F(x(t),x(t-\Delta),t) \;\; \text{a.e. } t \in [a,b], \\ x(t) \in C(t) \;\; \text{a.e. } t \in [a-\Delta,a), \quad \Delta > 0, \\ (x(a),x(b)) \in \Omega \subset X^2 \end{cases} \tag{61}$$

with an *Asplund* state space X. A specific feature of the delay system (61), which does not have any analogs for nondelayed systems, is the presence of *set-valued initial conditions* of the time $x(t) \in C(t)$ on $[a-\Delta,a)$, which particularly provides an additional source for optimization. The results obtained in [55] develop and extend those from [32,34] for the delay-differential problems under consideration, with deriving appropriate delay counterparts of conditions (59) and (60) as well as the new one corresponding to the multivalued "initial tail" part on $[a-\Delta,a)$.

9   Feedback Control of Constrained Parabolic Systems in Uncertainty Conditions

In the concluding section of the paper we discuss recent results by the author on *optimal control* and *feedback design* of *state-constrained parabolic systems* in *uncertainty conditions.* Problems of this type are among the most challenging and difficult in dynamic optimization for any kind of dynamical systems. The feedback design problem is formulated in the *minimax sense* to ensure *stabilization* of transients within the prescribed diapason and *robust stability* of the closed-loop control system under all feasible perturbations with *minimizing* an integral cost functional in the *worst* perturbation case.

The original motivation for our developments comes from practical design problems of automatic control of the soil groundwater regime in irrigation engineering networks functioning under uncertain weather and environmental conditions. In [41,42,44], we study such problems for parabolic systems with controls acting in boundary conditions of various types (Dirichlet, Neumann, Robin/mixed). In what follows we present the problem formulation and discuss the major results for the case of *Dirichlet boundary conditions*, which offer the *least regularity* properties for the parabolic dynamics and appear to be the *most challenging* in control theory for parabolic systems.

The system dynamics in the problem under consideration is given by the multidimensional *linear parabolic equation*

$$\begin{cases} \dfrac{\partial y}{\partial t} + Ay = w(t) & \text{a.e. in } Q := [0,T] \times \Omega, \\ y(0,x) = 0, & x \in \Omega, \\ y(t,x) = u(t), & (t,x) \in \Sigma := [0,T] \times \Gamma \end{cases} \tag{62}$$

with *controls* $u(\cdot)$ acting in the Dirichlet boundary conditions and distributed *perturbations* $w(\cdot)$ in the right-hand side of the parabolic equation. In (62), A is a *self-adjoint* and *uniformly strongly elliptic operator* on $L^2(\Omega)$ defined by

$$Ay := - \sum_{i,j=1}^{n} \frac{\partial}{\partial x_i}\left(a_{ij}(x)\frac{\partial y}{\partial x_j}\right) - cy, \tag{63}$$

where $\Omega \subset \mathbb{R}^n$ is an open bounded domain with the the boundary Γ that is supposed to be a sufficiently smooth $(n-1)$-dimensional manifold, and where $T > 0$ is a fixed time bound.

The sets of *admissible controls* U and *admissible perturbations* W are given by

$$U := \left\{ u \in L^\infty[0,T] \,\middle|\, -\alpha \leq u(t) \leq \alpha \text{ a.e. } t \in [0,T] \right\}, \tag{64}$$

$$W := \left\{ w \in L^\infty[0,T] \,\middle|\, -\beta \leq w(t) \leq \beta \text{ a.e. } t \in [0,T] \right\} \tag{65}$$

with some fixed bounds $\alpha, \beta > 0$ in the *pointwise/magnitude* constraints (64) and (65).

The underlying requirement on the system performance is to *stabilize* transients $y(t,x_0)$ near the initial equilibrium state $y(x,0) \equiv 0$ with a given accuracy $\eta > 0$ during the whole dynamic process. This is formalized via the *pointwise state constraints*

$$-\eta \leq y(t,x_0) \leq \eta \quad \text{a.e. } t \in [0,T]. \tag{66}$$

A characteristic feature of the dynamical process described by (62) is the *uncertainty* of perturbations $w \in W$: we can operate only with the bound β of the admissible region (65). Thus we can keep the system transients $y(t,x_0)$ within the prescribed stabilization region (66) only by using *feedback* boundary controls $u(\cdot)$ depending on the current state position $\xi = y(t, x_0)$ for each $t \in [0, T]$.

To formalize this description, consider a function $f \colon \mathbb{R} \to \mathbb{R}$ and construct boundary controls in (62) via the *feedback law*

$$u(t) := f(y(t, x_0)), \ t \in [0, T], \tag{67}$$

which defines a *feasible feedback regulator* if it generates controls $u(t)$ by (67) belonging to the admissible set U from (64) and keeps the corresponding transients $y(t, x_0)$ of (62) within the constraint area (66) for every admissible perturbation $w \in W$ from (65). We estimate the quality of feasible regulators $f = f(\xi)$ by the (energy-type) *cost functional*

$$J(f) := \max_{w \in W} \left\{ \int_0^T \left| f(y(t, x_0)) \right| dt \right\}. \tag{68}$$

The *maximum* operation in (68) reflects the required control energy needed to neutralize the adverse effect of the *worst perturbations* from (65) and to keep the state performance within the prescribed area (66). Finally, denote by $\mathscr{F}$ the set of all feasible feedback regulators and formulate the *minimax feedback control problem* as follows:

$$\text{minimize } J(f) \text{ over } f \in \mathscr{F}. \tag{69}$$

It has been well recognized in control theory and applications that *feedback* control problems are the most challenging and important for any type of dynamical systems, while PDE systems provide additional difficulties and much less investigated in comparison with the ODE dynamics. Furthermore, significant complications come from *pointwise/hard state constraints*, which are of high nontriviality even for open-loop control problems. We are not familiar with any constructive device applicable to the feedback control problem (P) under consideration among a variety of approaches and results available in the theories of differential games, H_∞-control, Riccati's feedback synthesis, and other developments in general settings; see more discussions and references in the aforementioned papers.

In these papers, we develop an approach to solving the feedback control problem (69), which is essentially based on certain underlying features of the parabolic dynamics, particularly on the *monotonicity property* of transients that is eventually related to the fundamental *Maximum Principle* for parabolic equations. Due to this property and the specific structures of the cost functional (68) and boundary controls in (62) and by employing the *convolution representation* of the transients obtained [53], we are able to select the *worst perturbations* in the area (65) for the class of *nonincreasing* and *odd feedbacks* (67). This allows us to study the corresponding *open-loop* optimal control problem with *pointwise state constraints* as a reaction of the parabolic system to the worst perturbations. Using the *spectral* Fourier-type representation of solutions to the parabolic system (62) and assuming the *positivity* of the *first eigenvalue* of the elliptic operator A in (63)—which is often the case— we observe the *dominance* of the *first term* in the exponential series representation of solutions to (62) as $t \to \infty$. In this way, we justify an efficient approximation of the open-loop optimal control problem for the parabolic system under consideration by that for the corresponding *ODE system* with state constraints on a sufficiently *large* time interval. Moreover, the approximating ODE optimal control problem is solved *exactly* by constructing *yet another approximation* of state constraints, employing the *Pontryagin maximum principle* that provides *necessary and sufficient* optimality conditions for the unconstrained approximating problems with both *bang-bang* and *singular modes* of optimal controls, and then by passing to the limit while meeting the state constraints. Furthermore, the *state constraints* occur to be a *regularization factor*, which simplifies the structure of optimal controls, especially when the time interval becomes bigger and bigger; this reveals the fundamental *turnpike property* of such dynamic systems expanding to the *infinite horizon*.

Thus using the ODE approximation described above, we justify an easily implemented *suboptimal* (or *near-optimal*) *structures* of optimal controls in both *open-loop* and *closed-loop* modes and then *optimize their parameters* along the *parabolic dynamics*. This allows us to arrive at a *three-positional feedback regulator* $f = f(\xi)$ in (67) acting via the Dirichlet boundary conditions of (62) that ensures the required state performance (66) under the fulfillments of all the constraints in (69) for *every feasible perturbation* from (65) providing a *near-optimal response* of the closed-loop control system in the case of *worst perturbations*.

The feedback control design constructed in this way leads us to the *highly nonlinear* closed-loop system (62) and (67), where $f(\xi)$ is a *discontinuous* three-positional regulator. The system may loose *robust stability* (in the large) and maintain the state performance (66) in an unacceptable *self-vibrating regime*. Developing a *variational approach* to robust stability that reduces the stability issue to a certain open-loop optimal control problem on the *infinite horizon*, we establish efficient conditions for robust stability of the closed-loop system whenever $t \geq 0$ in terms of the initial data of problem (69) and parameters of the three-positional feedback regulator. All the details can be found in [41,42,44].

References

1. Attouch, A., Buttazzo, G., Michaelle, G.: Variational Analysis in Sobolev and BV Spaces. In: Applications to PDEs and Optimization. MPS-SIAM Series on Optimization. SIAM Publications, Philadelphia (2006)
2. Bao, T.Q., Gupta, P., Mordukhovich, B.S.: Necessary conditions in multiobjective optimization with equilibrium constraints. J. Optim. Theory Appl. 135, 179–203 (2007)
3. Bao, T.Q., Mordukhovich, B.S.: Variational principles for set-valued mappings with applications to multiobjective optimization. Control Cybern. 36, 531–562 (2007)
4. Bao, T.Q., Mordukhovich, B.S.: Existence of minimizers and necessary conditions in multiobjective optimization with equilibrium constraints. Appl. Math. 26, 453–472 (2007)
5. Bao, T.Q., Mordukhovich, B.S.: Necessary conditions for super minimizers in constrained multiobjective optimization. J. Global Optim. (to appear)
6. Bao, T.Q., Mordukhovich, B.S.: Relative Pareto minimizers to multiobjective problems: existence and optimality conditions: Tech. Rep. 11, Department of Mathematics, Wayne State University (2007)
7. Borwein, J.M., Zhu, Q.J.: Techniques in Variational Analysis. CMS Books in Mathematics Series, vol. 20. Springer, New York (2005)
8. Borwein, J.M., Zhuang, D.M.: Super efficiency in vector optimization. Trans. Amer. Math. Soc. 338, 105–122 (1993)

9. Clarke, F.H.: Optimization and Nonsmooth Analysis. CMS Books in Mathematics Series, vol. 1. Wiley, New York (1983)
10. Crandall, M.G., Lions, P.-L.: Viscosity solutions to Hamilton-Jacobi equations. Trans. Amer. Math. Soc. 277, 1–42 (1983)
11. Dempe, S.: Foundations of Bilevel Programming. Kluwer Academic Publishers, Dordrecht (2002)
12. Dempe, S., Dutta, J., Mordukhovich, B.S.: New necessary optimality conditions in optimistic bilevel programming. Optimization 56, 577–604 (2007)
13. Donchev, T., Farkhi, E., Mordukhovich, B.S.: Discrete approximations, relaxation, and optimization of one-sided Lipschitzian differential inclusions in Hilbert spaces. J. Diff. Eq. 243, 301–328 (2007)
14. Dontchev, A.L., Lewis, A.S., Rockafellar, R.T.: The radius of metric regularity. Trans. Amer. Math. Soc. 355, 493–517 (2003)
15. Eberhard, A., Mordukhovich, B.S., Pearce, C.E.M.: On differentiability properties of prox-regular functions, preprint (2008)
16. Ekeland, I.: On the variational principle. J. Math. Anal. Appl. 47, 324–353 (1974)
17. Facchinei, F., Pang, J.-P.: Finite-Dimensional Variational Inequalities and Complementarity Problems. Series in Operations Research. Springer, New York (2003)
18. Geremev, W., Mordukhovich, B.S., Nam, N.M.: Coderivative calculus and metric regularity for constraint and variational systems. Nonlinear Anal. (to appear)
19. Göpfert, A., Riahi, H., Tammer, C., Zalinescu, C.: Variational Methods in Partially Ordered Spaces. CMS Books in Mathematics, vol. 17. Springer, New York (2003)
20. Ioffe, A.D.: Metric regularity and subdifferential calculus. Russian Math. Surv. 55, 501–558 (2000)
21. Jahn, J.: Vector Optimization: Theory, Applications and Extensions. Series in Operations Research. Springer, New York (2004)
22. Jeyakumar, V., Luc, D.T.: Nonsmooth Vector Functions and Continuous Optimization. Optimization and Its Applications Series, vol. 10. Springer, New York (2008)
23. Jourani, A., Thibault, L.: Coderivative of multivalued mappings, locally compact cones and metric regularity. Nonlinear Anal. 35, 925–945 (1999)
24. Klatte, D., Kummer, B.: Nonsmooth Equations in Optimization: Regularity, Calculus, Methods, and Applications. Kluwer Academic Publishers, Boston (2002)
25. Kruger, A.Y.: Weak stationarity: eliminating the gap between necessary and sufficient conditions. Optimization 53, 147–164 (2004)
26. Levy, A.B., Poliquin, R.A.: Characterizing the single-valuedness of multifunctions. Set-Valued Anal. 5, 351–364 (1997)
27. Mordukhovich, B.S.: Maximum principle in problems of time optimal control with nonsmooth constraints. J. Appl. Math. Mech. 40, 960–969 (1976)
28. Mordukhovich, B.S.: Metric approximations and necesary optimality conditions for general classes of nonsmooth extremal problems. Soviet Math. Dokl. 22, 526–530 (1980)
29. Mordukhovich, B.S.: Complete characterizations of openness, metric regularity, and Lipschitzian properties of multifunctions. Trans. Amer. Math. Soc. 340, 1–35 (1993)
30. Mordukhovich, B.S.: Discrete approximations and refined Euler-Lagrange conditions for nonconvex differential inclusions. SIAM Journal on Control and Optimization 33, 815–882 (1995)
31. Mordukhovich, B.S.: Variational Analysis and Generalized Differentiation, I: Basic Theory. Grundlehren Series (Fundamental Principles of Mathematical Sciences), vol. 330. Springer, Berlin (2006)
32. Mordukhovich, B.S.: Variational Analysis and Generalized Differentiation, II: Applications. Grundlehren Series (Fundamental Principles of Mathematical Sciences), vol. 331. Springer, Berlin (2006)

33. Mordukhovich, B.S.: Coderivative calculus and robust Lipschitzian stability for variational systems. J. Convex Anal. 13, 799–822 (2006)
34. Mordukhovich, B.S.: Variational analysis of evolution inclusions. SIAM J. Optim. 18, 752–777 (2007)
35. Mordukhovich, B.S.: Variational analysis in nonsmooth optimization and discrete optimal control. Math. Oper. Res. 26, 840–856 (2007)
36. Mordukhovich, B.S.: Necessary and sufficient conditions for linear suboptimality in constrained optimization. J. Global Optim. 40, 225–244 (2008)
37. Mordukhovich, B.S.: Methods of variational analysis in multiobjective optimization. Optimization (to appear)
38. Mordukhovich, B.S.: Characterizations of linear suboptimality for mathematical programs with equilibrium constraints. Math. Prog. (to appear)
39. Mordukhovich, B.S.: Multiobjective optimization problems with equilibrium constraints. Math. Prog. (to appear)
40. Mordukhovich, B.S.: Optimization and equilibrium problems with equilibrium constraints in infinite-dimensional spaces. Optimization (to appear)
41. Mordukhovich, B.S.: Suboptimal minimax design of constrained parabolic systems with mixed boundary control. Appl. Math. Comp. (to appear)
42. Mordukhovich, B.S.: Optimization and feedback design of state-constrained parabolic systems. Pacific J. Math. (to appear)
43. Mordukhovich, B.S.: Failure of metric regularity for major classes of variational systems. Nonlinear Anal. (to appear)
44. Mordukhovich, B.S.: Suboptimal feedback control of constrained parabolic systems in uncertainty conditions: Tech. Rep. 15, Department of Mathematics, Wayne State University (2006)
45. Mordukhovich, B.S., Nam, N.M.: Exact calculus for proximal subgradients with applications to optimization. ESAIM Proc. 17, 80–95 (2007)
46. Mordukhovich, B.S., Nam, N.M.: Variational analysis of extended generalized equations via coderivative calculus in Asplund spaces. J. Math. Anal. Appl. (to appear)
47. Mordukhovich, B.S., Nam, N.M., Phan, H.: Generalized differentiation of value functions in parametric optimization with applications to bilevel programming, preprint (2008)
48. Mordukhovich, B.S., Nam, N.M., Yen, N.D.: Fréchet subdifferential calculus and optimality conditions in nondifferentiable programming. Optimization 55, 685–708 (2006)
49. Mordukhovich, B.S., Nam, N.M., Yen, N.D.: Subgradients of marginal functions in parametric mathematical programming. Math. Prog. (to appear)
50. Mordukhovich, B.S., Outrata, J.V.: Coderivative analysis of quasivariational inequalities with applications to stability and optimization. SIAM J. Optim. 18, 389–412 (2007)
51. Mordukhovich, B.S., Outrata, J.V., Červinka, M.: Equilibrium problem with complementarity constraints: case study with applications to oligopolistic markets. Optimization 56, 479–494 (2007)
52. Mordukhovich, B.S., Pennanen, T.: Epiconvergent discretization of the generalized Bolza problem in dynamic optimization. Optimization Letters 1, 379–390 (2007)
53. Mordukhovich, B.S., Seidman, T.I.: Asymmetric games for convolution systems with applications to feedback control of constrained parabolic equations. J. Math. Anal. Appl. 333, 401–415 (2007)
54. Mordukhovich, B.S., Wang, B.: Generalized differentiation of parameter-dependent sets and mappings. Optimization 57, 17–40 (2008)
55. Mordukhovich, B.S., Wang, D., Wang, L.: Optimal control of delay-differential inclusions with multivalued initial conditions in infinite-dimensional spaces. Control Cybern. (to appear)

56. Outrata, J.V.: Mathematical programs with equilibrium constraints: theory and numerical methods. In: Haslinger, J., Stavroulakis, G.E. (eds.) Nonsmooth Mechanics of Solids. CISM Courses and Lecture Notes, vol. 485, pp. 221–274. Springer, New York (2006)
57. Robinson, S.M.: Generalized equations and their solutions. I: Basic theory. Math. Prog. Study 10, 128–141 (1979)
58. Rockafellar, R.T., Wets, R.J.-B.: Variational Analysis. Grundlehren Series (Fundamental Principles of Mathematical Sciences), vol. 317. Springer, Berlin (1998)
59. Schirotzek, W.: Nonsmooth Analysis. Series Universitext. Springer, Berlin (2007)
60. Vinter, R.B.: Optimal Control. Series Systems & Control: Foundations & Applications. Birkhäuser, Boston (2000)
61. Yao, J.C., Chadi, O.: Pseudomonotone complementarity problems and variational inequalities. In: Handbook of Generalized Convexity and Generalized Monotonicity, pp. 501–558. Springer, New York (2005)
62. Ye, J.J., Zhu, D.L.: Optimality conditions for bilevel programming problems. Optimization 33, 9–27 (1995)
63. Zheng, X.Y., Ng, K.F.: The Lagrange multiplier rule for multifunctions in Banach spaces. SIAM J. Optim. 17, 1154–1175 (2006)

Problems of Mathematical Finance by Stochastic Control Methods*

Łukasz Stettner

Institute of Mathematics Polish Academy of Sciences, Sniadeckich 8, 00-956 Warsaw and
Academy of Finance, Poland
stettner@impan.gov.pl

Abstract. The purpose of this paper is to present main ideas of mathematics of finance using the stochastic control methods. There is an interplay between stochastic control and mathematics of finance. On the one hand stochastic control is a powerful tool to study financial problems. On the other hand financial applications have stimulated development in several research subareas of stochastic control in the last two decades. We start with pricing of financial derivatives and modeling of asset prices, studying the conditions for the absence of arbitrage. Then we consider pricing of defaultable contingent claims. Investments in bonds lead us to the term structure modeling problems. Special attention is devoted to historical static portfolio analysis called Markowitz theory. We also briefly sketch dynamic portfolio problems using viscosity solutions to Hamilton-Jacobi-Bellman equation, martingale-convex analysis method or stochastic maximum principle together with backward stochastic differential equation. Finally, long time portfolio analysis for both risk neutral and risk sensitive functionals is introduced.

1 Pricing of Financial Derivatives

One of the fundamental problems of mathematics of finance is pricing of the derivative securities (shortly derivatives) i.e. securities the value of which depends on the basic securities such as stocks or bonds. In this section we restrict ourselves to stocks, although similar problems (unfortunately much harder) concern also derivatives of bonds.

1.1 Modeling of Asset Prices

We start with modeling of asset prices (stocks). We assume that we are given d assets on the market and denote the price of the i-th asset at time t by $S_i(t)$. We shall consider in parallel way two approaches: in discrete and in continuous time. We assume a given probability space $(\Omega, \mathscr{F}, (\mathscr{F}_t), P)$. In the case of discrete time the asset prices satisfy the relation

$$\frac{S_i(t+1)}{S_i(t)} = \zeta_i(t, z(t+1), \xi(t+1)), \tag{1}$$

* Research supported by MNiSzW grant no. 1 P03A 013 28.

A. Korytowski et al. (Eds.): System Modeling and Optimization, IFIP AICT 312, pp. 129–143, 2009.

for $i = 1, 2, \ldots, d$, where $(z(t))$ stands for the process of economic factors which have an impact on the prices, and $(\xi(t))$ is a noise process being a sequence of i.i.d. random variables such that $\xi(t)$ is independent of $z(s)$ for $s \leq t$.

In the case of continuous time model we assume that $S_i(t)$ is given by the following formula

$$S_i(t) = S_i(0)e^{\int_0^t a_i(s, z(s))ds + \int_0^t \sigma_i(s, z(s)) \cdot dL(s)}, \tag{2}$$

where the last term is the stochastic integral with respect to the process with stationary independent increments which is a Levy process or more specifically a Wiener process L. For the definition of stochastic integral we refer to [37].

1.2   Contingent Claims

A typical situation we encounter on financial markets is that at a given time T, called maturity we get a contingent claim the value of which is an $\mathscr{F}_T$ measurable random variable H. The problem is to determine its price at time 0. European options can serve as an example of contingent claims. Assume that we want to have a guarantee that at time T we shall buy one i-asset for the value of at most K (the so called striking price). For this purpose we buy a *European call option*. If the price $S_i(T)$ is greater than K we exercise this option buying one i-th asset for the price K. Otherwise, i.e. when $S_i(T) < K$ there is no reason to exercise this option - we simply buy the asset for the price $S_i(T)$. Consequently the value of this European call option is $(S_i(T) - K)^+$. Similarly, when we have an asset i we can guarantee to sell it at time T for the price K buying a *European put option*. In this case the value of the option is $(K - S_i(T))^+$. In both cases an important factor is the price for the option we have to pay at time 0. The terminal maturity time T is fixed and deterministic. One can consider contracts with random T. In the so called *American options* the buyer of an option is allowed to choose random maturity time τ, i.e. he chooses the time to exercise the option. Clearly such call or put options are more expensive since they offer a better bargain to the buyer.

1.3   Portfolio and Absence of Arbitrage

From stochastic control point of view the main decision we make on financial markets is the choice of a proper investment portfolio. We denote by $V(0) = v$ the initial value of the wealth process. We assume that we can invest v in the d-assets available on our market as well as in the savings account in a bank with a short term interest rate $r(t)$ at time t, which means that for one dollar invested at time 0 in the account we obtain $B_t = \exp\left(\int_0^t r_u du\right) = S_0(t)$ at time t. An investment strategy at each time t is a sequence $(N_0(t), N_1(t), \ldots, N_d(t))$ where $N_0(t)$ is the capital invested in our savings account, and $N_i(t)$ is the capital invested in the i-th asset at time t. The strategy can be also described as the sequence $\pi = (\pi_0(t), \pi_1(t), \ldots, \pi_d(t))$ where $\pi_i(t)$ is the portion of the wealth process $V(t)$ invested at time t in the i-th asset (for $i = 0$ in the savings account). Clearly $N_i(t) = \pi_i(t)V(t)$ for $i = 0, 1, \ldots, d$ and for discrete time model we have

$$V(t) = \sum_{j=0}^{d} N_j(t)S_j(t) = V(t-1) + \sum_{j=0}^{d} N_j(t-1)(S_j(t) - S_j(t-1))$$

$$= V(t-1) \sum_{j=0}^{d} \pi_j(t-1)\zeta_j(t,z(t),\xi(t)),$$

(3)

with $\zeta_0(t,z,\xi) = \exp\left(\int_t^{t+1} r_u du\right)$. The wealth process depends on the investment strategy π and initial wealth v. To point out this we shall denote the wealth process at time t by $V^{\pi,v}(t)$. We shall assume furthermore that the portfolio is *selffinancing* which means that no money is added or withdrawn i.e. we invest at each time t the value equal to our wealth $V^{\pi,v}(t)$. The fundamental assumption of mathematical finance is the so called *absence of arbitrage (AA)* in time T, which is the absence of the existence of portfolio π such that $V^{\pi,0}(T) \geq 0$ and $P\{V^{\pi,0}(T) > 0\} > 0$. Roughly speaking, it means that we are not able to create gain at time T without risk. The economical notion of the absence of arbitrage has an important analytical interpretation, which is very transparent in discrete time case.

We call a probability measure Q, which is equivalent to the original probability measure P, a martingale measure, whenever $(\frac{S_i(t)}{B_t})$ are martingales with respect to Q for $i = 1, 2, \ldots, d$ and $t = 0, 1, \ldots, T$. The martingale property means integrability of the ratios $\frac{S_i(t)}{B_t}$ plus the property

$$E\left\{\frac{S_i(t+1)}{B_{t+1}} | F_t\right\} = \frac{S_i(t)}{B_t},$$

(4)

for $t = 0, 1, \ldots, T - 1$. The following theorem (see [10] or [20,22] for more recent approaches) is fundamental in discrete time mathematics of finance.

Theorem 1. *(Dalang, Morton, Willinger, 1990) The absence of arbitrage is equivalent to the existence of a martingale measure Q.*

In what follows we shall denote by $\mathcal{Q}$ the family of all martingale measures. By Theorem 1 under (AA) we have $\mathcal{Q} \neq \emptyset$.

1.4 Pricing

We shall now consider the problem of pricing of contingent claim of value H at a given maturity time T. The buyer of it collects a gain which is a random variable H. We would like to evaluate the price of H at the initial time. One can look at this price from the perspective of the seller or the buyer. An acceptable price for the seller is such a price that for the amount he obtains at time 0 he is able (providing he invests it in an optimal way) to get at least the compensation for H (i.e. to hedge H), which he is supposed to deliver to the buyer. This investment forms an optimal stochastic control problem. We would like to find the smallest initial capital v, which invested in an optimal way gives us at least the value H at time T. Denote by $p_s(H)$ the minimal seller price. Consequently we can write the formula

$$p_s(H) = \inf\{v: \exists_\pi, V^{\pi,v}(T) \geq H\}.$$

(5)

The buyer price on the other hand is the value v such that if he starts with initial capital $-v$ and invests it in an optimal way, then at time T the value of his portfolio plus his gain H is nonnegative. Such maximal v forms the so called buyer price $p_b(H)$ and is the maximal price acceptable for the buyer. We have

$$p_b(H) = \sup \left\{ v : \exists_\pi, \ V^{\pi,-v}(T) + H \geq 0 \right\}. \tag{6}$$

Alternatively we can write that $p_b(H) = -p_s(-H)$. Under the absence of arbitrage assumption we clearly have that $p_b(H) \leq p_s(H)$. The interval $[p_b(H), p_s(H)]$ is called an absence of arbitrage interval and any price from this interval is acceptable in the sense that neither seller nor buyer is able to obtain a positive gain without risk at time T (which is an arbitrage). In particular situations, when $p_b(H) = p_s(H)$ for all bounded H, we say that the market is *complete* which in turn corresponds to the fair price or fair game between the seller and the buyer. By Theorem 1 it is clear that in the case of complete market the family $\mathscr{Q}$ is a singleton.

Assuming integrability of the contingent claim with respect to the set of all martingale measures $\mathscr{Q}$, i.e. $\sup_{Q \in \mathscr{Q}} E^Q |H| < \infty$ we have the following analytic representations for the seller and buyer prices called the fundamental asset pricing theorem.

Theorem 2. *Under (AA) we have*

$$p_s(H) = \sup_{Q \in \mathscr{Q}} E^Q \left\{ \frac{H}{B_T} \right\}, \tag{7}$$

and

$$p_b(H) = \inf_{Q \in \mathscr{Q}} E^Q \left\{ \frac{H}{B_T} \right\}. \tag{8}$$

The proof of this theorem (see [38]) is based on an important result from the theory of stochastic processes called optional decomposition.

Lemma 1. *(Föllmer, Kabanov [16]) If $(Y(t))$ is a Q supermartingale for each $Q \in \mathscr{Q}$, there is π and an adapted increasing process (d_t), $d_0 = 0$ such that $Y(T) = V^{\pi,Y(0)}(T) - d_T$.*

The absence of arbitrage interval $[p_b(H), p_s(H)]$ may be very large. Therefore one would like to find in this interval a proper price of the contingent claim. For this purpose we have to use other criteria: we choose martingale measures with minimal variance (see [38]) or minimal entropy (see [18]). Another approach is, instead of hedging with probability 1, to consider the so called quantile hedging under which we require inequalities in (5) and (6) to be satisfied with certain probability, e.g. $1 - \varepsilon$ (see [17]). Alternatively we can also use utility theory (see [18]) to price contingent claims. The theory described above was based on an assumption that there is no friction on our market, i.e. we do not pay transaction costs. The theory with proportional transaction costs (we pay costs proportional to the volume of transaction) is more complicated and still a subject of intensive studies (see e.g. [21,6,5] and references therein). In the case of large transactions one can expect to pay even smaller proportional transaction costs. These considerations lead to concave transaction costs. The problems are very complicated and at this moment some results are only available for simple Cox-Ross-Rubinstein model introduced below in Sect. 1.6 (see [39] and [26]).

In what follows we tacitly assume that there are no restrictions on trade and we have a competitive market: trader can sell unlimited quantities of securities without changing the security's price. Clearly this is a serious simplification. There are two approaches to handle the *liquidity risk* caused by large transactions: either we assume that the transaction can change the asset price (see [7]) or the price remains unchanged but we consider convex transaction costs (see [8]).

The theory we introduced so far concerned financial derivatives which are exercised exactly at time T. The American type options can be exercised at any time τ from the interval $[0, T]$, which is chosen by the buyer of the option. Consequently a contingent claim is an $(\mathscr{F}_t)$ adapted process $H = (H_t)$. In particular $H_t = ((S_1(t) - K)^+)$ or $H_t = ((K - S_1(t))^+)$ in the case of American call or put options. To avoid integrability problems assume that the process H_t is bounded. One case easily adjust the formulae (5) and (6) for the seller $p_s(H)$ and for the buyer $p_b(H)$ prices (the inequalities should be satisfied for each t from $[0, T]$, instead of T only). By an analogy to Theorem 2 we have (see [18])

Theorem 3. *Under (AA) we have*

$$p_s(H) = \sup_{Q \in \mathscr{Q}} \sup_{\tau} E^Q \left\{ \frac{H_\tau}{B_\tau} \right\}, \tag{9}$$

and

$$p_b(H) = \inf_{Q \in \mathscr{Q}} \sup_{\tau} E^Q \left\{ \frac{H_\tau}{B_\tau} \right\}, \tag{10}$$

where $\sup_\tau$ *denotes an optimal stopping problem: we maximize* $E^Q \left\{ \frac{H_\tau}{B_\tau} \right\}$ *over all stopping times* τ.

1.5 Continuous Time Markets

In this section we consider the case, when asset prices $(S(t))$ have dynamics of the form (2) and consequently we are allowed to change our portfolio at any moment. In such a case we have our portfolio at time t under the investment strategy π by anology to (3) in the form

$$V^{\pi,v}(t) = v + \int_0^t \pi(u) dS(u). \tag{11}$$

The notion of the absence of arbitrage (AA) remains the same. There is however a difference in the analytic form of the absence of arbitrage. First of all the notion of a martingale is replaced by a local martingale property. We say that $\frac{S_i(t)}{B_t}$ is a Q *local martingale* if there is a sequence of stopping times $\tau_n \to \infty$ such that $\forall n$ $\left(\frac{S(t \wedge \tau_n)}{B_{t \wedge \tau_n}} \right)$ is a Q martingale for $t \in [0, T]$. We shall denote by $\tilde{\mathscr{Q}}$ the set of all measures Q equivalent to P under which $\left(\frac{S(t)}{B_t} \right)$ is a local martingale.

Furthermore we introduce the so called *no free lunch with vanishing risk property* (NFLVR). It is satisfied when *there is no sequence of strategies* $(\pi^k(t))$ *such that*

- $\exists_{\alpha_k}, P\left\{V^{\pi^k,0}(t) \geq \alpha_k, t \in [0,T]\right\} = 1,$
- $\forall_k, \exists_{\delta_1,\delta_2>0}, P\left\{V^{\pi^k,0}(T) > \delta_1\right\} > \delta_2,$
- $V^{\pi^k,0}(T) \geq -\frac{1}{k}.$

We have (for the proof see [11])

Theorem 4. *(Delbaen, Schachermayer, 1994)*

$$(AA) \Longleftarrow (NFLVR) \Longleftrightarrow \tilde{\mathcal{Q}} \neq \emptyset. \tag{12}$$

Consequently we do not have full absence of arbitrage characterization as in discrete time.

1.6   Complete Markets

A particular situation arises, when $\mathcal{Q}$ or $\tilde{\mathcal{Q}}$ is a singleton. For discrete time model this is in the case when we have the so called *Cox-Ross-Rubinstein model* (CRR) (see [9]) consisting of the bank account $S_0(t)$ and an asset $S(t)$ with the dynamics

$$S_0(n+1) = (1+r)S_0(n), \tag{13}$$
$$S(n+1) = (1+\rho_n)S(n), \tag{14}$$

where r is deterministic and (ρ_n) is a sequence of i.i.d. random variables taking values u and d only, where $d < r < u$.

Notice that in this case given H which is $\mathcal{F}_T = \sigma\{\rho_i, i = 0, \ldots, T-1\}$ measurable, there is an investment strategy $\pi = (\pi(t))$ and $v \in R$ such that $V^{\pi,v}(T) = H$, and consequently we have a *replication* of the contingent claim H. For other markets, say multinomial markets we have only hedging property i.e. there are $(\pi(t))$ and $v \in R$ such that $V^{\pi,v}(T) \geq H$.

In the case of a continuous time market, $\tilde{\mathcal{Q}}$ is a singleton when we have the so called *Black and Scholes model* (see [4]). We assume that the short term interest rate $r_t = r$ is deterministic and besides of the bank account there is one asset with the dynamics of its price $S(t)$ of the form

$$S(t) = S(0)e^{\int_0^t a\,ds + \int_0^t \sigma\,dW(s)}. \tag{15}$$

where a and σ are deterministic. In this case we can obtain a famous analytic formula (called *Black-Scholes formula*) for the price of the European call option with the gain $H = (S(T) - K)^+$ at time T. We have (see [4])

Proposition 1

$$p_s(H) = p_b(H) = S(0)N(d_1(S(0),T)) - Ke^{rT}N(d_2(S(0),T)), \tag{16}$$

with $d_1(s,t) = \frac{\ln\frac{s}{K} + (r+0.5\sigma^2)t}{\sigma\sqrt{t}}, d_2(s,t) = d_1(s,t) - \sigma\sqrt{t}.$

Notice that the formula does not depend on a, since the drift part at is hidden in the local martingale measure.

1.7 Stochastic Control Methods (Summary)

We complete Sect. 1 with a summary of the stochastic control methods, used so far explicitly and implicitly. Generally speaking our main control problem was to find (minimal) v such that $V^{\pi,v}(T) \geq H$ i.e. find a minimal initial capital under which we can hedge the contingent claim H. Absence of arbitrage problem can be also formulated in these terms. In discrete time case such control problem can be solved using *backward dynamic programming* (see e.g. [36]). In continuous time case situation is more complicated. This is a stochastic version of optimal tracking problem which can be solved using controlled Backward Stochastic Differential Equations (BSDEs) (see [13]). The problem with BSDEs stimulated purely analytical approach based on martingale method. Although we have a nice formula for the pricing problem, the evaluation of the seller or buyer price is a rather difficult computational problem. Another problem appears in pricing of American options. We have to solve an optimal stopping problem which in the case of large markets (when the number of assets d is large) is not feasible.

2 Credit Risk

This section is devoted to special contingent claims which may default. In the previous section the contingent claim H was delivered at time T, in the case of European options and its value was not a subject to any change. For a given time horizon T a general *defaultable contingent claim* may be considered as a quintuple $(X, A, \tilde{X}, Z, \tau)$ consisting of the promised contingent claim X, promised dividends (A_t) payed up to default time, recovery claim $\tilde{X}$ i.e. the value of contingent claim we obtain if we have default before T and the recovery process Z, which is a certain recovery payoff at the time of default, payed when default occurs before or at the maturity T. We denote by τ *default time*, and by $H_t = 1_{\tau \leq t}$ the so called default process. The dividend process of the defaultable contingent claim is of the form

$$D_t = X^d(T)1_{t \geq T} + \int_{]0,t]} (1 - H_u)dA_u + \int_{]0,t]} Z_u dH_u, \qquad (17)$$

with $X^d(T) = X1_{\tau > T} + \tilde{X}1_{\tau \leq T}$. The problem is to price such dividend process. There are two approaches which vary depending on the form of the default time. If τ is a stopping time with respect to the available observation of the market i.e. first entry time of the company wealth process to the so called bankruptcy region, we have a *structural approach*. In this approach default time is predictable with respect to the available observation. Consequently pricing of defaultable contingent claims leads to stochastic control problems which can be solved for particular form of τ. An alternative approach called *intensity based approach* assumes that we are not able to predict τ. Our observation at time t is $\mathcal{G}_t = \mathcal{H}_t \vee \mathcal{F}_t$. We are not able to solve such problem using stochastic control methods. We consider therefore a martingale approach. It is not true in general that a $(\mathcal{G}_t)$ martingale measure is also an $(\mathcal{F}_t)$ martingale measure. To obtain an analytic formula we assume the so called *invariance property* which says that *for a $(\mathcal{G}_t)$ martingale measure Q^*: an $(\mathcal{F}_t)$ martingale is also a $(\mathcal{G}_t)$ martingale.* Let Q^* be a given $(\mathcal{G}_t)$ martingale measure. Under suitable integrability properties (see [2] for details) we have

Theorem 5. *The value of DCC* $(X, A, 0, Z, \tau)$ *at time* t *is equal to*

$$1_{\tau > t} G_t^{-1} B_t E^{Q^*} \left\{ \int_t^T B_u^{-1} (G_u dA_u - Z_u dG_u) + G_T B_T^{-1} X | \mathscr{F}_t \right\}, \tag{18}$$

with $G_t = P\{\tau > t | \mathscr{F}_t\}$.

3 Term Structure Modeling

In the theory developed in Sect. 1 we had an opportunity to invest in assets and savings account. We consider now investments in bonds also. By a *zero coupon bond* with maturity T we mean a financial instrument paying to its holder one unit of cash at time T. Let $B(t, T)$ be the price of such bond at time $t \leq T$. We shall use the following representation of $B(t, T)$

$$B(t, T) := \exp \left(- \int_t^T f(t, u) du \right), \tag{19}$$

where $f(t, T)$ is called an *instantaneous forward rate*. By the very definition clearly $B(T, T) = 1$. Assume that P^* is an equivalent measure to P such that $\frac{B(t,T)}{B_t}$ is a P^* martingale. From the martingale property we obtain

$$B(t, T) = E^{P^*} \left\{ e^{- \int_t^T r_u du} | F_t \right\}. \tag{20}$$

There are various ways to model *short term interest rate* r_t. In the so called *single factor models* it is modeled as a solution of a certain one-dimensional stochastic differential equation with one-dimensional Brownian motion as a single source of uncertainty. As an example of such models, Vasicek's model can serve in which

$$dr_t = (a - br_t)dt + \sigma dW_t, \tag{21}$$

or Cox-Ingersoll-Ross model (CIR) with

$$dr_t = (a - br_t)dt + \sigma \sqrt{r_t} dW_t. \tag{22}$$

We refer to [30] for more details. An alternative approach to term structure modeling is based on the so called Heath-Jarrow-Morton (HJM) methodology. We assume that the instantaneous forward rate is of the form

$$f(t, T) = f(0, T) + \int_0^t \alpha(u, T) du + \int_0^t \sigma(u, T) dW_u. \tag{23}$$

Consequently we can consider two kinds of martingale measures:

- *martingale measure forward* $\hat{P}$, which is equivalent to P and under $\hat{P}$, $\frac{B(t,T)}{B(t,T^*)}$ is a martingale for $T \leq T^*$;
- *martingale measure spot* P^*, which is an equivalent to P measure such that $\frac{B(t,T)}{B_t}$ is a P^* martingale.

The absence or arbitrage corresponds to the existence of martingale forward or spot measures (for details see [30]). In the case of bonds, similarly to the case of defaultable contingent claims, there is a problem to study pricing of financial derivatives, i.e. options based on bonds with the use of stochastic control. Therefore methods preferable in these cases are based on the martingale techniques.

4 Portfolio Analysis

Portfolio analysis is an important area of mathematics of finance. On the other hand this is a part of stochastic control theory. We shall consider two approaches: a static one in which we optimize our portfolio with respect to a certain functional on a one step horizon and a dynamic in which we have to find optimal portfolio for a longer time horizon.

4.1 Static Portfolios

Static portfolio is closely related to the so called *Markowitz theory*. Harry Markowitz (a Nobel prize winner in 1990) was the first economist who admitted the importance to study risk in portfolio analysis (see [28]) considered as a variance of the portfolio growth. Consider a discrete time asset model with prices such that its *random rate of return*

$$\zeta_i := \frac{S_i(1) - S_i(0)}{S_i(0)}. \tag{24}$$

Given portfolio $\pi = (\pi_1, \pi_2, \ldots, \pi_d)$ representing portions of the wealth process invested in assets and wealth process $V(0)$ at time 0, under prices $S(1)$, the wealth process $V(1)$ is such that the *portfolio rate of return R^π* is of the form

$$R^\pi := \frac{V^\pi(1) - V(0)}{V(0)} = \pi \cdot \zeta, \tag{25}$$

with $\cdot$ standing for scalar product. We are interested in maximization of the expected portfolio rate of return. On the other hand following Markowitz we would like to minimize the risk considered as the variance of the portfolio rate of return. One can notice that

$$Var(R(\pi)) = \pi^* \Sigma \pi, \tag{26}$$

where $*$ stands for the transpose of the vector π and Σ is the covariance matrix of the random vector ζ.

Consider now the so called *Markowitz order $\succeq$*: we say that portfolio rate of return R^π is better than $R^{\pi'}$ or in other words, the strategy π is better than π', we write $R^\pi \succeq R^{\pi'}$, whenever $E(R^\pi) \geq E(R^{\pi'})$ and $Var(R^\pi) \leq Var(R^{\pi'})$.

The maximal elements in this order form a so called *efficient frontier* on the plane $(ER^\pi, Var(R^\pi))$. To determine efficient frontier we maximize $Var(R^\pi)$ for fixed value of $E(R^\pi)$. This is a quadratic optimization problem with linear constraints. If we admit the so called short selling i.e. allow elements of π to be negative, such problem can be solved using standard Lagrange multiplier method (see [29] or [42]). There is a number of problems in implementation of the Markowitz theory. We need an estimate of the

- expected returns $E\{\zeta_i\}$
- covariance matrix $Cov(\zeta) = E\{(\zeta - E\zeta)(\zeta - E\zeta)^T\}$.

Although the covariance matrix can be well estimated based on historical data there is a problem with the estimation of the expected returns. To overcome this we use a

Black-Litterman (1990) approach based on Bayesian estimation and forecasted returns (see [3]). Variance of the portfolio as a measure of risk does not seem to be satisfactory. It penalizes in the same way positive events, when portfolio rate of return is above the expected value and negative events, when it is below the expected value. The following measure called *value at risk* $VaR_\alpha(R(\pi))$ was introduced to banking practice

$$VaR_\alpha(R(\pi)) = \inf\{x\colon P\{R(\pi)+x\leq 0\}\leq\alpha\}. \tag{27}$$

This is the smallest value which added to the portfolio rate of return allows nonpositive rate of return with probability below fixed level α. Although value at risk is very popular, still it has a number of deficiencies. First of all it measures only probability not the size of inconvenient for us situation. Therefore as a better measure of risk $CVaR_\alpha$, i.e. *conditional VaR* is considered called also expected shortfall

$$CVaR_\alpha(R(\pi)) = E\{-R(\pi)|R(\pi)+VaR_\alpha(R(\pi))\leq 0\}, \tag{28}$$

which is the expected value of $-R(\pi)$, given that $R(\pi)+VaR_\alpha(R(\pi))$ is nonpositive.

To evaluate *VaR* or *CVaR* we need to know a bit more about the distribution of the random rate of return ζ. Typical assumption that it is multinomial normal is not sufficient, since in practice the densities of such random variables have fat tails. To be more precise one can quote the result of Fergusson and Platen [14] saying that daily log-returns of the world stock market portfolio in 34 different currency denominations form t-Student distribution with 3.94 degrees of freedom. The proper class to study seems to be *elliptical distributions* introduced in [24]. We say that a d-dimensional random vector X is elliptic, whenever its density is of the form

$$f_X(x) = c_d|D|^{-\frac{1}{2}}g_d\left(\frac{1}{2}(x-\mu)^T D^{-1}(x-\mu)\right), \tag{29}$$

with suitable constant c_d, positive definite matrix D, its determinant $|D|$, vector μ, and a function g_d. This class includes: normal, t-Student, Laplace, and logistic distributions and is stable under linear transformations. Using a certain standardization procedure we can calculate *VaR* and *CVaR* for elliptically distributed vector ζ (see [27]). Although computation of the efficient frontier with *VaR* or *CVaR* as risk functions is more complicated, for elliptically distributed ζ this is a numerically feasible problem.

4.2 Dynamic Portfolios

The aspect of risk is also important in the dynamic case. We shall now consider another important optimization problem in mathematics of finance. We are interested to maximize a satisfaction measure as an expected value of a certain utility function of the wealth process. By *utility function* we mean a concave, increasing function U. A class of HARA (hyperbolic absolute risk aversion) utility functions consisting of $U(x) = x^\alpha$ with $\alpha \in [0,1)$ and $U(x) = \ln x$ can serve as an example. We maximize terminal utility

$$E\{B_T^{-1}U(V(T))\}, \tag{30}$$

or a functional with consumption $(c(t))$

$$E\left\{\sum_{t=0}^{T-1} B_t^{-1} U_1(c(t)) + B_T^{-1} U_2(V(T))\right\}, \tag{31}$$

where U_1, U_2 are utility functions, or we look for growth optimal portfolios (GOP) (see [35]):

$$E\{\ln(V(T))\}. \tag{32}$$

Maximization of the cost functionals (25)-(27) is a stochastic control problem, which in discrete time case can be solved using dynamic programming methods. The continuous time case strongly depends on the form of model for asset prices. The following three approaches can be considered

- solving Hamilton-Jacobi-Bellman equation,
- using convex analysis (duality approach),
- using stochastic maximum principle and BSDEs.

In the next subsections we sketch the main features of these approaches.

4.2.1 Hamilton-Jacobi-Bellman Equation

Let

$$w(t,v,z) = \sup_{\pi}\left\{e^{-\int_t^T r_u du} U(V^{\pi,v}(T))\right\} \tag{33}$$

be the value function of the terminal utility maximization. The wealth process dynamics under (2) is of the form

$$dV^{\pi,v}(t) = V^{\pi,v}\pi(t) \cdot \frac{dS(t)}{S(t-)} + V^{\pi,v}\pi_0(t) r_t dt. \tag{34}$$

The corresponding Hamilton-Jacobi-Bellman equation is of the form

$$0 = \sup_{\pi}\left[\frac{\partial w}{\partial t}(t,v,z) - r_t w(t,v,z) + \mathscr{L}^{\pi,v} w(t,v,z)\right], \tag{35}$$

with boundary condition $w(T,v,z) = U(v)$, and $\mathscr{L}^{\pi,v}$ being the generator of the pair $(V^{\pi,v}, z(t))$. Main problem we have is that the value function (28) is not in the domain of the generator $\mathscr{L}$. Typical verification theorem says that, if we are given a solution to Hamilton-Jacobi-Bellman equation (30), then it coincides with the value function (28). One can extend the notion of solution to (30) introducing the so called *viscosity solutions* and show that the value function w is a solution to (30) in this sense (for details see [15]).

4.2.2 Martingale – Convex Analysis Approach

Consider now a terminal utility maximization problem (25), assuming for simplicity $r_t = 0$ for $t \geq 0$ (i.e. $B_t = 1$) and that there are no economic factors $(z(t))$ in the asset dynamics. We assume furthermore that the market is complete and the wealth process $\{V^{\pi,v}(t)\}$ is a martingale under $dQ = \Lambda_T dP$, which is a given martingale measure. Clearly

$$E^Q\{V^{\pi,v}\} = v. \tag{36}$$

Let $\mathscr{H}(v)$ be the family of nonnegative random variables H such that $E\{\Lambda_T H\} = v$. Consider the following static optimization problem:

$$w(t,v) = \sup_{H \in \mathscr{H}(v)} E\{U(H)\}. \tag{37}$$

We want to find the solution H^* to the static problem and the strategy π^* such that $V^{\pi^*,v}(T) = H^*$. Assume additionally that the utility function U satisfies the so called *Inada conditions*: $U'(0) = \infty$ and $U'(\infty) = 0$. Let $I = (U')^{-1}$. Then

$$\sup_v [U(v) - vy] = U(I(y)) - yI(y). \tag{38}$$

Using (33) we therefore have

$$E\{U(H)\} \le E\{U(I(y\Lambda_T)\} - y(E[\Lambda_T I(y\Lambda_T)] - v). \tag{39}$$

If we choose y^* such that $E[\Lambda_T I(y^*\Lambda_T)] = v$, we obtain that

$$E\{U(H)\} \le E\{U(I(y^*\Lambda_T)\}.$$

Let $H^* = I(y^*\Lambda_T)$. One can show that $H^* \in \mathscr{H}(v)$. Consequently we have that $w(t,v) = E\{U(H^*)\}$ and H^* is a solution to the static problem (32). The optimal strategy π^* is determined from the martingale representation theorem as a strategy such that $V^{\pi^*,v}(T) = H^*$. For details see [23].

4.2.3 Stochastic Maximum Principle

We now consider another approach to terminal utility maximization (25). Let $S_i(t)$ be a solution to the following stochastic differential equation

$$dS(t) = S(t-)\left(\tilde{a}(t)dt + \sigma(t)dB(t) + \int_R \eta(t,x)\bar{N}(dt,dx)\right), \tag{40}$$

where $\bar{N}$ is the Poisson random measure N for large jumps (taking values outside of the unit ball) and coincides with the compensated Poisson measure $d\tilde{N}_j = dN_j - d\lambda_j dt$ for small jumps (with values from the unit ball). Under the strategy π the wealth process $V^{\pi,v}$ is a solution to the equation

$$V^{\pi,v}(t) = V^{\pi,v}(t-)(\pi_0(t)r_t + (1 - \pi_0(t))\pi(t) \cdot \tilde{a}(t)dt \\ + (1 - \pi_0(t))(\pi \cdot \sigma(t)dB(t) + \eta(t,x)\bar{N}(dt,dx)). \tag{41}$$

We define the *Hamiltonian* of the form

$$M(t,v,\pi,p,q,y) = v(\pi_0 \cdot r_t + (1 - \pi_0)\pi \cdot \tilde{a}(t)) \cdot p + (1 - \pi_0)vtr((\pi I\sigma(t))^T q)$$
$$+ (1 - \pi_0)v\int_{R^d}\left\{\sum_{j=1}^d (\eta_j(t,x)y_j(t,x)) + (\eta(t,x)p + vy)(I - Diag(\chi))\right\}d\lambda(x), \tag{42}$$

where I is the identity matrix and $Diag\chi$ is the diagonal matrix with χ_j entries such that $\chi_j = 0$ for large jumps and $\chi_j = 1$ for small jumps. Consider now the so called *adjoint equation* for (v, π), which is backward stochastic differential equation (BSDE) of the form

$$dp(t) = -D_v M(t, V(t), \pi(t), p_t, q_t, y(t)) dt + q_t dB_t$$
$$+ \int_{R^d} y(t-, x) \bar{N}(dt, dx), \tag{43}$$

with terminal condition $p(T) = D_v U(V(T))$ and D_v standing for the derivative with respect to v. By such solution we mean the triple (p, q, y) for which (38) holds. We have (see [19] and [32]).

Theorem 6. *(Stochastic maximum principle): If $(\hat{\pi}, \hat{V})$ is an admissible pair consisting of the strategy $\hat{\pi}$ and corresponding solution to (36) (we assume that there is a solution $\hat{V}$ for such $\hat{\pi}$) and for the triple $(\hat{p}, \hat{q}, \hat{y})$ being the solution to BSDE (38) we have*

$$M(t, \hat{V}(t), \hat{\pi}(t), \hat{p}(t), \hat{q}(t), \hat{y}(t)) = \sup_\pi M(t, \hat{V}(t), \pi, \hat{p}(t), \hat{q}(t), \hat{y}(t)), \tag{44}$$

for all $t \in [0, T]$ and $\hat{M}(v) := \sup_\pi M(t, v, \pi, \hat{p}(t), \hat{q}(t), \hat{y}(t))$ is concave in v, then under some regularity properties $(\hat{\pi}, \hat{V})$ is an optimal pair, i.e. $\hat{\pi}$ is an optimal control while $\hat{V}$ is the optimal wealth process.

4.3 Long Time Portfolio Functionals

In an independent section we consider a family of long time growth optimal portfolios (GOP). An infinite horizon analog of (27) is *risk neutral GOP* with the cost functional (introduced by Kelly in [25])

$$\liminf_{T \to \infty} \frac{1}{T} E \{\ln(V(T))\}. \tag{45}$$

An alternative is to study *risk sensitive GOP* (see [1,31,40]) with the cost functional

$$\limsup_{T \to \infty} \frac{1}{\gamma T} \ln E \{(V(T))^\gamma\}, \tag{46}$$

where $\gamma < 0$ is the so called risk factor. Such cost function is motivated by the Taylor expansion of the function $F(\gamma) = \ln E[e^{\gamma Y}]$. Namely, we have $F'(0) = EY$, $F''(0) = Var(Y)$ and

$$\frac{1}{\gamma} F(\gamma) \approx EY + \frac{1}{2} \gamma Var(Y). \tag{47}$$

Consequently, maximizing risk sensitive cost functional (41) we maximize the long run expected growth rate diminished by its variance (which is a measure of risk) with weight $-\gamma$. Risk sensitive GOP is a difficult problem, since we have to study a multiplicative cost functional. The risk factor γ is negative, however the theory with positive γ can be considered as a dual problem to certain long time maximization of the portfolio growth over a given benchmark (see [34] and [41]). Another important feature of the cost functional (41) is that it asymptotically, as $\gamma \to 0$ approximates risk neutral functional (40).

The same property can be proved for the optimal values of these functionals (see [12]). More generally, it can be shown (under certain assumptions) that risk neutral nearly optimal strategy is also nearly optimal for risk sensitive cost functional with risk factor γ close to 0. The study of the cost functionals (40) and (41) becomes even harder when we consider proportional, or fixed plus proportional transaction costs. From stochastic control point of view to study risk neutral GOP we have to solve additive Bellman equation, while in the case of risk sensitive GOP we have to solve multiplicative Bellman equations. If we additionally assume fixed costs our strategies are of impulse form (see [33]).

References

1. Bielecki, T.R., Pliska, S.: Risk sensitive dynamic asset management. JAMO 39, 337–360 (1999)
2. Bielecki, T.R., Rutkowski, M.: Credit Risk: Modelling, Valuation and Hedging. Springer, Heidelberg (2002)
3. Black, F., Litterman, R.: Global portfolio optimization. Financial Analysis J. 48, 28–43 (1992)
4. Black, F., Scholes, M.: The valuation of option contracts and a test of market efficiency. J. Finance 27, 399–417 (1972)
5. Bouchard, B.: No-arbitrage in discrete-time markets with proportional transaction costs and general information structure. Finance Stoch. 10, 276–297 (2006)
6. Campi, L., Schachermayer, W.: A super-replication theorem in Kabanov's model of transaction costs. Finance Stoch. 10, 579–596 (2006)
7. Çetin, U., Jarrow, R., Protter, P.: Liquidity risk and arbitrage pricing theory. Finance Stoch. 8, 311–341 (2004)
8. Çetin, U., Rogers, L.C.G.: Modeling liquidity effects in discrete time. Math. Finance 17, 15–29 (2007)
9. Cox, J.C., Ross, S.A., Rubinstein, M.: Option pricing: a simplified approach. J. Finan. Econom. 7, 229–263 (1979)
10. Dalang, R.C., Morton, A., Willinger, W.: Equivalent martingale measures and no-arbitrage in stochastic securities market model. Stochastics and Stochastics Rep. 29, 185–201 (1990)
11. Delbaen, F., Schachermayer, W.: A general version of the fundamental theorem of asset pricing. Math. Ann. 300, 463–520 (1994)
12. Di Masi, G.B., Stettner, Ł.: Remarks on risk neutral and risk sensitive portfolio optimization. In: Kabanov, Y., Liptser, R., Stoyanov, J. (eds.) From Stochastic Calculus to Mathematical Finance. The Shiryaev Festschrift, pp. 211–226. Springer, Heidelberg (2006)
13. El Karoui, N., Peng, S., Quenez, M.C.: Backward stochastic differential equations in finance. Math. Finance 7, 1–71 (1997)
14. Fegusson, K., Platen, E.: On the distributional characterization of log returns of a world stock index. Appl. Math. Finance 13, 19–38 (2006)
15. Fleming, W.H., Soner, H.M.: Controlled Markov Processes and Viscosity Solutions. Springer, Heidelberg (1992)
16. Föllmer, H., Kabanov, Y.M.: Optional decomposition and Lagrange multipliers. Finance Stoch. 2, 69–81 (1998)
17. Föllmer, H., Leukert, P.: Quantile hedging. Finance Stoch. 3, 1–25 (1997)
18. Föllmer, H., Schied, A.: Stochastic Finance. An Introduction in Discrete Time. de Gruyter (2002)
19. Framstad, N.C., Oksendal, B., Sulem, A.: Sufficient stochastic maximum principle for the optimal control of jump diffusions and applications to finance. JOTA 121, 77–98 (2004)

20. Kabanov, Y.M., Kramkov, D.O.: No-arbitrage and equivalent martingale measures: an elementary proof of the Harrison-Pliska theorem. Theory Probab. Appl. 39, 523–527 (1994)
21. Kabanov, Y.M., Rasonyi, M., Stricker, C.: On the closedness of sums of convex cones in L^0 and the robust no-arbitrage property. Finance Stoch. 7, 403–411 (2003)
22. Kabanov, Y.M., Stricker, C.: A teacher's note on no-arbitrage criteria. In: Sem. Probab. 35. Lecture Notes in Math., vol. 1755, pp. 149–152. Springer, Heidelberg (2001)
23. Karatzas, I.: Lectures on the Mathematics of Finance. AMS (1996)
24. Kelker, D.: Distribution theory of spherical distributions and location-scale parameter generalization. Sankhya 32, 419–430 (1970)
25. Kelly, J.L.: A New Interpretation of Information Rate. Bell System Technical Journal 35, 917–926 (1956)
26. Kociński, M.: Hedging in the CRR model under concave transaction costs. Demonstratio Math. 34, 497–512 (2001)
27. Landsman, Z., Valdez, E.A.: Tail Conditional Expectations for Elliptical Distributions. North American Actuarial Journal 7, 55–71 (2003)
28. Markowitz, H.: Portfolio selection. J. Finance 7, 77–91 (1952)
29. Markowitz, H.: Portfolio Selection Efficient Diversification of Investments. Wiley, Chichester (1959)
30. Musiela, M., Rutkowski, M.: Martingale Modelling in Financial Modelling. Springer, Heidelberg (2005)
31. Nagai, H.: Optimal strategies for risk sensitve portfolio opimization problems for a general factor models. SIAM J. Conrol Opim. 41, 1179–1800 (2003)
32. Oksendal, B., Sulem, A.: Applied Stochastic Control of Jump Diffusions. Springer, Heidelberg (2005)
33. Palczewski, J., Stettner, L.: Maximization of the portfolio growth rate under fixed and proportional transaction costs. Communications in Information and Systems 7, 31–58 (2007)
34. Pham, H.: A Large Deviations Approach to Optimal Long Term Investment. Finance Stoch. 7, 169–195 (2003)
35. Platen, E., Heath, D.: A Benchmark Approach to Quantitative Finance. Springer, Heidelberg (2006)
36. Pliska, S.R.: Introduction to Mathematical Finance. Dicrete Time Models. Blackwell, Malden (1997)
37. Protter, P.E.: Stochastic Integration and Differential Equations. Springer, Heidelberg (2004)
38. Shiryaev, A.N.: Essentials of Stochastic Finance. Facts, Models, Theory. World Scientific, Singapore (2001)
39. Stettner, L.: Discrete time markets with transaction costs. In: Yong, J. (ed.) Recent developments in mathematical finance, pp. 168–180. World Sci. Publ., River Edge (2002)
40. Stettner, L.: Risk sensitive portfolio optimization. Math. Meth. Oper. Res. 50, 463–474 (1999)
41. Stettner, L.: Duality and Risk Sensitvie Portfolio Optimization. Contemporary Mathematics 351, 333–347 (2004)
42. Szegö, G.P.: Portfolio Theory with Application to Bank Asset Management. Academic Press, London (1980)

Complete Shape Metric and Geodesic

Jean-Paul Zolésio

CNRS and INRIA, INRIA, 2004 route des Lucioles,
BP 93, 06902 Sophia Antipolis Cedex, France
jean-paul.zolesio@sophia.inria.fr

Abstract. We develop the framework for moving domain and geometry under minimal regularity (of moving boundaries). This question arose in shape control analysis and non cylindrical PDE analysis. We apply here this setting to the morphic measure between shape or images. We consider both regular and non smooth situations and we derive complete shape metric space with characterization of geodesic as being solution to Euler fluid-like equation. By the way, this paper also addresses the variational formulation for solution to the coupled Euler-transport system involving only condition on the convected terms. The analysis relies on compactness results which are the parabolic version to the Helly compactness results for the BV embedding in the linear space of integrable functions. This new compactness result is delicate but supplies to the lack of convexity in the convection terms so that the vector speed associated with the optimal tube (or moving domain), here the shape geodesic, should not be curl-free so that the Euler equation does not reduce to a classical Hamilton-Jacobi one. For topological optimization this geodesic construction is developed by level set description of the tube, and numerical algorithms are in the next paper of this book.

1 Introduction to Shape Metrics

The shape analysis arose in the early 70's from structural mechanics. The problem was to find a best shape which would minimize the *compliance* (the work of external forces in some elasticity modeling). Later this problem extended to optimal control-like situation in which the criteria to be extremized with respect to a geometrical shape had a more general form which implied the study of the so called material and shape derivatives for the solution of a partial differential equation with boundary conditions on the unknown part of boundary [8,12,16]. Very soon the concepts of topology on general shape families were introduced. The easiest one was the metric induced by the characteristic function of the shape (in this case the shape is just defined up to a zero measure subset). Besides this the thinner one was the *Courant metric*, see the book [6], which consists, very roughly speaking, in minimizing $||T - I_d|| + ||T^{-1} - I_d||$ for each application mapping a domain Ω_0 onto another Ω_1, the minimum being taken over the family of such invertible mappings T. Indeed this metric is not known to be differentiable and is very difficult to be computed in this very abstract and non geometrical form. Also by the class of the regularity imposed to the mappings T in the theory, it derives that the domains Ω_i, $i = 0, 1$ should be homeomorphic to one another and then should have the same topology. The aim of this work is to relax this metric definition in order to solve these two difficulties (i.e. give a geometrical interpretation with computational

A. Korytowski et al. (Eds.): System Modeling and Optimization, IFIP AICT 312, pp. 144–166, 2009.

algorithms using level set techniques and extend the metric to a larger class of domains having different topologies) but also and mainly to construct the geodesics. This last issue turns to have several applications in any kind of large deformation process but also in image analysis. Through a *Fully Eulerian* equivalent definition we shall characterize the geodesic tube as being built by solutions to a coupled incompressible Euler flow-transport equation (in case of given volume constraint); meanwhile we furnish a full mathematical result for such variational solution to the incompressible Euler flow which turns to be a new result concerning Euler equation. The new metric we present here, which in some sense is an extension of the Courant metric, is based on two main considerations: Shapes (or geometry) are elements of some set, say $\mathscr{F}$, and we consider all connecting tubes in $\mathscr{F}$. Then the metric is built on the shortest such tube which furnishes the geodesic, solution to some differentiable variational problem. Also we shall derive complete metric spaces. The concept of geodesic for usual metrics such as Hausdorff distance, or L^1 metric on characteristic functions makes no sense as there is obviously no hope to derive any local uniqueness for a shortest path. Here also we still have none such result (nor local stability for the geodesic) but this challenging question is hopeful as been formulated in term of local uniqueness for flow Euler-like equation to which we can add any viscosity perturbation. This paper follows [24,23] and the book [13]. The connecting tube concept arose in moving domain analysis and non cylindrical PDE study in the 90's, for example in [3,7,9,12,19,4,2,11,10,14,21,17].

2   Connecting Tubes

We consider the time interval as being $I = [0, 1]$ and D, a bounded domain in R^N with smooth boundary. We consider the set of characteristic functions

$$\mathscr{C} = \{\zeta = \zeta^2 \in L^1(I \times D)\}. \tag{1}$$

We consider the continuous elements

$$\mathscr{C}^0 = \mathscr{C} \cap C^0(I, L^1(D)). \tag{2}$$

Being given two measurable subsets $\Omega_i \subset D$, $i = 0, 1$, we consider the family of *connecting tubes*

$$\mathscr{T}^0(\Omega_0, \Omega_1) = \{\zeta \in \mathscr{C}^0 \text{ s.t. } \zeta(i) = \chi_{\Omega_i}, i = 0, 1\}. \tag{3}$$

2.1   Moving Domain

For any $\zeta \in \mathscr{T}^0(\Omega_0, \Omega_1)$ we consider the set $Q = \cup_{0 < t < 1} \{t\} \times \Omega_t \subset R^{N+1}$ such that $\zeta = \chi_Q$. This set Q is defined up to an $N+1$ dimensional zero measure set.

2.2   Generic Framework for Metric

The idea for constructing metrics is to consider in this set the infimum of some norm for the time derivative term $\frac{\partial}{\partial t}\zeta$. Indeed if such term is zero then ζ is not time depending. The general setting is to consider families of admissible tubes such that $\frac{\partial}{\partial t}\zeta \in$

$L^p(I, \mathcal{H}(D))$ for some Banach space $\mathcal{H}(D)$ of distributions over D, $\mathcal{H}(D) \subset \mathcal{D}'(D)$, and consider the following connecting tubes:

$$\mathcal{T}_{\mathcal{H}}^{0,p}(\Omega_0, \Omega_1) = \{\zeta \in \mathcal{T}^0(\Omega_0, \Omega_1), \text{ s.t. } \frac{\partial}{\partial t}\zeta \in L^p(I, \mathcal{H}(D))\}, \tag{4}$$

and for some $p \geq 1$, the metric in the following form:

$$d_{p,\mathcal{H}}(\Omega_0, \Omega_1) = Inf_{\zeta \in \mathcal{T}_{\mathcal{H}}^{0,p}(\Omega_0, \Omega_1)} \int_0^1 ||\frac{\partial}{\partial t}\zeta(t)||_{\mathcal{H}(D)}^p \, dt. \tag{5}$$

2.3   The Time L^p Regularity of $\frac{\partial}{\partial t}\zeta$ Implies $\zeta \in \mathcal{C}^0$

Let us define

$$\mathcal{C}_{\mathcal{H}}^{0,p} = \{\zeta \in \mathcal{C} \text{ s.t. } \frac{\partial}{\partial t}\zeta \in L^p(I, \mathcal{H}(D))\}. \tag{6}$$

Proposition 1. *Let $p \geq 1$, then $\mathcal{C}_{\mathcal{H}}^{0,p} \subset \mathcal{C}^0$.*

Proof. Obviously we have

$$\mathcal{C}_{\mathcal{H}}^{0,p} \subset W^{1,p}(I, \mathcal{H}(D)) \subset C^0(I, \mathcal{H}(D)) \subset C^0(I, \mathcal{D}'(D)). \tag{7}$$

So that from the following Lemma we get $\mathcal{C}_{\mathcal{H}}^{0,p} \subset C^0(I, L^1(D))$; then we see that the continuity property of the tube derives directly from $\zeta \in \mathcal{C}$ (that is $\zeta = \zeta^2$) and the weak regularity of the time derivative measure $\frac{\partial}{\partial t}\zeta$.   □

Lemma 1. *Let $\zeta = \zeta^2 \in L^1(I \times D) \cap C^0(I, \mathcal{D}'(D))$. Then $\zeta \in C^0(I, L^1(D))$.*

Proof. Notice that

$$||\zeta(t+s) - \zeta(t)||_{L^1(D)} = ||\zeta(t+s) - \zeta(t)||_{L^2(D)}^2. \tag{8}$$

Then it is enough to show that $\zeta \in C^0(I, L^2(D))$. We begin by establishing the weak $L^2(D)$ continuity: for any element $f \in L^2(D)$ consider

$$\int_D (\zeta(t+s)(x) - \zeta(t)(x))f(x)\,dx = \int_D (\zeta(t+s,x) - \zeta(t,x))\phi(x)\,dx$$
$$+ \int_D (\zeta(t+s,x) - \zeta(t,x))(f(x) - \phi(x))\,dx. \tag{9}$$

Let be given $\varepsilon > 0$, by the choice of $\phi \in \mathcal{D}'(D)$ (using here the density of $\mathcal{D}'(D)$ in $L^2(D)$), we have

$$|\int_D (\zeta(t+s,x) - \zeta(t,x))(f(x) - \phi(x))dx| \leq \int_D |f(x) - \phi(x)|dx \leq \varepsilon. \tag{10}$$

So we derive the continuity for the weak $L^2(D)$ topology. To reach the strong topology it is sufficient now to consider the continuity of the mapping

$$t \to \int_D |\zeta(t,x)|^2 dx = \int_D \zeta(t,x)\,dx = ((\zeta(t),1))_{L^2(D)}. \tag{11}$$

□

2.4   Metric and Pseudo Metric

We consider a set $\bar{\Omega} \subset D$ and the family of all subsets in D which are reachable in finite time from this $\bar{\Omega}$ by elements ζ, ζ describing the whole set $\mathscr{C}_{\mathscr{H}}^{0,p}$; more precisely:

$$\mathscr{O}_{\bar{\Omega}} = \{\Omega \subset D \text{ s.t. } \exists \zeta \in \mathscr{C}_{\mathscr{H}}^{0,p} \text{ with } \chi_\Omega = \zeta(1), \chi_{\bar{\Omega}} = \zeta(0)\}. \tag{12}$$

Notice that by construction any pair of elements in this family is connected:

$$\forall (\Omega_0, \Omega_1) \in (\mathscr{O}_{\bar{\Omega}})^2, \text{ the family } \mathscr{T}_{\mathscr{H}}^{0,p}(\Omega_0, \Omega_1) \text{ is not empty.} \tag{13}$$

Proposition 2. *For any $p \geq 1$, $d_{p,\mathscr{H}}$ is a quasi-metric in the following sense; for any elements $\Omega_i, i = 0,1,2$ in $\mathscr{O}_{\bar{\Omega}}$ we have:*

i) $d_{p,\mathscr{H}}(\Omega_0, \Omega_1) = 0$ iff $\chi_{\Omega_0} = \chi_{\Omega_1}$,
ii) $d_{p,\mathscr{H}}(\Omega_0, \Omega_1) = d_{p,\mathscr{H}}(\Omega_1, \Omega_0)$,
iii) $d_{p,\mathscr{H}}(\Omega_0, \Omega_2) \leq 2^{p-1}(d_{p,\mathscr{H}}(\Omega_0, \Omega_1) + d_{p,\mathscr{H}}(\Omega_1, \Omega_2))$.

For $p = 1$, $d_{1,\mathscr{H}}$ is a metric on $\mathscr{O}_{\bar{\Omega}}$.

Proof. 1a) When $\chi_{\Omega_1} = \chi_{\Omega_2}$ as elements in $L^1(D)$, we may choose $\zeta(t,x) = \chi_{\Omega_0}(x)$ so that the evolution domain Q is the cylinder $Q = I \times \Omega_0$ and $\frac{\partial}{\partial t}\zeta = 0$ realizes the minimum and leads to the null distance.

1b) Conversely for any $\varepsilon > 0$ there exists some admissible tube ζ^ε with $\zeta^\varepsilon(i) = \chi_{\Omega_i}$ and realizing the infimum up to ε. Then

$$\|\zeta^\varepsilon(0) - \zeta^\varepsilon(1)\|_{\mathscr{H}(D)} \leq \int_0^1 \|\frac{\partial}{\partial t}\zeta^\varepsilon\|_{\mathscr{H}} dt \leq (\int_0^1 \|\frac{\partial}{\partial t}\zeta^\varepsilon\|_{\mathscr{H}}^p dt)^{1/p} \leq \varepsilon^{1/p}. \tag{14}$$

We conclude $\chi_{\Omega_1} = \chi_{\Omega_0}$ as elements in $\mathscr{H}(D)$.

2) The symmetry is obviously realized by reversing the time variable. Indeed if ζ^ε realizes the infimum up to ε then we consider $\bar{\zeta}^\varepsilon(t,x) := \zeta^\varepsilon(1-t,x)$ and $\int_0^1 \|\frac{\partial}{\partial t}\bar{\zeta}^\varepsilon\|_{\mathscr{H}}^p dt = \int_0^1 \|\frac{\partial}{\partial t}\zeta^\varepsilon\|_{\mathscr{H}}^p dt$ so that the element $\bar{\zeta}^\varepsilon$ also approaches the infimum up to ε.

3) The triangle property derives from the following obvious generic construction: let us consider two connecting tubes $\zeta^{k,\varepsilon} \in \mathscr{T}_{\mathscr{H}(D)}^{0,p}(\Omega_{k-1}, \Omega_k), k = 1,2$ and realizing the infimum up to ε in the corresponding distances $d_{\mathscr{H}(D)}(\Omega_{k-1}, \Omega_k)$.

We introduce the new element $\bar{\zeta}^\varepsilon \in \mathscr{T}_{\mathscr{H}(D)}^{0,p}(\Omega_0, \Omega_1)$ piecewisely defined as follows:

$$\forall t \in [0, 1/2], \bar{\zeta}^\varepsilon(t) = \zeta^{1,\varepsilon}(2t); \forall t \in [1/2, 1], \bar{\zeta}^\varepsilon(t) = \zeta^{2,\varepsilon}(2t-1)$$
$$\int_0^1 \|\frac{\partial}{\partial t}\bar{\zeta}^\varepsilon\|_{\mathscr{H}}^p dt = \int_0^{1/2} \|\frac{\partial}{\partial t}\bar{\zeta}^\varepsilon\|_{\mathscr{H}}^p + \int_{1/2}^1 \|\frac{\partial}{\partial t}\bar{\zeta}^\varepsilon\|_{\mathscr{H}}^p dt. \tag{15}$$

Now

$$\forall t \in [0, 1/2], \frac{\partial}{\partial t}\bar{\zeta}^\varepsilon(t) = 2\frac{\partial}{\partial t}\zeta^{1,\varepsilon}(2t); \forall t \in [1/2, 1], \frac{\partial}{\partial t}\bar{\zeta}^\varepsilon(t) = 2\frac{\partial}{\partial t}\zeta^{2,\varepsilon}(2t-1). \tag{16}$$

Then

$$\int_0^1 \|\frac{\partial}{\partial t}\bar{\zeta}^\varepsilon\|_{\mathscr{H}}^p dt = 2^p \int_0^{1/2} \|\frac{\partial}{\partial t}\zeta^{1,\varepsilon}(2t)\|_{\mathscr{H}}^p dt + 2^p \int_{1/2}^1 \|\frac{\partial}{\partial t}\zeta^{2,\varepsilon}(2t-1)\|_{\mathscr{H}}^p dt. \tag{17}$$

By respective changes of variables $s = 2t$ and $s = 2t - 1$ we get

$$\int_0^1 \|\frac{\partial}{\partial t} \tilde{\zeta}^\varepsilon\|_{\mathcal{H}}^p dt = 2^{p-1} \int_0^1 \|\frac{\partial}{\partial t} \zeta^{1,\varepsilon}(s)\|_{\mathcal{H}}^p ds + 2^{p-1} \int_0^1 \|\frac{\partial}{\partial t} \zeta^{2,\varepsilon}(s)\|_{\mathcal{H}}^p ds. \quad (18)$$

So that $\forall \varepsilon > 0$ we have:

$$d_{p,\mathcal{H}}(\Omega_0, \Omega_2) \leq \int_0^1 \|\frac{\partial}{\partial t} \tilde{\zeta}^\varepsilon\|_{\mathcal{H}}^p dt \leq 2^{p-1}(d_{p,\mathcal{H}}(\Omega_0, \Omega_1) + d_{p,\mathcal{H}}(\Omega_1, \Omega_2) + 2\varepsilon). \quad (19)$$

$\square$

2.5 Banach Space of Bounded Measures

We make the choice, as Banach space of measures $\mathcal{H}(D)$, of the space of bounded measure $M^1(D)$ and set

$$p \geq 1, \quad \mathscr{C}^p = \{\zeta \in \mathscr{C} \text{ s.t. } \frac{\partial}{\partial t} \zeta \in L^p(I, M^1(D))\} \quad (20)$$

that is

$$\mathscr{C}^p = \mathscr{C} \cap L^p(I, BV(D)). \quad (21)$$

From the previous considerations we get $\mathscr{C}^p \subset C^0(I, L^1(D))$, so that

$$p \geq 1, \quad \mathscr{C}^p = \{\zeta \in \mathscr{C}^0 \text{ s.t. } \frac{\partial}{\partial t} \zeta \in L^p(I, M^1(D))\}. \quad (22)$$

The set of *connecting tubes* is then:

$$\mathscr{T}^p(\Omega_0, \Omega_1) = \{\zeta \in \mathscr{C}^p \text{ s.t. } \zeta(i) = \chi_{\Omega_i}, \ i = 0, 1\}. \quad (23)$$

Corollary 1. *Let $p \geq 1$, then*

$$d_p(\Omega_0, \Omega_1) = Inf_{\zeta \in \mathscr{T}^p(\Omega_0, \Omega_1)} \int_0^1 \|\frac{\partial}{\partial t} \zeta(t)\|_{M^1(D)}^p dt \quad (24)$$

is a quasi metric. When $p = 1$, d_1 is a metric.

2.6 Smooth Domains

When a tube $\zeta = \chi_Q$ is smooth, $Q = \cup_{0<t<1}\{t\} \times \Omega_t$, with *lateral boundary* $\Sigma = \cup_{0<t<1}\{t\} \times \partial\Omega_t$ being a C^k manifold in $I \times D \subset R^{N+1}$, with the integer $k \geq 1$, there exists a vector field $V \in C^0(\bar{I}, C^k(D, R^N))$ with $< V(t,x), n_{\partial D} >= 0$ such that Ω_t is built by the flow mapping of V, that is $\Omega_t = T_t(V)(\Omega_0)$.

For example, when $k = 2$ the oriented distance function $b_{\Omega_t} = d_{\Omega_t} - d_{\Omega_t^c} \in C^2(\mathcal{U})$ where $\mathcal{U}$ is some tubular neighborhood of the boundary $\partial\Omega_t$, and we may choose any extension of $\nabla b_{\Omega_t}(x)v(t)op_t(x)$ as speed vector $V(t,x)$, where the normal field is $n_t(x) = \nabla b_{\Omega_t}(x), x \in \Gamma_t = \partial\Omega_t$, the projection p_t onto Γ_t being defined in $\mathcal{U}$ by $p_t(y) = y - b_{\Omega_t}(y)\nabla b_{\Omega_t}(y)$ (we recall that $\nabla b_{\Omega_t}op_t(y) = \nabla b_{\Omega_t}(y)$ for any $y \in \mathcal{U}$).

In the smooth situation the tube characteristic function ζ verifies the classical convection problem (in weak sense):

$$\zeta^2 = \zeta \in L^1(I \times D), \quad \frac{\partial}{\partial t}\zeta + \nabla\zeta.V = 0, \quad \zeta(0) = \chi_{\Omega_0}. \tag{25}$$

Then, without any restriction, we consider smooth domains generated from Ω_0 by the flow mapping $T_t(V)$ of smooth vector fields $V(t,x)$, $V \in E^k$ with:

$$E^k := \{V \in C^0(\bar{I}, C^k(\bar{D}, R^N)) \text{ s.t. } \forall t \in \bar{I}, \ \langle V(t), n_{\partial D}\rangle = 0\}. \tag{26}$$

The *connecting* condition is then: $\Omega_1 = T_1(V)(\Omega_0)$, where $T_t(V)$ is the flow mapping of V at time $t \in [0, 1]$. We set $\Omega_t := T_t(V)(\Omega_0)$ and $\zeta(t,.) = \chi_{\Omega_t}$ is an admissible connecting tube, moreover we have:

$$\|\frac{\partial}{\partial t}\zeta(t)\|_{M^1(D)} = \int_{\partial\Omega_t} |\langle V(t,x)), n_t(x)\rangle| \, d\Gamma_t(x) \tag{27}$$

and the metric would turn to be

$$d_{k,p}(\Omega_0, \Omega_1) = Inf_{V \in \mathcal{V}_k(\Omega_0, \Omega_1)} \int_0^1 (\int_{\partial\Omega_t(V)} |\langle V(t,x), n_t(x)\rangle| \, d\Gamma_t(x))^p dt \tag{28}$$

where $\mathcal{V}_k(\Omega_0, \Omega_1)$, defined below, stands for the family of connecting vector fields in E^k, $k \geq 1$. As the time regularity required for the classical flow analysis is just time continuity (in the very definition of E^k) this connecting family turns to be stable through the *generic construction* of connecting vector field $\bar{V}$ similar to the point 3 in the proof of Proposition 2.

Proposition 3. *Let $k \geq 1$ and $\bar{\Omega}$ be open domain in $D \subset R^N$ whose boundary $\bar{\Gamma}$ is a C^k manifold. We consider the family of smooth domains*

$$\mathcal{O}_k = \{\Omega \subset D \text{ s.t. } \exists V \in E^k, \Omega = T_1(V)(\bar{\Omega})\}. \tag{29}$$

For any pair of elements Ω_i, $i = 0, 1$ in this family, the set of connecting fields

$$\mathcal{V}_k := \{V \in E^k \text{ s.t. } T_1(V)(\Omega_0) = \Omega_1\} \tag{30}$$

is never empty. Equipped with $d_{p,k}$, the family $\mathcal{O}_k$ is a p-quasi-metric space (and a metric space when $p = 1$).

An important point here is that in this family $\mathcal{O}_k$, $k \geq 1$, all domains are homeomorphic to the domains $\bar{\Omega}$ so that we cannot evaluate distance between domains with different topologies, even when they are smooth. In order to escape from that classical difficulty we shall develop two classes of issues. The first one is based on time piecewise regularity of domains leading to a good modeling for classical topological changes such as holes collapse or holes creation (at a given time t_0), and topological separations. The second one is based on completing different approach relying on the fully eulerian description of tubes with non smooth vector fields V.

2.6.1   The Piecewise Smooth Situation

In some applications we shall consider the situation in which the time interval can be decomposed in a finite number of time intervals of smoothness for the lateral boundaries: we consider tubes such that there exists an integer K (tube dependent) and time partitions t_k such that

$$I = \cup_{1 \le k \le K} \bar{I}_k, I_k =]t_{k-1}, t_k[. \tag{31}$$

We assume that for $t \in I_k$ the lateral boundary Σ_k of the set $Q_k = \cup_{t \in I_k} \{t\} \times \Omega_t$, $\Sigma_k = \cup_{t \in I_k} \{t\} \times \partial\Omega_t$, is a C^1 manifold in R^{N+1}. We consider the unit normal field ν_k to Σ_k, out going to Q_k on Σ_k. It can be uniquely written in the form

$$\forall t \in I_k, \forall x \in \Gamma_k, \nu_k(t,x) = \frac{1}{\sqrt{1 + v_k(t,x)^2}}(-v_k(t,x), n_t(x)) \in R_t \times R_x^N. \tag{32}$$

The term $v_k(t,.)$ is called the normal speed of the moving boundary Γ_t. Obviously we have

$$\forall t \in I_k, \forall \phi \in \mathcal{D}(D), \left\langle \frac{\partial}{\partial t} \zeta, \phi \right\rangle = \int_{\Gamma_t} v_k(t,x)\phi(x) \, d\Gamma_k(x). \tag{33}$$

2.6.2   Behavior of the Normal Speed at $t = t_i$

To discuss the global regularity of $\frac{\partial}{\partial t}\zeta$ we must choose the regularity of v at the junction times t_k. Consider

$$\left\langle \zeta, -\frac{\partial}{\partial t}\Phi \right\rangle_{L^p(I,L^2(D)) \times L^q(I)} = \int_0^1 \int_{\Omega_t} -\frac{\partial}{\partial t}\Phi(t,x) \, dt dx$$
$$= \Sigma_{k=1,\dots,K} \int_{t_{k-1}}^{t_k} \int_{\Omega_t} -\frac{\partial}{\partial t}\Phi(t,x) \, dt dx \tag{34}$$

But

$$\forall t \in I_k, \frac{\partial}{\partial t}\int_{\Omega_t} \Phi(t,x) \, dt dx = \int_{\Gamma_t^k} \Phi(t,x)v_k(t,x)d\Gamma_t^k(x) + \int_{\Omega_t} \frac{\partial}{\partial t}\Phi(t,x) \, dx. \tag{35}$$

So that assume that $v_k \in L^1(\Sigma_k)$ and as $\zeta \in C^0(I,L^1(D))$ we have

$$\int_{t_{k-1}+\varepsilon}^{t_k-\varepsilon} \int_{\Omega_t} \frac{\partial}{\partial t}\Phi(t,x) dt dx = \int_{t_{k-1}+\varepsilon}^{t_k-\varepsilon} \int_{\Gamma_t} |v_k(t,x)| \, d\Gamma_k(x) dt$$
$$+ \int_D (\zeta(t_{k-1}+\varepsilon,x)\Phi(t_{k-1}+\varepsilon,x) - \zeta(t_k-\varepsilon,x)\Phi(t_k-\varepsilon,x)) dx$$
$$\to_{\varepsilon \to 0} \int_{t_{k-1}}^{t_k} \int_{\Gamma_t} |v_k(t,x)| \, d\Gamma_k(x) dt + \int_D (\zeta(t_{k-1},x)\Phi(t_{k-1},x) - \zeta(t_k,x))\Phi(t_k,x)) dx \tag{36}$$

Finally we get $\forall \Phi \in \mathcal{D}(I \times D)$,

$$\left\langle \zeta, \frac{\partial}{\partial t}\Phi \right\rangle = \lim_{\varepsilon \to 0} \Sigma_{1 \le k \le K} \int_{t_{k-1}+\varepsilon}^{t_k-\varepsilon} \int_D \zeta(t,x)\Phi(t,x) dx$$
$$= \int_0^1 \int_{\partial\Omega_t} v(t,x)\Phi(t,x) \, d\Gamma_t(x) dx \tag{37}$$

This expression continuously extends for any $\Phi \in C_c^0(I \times D)$ (with compact support) and we get

$$\left\|\frac{\partial}{\partial t}\zeta(t)\right\|_{L^1(I,M^1(D))} = \int_0^1 \int_{\Gamma_t} |v(t,x)| d\Gamma_t(x),$$ (38)

and we have

$$\int_0^1 \left\|\frac{\partial}{\partial t}\zeta(t)\right\|_{M^1(D)}^p dt = \int_0^1 \left(\int_{\Gamma_t} |v(t,x)| d\Gamma_t(x)\right)^p dt.$$ (39)

2.6.3 "Piecewise Metric"

Proposition 4. *Let $\bar{\Omega}$ be a smooth subset in D, $k \geq 1$, $p \geq 1$. We consider the family $\mathcal{O}_{pwk}(\bar{\Omega})$ of all subsets connected to $\bar{\Omega}$ by piecewise C^k tubes in the previous sense and verifying the following qualification condition:*

$$\int_0^1 \int_{\Gamma_t} |v(t,x)| d\Gamma_t(x) dt < \infty.$$ (40)

Then equipped with

$$\delta_{pwk}^p(\Omega_0, \Omega_1) = \text{Inf}_{\zeta \in \mathscr{T}_{pwk}^p(\Omega_0,\Omega_1)} \int_0^1 \left(\int_{\Gamma_t} |v(t,x)| d\Gamma_t(x)\right)^p dt,$$ (41)

the family $\mathcal{O}_{pwk}^p(\bar{\Omega})$ is a p-quasi-metric space. For $p = 1$, the family $\mathcal{O}_{pwk}^1(\bar{\Omega})$ equipped with δ_{pwk}^1, is a metric space.

Notice that a sufficient condition for deriving the condition (40) is that the lateral surface Σ would have a finite $\mathscr{H}^{n-1}$ Hausdorff measure (that is to say that the tube Q has a finite perimeter in $I \times D$). Indeed we have:

$$P_{I \times D}(Q) = \int_0^1 \int_{\Gamma_t} \sqrt{1+v^2} d\Gamma_t(x) dt \geq \int_0^1 \int_{\Gamma_t} |v| d\Gamma_t(x) dt.$$ (42)

2.6.4   Level Set Formulation

Let $\Psi(t,x) \in C^1(\bar{I} \times \bar{D})$ and consider

$$\forall t \in I, \Omega_t = \{x \in D \text{ s.t. } \Psi(t,x) > 0\}, \Gamma_t = \{x \in D \text{ s.t. } \Psi(t,x) = 0\}.$$ (43)

An important case is when the function has the following form

$$\Psi(t,x) = \Phi(x) - t \text{ then } \Omega_t = \{x \in D \text{ s.t. } \Phi(x) > t\}.$$ (44)

In this very situation, from Sard's theorem we know that for almost every t in I the manifold Γ_t is of class C^1 which does not insure the tube associated to Ψ to be pwk (even for $k = 1$).

In the general setting the qualification condition (40) would write

$$\int_0^1 \int_{\Gamma_t} \left(\frac{|\frac{\partial}{\partial t}\Psi|}{\|\nabla_x \Psi\|}\right)(t,x) d\Gamma_t(x) < \infty.$$ (45)

We shall restrict our study to the pwk level set tubes, i.e. functions $\Pi(t,x)$ such that the generated tubes verify the previous pw1 condition: $\exists t_k, t_0 = 0 < t_1 < \ldots < t_K = 1$ such that on each open interval $I_k =]t_{k-1}, t_K[$,

$$\exists \alpha_k(.) \in C^0(I_k) \text{ s.t. } \forall x \in \Gamma_t, \|\nabla_x \Psi(t,x)\| \geq \alpha_k(t) > 0. \tag{46}$$

In this class the previous piecewise tubes analysis applies and we get an associated metric in terms of level sets. In the proof of the following result the only main point is to verify that in the generic construction for the triangle axiom (point 3 in the proof of Proposition 2) the connecting element $\bar{\zeta}^\varepsilon$ piecewisely defined is still in the class. Indeed $\bar{\zeta}^\varepsilon$ is associated to the function

$$\bar{\Psi}^\varepsilon(t,x) = \begin{cases} \Psi^{1,\varepsilon}(2t,x), & \text{if } 0 < t < 1/2, \\ \Psi^{2,\varepsilon}(2t-1,x), & \text{if } 1/2 < t < 1. \end{cases} \tag{47}$$

Obviously this element $\bar{\Psi}^\varepsilon$ verifies the two conditions (45) and (46) if the element $\zeta^{i,\varepsilon}$, $i = 0,1$ does.

Proposition 5. *Let $\bar{\Omega} \subset D$ be a C^1 domain. We consider the family*

$$\mathscr{P}_{pw1} = \{\Psi \in C^1(\bar{I} \times \bar{D}), \text{ s.t. } \exists t_k, 1 \leq k \leq K_\Psi, \text{ s.t. }$$
$$\Sigma_k = \{(t,x) | t \in I_k, \Psi(t,x) = 0\} \tag{48}$$
$$\text{is a } C^1 \text{ manifold in } R^{N+1} \text{ and } \Psi \text{ verifying } (45), (46)\}.$$

We also consider the family generated by this class of piecewise C^1 (pw1) functions:

$$\mathscr{O}_{LS} = \{\Omega = \{x \in D | \Psi(1,x) > 0\}, \Psi \in \mathscr{P}_{pw1}\}. \tag{49}$$

Obviously two elements Ω_i, $i = 0,1$ in this family are connected by tube in the form of (47) and we denote

$$\mathscr{T}_{LS}(\Omega_0, \Omega_1) = \{\Psi \in \mathscr{P}_{pw1} \text{ s.t. } \Omega_i = \{x \in D | \Psi(i,x) > 0\}\}. \tag{50}$$

We set

$$\delta_{LS}(\Omega_0, \Omega_1) = Inf_{\Psi \in \mathscr{T}_{LS}(\Omega_0, \Omega_1)} \int_0^1 \int_{\Psi(t)^{-1}(0)} |\frac{\partial}{\partial t}\Psi(t,x)| \, \|\nabla_x \Psi(t,x)\|^{-1} \, d\Gamma_t(x) \, dt. \tag{51}$$

Then equipped with δ_{LS} the family $\mathscr{O}_{LS}$ is a metric space.

2.6.5 Submetrics

In the level set setting it is easy to describe some connecting elements. Assume that $\Omega_i = \{x \in D | \Phi_i(x) > 0\}$, $i = 0,1$. Then let

$$\Psi(t,x) = \rho(t)\Phi_1(x) + (1 - \rho(t))\Phi_0(x), \text{ with } \rho \in C^1(\bar{I}), \rho(i) = i, i = 0,1, \tag{52}$$

and we could consider the "submetric" associated to these connections, for different admissible such functions ρ.

2.6.6 Level Set Metric Associated to Subspace

In the definition (48) of the set of "potential" functions Ψ we can limit to a given subspace of functions in the following way: let E be a closed subspace in $C^1(\bar{I} \times \bar{D})$, we consider

$$\mathscr{P}_{pwE} = \mathscr{P}_{pw1} \cap E \tag{53}$$

As $\mathscr{P}_{pwE} \subset \mathscr{P}_{pw1}$ we get the similar inclusions $\mathscr{O}_{LSE} \subset \mathscr{O}_{SL}$, $\mathscr{T}_{LSE}(\Omega_0, \Omega_1) \subset \mathscr{T}_{LS}$ (Ω_0, Ω_1) and the family $\mathscr{O}_{LSE}$ is equipped with the metric $\delta_{LSE} \leq \delta_{LS}$.

In the specific situation where the Banach space is of finite dimension we consider the Galerkin-like construction. Let $E_1, \dots E_M$ be M given elements in $C^1(\bar{D})$ and consider

$$E = \{e(t,x) = \Sigma_{1 \leq m \leq M} \lambda_m(t) E_m(x) \mid \lambda \in C^1(I)^M \}. \tag{54}$$

When the elements $E_m(x)$ are chosen as polynomial functions the surfaces Γ_t are algebraic surfaces (or curves) in D and it is an open question to characterize conditions on the coefficients λ in order that the tube satisfies (45) and (46). Nevertheless in applications it seems difficult to violate them.

3 Complete Metric: Existence of Geodesic

We address now the question concerning the infima in the previous metrics (or pseudo metrics) we described in the previous sections. Let ζ^n be a minimizing sequence in (24). The element $\frac{\partial}{\partial t}\zeta^n$ remains bounded in $L^p(I, M^1(D))$. Then when $p > 1$, there exists a subsequence, still denoted $\frac{\partial}{\partial t}\zeta^n$ and weakly converging to an element $\omega \in L^p(I, M^1(D))$. The difficulty is now to get ω in the form $\omega = \frac{\partial}{\partial t}\zeta^*$ for some admissible ζ^*. As $\zeta^n \in \mathscr{C}^0$, it remains bounded as an element of $\mathscr{C}$ in $L^r(I \times D)$, and this for any $1 \leq r \leq \infty$. Let us consider a subsequence ζ^n weakly convergent to some element ρ. By continuity of the derivative in weak topologies we derive that $\omega = \frac{\partial}{\partial t}\rho$ but a priori the element ρ is not a characteristic function. Indeed we shall have $\rho \in \mathscr{C}$, that is $\rho^2 = \rho$, if and only if ρ^n strongly converges to ρ in $L^1(I \times D)$. Nevertheless, this strong $L^1(I \times D)$ convergence would not imply the limiting element ρ to be in $\mathscr{C}^0$. Now, if this element is not time continuous the connection makes no sense and it could not be a candidate for geodesic.

3.1 Compacity Arguments and Complete Metric

3.1.1 Surface Tension-Like Term

We shall propose now several changes in the metric (or p-quasi-metric) to derive this strong convergence. First of all let us denote that if we complete in (24) the metric by the following, with $\sigma > 0$ (a surface tension term)

$$d(\Omega_0, \Omega_1) = Inf_{\zeta \in \mathscr{T}(\Omega_0, \Omega_1)} L_\sigma(\zeta) \tag{55}$$

with

$$L_\sigma = \int_0^1 \|\frac{\partial}{\partial t}\zeta\|_{M^1(D)} dt + \sigma \int_0^1 \|\nabla_x \zeta(t)\|_{M^1(D,R^N)} dt, \tag{56}$$

then we could derive, for any *smooth minimizing sequence*, $\zeta^n(t,x) = \chi_{Q^n}$, the tubes with bounded perimeter in $I \times D$ as we have.

3.1.2   Boundedness of the Perimeter in $I \times D \subset R^{N+1}$

Proposition 6. *Assume the evolution domain Q to be smooth, then*

$$P_{I \times D}(Q) \leq \int_0^1 ||\frac{\partial}{\partial t}\zeta||_{M^1(D)}\,dt + \int_0^1 ||\nabla_x\zeta(t)||_{M^1(D,R^N)}\,dt. \tag{57}$$

Proof

$$P_{I \times D}(Q) = \int_\Sigma d\Sigma = \int_0^1 \int_{\Gamma_t} \sqrt{1+v^2}\,d\Gamma_t(x)\,dt$$

$$P_{I \times D}(Q) \leq \int_0^1 \int_{\Gamma_t}(1+|v|)\,d\Gamma_t(x)\,dt = \int_0^1 (P_D(\Omega_t) + \int_{\Gamma_t}|v|\,d\Gamma_t)\,dt, \tag{58}$$

but

$$P_D(\Omega_t) = ||\nabla\zeta(t)||_{M^1(D,R^N)}, \int_{\Gamma_t}|v|\,d\Gamma_t(x) = ||\frac{\partial}{\partial t}\zeta||_{M^1(D)}. \tag{59}$$

So that (57) is true when the domain is smooth.   □

3.1.3   Metric on the Closure of Smooth Tubes Would Fail

We could hope that (57), by some density arguments, extends for all tubes $\zeta \in L^1$ $(I, BV(D)) \cap W^{1,1}(I, M^1(D))$ (which is an open question) or define the metric as follows. Introducing the family of smooth tubes, say $\mathscr{C}^\infty$ (elements $\zeta = \chi_Q$ with lateral boundary being a C^∞ manifold in $I \times D \subset R^{N+1}$), set

$$d_\sigma^\infty(\Omega_0, \Omega_1) = Inf_{\zeta \in \mathscr{C}^\infty} L_\sigma(\zeta). \tag{60}$$

Any minimizing sequence, from (57) would remain bounded in $BV(I \times D)$ and then there shall exist a subsequence strongly converging in $L^1(I \times D)$ so that the limiting element will be $\zeta \in \mathscr{C}$ with

$$||\zeta||_{BV(I \times D)} \leq liminf_{n \to \infty} ||\zeta_n||_{BV(I \times D)}, \tag{61}$$

and by similar weak l.s.c. arguments on each term of L_σ we would see that the limiting element $\zeta \in \mathscr{C}$ would be a minimizer of L_σ on some closure of $\mathscr{C}^\infty$. Nevertheless this element would not belong to $\mathscr{C}^0$, and being not continuous in time the connection property $\zeta \in \mathscr{T}(\Omega_0, \Omega_1)$ could be lost and this candidate for metric would fail, while having a minimizer. Finally we understand that even if the inequality (57) extends to a more general family of tubes it would not help for deriving metric with geodesic.

An important point here is that any expression in the form of

$$\tilde{d}_\sigma(\Omega_0, \Omega_1) = Inf_{\zeta \in \mathscr{T}} L_\sigma(\zeta) \tag{62}$$

would fail to be a metric because *it violates the first metric axiom*. Indeed the new perimeter term $\sigma \int_0^1 P_D(\Omega_t)\,dt$ cannot be zero.

3.2   Compactness Results

We have seen that the compactness result deriving from the boundedness of L_σ, i.e. boundedness in the Banach space

$$\mathscr{B}^1 = L^1(I, BV(D)) \cap W^{1,1}(I, M^1(D)) \subset C(I, M^1(D)) \tag{63}$$

is not enough.

Proposition 7. *Consider ζ_n bounded in $L^1(I, BV(D))$, together with $\frac{\partial}{\partial t}\zeta_n$ bounded in $L^p(I, M^1(D))$ for some $p > 1$. Then there exists a subsequence, still denoted ζ_n, and an element $\zeta \in L^1(I, BV(D)) \cap W^{1,1}(I, M^1(D)) \subset C^0(I, M^1(D))$ such that ζ_n strongly converges to ζ in $L^1(I, L^1(D))$ with $\frac{\partial}{\partial t}\zeta \in L^p(I, M^1(D))$ verifying $\|\zeta\|_{L^1(I, BV(D))} \le liminf\|\zeta_n\|_{L^1(I, BV(D))}$ and $\|\frac{\partial}{\partial t}\zeta\|_{L^p(I, M^1(D))} \le liminf\|\frac{\partial}{\partial t}\zeta_n\|_{L^p(I, M^1(D))}$.*
Continuity $\zeta \in W^{1,1}(I, M^1(D))$ implies $\zeta \in C^0(I, L^1(D))$. Moreover $\zeta(t,x) = \zeta^2(t,x)$, a.e.$(t,x) \in I \times D$ and $\zeta \in C^0(I, L^1(D))$ implies that the mapping:

$$t \in \bar{I} \rightarrow p(t) := \|\nabla_x \zeta(t)\|_{M^1(D, R^N)} \text{ is s.c.i.} \tag{64}$$

Proof. See [24,13].   □

Also a similar compacity result can be derived with $p = 1$, leading to a metric, but assuming some *uniform integrability* for the $\|\frac{\partial}{\partial t}\zeta\|_{M^1(D)}$ term.

Proposition 8. *Consider ζ_n bounded in $L^1(I, BV(D))$ together with $\frac{\partial}{\partial t}\zeta_n$ bounded in $L^1(I, M^1(D))$, and assume there exists an element $\theta \in L^1(I)$ such that*

$$a.e.\ t \in I,\ \|\frac{\partial}{\partial t}\zeta_n\|_{M^1(D)} \le \theta(t). \tag{65}$$

Then there exists a subsequence, still denoted ζ_n, and an element $\zeta \in L^1(I, BV(D)) \cap W^{1,1}(I, M^1(D)) \subset C^0(I, M^1(D))$ such that ζ_n strongly converges to ζ in $L^1(I, L^1(D))$ with $\frac{\partial}{\partial t}\zeta \in L^p(I, M^1(D))$ verifying $\|\zeta\|_{L^1(I, BV(D))} \le liminf\|\zeta_n\|_{L^1(I, BV(D))}$ and $\|\frac{\partial}{\partial t}\zeta\|_{L^p(I, M^1(D))} \le liminf\|\frac{\partial}{\partial t}\zeta_n\|_{L^p(I, M^1(D))}$.
Continuity $\zeta \in W^{1,1}(I, M^1(D))$ implies $\zeta \in C^0(I, L^1(D))$. Moreover $\zeta(t,x) = \zeta^2(t,x)$, a.e. $(t,x) \in I \times D$ and $\zeta \in C^0(I, L^1(D))$ implies that the mapping:

$$t \in \bar{I} \rightarrow p(t) := \|\nabla_x \zeta(t)\|_{M^1(D, R^N)} \text{ is s.c.i.} \tag{66}$$

Proof. See [21,13].   □

3.3   Use of Compactness

The idea is to consider the following expression for the shape metric defined by (24):

$$\bar{d}^p(\Omega_0, \Omega_1) = Inf_{\zeta \in \mathscr{T}^p(\Omega_0, \Omega_1)} \int_0^1 \|\frac{\partial}{\partial t}\zeta(t)\|_{M^1(D)} dt + " \int_0^1 |p'_\zeta(t)|^p dt". \tag{67}$$

Indeed the last term is not finite in general as it would imply $p(t)$ to be time continuous which is known to be false (the perimeter is *l.s.c.* only and may "jump down" as in the celebrate "*Camembert entamé*" example: take a circular cheese Camembert with radius R and subtract a radial triangular part with angle α, the perimeter is $p(\alpha)$ and $liminf_{\alpha \to 0} p(\alpha) = (2\pi + 2)R > p(0) = 2\pi R$).

We relax this term by introducing (see [23]) the "time capacity" term

$$\theta^p(\zeta) = Inf_{\mu \in K^p(\zeta)} \int_0^1 |\mu'(t)|^p dt, \tag{68}$$

with the closed convex set

$$K^p(\zeta) = \{\mu \in W^{1,p}(I) \text{ s.t. } ||\nabla_x \zeta(t)||_{M^1(D,R^N)} \le \mu(t) \text{ a.e. } t \in I\}. \tag{69}$$

Then the metric is

$$\bar{d}^p(\Omega_0, \Omega_1) := Inf_{\zeta \in \mathcal{T}^p(\Omega_0, \Omega_1)} \int_0^1 ||\frac{\partial}{\partial t}\zeta(t)||_{M^1(D)} dt + \theta^p(\zeta). \tag{70}$$

3.4   Complete Quasi-Metric by Level Set Formulation

Let $p > 1$ and Ω_i, $i = 1, 2$ be two arbitrary measurable subsets in D. Let

$$K(\Omega_1, \Omega_2) = \{\phi \in L^2(I, H^1(D)) \cap W^{1,1}(I, L^2(D)),$$

$$\frac{\partial}{\partial t}\phi \in L^p(I, L^2(D)), \tag{71}$$

$$\Omega_1 = \{\Phi(0,.) > 0\}, \Omega_2 = \{\Phi(1,.) > 0\}\}.$$

Notice that $K(\Omega_1, \Omega_2) \subset C^0(\bar{I}, L^2(D))$. We set

$$d_{LS,p}(\Omega_1, \Omega_2) := Inf_{\phi \in K(\Omega_1, \Omega_2)} \int_0^1 \left(\alpha ||\phi(t)||_{H^1(D)}^2 + ||\frac{\partial}{\partial t}\phi(t)||_{L^2(D)}^p\right) dt. \tag{72}$$

Proposition 9. *Let $1 < p \le 2$. Equipped with $d_{LS,p}$, the family of measurable subsets in D is a complete quasi-metric space.*

4   Fully Eulerian Metric d_e

For non smooth vector fields, being given the element Ω_0 in D the problem (25) may have no solution or several solutions (depending on the weak regularity of the speed vector field V). As soon as V satisfies the minimal regularity $V \in \mathcal{V}^p$ where

$$p \ge 1, \mathcal{V}^p = \{V \in L^p(I \times D, R^N), divV \in L^p(I \times D),$$

$$\langle V, n_{\partial D}\rangle = 0 \text{ in } W^{-1,p}(\partial D)\}, \tag{73}$$

the following classical convection problem

$$\zeta \in L^1(I \times D), \frac{\partial}{\partial t}\zeta + \nabla\zeta.V = 0, \zeta(0) = \chi_{\Omega_0} \tag{74}$$

possesses solution (the proof is classically done by the Galerkin method when $V \in L^2(I \times D)$ and $(divV)^+ \in L^\infty(I, L^2(D))$, see [20], and there is no uniqueness result for the solution, which a priori is not an element in $\mathscr{C}$, nor in $\mathscr{C}^0$). The element ζ is not a characteristic function but is time continuous, $\zeta \in C^0([0,1], H^{-1/2}(D))$. Indeed we consider weak solutions to problems (25) and (74), in the following sense:

$$\forall \phi \in C^\infty(\bar{I} \times \bar{D}) \text{ s.t. } \phi(1,.) = 0,$$

$$\int_0^1 \int_D \zeta(-\frac{\partial}{\partial t}\phi - div(\phi V))dxdt = \int_D \chi_{\Omega_0}\phi(0,x)dx. \tag{75}$$

The time derivative, for any solution to (74) (then to (25)) verifies:

$$\frac{\partial}{\partial t}\zeta = div(\zeta V) - \zeta\,divV \in L^p(I, W^{-1,p}(D)), \tag{76}$$

so that

$$\zeta \in W^{1,p}(I, W^{-1,p}(D)) \subset C^0(\bar{I}, W^{-1,p}(D)). \tag{77}$$

Notice that weak solutions to (74) can also be obtained by the following technique, without any L^∞ requirement on the divergence:

Proposition 10. *Let $p > 1$ and $V \in \mathscr{V}^p$ defined in (73), let $V_n \in \mathscr{V}^p \cap C^\infty(\bar{I} \times \bar{D}, R^N))$ such that $V_n \to V$ strongly in $\mathscr{V}^p$. Consider the element $\zeta_n(t) = \chi_{\Omega_0} o T_t(V_n)^{-1} \in \mathscr{C}^0$, a unique solution to the characteristic convection problem (25). There exists a subsequence, still denoted ζ_n which weakly converges in $L^p(I \times D)$ to an element $\rho \in L^p(I \times D) \cap W^{1,1}(I, W^{-1,p}(D)) \subset C^0(\bar{I}, W^{-1,p}(D))$, a solution to the convection problem (74) or (75).*

Proof. We pass to the limit in the weak form (75). □

The concept of distance between the two sets $\Omega_i, i = 0, 1$ is associated to the "shortest path", that is now introduced through the Euler description using **the product space approach** which is described in [23] and [24]. Let us consider the *eulerian* connecting tubes defined as set of couples (ζ, V) solving the convection equation:

$$\mathscr{T}_e^p(\Omega_0, \Omega_1) = \{(\zeta, V) \in \mathscr{C} \times \mathscr{V}^p \text{ solving (25) with } \zeta(i) = \chi_{\Omega_i}, i = 0, 1\}. \tag{78}$$

4.1 Eulerian Metrics

Let

$$d_p^e(\Omega_0, \Omega_1) := Inf_{(\zeta,V)\in\mathscr{T}^p(\Omega_0,\Omega_1)} \int_0^1 (\|V(t)\|_{L^p(D,R^N)}^p + |divV(t)|_{L^p(D)}^p)dt, \tag{79}$$

and

$$\bar{d}_e^p(\Omega_0, \Omega_1) := Inf_{(\zeta,V)\in\mathscr{T}_e^p(\Omega_0,\Omega_1)} \int_0^1 \int_D (|V|^p + (divV)^p)dxdt + \theta^p(\zeta). \tag{80}$$

Proposition 11. *For $p \geq 1$, d_p^e is a quasi-metric in the following sense:*

i) $d_p^e(\Omega_0, \Omega_1) = 0$ iff $\chi_{\Omega_0} = \chi_{\Omega_1}$,
ii) $d_p^e(\Omega_0, \Omega_1) = d_p^e(\Omega_1, \Omega_0)$,
iii) $d_p^e(\Omega_0, \Omega_2) \leq 2^{p-1}(d_p^e(\Omega_0, \Omega_1) + d_p^e(\Omega_1, \Omega_2))$.

Moreover, equipped with $\bar{d}_p^e$ the family $\mathcal{O}_{\hat{\Omega}}^e$ is a complete quasi-metric space and for $p = 1$, equipped with d_1^e, it is a complete metric space.

4.2   BV **Regularity of the Field** V

When the speed vector field V verifies some BV properties, $V \in L^2(I, BV(D)^N)$ ([1,23,24]), there is a unique tube associated to V, then we do have an application $V \to \zeta_V$ and with such regularity on V we can revisit the complete metric d being completely delivered of the non differential perimeter and curvature terms that we were obliged to introduce in order to apply the compacity theorems. From the tube analysis we consider several interesting choices for the space regularity of the speed vector field (together with its divergence field). Let

$$\mathcal{E}^{1,1} = \{V \in L^1(I \times D, R^N) \text{ s.t. } divV \in L^1(D), V.n_D \in W^{-1,1}(\partial D)\}, \qquad (81)$$

and let E be a closed subspace in $BV(D) \cap \mathcal{E}^{1,1}$ such that any element $V \in E$ verifies the required assumptions. A first example is, when working with prescribed volume for the moving domain

$$E_0 = \{V \in BV(D, R^N) \cap \mathcal{E}^{1,1}, \text{ s.t. } divV = 0 \text{ a.e. } (t,x) \in I \times D\} \qquad (82)$$

V be a divergence-free vector field with $divV = 0$, $V \in L^1(I, E_0)$, where $E = BV(D, R^N)$ or any closed subspace (for example $E = \{V \in H_0^1(D, R^N), \text{ s.t. } divV = 0\}$). An *obvious* metric is to consider the set

$$\mathcal{V}(\Omega_1, \Omega_2) = \{V \in \mathcal{E}^{1,1} \text{ s.t. } V, divV \in L^p(I, E_0), \text{ s.t. } \zeta_0 = \chi_{\Omega_1}, \zeta(1) = \chi_{\Omega_2}\}$$
$$\delta_{E_0}(\Omega_1, \Omega_2) = Inf_{V \in \mathcal{V}(\Omega_1, \Omega_2)} \int_0^1 ||V(t)||_{E_0} dt. \qquad (83)$$

As V is divergence-free the previous boundedness assumption on the divergence is verified and to each V a tube ζ_V is associated through the convection. Then we get:

Proposition 12. *Let E be any subspace of $BV(D, R^N) \cap \mathcal{E}^{1,1}$ such that any element V satisfies assumptions of Theorem 2.12 of [24], for example $E = E_0$. Then equipped with δ_E, the family $\mathcal{O}_{\Omega_0}^E$ is a metric space.*

$$p > 1, d_{E_0}(\Omega_1, \Omega_2) = Inf_{V \in \mathcal{V}(\Omega_1, \Omega_2)} ||V||_{L^p(I, E_0)} + ||\frac{\partial}{\partial t} V||_{L^1(I, M^1(D, R^N))}. \qquad (84)$$

Proposition 13. *Let E be any subspace of $BV(D, R^N) \cap \mathcal{E}^{1,1}$, such that any element V whose divergence satisfies assumptions of Theorem 2.12 of [24]. Then equipped with d_E the family $\mathcal{O}_{\Omega_0}^E$ is a complete quasi-metric space.*

4.2.1 Geodesic Characterization via Transverse Field Z

In that setting we are concerned with smooth vector fields $Z(s,t,x) \in R^N$ such that $Z(s,0,x) = Z(s,1,x) = 0$ so that the extremities (for $t = 0, t = 1$) of the perturbed tube $Q_s := T_s(Z(s))(Q)$ are preserved. The parameter s appears here as a *perturbation* parameter. Indeed the family of connecting tubes $\mathcal{T}_e^P(\Omega_0, \Omega_1)$ is not a linear space nor equipped with any manifold structure. Nevertheless we can describe some tangential space $\mathbf{T}_{(\zeta, V)} \mathcal{T}_e^P(\Omega_0, \Omega_1)$ at any element (tube) (ζ, V): if $(\zeta, V) \in \mathcal{T}$ then $\zeta \circ T_s(Z(t))^{-1}, V^s) \in \mathcal{T}$ where $[\frac{d}{ds} V^s]_{s-0} = Z_t + [Z, V]$. The previous study for the transverse field [18,13,11] implies that for given such a vector field Z, with $div_x Z(s,t,x) = 0$ we get the admissible perturbation of the field V in the form $V + sW(s,t,x)$, with

$$W(s,t,x) = \frac{\partial}{\partial t} Z(s,t,x) + [Z, V]. \tag{85}$$

More precisely, define the Lipschitz-continuous connecting set:

$$\mathcal{V}^{1,\infty}(\Omega_1, \Omega_2) = \{V \in L^1(I, W^{1,\infty}) \cap \mathcal{E}^{1,1}, \text{ s.t. } \zeta_V \in \bar{\mathcal{T}}(\Omega_1, \Omega_2)\} \tag{86}$$

and the set of smooth transverse vector fields:

$$\mathcal{Z} = \{Z(t,x) \in C_{comp}^\infty(I \times D, R^N)\}. \tag{87}$$

Notice that such Z verifies $Z(0,.) = Z(1,.) = 0$ on D.

Proposition 14. *Let $V \in \mathcal{V}(\Omega_1, \Omega_2)$ and $Z(t,x) \in \mathcal{Z}$. The transformation $\mathcal{T} = T_s(Z) \circ T_t(V)$ maps $\Omega_t(V)$ onto $\Omega_t^s := T_s(Z)(\Omega_t(V))$ so that*

$$\forall s, \forall Z, V^s(t,x) = \frac{\partial}{\partial t} \mathcal{T} \circ \mathcal{T}^{-1}$$

$$= (\frac{\partial}{\partial t} T_s(Z(t)) + DT_s(Z(t)).V(t)) \circ T_s(Z(t))^{-1} \in \mathcal{V}^{1,\infty}(\Omega_1, \Omega_2). \tag{88}$$

Lemma 2

$$\frac{\partial}{\partial s} V^s(t,x)|_{s=0} = \frac{\partial}{\partial t} Z(t) + [Z(t), V(t)]. \tag{89}$$

Corollary 2. *Consider a functional $\mathcal{J}(V) = j(\zeta_V)$ and let $\bar{V}$ be a minimizing element of $\mathcal{J}$ on $\mathcal{V}(\Omega_1, \Omega_2)$. Then we have*

$$\forall Z \in \mathcal{Z}, \frac{\partial}{\partial s} \mathcal{J}(\bar{V}^s)_{s=0} = J'(\bar{V}; (\frac{\partial}{\partial s} V^s)_{s=0}) = \mathcal{J}'(\bar{V}; \frac{\partial}{\partial t} Z(t) + [Z(t), V(t)]) \geq 0. \tag{90}$$

That variational principle extends to vector field $V \in E$ for which the flow mapping $T_t(V)$ is poorly defined. The element $\zeta_V \in \mathcal{H}^c$ is uniquely defined. For any $Z \in \mathcal{Z}$ the perturbed $\zeta_V^s := \zeta_V \circ T_s(Z)^{-1} \in \bar{\mathcal{T}}(\Omega_1, \Omega_2)$; on the other hand the following result is easily verified.

Proposition 15. $\zeta_V^s = \zeta_{V^s}$ *with*

$$V^s(t,.) := -DT_s^{-1}(-Z(t)).(V(t) \circ T_s(Z(t))^{-1}) - \frac{\partial}{\partial t} T_s(-Z(t)) \tag{91}$$

In other words:

$$\frac{\partial}{\partial t}\zeta + \nabla\zeta.V = 0 \text{ implies } \frac{\partial}{\partial t}(\zeta o T_s(Z(t))^{-1}) + \nabla(\zeta o T_s(Z(t))^{-1}).V^s = 0. \quad (92)$$

It can also be verified that the expression (89) for the derivative of the field still holds true so that the variational principle (90) is valid for any functional $\mathscr{J}$ minimized over the lipschitzian connecting family $\mathscr{V}^{1,\infty}(\Omega_1, \Omega_2)$. And more generally, without assuming V in E we have:

Proposition 16. *Let* $(\zeta, V) \in \mathscr{T}^{p,q}(\Omega_1, \Omega_2)$, *then for all* $s > 0$ *and* $Z \in \mathscr{Z}$ *we have:*

$$(\zeta o T_s(Z)^{-1}, V^s) \in \mathscr{T}^{p,q}(\Omega_1, \Omega_2). \quad (93)$$

In order to get a differentiable metric, we could consider

$$\tilde{d}(\Omega_1, \Omega_2) = Inf_{V \in \mathscr{V}(\Omega_1, \Omega_2)} \int_0^1 (\|V(t)\|_{H_0^1 \cap E_0} + \|\frac{\partial}{\partial t}V\|_{L^2(D)})dt. \quad (94)$$

Equipped with $\tilde{d}$, $\mathscr{O}_{\Omega_0}$ would be a complete metric space but $\tilde{d}$ fails to be a metric because of the triangle axiom. The advantage is that now the associated functional is differentiable with repect to V, then we can apply the previous variational principle with transverse vector field Z. Let $\bar{V}$ be a minimizer in $\mathscr{V}(\Omega_1, \Omega_2)$ for $\tilde{d}(\Omega_1, \Omega_2)$. Then $\forall Z \in \mathscr{Z}$ we have:

$$\int_0^1 \{\|V(t)\|^{-1}\langle V(t), Z_t + [Z,V]\rangle + |V'(t)|^{-1}((V'(t)(Z_t + Z,V)'))\}dt = 0, \quad (95)$$

where $\langle,\rangle$ is the $H_0^1(D, R^N)$ inner product while $((,))$ is the $L^2(D, R^N)$ one. In order to recover a differentiable complete metric, we introduce again the constraint on the perimeter as in the beginning and set

$$\delta_{H^1}(\Omega_1, \Omega_2) = Inf_{V \in \mathscr{V}(\Omega_1, \Omega_2)} \int_0^1 \|V(t)\|_{H_0^1 \cap E_0} dt. \quad (96)$$

The optimality condition is: $\forall Z \in \mathscr{Z}$

$$\text{s.t. } \int_0^1 \int_{\Gamma_t} H(t)\langle Z(t), n_t\rangle d\Gamma_t dt = 0,$$

$$\int_0^1 \|V(t)\|^{-1}\langle V(t), Z_t + [Z,V]\rangle dt = 0. \quad (97)$$

4.2.2   Euler Equation for Geodesics

$$\exists c(t), P \text{ s.t. } \frac{\partial}{\partial t}(\|V(t)\|^{p-2}V(t)) + \|V(t)\|^{p-2}(DV(t).V + D^*V.V(t))$$
$$= \nabla P + c\chi_{\Gamma_t} div_{\Gamma_t}(n_t) n_t. \quad (98)$$

That is,

$$(p-2)||V||^{p-4}((V,\frac{\partial}{\partial t}V)) + ||V(t)||^{p-2}(\frac{\partial}{\partial t}V + DV(t).V + D^*V.V(t))$$
$$= c\,\chi_{\Gamma_t}\,div_{\Gamma_t}(n_t)\,n_t, \tag{99}$$

which can be written as (with $\bar{V} = ||V||^{-1}V$, $\Pi = P - 1/2|V|^2$):

$$divV = 0,$$

$$\frac{\partial}{\partial t}V + (p-2)((\frac{\partial}{\partial t}V, \bar{V}))\,\bar{V} = DV.V = \nabla\Pi + c(t)||V||^{2-p}\chi_{\Gamma_t}\,div_{\Gamma_t}(n_t)\,n_t. \tag{100}$$

5 Variational Formulation for Euler Flow

As an application of the previous results we give a variational formulation for Euler incompressible flow with tube boundary condition. We consider two non miscible fluids and the tube describes the densities distribution. For shortness in this section we assume $p = 2$ and we consider the quadratic situation with divergence-free vector fields. Then we consider the Hilbert space

$$H = \{V \in L^2(I \times D, R^N) \text{ s.t. } divV = 0, V.n_D = 0\}. \tag{101}$$

We consider any Banach space $E_1 \subset L^1(D, R^N)$ with continuous and compact inclusion mapping.

Examples are $E_1 = BV(D, R^N)$ or $E_1 = W^{\varepsilon,p}(D, R^N)$, for $\varepsilon < 1/p$, $1 \leq p < \infty$, which is, for $p = 2$, the Hilbert space $E_1 = H^\varepsilon(D, R^N)$, for $\varepsilon < 1/2$.

The set of tubes under consideration is then

$$\mathcal{T} = \{(\zeta, V) \in L^2(I \times D) \times H, \text{ s.t. } \zeta = \zeta^2,$$
$$\nabla\zeta \in L^1(I, E_1), \zeta_t + \nabla\zeta.V = 0, \zeta(\tau) = \chi_{\Omega_1}\}. \tag{102}$$

Notice that the convection equation implies that as $\zeta_t = div(-\zeta V)$, then we get:

$$\zeta \in C^0(I, W^{-1,1}(D, R^N)). \tag{103}$$

Proposition 17. *The set $\mathcal{T}$ of tubes is non empty.*

5.1 Tube-Variational Principle

We introduce the optimal control view point: the state equation will be the convection equation (102) while we shall minimize a "Tube-Energy" cost functional governed by this equation. The regularizing term is a surface tension-like term. As in the previous sections this term will be needed in order to make use of the previous parabolic compactness of tubes. Indeed we shall introduce a kind of "density" perimeter θ_h associated with $L^1(I, H^\varepsilon(D))$ norm of the tube ζ, which turns to be differentiable under smooth transverse fields perturbations ζ_s.

5.1.1 Given Initial Domain $\Omega_0 \subset D$

Being given $\alpha > 0, \beta > 0$, we consider the following Tube-Energy functional:

$$\mathscr{E}(\zeta, V) = 1/2 \int_0^\tau \int_D (\alpha \zeta(t,x) + \beta) |V(t,x)|^2 dxdt + \sigma \int_0^\tau \|\zeta(t)\|_{E_1} dt$$
$$- \int_0^\tau \int_D V_0(x).V(t,x) dxdt. \tag{104}$$

Theorem 1. *The functional $\mathscr{E}$ reaches its minimum on the set $\mathscr{T}$ of tubes.*

Proof. We consider a minimizing sequence $(\zeta_n, V_n) \in \mathscr{T}$. There exist subsequences such that $V_n \rightharpoonup V$, weakly in $L^2(I \times D)$ and $\zeta_n \to \zeta$ strongly in $L^1(I \times D)$. Effectively as $(\zeta_n)_t = div(-\zeta_n V_n)$, we have:

$$\|\zeta_n\|_{L^1(I,E_1)} \le M_1, \quad \|(\zeta_n)_t\|_{L^2(I,W^{-1,1}(D,R^N))} \le M_2. \tag{105}$$

The conclusion derives from the compacity result. From this strong L^1 convergence we derive that $\zeta^2 = \zeta$. We consider the weak formulation for the convection problem (102):

$$\forall \psi \in C^1(I \times \bar{D}, R^N), \ \psi(0,.) = 0,$$
$$\int_0^\tau \int_D \zeta_n(-\psi_t - \nabla\psi.V_n) dxdt = -\int_{\Omega_1} \psi(0,x) dx, \tag{106}$$

in which we can pass to the limit and conclude that $(\zeta, V) \in \mathscr{T}$. Moreover, the element (ζ, V) is classically a minimizer as the three terms are weakly lower semi- continuous, respectively for each weak topology. $\square$

5.2 Euler Equation Solved by the Minimizer

In order to analyze the necessary conditions associated with any minimizer of $\mathscr{E}$ over the set $\mathscr{T}$ we introduce transverse transformations of the tube.

5.2.1 Transverse Field

Let us consider a perturbation parameter $s \ge 0$ and any smooth horizontal non autonomous vector field over R^{N+1} (s being the evolution parameter for a dynamic in R^{N+1})

$$\mathscr{Z}(s,t,x) = (0, z(s,t,x)) \in R_r \times R^N, \tag{107}$$

such that $\mathscr{Z}(s,0,x) = 0$.

For any element $(\zeta, V) \in \mathscr{T}$ we consider the perturbed tube (ζ^s, V^s), where:

$$\zeta^s(t,x) := \zeta o T_s(\mathscr{Z}^t)(x))^{-1}$$
$$V^s(t,x) = (D(T_s(\mathscr{Z}^t)^{-1})^{-1}.(V(t) o T_s(\mathscr{Z}^t)^{-1} - \frac{\partial}{\partial t}(T_s(\mathscr{Z}^t)^{-1})). \tag{108}$$

Indeed we can show

Proposition 18. $\forall(\zeta, V) \in \mathscr{T}, \forall \mathscr{Z}$, *the previously defined elements* $(\zeta^s, V^s) \in \mathscr{T}$.

5.2.2 Transverse Derivative

Assume that $div_x \mathscr{Z}^t = 0$, then:

$$\int_D (\alpha \zeta^s(t,x) + \beta)|V^s(t,x)|^2 \, dx = \int_D (\alpha \zeta(t,x) + \beta)|V^s(t) o T_s(\mathscr{Z}^t)(x)|^2 \, dx \qquad (109)$$

So that the optimality of the element (ζ, V) writes:

$$1/s(\mathscr{E}(\zeta^s, V^s o T_s) - \mathscr{E}(\zeta, V)) \geq 0. \qquad (110)$$

Now the following quotient has a strong limit in $L^2(I \times D)$:

$$\begin{aligned}
\frac{V^s o T_s - V}{s} &= \frac{d}{ds}[V^s o T_s(\mathscr{Z}^t)]_{s=0} \\
&= \frac{d}{ds}[(D(T_s(\mathscr{Z}^t)^{-1})^{-1}.(V(t) - \frac{\partial}{\partial t}(T_s(\mathscr{Z}^t)^{-1}) o T_s(\mathscr{Z}^t))]_{s=0} \\
&= \frac{d}{ds}[(D(T_s(\mathscr{Z}^t) o T_s(\mathscr{Z}^t)^{-1}.(V(t) - \frac{\partial}{\partial t}(T_s(\mathscr{Z}^t)^{-1}) o T_s(\mathscr{Z}^t))]_{s=0} \\
&= \frac{\partial}{\partial t}Z(t) + DZ(t).V(t) \in L^2(I \times D, R^N),
\end{aligned} \qquad (111)$$

where we always denote $Z(t)(x) = Z(t,x) := \mathscr{Z}^t(0,x)$ (that is at $s = 0$). Indeed we know that if V was smoother, say $V \in L^2(H^1(\Omega))$, we would have:

$$\frac{\partial}{\partial s}[V^s]_{s=0} = Z_t + [Z(t), V(t)] := H_V.Z, \qquad (112)$$

where the Lie bracket is $[Z,V] = DZ.V - DV.Z$, so we would get the previous expression for the derivative of $V^s o T_s(\mathscr{Z}^t)$, as $(V^s o T_s)_s = (V^s)_s + DV^s.DZ(t)$.

5.3 Necessary Condition

5.3.1 Quadratic Term of $\mathscr{E}$

The quadratic term may be decomposed as follows:

$$\begin{aligned}
&\int_0^\tau \int_D ((\alpha \zeta^s + \beta)|V^s|^2 - (\alpha \zeta + \beta)|V|^2)/s \, dx dt \\
&= \int_0^\tau \int_D ((\alpha \zeta + \beta)(|V^s o T_s|^2 - |V|^2)/s \, dx dt \\
&= \int_0^\tau \int_D ((\alpha \zeta + \beta)(V^s o T_s + V)(V^s o T_s - V)/s \, dx dt \\
&\to 2 \int_0^\tau \int_D ((\alpha \zeta + \beta)V.(\frac{\partial}{\partial t}Z(t) + DZ(t).V(t)) \, dx dt \\
&= -2 \left\langle \frac{\partial}{\partial t}((\alpha \zeta + \beta)V) + "D((\alpha \zeta + \beta)V).V", Z \right\rangle_{\mathscr{D}' \times \mathscr{D}} \\
&\quad + \int_D (\alpha \chi_{\Omega_\tau} + \beta)V(\tau).Z(\tau) \, dx - \int_D (\alpha \chi_{\Omega_0} + \beta)V(0).Z(0) \, dx,
\end{aligned} \qquad (113)$$

where

$$"D((\alpha \zeta + \beta)V).V_i" = \partial_j((\alpha \zeta + \beta)V_i V_j) \in W^{-1,1}(D). \qquad (114)$$

In fact we shall consider Z such that $Z(\tau,.) = 0$ over D.

5.3.2 The Linear Term
Let V_0 be any given element in R^N. We have:

$$\int_0^\tau \int_D V_0.(V^s(t,x) - V(t,x))/s\,dxdt$$

$$= \int_0^\tau \int_D V_0.(V^s(t)oT_s(\mathscr{Z}^t)(x) - V(t,x))/s\,dxdt$$

$$+ \int_0^\tau \int_D V_0.(V^s(t)oT_s(\mathscr{Z}^t)(x) - V(t,x))/s\,dxdt \tag{115}$$

$$\rightarrow \int_0^\tau \int_D V_0.(Z_t(t,x) + DZ(t,x).V(t,x))\,dxdt$$

$$= \int_0^\tau \frac{\partial}{\partial t}(\int_D V_0.Z(t,x)\,dx)\,dt = \int_D V_0.Z(\tau,x)\,dx - \int_D V_0.Z(0,x)\,dx.$$

5.3.3 "Perimeter" Term in $\mathscr{E}$
Assume formally that the minimizer element ζ is smooth enough, so that with the choice $E_1 = BV(D, R^N)$ we have the surface tension term in the classical form:

$$\sigma \int_0^\tau \|\nabla\zeta\|_{M^1(D,R^N)}dt = \sigma \int_0^\tau P_D(\Omega_t)dt. \tag{116}$$

We would obtain as derivative with respect to s:

$$\sigma \int_0^\tau \int_{\Gamma_t} \Delta b_{\Omega_t} \langle Z(t), n_t \rangle \, d\Gamma_t\, dt. \tag{117}$$

In the interesting case where $E_1 = W^{\varepsilon,p}(D)$ we introduce the term, for any given "small" $h > 0$:

$$\theta_h(\zeta) := \int_0^\tau (\int_{D \times D} \rho_h(\|x - y\|) \frac{|\zeta(x) - \zeta(y)|^p}{\|x - y\|^{N+\varepsilon p}} dxdy)\,dt, \tag{118}$$

where ρ_h is any smooth positive function such that $\rho(z) = 0$ for $|z| \geq 2h$, and $\rho(z) = 1$ for $|z| \leq h$.

As a result we have

Lemma 3

$$\int_0^\tau \|\zeta(t)\|_{W^{\varepsilon,p}(D)}\,dt \leq \tau(meas(D) + \frac{1}{h^{N+\varepsilon p}}meas(D)^2) + \theta_h(\zeta). \tag{119}$$

So that it is enough to choose the surface tension term in the form $\sigma\,\theta_h(\zeta)$. This term turns to be always differentiable with respect to the transverse perturbations as follows:

$$\theta_h(\zeta oT_s(\mathscr{Z})^{-1})$$

$$= \int_0^\tau \int_{D \times D} \rho_h(\|T_s(\mathscr{Z})(x) - T_s(\mathscr{Z})(y)\|) \frac{|\zeta(x) - \zeta(y)|^p}{\|T_s(\mathscr{Z})(x) - T_s(\mathscr{Z})(y)\|^{N+\varepsilon p}} dxdydt \tag{120}$$

Part II

Regular Papers

Galerkin Strategy for Level Set Shape Analysis: Application to Geodesic Tube

Louis Blanchard[1] and Jean-Paul Zolésio[2]

[1] INRIA, Sophia Antipolis, France
louis.blanchard@sophia.inria.fr
[2] CNRS and INRIA, 2004 route des Lucioles, BP 93,
06902 Sophia Antipolis Cedex, France
jean-paul.zolesio@sophia.inria.fr

Abstract. In this paper, we consider the geodesic tube characterization using a Galerkin-Level Set strategy. The first section is devoted to the analysis of a geodesic tube construction between two sets through the definition of the shape metric. In the second section, we define the Galerkin-Level Set strategy in shape analysis. This new variational formulation associated to a Hilbert space metric for shape identification problem consists in parameterizing the level set function in a finite dimensional subspace spanned by linear independent functions. Consequently, this method is more focused on topological changes than on high accuracy for the boundary evaluation as in a traditional level set formulation. In the third section, we use the Galerkin-Level Set formulation applied to a geodesic tube construction between two sets, through the calculus of the shape derivative of the normal speed. Finally, this geodesic tube construction is validated by a numerical experiment.

1 Tube Formulation Using Moving Domain

In this section, we briefly recall the concept of connecting tube, introduced in [6]. Let us consider $\mathbb{D}$ as a bounded universe in $\mathbb{R}^n$ and two open sets domains $\Omega_0, \Omega_1 \subset \mathbb{D}$. We denote the initial domain by Ω_0 and the final domain by Ω_1, and consider the tube connecting Ω_0 with Ω_1 defined by the $n+1$ dimensional graph of an n-dimensional moving domain: see Fig. 1. Consequently, considering the time interval $I = [0,1]$, we define the tube evolution Q by product space, using the cylinder $I \times \Omega$ as follows:

$$Q = \bigcup_{0 \le t \le 1} \{t\} \times \Omega_t \tag{1}$$

Moreover, we denote by Σ the lateral boundary of the tube, defined by the following expression: $\Sigma = \bigcup_{0 \le t \le 1} \{t\} \times \Gamma_t$, where Γ_t denotes the boundary of Ω_t. The characteristic function of the tube is defined by $\zeta(t,x) \overset{def}{=} \chi_{\Omega_t}(x)$ and verifies $\zeta^2 = \zeta$. Following [4,5], the set of connecting tubes between Ω_0 and Ω_1 is defined by:

$$\mathscr{T}(\Omega_0, \Omega_1) = \left\{ \zeta \in L^\infty(I \times \mathbb{D}) \text{ and piecewise } C^1, \begin{bmatrix} \zeta(0) = \chi_{\Omega_0} \\ \zeta(1) = \chi_{\Omega_1} \end{bmatrix} \right\} \tag{2}$$

A. Korytowski et al. (Eds.): System Modeling and Optimization, IFIP AICT 312, pp. 169–184, 2009.
© IFIP International Federation for Information Processing 2009

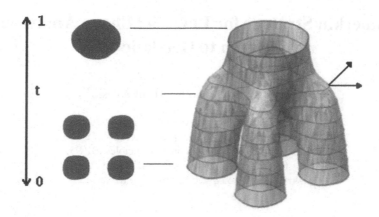

Fig. 1. Continuous tube between Ω_0 and Ω_1

The outgoing unitary normal vector field on the lateral boundary of the tube Σ is defined by

$$v(t,x) = \frac{1}{\sqrt{1+v(t,x)^2}} \begin{pmatrix} -v(t,x) \\ \mathbf{n}(t,x) \end{pmatrix} \tag{3}$$

where $\mathbf{n}(t,x)$ is the normal field to Γ_t and $v(t,x)$ is an intrinsic geometric entity called the normal speed of the boundary Γ_t.

Definition 1. *In order to characterize the minimal tube path between Ω_0 and Ω_1, we introduce the function:*

$$d(\Omega_0, \Omega_1) = \inf_{\zeta \in \mathscr{T}(\Omega_0, \Omega_1)} \int_0^1 \int_{\Gamma_t} |v(t,x)| \, d\Gamma(x) \, dt \tag{4}$$

Lemma 1. *The function $d(\Omega_0, \Omega_1)$ is a metric.*

Proof. We have to prove that the function $d(\Omega_0, \Omega_1)$ satisfies:

I. (Identity of indiscernibles) $d(\Omega_0, \Omega_1) = 0 \Leftrightarrow \Omega_0 = \Omega_1$.
  - If $\Omega_0 = \Omega_1$ then $v = 0$ and $d(\Omega_0, \Omega_1) = 0$.
  - If $d(\Omega_0, \Omega_1) = 0$ that implies $\forall t \in [0,1]$, $v(t,.) = 0$ and the time space normal (3) is $v(t,.) = (0, \mathbf{n}(t,.))$. Then the tube is a cylinder and the domain Ω_t does not depend on time, consequently $\Omega_0 = \Omega_1$.

II. (Symmetry) $d(\Omega_0, \Omega_1) = d(\Omega_1, \Omega_0)$.
  - If we consider the backward tube $\hat{\zeta}(t) = \zeta(1-t) \in \mathscr{T}(\Omega_1, \Omega_0)$, that implies $\hat{v}(t,.) = -v(1-t,.)$, and consequently $d(\Omega_0, \Omega_1) = d(\Omega_1, \Omega_0)$.

III. (Triangle inequality) $d(\Omega_0, \Omega_2) \leq d(\Omega_0, \Omega_1) + d(\Omega_1, \Omega_2)$.
  - We consider three open sets domains in $\mathbb{D}$: $\Omega_i \,\forall i \in [0,2]$. We denote by $\zeta_1 \in \mathscr{T}(\Omega_0, \Omega_1)$ the tube connecting Ω_0 to Ω_1, and by $\zeta_2 \in \mathscr{T}(\Omega_1, \Omega_2)$ the tube

connecting Ω_1 to Ω_2. Let us consider the piecewise C^1 tube defined, through its characteristic function $\hat{\zeta}$ as follows:

$$\hat{\zeta}(t,x) = \begin{cases} \zeta_1(2t,x) & \text{if } 0 \leq t \leq \frac{1}{2} \\ \zeta_2(2t-1,x) & \text{if } \frac{1}{2} \leq t \leq 1 \end{cases} \tag{5}$$

Consequently, the normal speed on the boundary Γ_t is given by:

$$\hat{v}(t,x) = \begin{cases} 2v_1(2t,x) & \text{if } 0 \leq t \leq \frac{1}{2} \\ 2v_2(2t-1,x) & \text{if } \frac{1}{2} \leq t \leq 1 \end{cases} \tag{6}$$

Now by construction $\hat{\zeta} \in \mathscr{T}(\Omega_0,\Omega_2)$ is a tube connecting Ω_0 to Ω_2 and we get:

$$\begin{aligned}
d(\Omega_0,\Omega_2) &\leq \int_0^{\frac{1}{2}} \int_{\Gamma_t} |\hat{v}(t,x)| \, d\Gamma(x) \, dt + \int_{\frac{1}{2}}^1 \int_{\Gamma_t} |\hat{v}(t,x)| \, d\Gamma(x) \, dt \\
&\leq \int_0^{\frac{1}{2}} \int_{\Gamma_t} |2v_1(2t,x)| \, d\Gamma(x) \, dt + \int_{\frac{1}{2}}^1 \int_{\Gamma_t} |2v_2(2t-1,x)| \, d\Gamma(x) \, dt \\
&\leq \int_0^1 \int_{\Gamma_t} |v_1(r,x)| \, d\Gamma(x) \, dr + \int_0^1 \int_{\Gamma_t} |v_2(u,x)| \, d\Gamma(x) \, du
\end{aligned} \tag{7}$$

and as v_1 (resp. v_2) is the infimum in the definition of $d(\Omega_0,\Omega_1)$ (resp. $d(\Omega_1,\Omega_2)$) up to $\varepsilon > 0$, then $\forall \varepsilon \in \mathbb{R}_+^*$ we get:

$$d(\Omega_0,\Omega_2) \leq d(\Omega_0,\Omega_1) + d(\Omega_1,\Omega_2) + 2\varepsilon \tag{8}$$

$\square$

1.1 Tube Formulation Using a Level Set Method

In this paper, we use a level set parameterization for the domain evolution. In this method the moving domain Ω_t is defined by the set of points in $\mathbb{D}$ for which the level set function Φ is positive:

$$\Omega_t = \left\{ x \in \mathbb{D} \mid \Phi(t,x) > 0 \right\} \tag{9}$$

We denote by Φ_0 the level set function of the domain Ω_0, and by Φ_1 the level set function of the domain Ω_1:

$$\Omega_0 = \left\{ x \in \mathbb{D} \mid \Phi_0(x) > 0 \right\}, \ \Omega_1 = \left\{ x \in \mathbb{D} \mid \Phi_1(x) > 0 \right\} \tag{10}$$

Using the level set formulation, the set of connecting tubes between the initial domain Ω_0 and the final domain Ω_1 becomes:

$$\mathscr{T}_{LS}(\Omega_0,\Omega_1) = \left\{ \begin{array}{l} \Phi(t,x) \in L^1(I,C^0(\bar{\mathbb{D}})) \\ \chi_{\Omega_t} \in C^0(\bar{I},L^1(\mathbb{D})) \\ \Phi \text{ piecewise } C^0 \end{array}, \begin{bmatrix} \Phi(0,x) = \Phi_0(x) \\ \Phi(1,x) = \Phi_1(x) \end{bmatrix} \right\} \tag{11}$$

We consider a decomposition of the time interval I into a finite number of time intervals in which the level set function Φ is continuous. Therefore Φ is piecewise C^0, which means that there exists an integer N and an increasing sequence: $(t_0 = 0 < t_1 < \cdots < t_N = 1)$ with a decomposition of the time interval as follows: $I = \bigcup_{1 \leq k \leq N} \bar{I}_k$ where $I_k =]t_k, t_{k+1}[$, such that:

$$\forall k \in [1, N], \ \Phi(t, .)\big|_{I_k} \in C^0(I_k) \tag{12}$$

Definition 2. *The metric d defined by the equation (4) can be expressed, in term of the level set function Φ as follows:*

$$d(\Omega_0, \Omega_1) = \inf_{\Phi \in \mathcal{S}_{LS}(\Omega_0, \Omega_1)} \int_0^1 \int_{\Gamma_t = \Phi^{-1}(0)} \frac{|\partial_t \Phi(t, x)|}{\|\nabla \Phi(t, x)\|} \, d\Gamma(x) \, dt \tag{13}$$

Indeed, using the level set formulation we have the relations:

$$\mathbf{n}(t, x) = \frac{-\nabla \Phi(t, x)}{\|\nabla \Phi(t, x)\|}, \quad \mathbf{V}(t, x) = -\partial_t \Phi(t, x) \frac{\nabla \Phi(t, x)}{\|\nabla \Phi(t, x)\|} \tag{14}$$

Then the normal speed of the boundary Γ_t turns into:

$$v(t, x) = \langle \mathbf{V}(t, x), \mathbf{n}(t, x) \rangle_{\mathbb{R}^n} = \frac{\partial_t \Phi(t, x)}{\|\nabla \Phi(t, x)\|} \tag{15}$$

where $\langle ., . \rangle_{\mathbb{R}^n}$ denotes the inner product in $\mathbb{R}^n$.

Assumption 1. *The function $d(\Omega_0, \Omega_1)$ expressed in term of the level set function Φ, is also a metric.*

1.2   Tube Formulation Using the Federer Theorem

In this section, we consider the tube formulation through the level set method described previously, and we consider an approximation of the metric d using the Federer measure decomposition theorem.

Theorem 1   (Federer measure decomposition). *Let us consider a functional $F \in L^1(\mathbb{D})$, and $\forall h > 0$ the domain*

$$U_h(\Gamma_t) = \left\{ x \in \mathbb{D} \mid \|\Phi(t, x)\| < h \right\} \tag{16}$$

Then we have:

$$\int_{U_h(\Gamma)} F(x) \, dx = \int_{-h}^{+h} \left(\int_{\Phi^{-1}(z)} \frac{F(x)}{\|\nabla_x \Phi(x)\|} \, d\Gamma(x) \right) dz \tag{17}$$

Corollary 1. *Assuming the mapping:*

$$z \in [-h, +h] \rightarrow \int_{\Phi^{-1}(z)} \frac{F(x)}{\|\nabla_x \Phi(x)\|} \, d\Gamma \tag{18}$$

to be continuous, we obtain:

$$\int_\Gamma \frac{F(x)}{\|\nabla_x \Phi(x)\|} \, d\Gamma(x) = \frac{1}{2h} \int_{U_h(\Gamma)} F(x) \, dx + o(1), \ h \to 0 \tag{19}$$

Definition 3. *Using the Federer measure decomposition theorem and according to the previous corollary, we consider an approximation of the metric d denoted by d_h and defined as follows:*

$$d_h(\Omega_0, \Omega_1) = \inf_{\Phi \in \mathscr{T}_{LS}(\Omega_0, \Omega_1)} \int_0^1 \frac{1}{2h} \int_{U_h(\Gamma_t)} |\partial_t \Phi(t, x)| \, dx \, dt \qquad (20)$$

Lemma 2. *The approximation of the metric $d(\Omega_0, \Omega_1)$, denoted $d_h(\Omega_0, \Omega_1)$ is also a metric.*

Proof. We have to prove that the function $d_h(\Omega_0, \Omega_1)$ satisfies:

I. (Identity of indiscernibles) $d_h(\Omega_0, \Omega_1) = 0 \Leftrightarrow \Omega_0 = \Omega_1$.
   - If $\Omega_0 = \Omega_1$ then $\partial_t \Phi = 0$ and $d_h(\Omega_0, \Omega_1) = 0$.
   - If $d_h(\Omega_0, \Omega_1) = 0$ that implies $\forall t \in [0, 1]$, $\partial_t \Phi(t, .) = 0$ in $\mathscr{D}_h = \bigcup_{0 \le t \le 1} \{t\} \times U_h(\Gamma_t)$ and that implies $\Phi = \Phi(x) \in \mathscr{D}_h$. Consequently, the boundary $\Gamma_t \stackrel{def}{=} \left\{ x \in \mathbb{D} \mid \Phi(x) = 0 \right\}$ does not depend on time, and $\Omega_0 = \Omega_1$.

II. (Symmetry) $d_h(\Omega_0, \Omega_1) = d_h(\Omega_1, \Omega_0)$.
   - If we consider the backward tube $\hat{\Phi}(t, .) = \Phi(1 - t, .) \in \mathscr{T}(\Omega_1, \Omega_0)$, that implies $\partial_t \hat{\Phi}(t, .) = -\partial_t \Phi(1 - t, .)$, and $d_h(\Omega_0, \Omega_1) = d_h(\Omega_1, \Omega_0)$.

III. (Triangle inequality) $d_h(\Omega_0, \Omega_2) \le d_h(\Omega_0, \Omega_1) + d_h(\Omega_1, \Omega_2)$.
   - We assume three open sets domains in $\mathbb{D}$: $\Omega_i \; \forall i \in [0, 2]$. We denote by $\Phi_1 \in \mathscr{T}_{LS}(\Omega_0, \Omega_1)$ the tube connecting Ω_0 to Ω_1, and by $\Phi_2 \in \mathscr{T}_{LS}(\Omega_1, \Omega_2)$ the tube connecting Ω_1 to Ω_2. Let us consider the piecewise C^1 tube defined, through its level set function Φ as follows:

$$\bar{\Phi}(t, x) = \begin{cases} \Phi_1(2t, x) & \text{if } 0 \le t \le \frac{1}{2} \\ \Phi_2(2t - 1, x) & \text{if } \frac{1}{2} \le t \le 1 \end{cases} \qquad (21)$$

Consequently, the time derivative of level set function Φ on the domain $U_h(\Gamma_t)$ is given by:

$$\partial_t \bar{\Phi}(t, x)(t, x) = \begin{cases} 2 \partial_t \Phi_1(2t, x) & \text{if } 0 \le t \le \frac{1}{2} \\ 2 \partial_t \Phi_2(2t - 1, x) & \text{if } \frac{1}{2} \le t \le 1 \end{cases} \qquad (22)$$

Now by construction $\bar{\Phi} \in \mathscr{T}_{LS}(\Omega_0, \Omega_2)$ is a tube connecting Ω_0 to Ω_2 and we get:

$$d_h(\Omega_0, \Omega_2) \le \int_0^{\frac{1}{2}} \frac{1}{2h} \int_{U_h(\Gamma_t)} |\partial_t \bar{\Phi}(t, x)| \, dx \, dt + \int_{\frac{1}{2}}^1 \frac{1}{2h} \int_{U_h(\Gamma_t)} |\partial_t \bar{\Phi}(t, x)| \, dx \, dt$$

$$\le \frac{1}{2h} \left[\int_0^{\frac{1}{2}} \int_{U_h(\Gamma_t)} |2 \partial_t \Phi_1(2t, x)| \, dx \, dt + \int_{\frac{1}{2}}^1 \int_{U_h(\Gamma_t)} |2 \partial_t \Phi_2(2t - 1, x)| \, dx \, dt \right]$$

$$\le \int_0^1 \frac{1}{2h} \int_{U_h(\Gamma_t)} |\partial_t \Phi_1(r, x)| \, dx \, dr + \int_0^1 \frac{1}{2h} \int_{U_h(\Gamma_t)} |\partial_t \Phi_2(u, x)| \, dx \, du$$

$$\tag{23}$$

and as Φ_1 (resp. Φ_2) is the infimum in the definition of $d_h(\Omega_0, \Omega_1)$ (resp. $d_h(\Omega_1, \Omega_2)$) up to $\varepsilon > 0$. Then $\forall \varepsilon \in \mathbb{R}_+^*$, we get

$$d_h(\Omega_0, \Omega_2) \leq d_h(\Omega_0, \Omega_1) + d_h(\Omega_1, \Omega_2) + 2\varepsilon \qquad (24)$$

□

2   Shape Identification Problem

We address the question concerning the shape identification of a given smooth domain. A commonly used approach in shape analysis consists in choosing a level set formulation for the evolution of moving domain. The main advantage of a level set formulation concerns the easy generation of topological changes during the evolution process.

2.1   Shape Identification Using a Level Set Method

Let us denote by $\Omega_* \in \mathbb{D}$ a smooth domain to identify and by χ_{Ω_*} its characteristic function satisfying: $\chi_{\Omega_*} \in H^s(\mathbb{D})$, $0 < s < \frac{1}{2}$. Following [1], the evaluation of the distance between the given domain Ω_* and the moving domain Ω_t is made by the use of a metric associated to the Hilbert space H^s denoted $\delta_s(\Omega, \Omega_*)$ and defined by:

$$\forall s \in]0, \frac{1}{2}[, \quad \delta_s(\Omega, \Omega_*) = \| \chi_\Omega - \chi_{\Omega_*} \|_{H^s(\mathbb{D})}$$
$$= \| \chi_\Omega - \chi_{\Omega_*} \|_{L^2(\mathbb{D})} + \| \chi_\Omega - \chi_{\Omega_*} \|_s \qquad (25)$$

where

$$\| \chi_\Omega \|_s^2 = \int_\mathbb{D} \int_\mathbb{D} |\chi_\Omega(x) - \chi_\Omega(y)|^2 G(x,y) \, dx \, dy \qquad (26)$$

and where the kernel function defined by: $G(x,y) = |x-y|^{-(n+2s)}$ is singular on the diagonal $\Delta = \{ (x,x) \subset \mathbb{D} \times \mathbb{D}, x \in \mathbb{D} \}$.

2.1.1   Shape Analysis via the Speed Method

Finally, we use the concept of speed method from shape analysis [3] to compute the shape derivative of the metric $\delta_s(\Omega, \Omega_*)$ which corresponds to a gradient direction for the underling shape optimization problem:

$$\min_{\Omega \in \mathbb{D}} \delta_s(\Omega, \Omega_*) \qquad (27)$$

Definition 4. *Let us consider an open set domain Ω where $\Gamma = \partial \Omega$ is of class C^1. We define the eulerian derivative of the functional J in the direction of a perturbation vector field $\mathbf{W} \in C_0^1(\mathbb{D}; \mathbb{D})$ by*

$$dJ(\Gamma_t, \mathbf{W}) = \frac{\partial J(\Gamma(t+\varepsilon))}{\partial \varepsilon} \Big|_{\varepsilon=0} \qquad (28)$$

Lemma 3. *The funtional $\delta_s(\Omega, \Omega_*)$ is shape derivative for perturbation vector fields $\mathbf{V} \in C_0^1(\mathbb{D}, \mathbb{D})$, and expressed as follows:*

$$d\,\delta_s\big((\Omega,\Omega_*);\mathbf{V}\big) = \int_\Gamma F(x)\,\langle\mathbf{V}(0,x),\mathbf{n}(x)\rangle_{\mathbb{R}^n}\,d\Gamma(x) \tag{29}$$

where $d\Gamma$ is the arclength measure on Γ and where:

$$F(x) = \frac{1-2\chi_{\Omega_*}(x)}{2\,\|\chi_\Omega - \chi_{\Omega_*}\|_{L^2(\mathbb{D})}} + \frac{\int_{\mathbb{D}}\big[1-2\chi_{\Omega_t}(y) + 2\,[\chi_{\Omega_*}(y)-\chi_{\Omega_*}(x)]\big]\,G(x,y)\,dy}{\|\chi_\Omega - \chi_{\Omega_*}\|_s} \tag{30}$$

Proof. See [1].   □

2.2   Shape Identification Using a Galerkin-Level Set Strategy

Generally, the parameterization of the level set function Φ is done by the *oriented distance function* denoted b_{Ω_t}, see [2,3] for references:

$$\Phi(t,x) = -b_{\Omega_t}(x) \tag{31}$$

where $b_\Omega(x)$ is also called *signed distance function* and is defined as follows:

$$b_\Omega(x) = d_\Omega(x) - d_{\complement\Omega}(x) \quad \text{with} \quad d_A(x) = \inf_{y\in A}|y-x| \tag{32}$$

The choice of the *oriented distance function* for the parameterization of the level set function can be necessary for having a high accuracy of the boundary approximation. However, the choice of the *oriented distance function* implies an expansive computational cost owing to the complexity of its evaluation and imposes a reinitialization during the evolution process. Consequently, according to the fact that in this paper we focus on topological changes without considering the approximation of the boundary as an essential point, we use a new approach called Galerkin-Level Set method.

2.2.1   Galerkin-Level Set Strategy

The Galerkin-Level Set strategy consists in parameterizing the level set function in a finite dimensional subspace $\mathcal{E}$, spanned by linear independent functions defined over $\mathbb{D}$: $\mathcal{E} = \{E_1,\dots,E_m\}$. We denote by $\Lambda(t) = \big(\lambda_1(t),\dots,\lambda_m(t)\big)$ the parameter vector of the Galerkin decomposition of Φ in the basis $\mathcal{E}$:

$$\Phi(t,x) = \sum_{k=1}^{m} \lambda_k(t)\,E_k(x) \tag{33}$$

Consequently, using the Galerkin decomposition of the level set function, the parameterization of the moving domain $\Omega(t)$ is defined as follows:

$$\Omega(t) = \big\{x \in \mathbb{D} \mid \Phi(t,x) = \sum_{k=1}^{m} \lambda_k(t)\,E_k(x) > 0\big\} \tag{34}$$

2.2.2   Level Set Equation

In a level set formulation, the moving domain evolves by advecting the level set function Φ following the flow of the shape gradient. Then, in a traditional level set formulation, the transport equation is a Partial Differential Equation (PDE) of Hamilton-Jacobi type:

$$\begin{cases} \partial_t \Phi(t,x) + \rho F(x) \|\nabla \Phi(t,x)\| = 0, & \rho > 0 \\ \Phi(0,x) = \Phi_0(x), & (t,x) \in [0,\tau] \times \Omega_t \end{cases} \tag{35}$$

Remark 1. *The main advantage of the Galerkin-Level Set method compared to the traditional level set formulation concerns the level set equation that turns, in the Galerkin-Level Set method, into a system of ordinary differential equations.*

Lemma 4. *Using the Galerkin-Level Set strategy (34), the level set equation turns into a system of m ordinary differential equations:*

$$\begin{cases} \partial_t \Lambda(t) + \rho \mathscr{F}(t,x) = 0, & \rho > 0 \\ \Lambda(0) = \Lambda_0, & (t,x) \in [0,\tau] \times \Omega_t \end{cases} \tag{36}$$

where

$$\mathscr{F}(t,x) = \left(\int_{\Gamma_t} \frac{F(x)}{\|\nabla \Phi(t,x)\|} E_1(x) d\Gamma(x), \ldots, \int_{\Gamma_t} \frac{F(x)}{\|\nabla \Phi(t,x)\|} E_m(x) d\Gamma(x) \right) \tag{37}$$

Proof. According to the Galerkin-Level Set strategy, consisting in the decomposition of function Φ (33), the shape derivative of the functional $\delta_s(\Omega, \Omega_*)$ with respect to the vector of parameters $\Lambda(t)$ turns into:

$$d\, \delta_s\big((\Omega, \Omega_*); \mathbf{V}\big) = \sum_{k=1}^m \partial_t \lambda_k(t) \int_{\Gamma_t} F(x) \frac{E_k(x)}{\|\nabla \Phi(t,x)\|} d\Gamma(x) \tag{38}$$

where only the vector of parameters $\Lambda(t)$ depends on time. A sufficient condition to decrease the shape gradient is to choose:

$$\forall k \in [1,m], \ \forall \rho \in \mathbb{R}_+^*, \quad \partial_t \lambda_k(t) = -\rho \int_{\Gamma_t} F(x) \frac{E_k(x)}{\|\nabla \Phi(t,x)\|} d\Gamma(x) \tag{39}$$

Finally, considering the level set equation we obtain a system of m ordinary differential equations (36). $\qquad\square$

Corollary 2. *Substituting the approximation of the boundary integral calculus from the equation (19), into the system of m ordinary differential equations (37), we obtain an approximation of the vector $\mathscr{F}$ defined as follows:*

$$\tilde{\mathscr{F}}(t,x) = \left(\frac{1}{2h} \int_{U_h(\Gamma_t)} F(x) E_1(x)\, dx, \ldots, \frac{1}{2h} \int_{U_h(\Gamma_t)} F(x) E_m(x)\, dx \right) \tag{40}$$

Note that in this new formulation the main advantage is that the denominator term $\|\nabla \Phi(x)\|$ has been eliminated.

From now, we use the previous corollary for the level set equation and we consider the following algorithm.

Algorithm 1

I. **Initialization:** Choose an initial vector of parameters $\Lambda_0 = (\lambda_1^0, \ldots, \lambda_m^0)$. Initialize the level set function $\Phi_0(x) = \sum_{l=1}^m \lambda_l^0 E_l(x)$. Set $k = 0$.

II. **Shape gradient direction:** Find the tubular neighborhood $U_h(\Gamma_{t_k})$ of the zero level set Γ_{t_k} of the actual level set function $\Phi(t_k, x)$. Compute $\tilde{\mathscr{F}}(t_k, x)$ from the equation (40).

III. **Update:** Perform a time step in the level set equation (36) to update $\Lambda(t_k)$. Let $\Lambda(t_{k+1})$ denote this update: $\Lambda(t_{k+1}) = \Lambda(t_k) - \rho\,\tilde{\mathscr{F}}(t_k, x)$, $\rho > 0$. Update the function $\Phi(t_{k+1}, x) = \sum_{l=1}^m \lambda_l(t_{k+1}) E_l(x)$. Set $k = k + 1$ and go to 2.

2.3 Numerical Experiment

We present a numerical experiment based on the algorithm 1 for a 3D shape identification problem using the Galerkin-Level Set method described in the previous section. In this numerical experiment, the given domain Ω_* to identify is the gray matter of a human brain. We consider a Galerkin-Level Set expansion of the level set function Φ in Fourier series of dimension $m = 25^3$; note that in this 3D case the level set function Φ is in $\mathbb{R}^4$. We start with a smooth initial domain $\Omega_{t=0}$ corresponding to the lower frequency of the Fourier series: see left-hand picture in Fig. 2. The algorithm detects the contour of the human brain after only 8 iterations.

3 Geodesic Tube Formulation Using Moving Domain

3.1 Tube Formulation Using a Galerkin Strategy

The tube path between Ω_0 and Ω_1 is made by a Galerkin-Level Set approach. The moving domain Ω_t of the tube evolution defined by the equation (1) is parameterized by the Galerkin-Level Set formulation and defined as follows:

$$\Omega(t) = \left\{ x \in \mathbb{D} \mid \Phi(t, x) = \sum_{k=1}^m \lambda_k(t) E_k(x) > 0 \right\} \tag{41}$$

where $\Lambda(t) = (\lambda_1(t), \ldots, \lambda_m(t)) \in \mathbb{R}^m$ is the vector of parameters in the Galerkin expansion of the level set function. The first step consists in identifying the initial domain Ω_0 and the final domain Ω_1 through the research of the parameters $\Lambda_0 = (\lambda_1^0, \ldots, \lambda_m^0) \in \mathbb{R}^m$ and $\Lambda_1 = (\lambda_1^1, \ldots, \lambda_m^1) \in \mathbb{R}^m$ which satisfy the equations:

$$\Phi_0(x) = \sum_{k=1}^m \lambda_k^0 E_k(x), \quad \Phi_1(x) = \sum_{k=1}^m \lambda_k^1 E_k(x) \tag{42}$$

Thus, the feasible set of connecting tubes between Ω_0 and Ω_1 through the Galerkin-Level Set formulation turns into:

$$\mathscr{T}_\Lambda(\Omega_0, \Omega_1) = \left\{ \Lambda(t) \in \left(L^2(I)\right)^m, \begin{bmatrix} \Lambda(0) = \Lambda_0 \\ \Lambda(1) = \Lambda_1 \end{bmatrix} \right\} \tag{43}$$

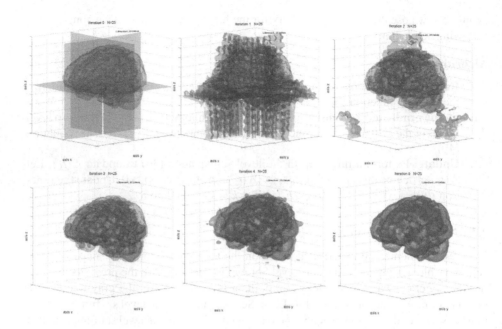

Fig. 2. Shape identification of gray matter of human brain using a Galerkin-Level Set method

Remark 2. *The feasible set of connecting tubes between Ω_0 and Ω_1 is not empty. Indeed, if we consider the vector of parameters $\Lambda(t)$ as a convex combination of Λ_0 and Λ_1: $\Lambda(t) = \Lambda_1 t + \Lambda_0 (1-t)$, we have $\Lambda(t) \in \mathscr{T}_\Lambda(\Omega_0, \Omega_1)$. Moreover, the parameters $\Lambda(t)$ defined as a convex combination of Λ_0 and Λ_1 generate an admissible tube that we use for the initialization during the tube optimization process.*

3.2 Geodesic Tube Construction between Two Domains

We focus on the construction of an optimal tube connecting the initial domain Ω_0 to the final domain Ω_1, this optimal tube is also called a geodesic tube. The question is to determine, through the use of shape metrics $d(\Omega_0, \Omega_1)$ and $d_h(\Omega_0, \Omega_1)$, which tube is an optimal tube among all those tubes in the admissible set (see Fig. 3).

Let us consider the metrics d and d_h defined by (4) and (20) that we can rewrite as follows:

$$
\begin{aligned}
d(\Omega_0, \Omega_1) &= \inf_{\Phi \in \mathscr{T}_{LS}(\Omega_0, \Omega_1)} \int_0^1 J(\Gamma_t)\, dt \\
d_h(\Omega_0, \Omega_1) &= \inf_{\Phi \in \mathscr{T}_{LS}(\Omega_0, \Omega_1)} \int_0^1 J_h(\Gamma_t)\, dt
\end{aligned}
\tag{44}
$$

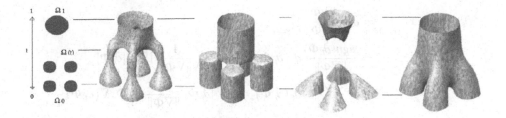

Fig. 3. Different continuous tubes between Ω_0 and Ω_1

where the functionals $J\left(\Gamma_t\right)$ and $J_h\left(\Gamma_t\right)$ are defined by:

$$J(\Gamma_t) = \int_{\Gamma_t} |v(t,x)|\, d\Gamma(x) = \int_{\Gamma_t} \frac{|\partial_t \Phi(t,x)|}{\|\nabla\Phi(t,x)\|}\, d\Gamma(x)$$

$$J_h(\Gamma_t) = \frac{1}{2h}\int_{U_h(\Gamma_t)} |\partial_t \Phi(t,x)|\, dx \tag{45}$$

Then, in order to solve the problem concerning the geodesic tube, that is to say to compute the metrics d or d_h defined by (44), we use a gradient method based on the computation of the shape derivative.

Lemma 5. *According to (28), the eulerian derivative of the functional J in the direction of a perturbation vector field* $\mathbf{W} \in C_0^1(\mathbb{D};\mathbb{D})$ *is:*

$$dJ(\Gamma_t,\mathbf{W}) = \int_{\Gamma_t} \partial_\varepsilon |v(t+\varepsilon,x)|\Big|_{\varepsilon=0}\, d\Gamma(x)$$

$$+ \int_{\Gamma_t} \left[\frac{\partial |v(t,x)|}{\partial n} + H(t,x)\,|v(t,x)|\right] \langle \mathbf{W}(t,x), \mathbf{n}(t,x) \rangle_{\mathbb{R}^n}\, d\Gamma(x) \tag{46}$$

where H is the mean curvature. Using the level set formulation the eulerian derivative of the functional J turns into:

$$dJ(\Gamma_t,\mathbf{W}) = \int_{\Gamma_t} \left[\frac{sign(\partial_t \Phi)}{\|\nabla\Phi\|}\partial_\varepsilon(\partial_t \Phi)\Big|_{\varepsilon=0} - |\partial_t \Phi|\frac{1}{\|\nabla\Phi\|^3}\nabla\Phi.\nabla(\partial_\varepsilon \Phi)\Big|_{\varepsilon=0} \right.$$

$$+ \left(-sign(\partial_t \Phi)\frac{\nabla\Phi}{\|\nabla\Phi\|}\cdot\frac{\nabla(\partial_t \Phi)}{\|\nabla\Phi\|} + 2|\partial_t \Phi|\frac{\nabla\Phi}{\|\nabla\Phi\|}\cdot\left[\frac{D^2\Phi}{\|\nabla\Phi\|^2}\cdot\frac{\nabla\Phi}{\|\nabla\Phi\|}\right] - \right.$$

$$\left. \left. - \frac{|\partial_t \Phi|}{\|\nabla\Phi\|^2}\Delta\Phi \right)\frac{\partial_\varepsilon \Phi}{\|\nabla\Phi\|} \right] d\Gamma(x) \tag{47}$$

Proof. According to the equation (46), the eulerian derivative of the functional J_{ls} in the direction of a perturbation vector field $\mathbf{W}$ turns into:

$$\partial_\varepsilon |v(t+\varepsilon,x)|\Big|_{\varepsilon=0} = \partial_\varepsilon \left(\frac{|\partial_t \Phi|}{\|\nabla \Phi\|}\right)\Big|_{\varepsilon=0}$$

$$= \frac{sign(\partial_t \Phi)}{\|\nabla \Phi\|} \partial_\varepsilon (\partial_t \Phi)\Big|_{\varepsilon=0} + \partial_\varepsilon \left(\frac{1}{\|\nabla \Phi\|}\right)\Big|_{\varepsilon=0} |\partial_t \Phi|$$

$$= \frac{sign(\partial_t \Phi)}{\|\nabla \Phi\|} \partial_\varepsilon (\partial_t \Phi)\Big|_{\varepsilon=0} - |\partial_t \Phi| \frac{1}{\|\nabla \Phi\|^3} \nabla \Phi . \nabla (\partial_\varepsilon \Phi)\Big|_{\varepsilon=0}$$

$$(48)$$

and

$$\frac{\partial |v(t,x)|}{\partial n} + H |v(t,x)| = \frac{-\nabla \Phi}{\|\nabla \Phi\|} . \nabla \left(\frac{|\partial_t \Phi|}{\|\nabla \Phi\|}\right) + \nabla . \left(\frac{-\nabla \Phi}{\|\nabla \Phi\|}\right) \frac{|\partial_t \Phi|}{\|\nabla \Phi\|} \qquad (49)$$

$$\frac{-\nabla \Phi}{\|\nabla \Phi\|} . \nabla \left(\frac{|\partial_t \Phi|}{\|\nabla \Phi\|}\right) + \left[-\nabla \left(\frac{1}{\|\nabla \Phi\|}\right) . \nabla \Phi - \frac{\Delta \Phi}{\|\nabla \Phi\|}\right] \frac{|\partial_t \Phi|}{\|\nabla \Phi\|}$$

$$= \frac{-\nabla \Phi}{\|\nabla \Phi\|} . \frac{\nabla (|\partial_t \Phi|)}{\|\nabla \Phi\|} - 2 |\partial_t \Phi| \frac{\nabla \Phi}{\|\nabla \Phi\|} . \nabla \left(\frac{1}{\|\nabla \Phi\|}\right) - \frac{|\partial_t \Phi|}{\|\nabla \Phi\|^2} \Delta \Phi$$

$$(50)$$

$$= -sign(\partial_t \Phi) \frac{\nabla \Phi}{\|\nabla \Phi\|} . \frac{\nabla (\partial_t \Phi)}{\|\nabla \Phi\|} +$$

$$+ 2 |\partial_t \Phi| \frac{\nabla \Phi}{\|\nabla \Phi\|} . \left[\frac{D^2 \Phi}{\|\nabla \Phi\|^2} . \frac{\nabla \Phi}{\|\nabla \Phi\|}\right] - \frac{|\partial_t \Phi|}{\|\nabla \Phi\|^2} \Delta \Phi$$

$$\square$$

Lemma 6. *According to (28), the eulerian derivative of the functional J_h in the direction of a perturbation vector field* $\mathbf{W} \in C_0^1(\mathbb{D}; \mathbb{D})$ *is:*

$$dJ_h(\Gamma_t, \mathbf{W}) = \frac{1}{2h} \int_{\mathbb{D}} \left[\partial_\varepsilon (|\partial_t \Phi(t+\varepsilon,x)|)\Big|_{\varepsilon=0} \rho_h \circ b_{\Omega_t}(x)\right] dx$$

$$+ \frac{1}{2h} \int_{\mathbb{D}} \left[|\partial_t \Phi(t,x)| \partial_\varepsilon \left(\rho_h \circ b_{\Omega_{t+\varepsilon}}(x)\right)\Big|_{\varepsilon=0}\right] dx \qquad (51)$$

where the function ρ_h is defined by: $\rho_h(x) = \begin{cases} \frac{x}{h}+1 & if x \in [-h,0] \\ \frac{-x}{h}+1 & if x \in [0,h] \\ 0 & if x \in \mathbb{R} \setminus [-h,h] \end{cases}$.

Proof. Due to the fact that $\rho_h \circ b_{\Omega_t}(x)\Big|_{\Gamma_t} = 1$, and using the fact that $supp(\rho_h \circ b_{\Omega_t}) \subseteq U_h(\Gamma_t)$ we can rewrite the functional J_h as follows

$$J_h(\Gamma_t) = \frac{1}{2h} \int_{\mathbb{D}} |\partial_t \Phi(t,x)| \rho_h \circ b_{\Omega_t}(x) dx \qquad (52)$$

Consequently, the eulerian derivative of the functional J_h turns into the equation (51) where:

$$\partial_\varepsilon (|\partial_t \Phi(t+\varepsilon,x)|)\Big|_{\varepsilon=0} = sign(\partial_t \Phi) \partial_\varepsilon (\partial_t \Phi(t,x))\Big|_{\varepsilon=0}. \qquad (53)$$

Using $\partial_\varepsilon\left(b_{\Omega_t}(x)\right) + \nabla b_{\Omega_t}(x).\mathbf{W}\circ p = 0$ and $\nabla b_{\Omega_t}(x) = \mathbf{n}(t,x)$, we get

$$
\begin{aligned}
\partial_\varepsilon\left(\rho_h\circ b_{\Omega_t}(x)\right)\Big|_{\varepsilon=0} &= \rho_h'\circ b_{\Omega_t}(x)\,\partial_\varepsilon\left(b_{\Omega_t}(x)\right) \\
&= -\rho_h'\circ b_{\Omega_t}(x)\,\langle\mathbf{W},\mathbf{n}\rangle_{\mathbb{R}^n} \\
&= -\rho_h'\circ b_{\Omega_t}(x)\,\frac{\partial_\varepsilon\Phi}{\|\nabla\Phi\|}
\end{aligned}
\tag{54}
$$

The derivative of the function $\rho_h(x)$ is defined by:

$$
\rho_h'(x) = \begin{cases} \frac{1}{h} & \text{if } x \in [-h,0] \\ \frac{-1}{h} & \text{if } x \in [0,h] \\ 0 & \text{if } x \in \mathbb{R}\setminus[-h,h] \end{cases} = \begin{cases} \frac{1}{h}\left(1 - 2\chi_{\Omega_t}(x)\right) & \text{if } x \in U_h(\Gamma_t) \\ 0 & \text{if } x \in \mathbb{R}\setminus[-h,h] \end{cases}
$$

Finally, we get for the eulerian derivative of the functional J_h in the direction of a perturbation vector field $\mathbf{W} \in C_0^1(\mathbb{D};\mathbb{D})$:

$$
\begin{aligned}
dJ_h(\Gamma_t,\mathbf{W}) =\ & \frac{1}{2h}\int_{\mathbb{D}}\left[\text{sign}(\partial_t\Phi)\,\partial_\varepsilon\left(\partial_t\Phi(t,x)\right)\Big|_{\varepsilon=0}\rho_h\circ b_{\Omega_t}(x)\right]dx \\
& - \frac{1}{2h}\int_{\mathbb{D}}\left[|\partial_t\Phi(t,x)|\rho_h'\circ b_{\Omega_t}(x)\,\frac{\partial_\varepsilon\Phi}{\|\nabla\Phi\|}\right]dx
\end{aligned}
\tag{55}
$$

$\square$

3.2.1   Polynomial Decomposition of the Parameter $\Lambda(t)$

We continue the study of a geodesic tube through a tube formulation using a Galerkin-Level set strategy. Consequently, $\Lambda(t) \in \mathscr{T}_\Lambda(\Omega_0,\Omega_1)$ represents the parameters of the optimization process. For complexity reason, we consider a polynomial decomposition of the parameter $\Lambda(t)$ as follows:

$$
\Lambda(t) = P_\alpha(t)\Lambda_1 + \left(1 - P_\alpha(t)\right)\Lambda_0\,, \quad P_\alpha(t) = \sum_{i=1}^M \alpha_i\,e_i(t)
\tag{56}
$$

where $\alpha = (\alpha_1,\ldots,\alpha_M)$ are the coefficients of the decomposition of the polynomial $P_\alpha(t)$ in the basis $\{e_1(t),\ldots,e_M(t)\}$. Consequently, the feasible set of connecting tubes defined by (43) with initial and final conditions on $\Lambda(t)$ turns into a feasible set with initial and final conditions on the polynomial P_α defined as follows:

$$
\mathscr{T}_\alpha(\Omega_0,\Omega_1) = \left\{\alpha \in \mathbb{R}^M\,,\ \begin{bmatrix} P_\alpha(0) = 0 \\ P_\alpha(1) = 1 \end{bmatrix}\right\}
\tag{57}
$$

Let us consider the metrics d and d_h defined by (4) and (20) that we can rewrite as follows:

$$
\begin{aligned}
d(\Omega_0,\Omega_1) &= \inf_{\alpha\in\mathscr{T}_\alpha(\Omega_0,\Omega_1)}\int_0^1 \tilde{J}(\Gamma_t)\,dt \\
d_h(\Omega_0,\Omega_1) &= \inf_{\alpha\in\mathscr{T}_\alpha(\Omega_0,\Omega_1)}\int_0^1 \tilde{J}_h(\Gamma_t)\,dt
\end{aligned}
\tag{58}
$$

where the functionals $\tilde{J}(\Gamma_t)$ and $\tilde{J}_h(\Gamma_t)$ are defined by:

$$\tilde{J}(\Gamma_t) = |\dot{P}_\alpha(t)| \int_{\Gamma_t} \frac{|\Phi_1(x) - \Phi_0(x)|}{\|\nabla\Phi\|} d\Gamma(x)$$

$$\tilde{J}_h(\Gamma_t) = \frac{|\dot{P}_\alpha(t)|}{2h} \int_{U_h(\Gamma_t)} |\Phi_1(x) - \Phi_0(x)| dx \qquad (59)$$

Then, in order to solve the problem concerning the geodesic tube, that is to say to compute the metrics d or d_h defined by (58), we use a gradient method based on the computation of the shape derivative.

Assumption 2. *The shape derivative of the functional J defined by (46) can be rewritten as follows:*

$$dJ(\Gamma_t, \mathbf{W}) = \left.\frac{\partial J(\alpha + \varepsilon h)}{\partial \varepsilon}\right|_{\varepsilon=0} = \langle h, \nabla J(\Gamma_t) \rangle_{\mathbb{R}^M} \qquad (60)$$

where $\forall i \in [1, M]$:

$$(\nabla J(\Gamma_t))_i = \dot{e}_i(t)\, sign(\dot{P}_\alpha(t)) \int_{\Gamma_t} \frac{|\Phi_1(x) - \Phi_0(x)|}{\|\nabla\Phi\|} d\Gamma(x)$$

$$+ e_i(t)|\dot{P}_\alpha(t)| , \int_{\Gamma_t} \frac{|\Phi_1(x) - \Phi_0(x)|}{\|\nabla\Phi\|} K(t,x)\, d\Gamma(x) \qquad (61)$$

and

$$K(t,x) = \left[-2\frac{\nabla\Phi . (\nabla\Phi_1(x) - \nabla\Phi_0(x))}{\|\nabla\Phi\|^2} + \right.$$

$$+ 2(\Phi_1(x) - \Phi_0(x)) \frac{\nabla\Phi}{\|\nabla\Phi\|} \cdot \left[\frac{D^2\Phi}{\|\nabla\Phi\|^2} \cdot \frac{\nabla\Phi}{\|\nabla\Phi\|} \right] \qquad (62)$$

$$\left. - (\Phi_1(x) - \Phi_0(x)) \frac{\Delta\Phi}{\|\nabla\Phi\|^2} \right]$$

Assumption 3. *The shape derivative of the functional J_h defined by (51) can be rewritten as follows:*

$$dJ_h(\Gamma_t, \mathbf{W}) = \left.\frac{\partial J_h(\alpha + \varepsilon h)}{\partial \varepsilon}\right|_{\varepsilon=0} = \langle h, \nabla J_h(\Gamma_t) \rangle_{\mathbb{R}^M} \qquad (63)$$

where $\forall i \in [1, M]$:

$$(\nabla J_h(\Gamma_t))_i = \frac{\dot{e}_i(t)}{2h}\, sign(\dot{P}_\alpha(t)) \int_{U_h(\Gamma_t)} |\Phi_1(x) - \Phi_0(x)|\, dx$$

$$- \frac{e_i(t)}{2h^2}|\dot{P}_\alpha(t)| \int_{U_h(\Gamma_t)} (1 - 2\chi_{\Omega_t}(x)) |\Phi_1(x) - \qquad (64)$$

$$\Phi_0(x)| \frac{(\Phi_1(x) - \Phi_0(x))}{\|\nabla\Phi\|}\, dx$$

Algorithm 2

I. **Initialization:** Choose an initial vector of parameters $\Lambda(t)$ defined by (56) which generate an admissible connecting tube between Ω_0 and Ω_1 through the choice of the parameter α. Initialize the level set function
$\Phi(t,x) = P_\alpha(t)\,\Phi_1(x) + (1 - P_\alpha(t))\,\Phi_0(x)$. Set $k = 0$.

II. **Shape gradient direction:** For every $t \in I$, find the tubular neighborhood $U_h(\Gamma_t)$ of the zero level set Γ_t of the actual level set function $\Phi(t,x)$. Compute $\nabla J_h(\Gamma_t)$ from the equation (64).

III. **Update:**
- Perform a time step to update α.
- Let α^+ denote this update: $\alpha^+ = \alpha - \rho \int_0^1 \nabla J_h(\Gamma_t)\,dt$, $\rho > 0$.
- Update the function $\Phi^+(t,x) = P_{\alpha^+}(t)\,\Phi_1(x) + (1 - P_{\alpha^+}(t))\,\Phi_0(x)$.
- Set $k = k+1$ and go to (2).

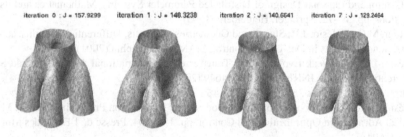

iteration 0 : J = 157.9299   iteration 1 : J = 148.3238   iteration 2 : J = 140.6641   iteration 7 : J = 129.2464

Fig. 4. Tube optimization using the metric $d_h(\Omega_0, \Omega_1)$

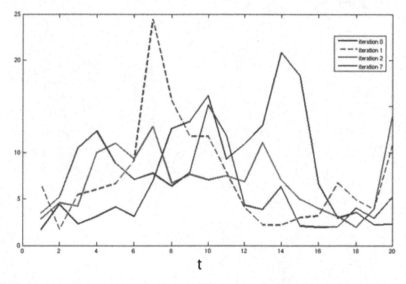

Fig. 5. Distribution of the functional values $J_h(\Gamma_t)$ for tube obtained during the optimization process of Fig. 4

3.3 Numerical Experiment of a Geodesic Tube Construction

We present a numerical experiment based on the algorithm 2 for a 3D tube optimization. Fig. 4 shows tubes obtained during the optimization process for different iterations. From Fig. 5 we can see that the tube obtained after seven iterations has a more homogeneous distribution of the functional values $J_h(\Gamma_t)$ compared to the initial tube. The result of this optimization process is the construction of a smoother tube than the initial tube.

References

1. Blanchard, L., Zolésio, J.-P.: Morphing by moving. Shape modeling with Galerkin approximation. In: Sivasundaram, S. (ed.) International Conference on Nonlinear Problems in Aviation and Aerospace (to appear)
2. Delfour, M.C., Zolésio, J.-P.: Oriented distance functions in shape analysis and optimization. In: Control and Optimal Design of Distributed Parameter Systems. Mathematics and its Applications, pp. 39–72. Springer, Heidelberg (1995)
3. Delfour, M.C., Zolésio, J.P.: Shapes and Geometries. Analysis, Differential Calculus, and Optimization. Advances in Design and Control. SIAM, Philadelphia (2001)
4. Zolésio, J.-P.: Set Weak Evolution and Transverse Field, Variational Application and Shape Differential Equation. INRIA report RR-4649 (2002),
   http://www.inria.fr/rrrt/rr-4649.html
5. Zolésio, J.-P.: Shape Topology by Tube Geodesic. In: Information Processing: Recent Mathematical Advances in Optimization and Control, pp. 185–204. Presse de l'Ecole des Mines de Paris (2004)
6. Zolésio, J.-P.: Optimal tubes: geodesic metric, Euler flow, moving domain. In: Free and Moving Boundaries: Analysis, Simulation and Control, vol. 252, pp. 203–213. Taylor & Francis, CRC Press (2005)

Remarks on 0-1 Optimization Problems with Superincreasing and Superdecreasing Objective Functions

Marian Chudy

Military University of Technology, Faculty of Cybernetics, Warsaw, Poland
mchudy@wat.edu.pl

Abstract. The set of particular 0-1 optimization problems solvable in polyno-mial time has been extended. This becomes when the coefficients of the ob-jective function belong to the set of superincreasing or superdecreasing types of sequence. We have defined special superincreasing sequences which we call the nearest up and nearest down to the sequence (c_j) of objective function co-efficients. They are applied to calculate the upper and lower bound of optimal objective function value. When the problem needs to compute the minimum of objective function with the superdecreasing sequence (c_j), two cases are con-sidered. Firstly, we have described a type of problem when optimal solution can be obtained directly using a polynomial procedure. The second case needs two phases to calculate an optimal solution. The second phase relies on improving a feasible solution. The complexities of all the presented procedures are given.

1 Introduction

The most frequently met formulation of 0-1 optimization problem (PLB) is:

$$\max \sum_{j=1}^{n} c_j x_j \tag{1}$$

subject to

$$\sum_{j \in N_i} a_{ij} x_j \le d_i, \quad i = \overline{1, m} \tag{2}$$

$$x_j \in \{0, 1\}, \quad j \in N = \{1, 2, \ldots, n\}, \quad N_i \subset N \tag{3}$$

To refer to the title of paper we are reminded that the sequence (c_j) is called superin-creasing when

$$\sum_{i=1}^{j-1} c_i < c_j \quad \text{for } j = 2, 3, \ldots \tag{4}$$

We will consider sequences containing n elements only and assume that for $n = 1$ a sequence is a superincreasing one.

A. Korytowski et al. (Eds.): System Modeling and Optimization, IFIP AICT 312, pp. 185–195, 2009.
© IFIP International Federation for Information Processing 2009

For $m = 1$ and positive a_{ij}, d_i, c_j the problem (1)-(3) becomes knapsack one and its special case with superincreasing parameters was effectively applied in the knapsack-type public key cryptosystem [5]. The decision type of knapsack problem with super-increasing parameters was shown in [4] to be P-complete. Short reference to the 0-1 optimization problems with superincreasing objective functions was presented in [3] and [1].

For two-constraint 0-1 knapsack problem, an exact algorithm was described in [6]. A general knapsack problem can be solved using Sbihi's new algorithm [7].

To the best of our knowledge, very few algorithms solving also PLB, are available. We will focus on these kinds of PLB which are solvable in polynomial time or their solutions can be estimated in polynomial time. Some of these cases were also presented in [2]. Now we attempt to extend this class.

2   Superincreasing Sequence and 0-1 Optimization Problem. General Remarks

For further considerations we will enumerate a few useful properties of superincreasing sequences:

I. Each subsequence of a superincreasing sequence is a superincreasing one,
II. Each increasing sequence containing only negative elements is a superincreasing one,
III. Each nondecreasing sequence (c_j) such that $c_1 \neq c_2$ containing only negative elements is a superincreasing one,
IV. Nondecreasing finite sequence of nonnegative elements contains some superincreasing subsequence.

Proposition 1. *If the problem (1)-(3) satisfies the following assumptions:*

- *the sequence (c_j) is superincreasing and nonnegative,*
- *elements a_{ij} are nonnegative $(a_{ij} \geq 0)$,*

then the optimal solution of the problem (1)-(3) is given by the following procedure

$$
x_j^* = \begin{cases} 1 & when \begin{cases} a_{1j} \leq d_1 - \sum_{k \in N_j^+} a_{1k} \\ a_{2j} \leq d_2 - \sum_{k \in N_j^+} a_{2k} \\ \cdot \\ \cdot \\ \cdot \\ a_{mj} \leq d_m - \sum_{k \in N_j^+} a_{mk} \end{cases} & j = n, n-1, ..., 1 \\ 0 & otherwise \end{cases} \tag{5}
$$

where a_j is the j-th column of the constraint matrix (2)

$$
d = (d_1, d_2, ..., d_m)^T, N_n^+ = \phi
$$

$$
N_j^+ = \{k : x_k^* = 1, \ k \in \{n, n-1, ..., j+1\}\}.
$$

The proof results from (4) and assumptions.

The complexity of procedure (5) is equal to $f_5(n) \in O(n^3)$. To calculate this function one should observe that calculating each element of x_j^* for $j = n, n-1, ..., 1$ needs $n, 2n, ..., n \times n$ basic operations, respectively. The total sum of these operations gives us that function.

Proposition 1 allows us to solve 0-1 optimization problem in polynomial time, when the assumptions it needs are satisfied.

Remark. The following example shows that the assumption $a_{ij} \geq 0$ is significant.

Example 1

$$\max \ x_1 + 2x_2 + 5x_3 + 10x_4$$

subject to

$$x_1 - 2x_2 + x_3 + 8x_4 \leq 6$$
$$2x_1 - x_2 - x_3 + 3x_4 \leq 2$$
$$x_j \in \{0,1\}, \ j = \overline{1,4}$$

The coefficients c_j form the superincreasing sequence $(c_j) = (1, 2, 5, 10)$. The optimal solution is $x^* = (0, 1, 0, 1)$ and optimal objective function value is equal to 12. On the other hand, using procedure (5) we obtain vector $\bar{x} = (1, 1, 1, 0)$ and the objective function value is equal to 8.

Remark. A simple example shows that assumption $c_j \geq 0$, $j = \overline{1,n}$ is also important.

Example 2

$$\max \ -2x_1 + 3x_2 + 3x_3 + 5x_4$$

subject to

$$x_1 + 2x_2 + 4x_3 + 6x_4 \leq 6$$
$$4x_1 + x_2 + x_3 + 5x_4 \leq 5$$
$$x_j \in \{0,1\}, \ j = \overline{1,4}$$

One can observe that the sequence $(c_j) = (-2, 3, 3, 5)$ is superincreasing. Procedure (5) computes the vector $x = (0, 0, 0, 1)$ and we obtain an objective function value of $(c|x) = 5$. However, the optimal solution is $x^* = (0, 1, 1, 0)$ and the optimal objective function value is equal to $(c|x^*) = 6$.

To continue our considerations we renumber, if necessary, all variables of the 0-1 problem and assume that the sequence (c_j) is **integer, nonnegative** and **non decreasing**.

Let us start with the following example.

Example 3

$$\max \ x_1 + 2x_2 + 3x_3 + 5x_4 + 5x_5$$

subject to

$$x_1 + x_3 + x_4 + x_5 \leq 2$$
$$x_2 + x_4 + x_5 \leq 2$$
$$x_1 + x_3 + x_4 \leq 3$$
$$x_j \in \{0,1\}, \ j = \overline{1,5}$$

The optimal solution is $x^* = (0,1,1,1,0)$. The sequence $(c_j) = (1,2,3,5,5)$ is not superincreasing but we can indicate some superincreasing subsequence of (c_j) which corresponds to some feasible solution. This feasible solution can be obtained by correctly selecting elements of vector $x^* = (0,1,1,1,0)$ which are equal to one.

The vectors we mentioned above are: $x^1 = (0,1,1,0,0)$, $x^2 = (0,1,0,1,0)$, $x^3 = (0,0,1,1,0)$. They correspond to superincreasing subsequence $(c_2,c_3) = (2,3)$, $(c_2,c_4) = (2,5)$, $(c_3,c_4) = (3,5)$ respectively. It leads to the following proposition.

Proposition 2. *If the coefficients of problem (1)-(3) are nonnegative, i.e., $c_j \geq 0$, $a_{ij} \geq 0$ for all i,j and there exists a feasible solution $x \neq 0 = (0,0,...,0)$, then there exists a feasible solution x^1 having elements $x_{j_r}^1 = 1$, $j_r \in N^1 \subset N^+ = \{j : x_j = 1\}$, $x_{j_r}^1 = 0$ $j_r \in N \backslash N^1$ that correspond to the superincreasing subsequence (c_{j_r}) of the sequence (c_j) which satisfies property IV.*

The proof results from (4), assumptions of this proposition and property IV.
Let us consider two superincreasing subsequences of sequence (c_j):

I. Subsequence $(c_{i_1}, c_{i_2}, ..., c_{i_r})$ which corresponds to a feasible solution x

$$x_j = \begin{cases} 1 & \text{for } j = i_k \\ 0 & \text{for } j \neq i_k \end{cases} \quad k = \overline{1,r}, \tag{6}$$

II. Subsequence $(c_{j_1}, c_{j_2}, ..., c_{j_r})$ which corresponds to a feasible solution y

$$y_j = \begin{cases} 1 & \text{for } j = j_k \\ 0 & \text{for } j \neq j_k \end{cases} \quad k = \overline{1,r}. \tag{7}$$

Proposition 3. *If $c_{i_r} > c_{j_r}$, then such inequality holds $\sum_{j=1}^{n} c_j x_j > \sum_{j=1}^{n} c_j y_j$ i.e., solution x is better than solution y. The proof results directly from the definition of the superincreasing sequence.*

One can describe this dependence using subsequences (c_3,c_4), (c_2,c_3) from example 3.

We observe that Propositions 2 and 3 can be applied to improve some given feasible solution.

It results in the fact that for many 0-1 optimization problems we can construct suitable 0-1 optimization problems with superincreasing objective functions and use them to compute, in polynomial time, high quality upper and lower bounds of optimal objective function values.

3   Superincreasing Sequence and Upper Bound

To obtain the upper bound of optimal objective function value, we have to introduce several new objects. Denote by:

- H^n – the set of all finite superincreasing integer sequences (h_j), $j = \overline{1,n}$,
- $A^n = \{h \in H^n : h_j \geq c_j, j = \overline{1,n}\}$ – the set of finite superincreasing sequences with integer elements no smaller than suitable elements of the sequence (c_j).

Remembering that (c_j) is nondecreasing we form the following definition.

Definition 1. *A superincreasing sequence $h^* = (h_j^*)$ is called the nearest up to the sequence (c_j) when*

$$h^* \in A^n \text{ and } \|c - h^*\| = \min_{h \in A^n} \|c - h\| = \min_{h \in A^n} \sum_{j=1}^{n} |c_j - h_j| \qquad (8)$$

For a given (c_j) we can compute the sequence $h^* = (h_j^*)$ in the following way:

$$h_1^* = c_1 \qquad (9)$$

and for $j = \overline{2, n}$

$$h_j^* = \sum_{k=1}^{j-1} h_k^* + 1 \quad \text{when} \quad c_j \le \sum_{k=1}^{j-1} h_k^* \qquad (10)$$

$$h_j^* = c_j \qquad \text{when} \quad c_j > \sum_{k=1}^{j-1} h_k^*. \qquad (11)$$

We notice that $(h_j^*) = (c_j)$ when (c_j) is a superincreasing sequence.

To compute all elements of h^*, the following numbers of basic operations are needed: 0 for $j = 1$, 2 for $j = 2$, 3 for $j = 3$, ..., n for $j = n$, respectively. Hence, the complexity of the procedure (9)-(11) is equal to $f_{h^*}(n) \in O(n^2)$.

The upper bound of optimal objective function value for the PLB is given by

$$\sum_{j=1}^{n} h_j^* x_j \ge \sum_{j=1}^{n} c_j x_j^* \qquad (12)$$

where:

- $x = (x_j), j = \overline{1,n}$ denotes a feasible solution computed by procedure (5) when we set the sequence (h_j^*) instead of the sequence (c_j) in PLB,
- $x^* = (x_j^*), j = \overline{1,n}$ denotes an optimal solution of the problem (1)-(3), under the assumption $a_{ij} \ge 0, c_j \ge 0$.

Example 4

$$\max \ x_1 + 4x_2 + 5x_3 + 6x_4$$

subject to

$$x_1 + x_2 + x_3 + 6x_4 \le 6$$
$$2x_1 + x_2 + x_3 + x_4 \le 7$$
$$x_j \in \{0,1\}, \quad j = \overline{1,4}$$

The vector $x^* = (1,1,1,0)$ is the optimal solution of this problem and the optimal objective function value $(c|x^*)$ is equal to 10.

According to procedure (9)-(11) the vector $h^* = (h_1^*, h_2^*, h_3^*, h_4^*) = (1,4,6,12)$ is superincreasing and the nearest up to the vector $c = (c_1, c_2, c_3, c_4) = (1,4,5,6)$ which is not superincreasing.

Setting the sequence (h_j^*) instead of (c_j) and applying procedure (5) we obtain a feasible solution of $x = (0,0,0,1)$ and objective function value of $(c|x) = 6$.

The upper bound of the optimal value $(c|x^*) = 10$, based on (12), is equal to $(h^*|x) = 12$.

At this point we should underline a very important fact: procedure (5), in every case, produces an upper bound of optimal function value when we use vector h^* instead of vector c. But procedure (5) cannot compute the upper bound without setting h^* instead of c. From example 4 results we obtain the same vector $x = (0,0,0,1)$ as when we use procedure (5) and keep the vector $c = (c_1, c_2, c_3, c_4) = (1,4,5,6)$.

4   Superincreasing Sequence and Lower Bound

To improve an assessment of optimal objective function value we propose to compute a lower bound of it.

Definition 2. *Let (c_j) be a non decreasing integer sequence.*
A superincreasing sequence $h^o = (h_j^o)$ is called the nearest down to the sequence $c = (c_j)$ when

$$\|c - h^o\| = \min_{h \in B^n} \|c - h\| = \min_{h \in B^n} \sum_{j=1}^{n} |c_j - h_j| \tag{13}$$

where

$$B^n = \{h \in H^n : \; h_j \le c_j, \; j = \overline{1,n}\}$$

For the given $c = (c_j)$ a sequence $h^o = (h_j^o)$ can be computed according to the following procedure:

I. For $n = 1$, $h_1^o = c_1$,
II. For $n = 2$, ; $h_2^o = c_2$,

$$h_1^o = \begin{cases} c_1 & \text{if } \; c_1 < h_2^o = c_2 \\ c_1 - 1 & \text{if } \; c_1 = c_2 = h_2^o \end{cases} \tag{14}$$

III. For $n \ge 3$, ; $h_j^o = c_j$, ; $j = \overline{n,2}$ and first element h_1^o needs a special recurrence formula to compute:

$$h_1^{k-2} = \begin{cases} c_1 & \text{for } k = n \\ h_1^{k-1} & \text{if } \; h_1^{k-1} + \sum_{i=2}^{k} c_i < h_{k+1}^o = c_{k+1}, \quad k = \overline{n-1,2} \\ h_1^{k-1} - A & \text{if } \; h_1^{k-1} + \sum_{i=2}^{k} c_i \ge h_{k+1}^o = c_{k+1} \end{cases} \tag{15}$$

where $A = (c_{k+1} - (h_1^{k-1} + \sum_{i=2}^{k} c_i)) + 1$

Example 5

$$1. \begin{cases} c = (1,1,1,1,4) \\ h^o = (-2,1,1,1,4) \end{cases} \quad 2. \begin{cases} c = (-1,-1,-1,-1,4) \\ h^o = (-2,-1,-1,-1,4) \end{cases}$$

$$3. \begin{cases} c = (-3,-1,3,4,4) \\ h^o = (-3,-1,3,4,4) \end{cases} \quad 4. \begin{cases} c = (-1,2,5,7,8,8) \\ h^o = (-15,2,5,7,8,8) \end{cases}$$

To evaluate the complexity of computing (h_j^o) for the given (c_j), it will be enough to take into account expression (15). The recurrence structure of this formula leads us to the following evaluation of the complexity: $f_{h^o}(n) \in O(n^2)$.

5   Some Useful Properties of (h_j^*) and (h_j^o)

I. From Definitions 1 and 2 it follows that if a sequence (c_j) is superincreasing then

$$c_j = h_j^* = h_j^o, \quad j = \overline{1,n}. \tag{16}$$

II. For each non decreasing (c_j) the following inequalities hold

$$h_j^* \geq c_j \geq h_j^o, \quad j = \overline{1,n}. \tag{17}$$

III. If some non decreasing sequence (c_j) satisfies

$$h_j^* = h_j^o, \quad j = \overline{1,n} \tag{18}$$

then (c_j) is superincreasing.

IV. For each vector x such that $x \in S = \{x \in E^n : ; (2), (3) \text{hold}\}$ we can obtain from (17) the following evaluation

$$(h^*|x) \geq (c|x) \geq (h^o|x). \tag{19}$$

V. The previous properties allow us to formulate

$$(c|x^*) \geq (c|x) \geq (h^o|x) \quad x^*, x \in S. \tag{20}$$

It means that value $(c|x), x \in S$ is not worse a lower bound of $(c|x^*)$ than $(h^o|x)$.

6   Superdecreasing Sequence and 0-1 Optimization Problem

Some 0-1 optimization problems have the following form:

$$\min \sum_{j=1}^n c_j x_j = \min(c|x) \tag{21}$$

subject to

$$\sum_{j \in N_i} a_{ij} x_j \geq d_i, \quad i = \overline{1,m} \tag{22}$$

$$x_j \in \{0,1\}, \quad j \in N = \{1,2,...,n\}, \quad N_i \subset N \tag{23}$$

This problem needs a different approach than the approach to (1)-(3).

Definition 3. *A sequence* (c_j) *is called a superdecreasing one when*

$$c_j > \sum_{i=j+1}^{n} c_i, \quad j = 1,...,n-1 \tag{24}$$

and for $n = 1$, (c_j) *is superdecreasing.*

The following properties of a superdecreasing sequence take place:

I. Each subsequence of the superdecreasing sequence (c_j) is superdecreasing,
II. Each of the decreasing sequence (c_j) containing only negative elements is superdecreasing,
III. Each of the non increasing sequence (c_j) containing only negative elements and satisfying $c_1 \neq c_2$ is superdecreasing,
IV. Each of the non increasing sequence (c_j) with only negative elements contains a superdecreasing subsequence.

We are able to select several cases when problems (21)-(23) are easily solvable.

Proposition 4. *Consider the problem (21)-(23) with superdecreasing* (c_j) *and let the following conditions hold:* $c_j \geq 0$, $a_{ij} \geq 0$ *and there exists* j *such that* $a_{ij} \geq d_i$, $i = \overline{1,m}$, *then an optimal solution has the form:*

$$x_{j_*}^* = 1 \text{ and } x_j^* = 0 \text{ for } j \neq j_*$$

when there exists i such that

$$\sum_{j=j_*+1}^{n} a_{ij} < d_i, \quad i \in \{1,2,...,m\} \text{ and } j_* < n$$

$$j_* = \max\{j : a_{ij} \geq d_i, \ i = \overline{1,m}\} \tag{25}$$

or

$$j_* = n$$

The proof results from Definition 3 and the conditions presented above.

Example 6

$$\min \ 10x_1 + 5x_2 + 2x_3 + x_4$$

subject to

$$x_1 + 2x_2 + 7x_3 + 8x_4 \geq 6$$
$$2x_1 + x_2 + 5x_3 + 2x_4 \geq 4$$
$$x_j \in \{0,1\}, \quad j = \overline{1,4}$$

In this problem we have $j_* = 3$, $x^* = (0,0,1,0)$ which satisfy all the conditions that Proposition 4 requires.

To obtain an optimal solution using the procedure described in Proposition 4, we need to execute the following numbers of basic operations:

I. At most n^2 to compute j_*,

II. To check if there exists I such that $]; \sum_{j=j_*+1} a_{ij} < d_i, \ i \in \{1,2...,m\}$, the worst case takes place for $j_* = 1$ and the number of basic operations equals at most $(n-1)n$.

Hence, the complexity of the procedure is equal to $f_{\min}(n) = O(n^2)$.

Proposition 5. *Let us consider the problem (21)-(23) with a superdecreasing (c_j) and $c_j \geq 0, \ a_{ij} \geq 0$.*

Assume that there is no j such that $a_{ij} \geq d_i, \ i = \overline{1,m}$. Then the following expressions

$$x_j^o = \begin{cases} 1 & for \ j = n, n-1, ..., j_o \\ 0 & otherwise \end{cases} \tag{26}$$

$$j_o = \max \left\{ j : \sum_{k=j}^{n} a_{ik} \geq d_i, \ i = \overline{1,m} \right\} \tag{27}$$

give us the upper bound $(c|x^o)$ of optimal objective function value $(c|x^*)$ of the problem (21)-(23). The proof we obtain from (26) and Definition 3.

The complexity of this procedure is determined in (27). In the worst case it needs at most n^3 basic operations. Hence, the complexity equals $f_{up}(n) \in O(n^3)$.

Example 7

$$\min \ 10x_1 + 5x_2 + 2x_3 + x_4$$

subject to

$$x_1 + 2x_2 + x_3 + 8x_4 \geq 6$$
$$2x_1 + 2x_2 + x_3 + 2x_4 \geq 4$$
$$x_j \in \{0,1\}, \ j = \overline{1,4}$$

The sequence $(c_j) = (10,5,2,1)$ is superdecreasing.

Using procedure (26), (27) we obtain $j_o = 2$ and the feasible solution $x = (0,1,1,1)$ that gives $(c|x) = 8$. This is not an optimal solution. The optimal solution is $x^* = (0,1,0,1)$ and gives $(c|x^*) = 6$. We can improve the feasible solution $x = (0,1,1,1)$ applying the procedure given below.

Proposition 6. *Let $x^o = (x_j^o), \ j = \overline{1,n}$ be the feasible solution of problem (21)-(23) which was obtained using procedure (26), (27) under the assumptions: (c_j) is superdecreasing, $c_j \geq 0, \ a_{ij} \geq 0$.*

Defining auxiliary parameters:

$$N_o^+ = \left\{ j : x_j^o = 1, \ x_j^o \ that \ satisfies \ (26),(27) \right\} = \{n, n-1, ..., j_o\},$$

$$N_{j_o+1}^- = \begin{cases} \phi & when \ \sum_{k \in N_o^+ \setminus \{j_o+1\}} a_{ik} < d_i, \ i = \overline{1,m} \\ \{j_o+1\} & when \ \sum_{k \in N_o^+ \setminus \{j_o+1\}} a_{ik} \geq d_i, \ i = \overline{1,m} \end{cases}$$

$$N_j^- = \begin{cases} N_{j-1}^- & when \; \sum_{k\in N_o^+ \setminus (N_{j-1}^- \cup \{j\})} a_{ik} < d_i, \; i=\overline{1,m} \\ N_{j-1}^- \cup \{j\} & when \; \sum_{k\in N_o^+ \setminus (N_{j-1}^- \cup \{j\})} a_{ik} \geq d_i, \; i=\overline{1,m} \end{cases} \quad j=\overline{j_o+2,n}$$

the optimal solution can be expressed in the following way

$$x_j^* = \begin{cases} 0 & when \; \sum_{k\in N_o^+ \setminus \{j\}} a_{ik} \geq d_i, \; i=\overline{1,m} \\ 1 & otherwise \end{cases} \quad j=j_o+1 \qquad (28)$$

$$x_j^* = \begin{cases} 0 & when \; \sum_{k\in N_o^+ \setminus N_j^-} a_{ik} \geq d_i, \; i=\overline{1,m} \\ 1 & otherwise \end{cases} \quad j=\overline{j_o+2,n} \qquad (29)$$

$$x_j^* = 0 \; for \; j = \overline{1,j_o-1}. \qquad (30)$$

The essence of the procedure (28)-(30) relies on reducing to zero, if possible, these elements x_j^o of vector x^o which are equal to one and c_j is large. It is also essence of the proof.

The similarity between (26) and (28), (29) allows us to write $f_{cor}(n) \in O(n^3)$ as the complexity of the vector x^o improving .

In example 7, we can correct $x = (0,1,1,1)$ to the form $x^* = (0,1,0,1)$ using procedure (28)-(30), because the sum of the second and fourth column satisfies (28), i.e.

$$\begin{bmatrix} 2 \\ 2 \end{bmatrix} + \begin{bmatrix} 8 \\ 2 \end{bmatrix} \geq \begin{bmatrix} 6 \\ 4 \end{bmatrix}.$$

7   Conclusions

The results we have obtained are applicable to:

- solving 0-1 optimization problems with a superincreasing and superdecreasing objective function, if the indicated assumptions hold,
- computing upper bounds and lower bounds of optimal objective function value for 0-1 optimization problems under suitable assumptions,
- improving given feasible solution of 0-1 optimization problem using some properties of superincreasing and superdecreasing sequences.

It is worth underlining that all of these procedures are polynomial. The practical application area of 0-1 optimization problems is very broad. There are no reasons to exclude these results from this area.

References

1. Alfonsin, R.: On variations of the subset sum problem. Discrete Applied Mathematics 81, 1–7 (1998)
2. Chudy, M.: Some properties of 0-1 optimization problems. In: Proceedings of the 11th International Conference on System Modelling Control (SMC), pp. 61–67. EXIT (2005)
3. Jenner, B.: Knapsack problems for NL. Information Processing Letters 54, 167–174 (1995)

4. Karloff, H.J., Ruzzo, W.L.: The iterated mod problem. Information and Computation 80, 193–204 (1989)
5. Koblitz, N.: A Course in Number Theory and Cryptography. Springer, New York (1994)
6. Martello, S., Toth, P.: An exact algorithm for two-constraint 0-1 knapsack problem. Operations Research 51, 826–835+837 (2003)
7. Sbihi, A.: A best first search algorithm for the multiple-choice multidimensional knapsack problem. Journal of Combinatorial Optimization 13, 337–351 (2007)

A Hierarchical Multiobjective Routing Model for MPLS Networks with Two Service Classes*

José Craveirinha[1], Rita Girão-Silva[1], João Clímaco[2], and Lúcia Martins[1]

[1] Department of Electrical Engineering Science and Computers of the University of Coimbra
Pólo II, Pinhal de Marrocos, P-3030-290 Coimbra, Portugal Institute of Computers and Systems
Engineering of Coimbra (INESC-Coimbra) R. Antero de Quental,
199, P-3000-033 Coimbra, Portugal
`jcrav@deec.uc.pt, rita@deec.uc.pt, lucia@deec.uc.pt`
[2] Faculty of Economics of the University of Coimbra Av. Dias da Silva, 165, P-3004-512
Coimbra, Portugal Institute of Computers and Systems Engineering of Coimbra
(INESC-Coimbra) R. Antero de Quental, 199, P-3000-033 Coimbra, Portugal
`jclimaco@inescc.pt`

Abstract. This work presents a model for multiobjective routing in MPLS networks formulated within a hierarchical network-wide optimization framework, with two classes of services, namely QoS and Best Effort (BE) services. The routing model uses alternative routing and hierarchical optimization with two optimization levels, including fairness objectives. Another feature of the model is the use of an approximate stochastic representation of the traffic flows in the network, based on the concept of effective bandwidth. The theoretical foundations of a heuristic strategy for finding "good" compromise solutions to the very complex bi-level routing optimization problem, based on a conjecture concerning the definition of marginal implied costs for QoS flows and BE flows, will be described. The main features of a first version of this heuristic based on a bi-objective shortest path model and some preliminary results for a benchmark network will also be revealed.

1 Introduction and Motivation

Modern multiservice network routing functionalities have to deal with multiple, heterogeneous and multifaceted QoS (Quality of Service) requirements. This led to routing models, the aim of which is the calculation and selection of one (or more) sequences of network resources (designated as routes, which correspond to loopless paths in the network representation) satisfying certain QoS constraints and the optimization of route related metrics. Therefore there are potential advantages in formulating important routing problems in these types of networks as multiple objective optimization problems. These formulations enable the trade-offs among distinct performance metrics and other network cost function(s) to be pursued in a consistent manner. Note that the definition of the objective functions and constraints depends strongly on the nature of the considered routing principles, the type of network technological platform and the features of the offered traffic flows associated with different service types.

* Work partially supported by programme POSI of the EC programme cosponsored by FEDER and national funds.

A. Korytowski et al. (Eds.): System Modeling and Optimization, IFIP AICT 312, pp. 196–219, 2009.
© IFIP International Federation for Information Processing 2009

In these networks, connection-oriented services, namely with guaranteed QoS levels, may be implemented. The traffic flows are composed (at the physical level) of packet streams that are forwarded from node to node, according to some specific technical rules. When the packets enter the network, they are grouped in different FECs (Forward Equivalence Classes) according to specific criteria, namely the originating node, the destination node and the grade of service that has to be provided. The 'traffic trunks' are an aggregation of flows of a certain class and can be characterized by the ingress and egress nodes, the FEC they are associated with, and a set of parameters/attributes with impact on the traffic engineering schemes, which define some essential requirements of the routing models. The routing mechanism for packets used in the MPLS networks is based on the establishment of the so-called LSPs (Label Switched Paths). At the ingress node, a label containing information on the FEC is associated with the packets. At each intermediate node, the LSRs (Label Switching Routers) forward the packets using a specific label switching technique: the label is an index into a routing table with information on the next network arc (or hop) and the next label to be assigned to the packet. Therefore, end-to-end "explicit routes" may be established in association with the implementation of advanced QoS-based routing mechanisms. In particular explicit routes enable source routing mechanisms, characterized by the fact that the route followed by each packet stream (of a given connection) is entirely determined by the ingress router. This is an inherent advantage by comparison with the hop-by-hop (i.e. node by node) routing method typical of IP routing. Details on traffic engineering-related concepts in MPLS networks relevant in the present context are described in [2,3,33]. The described features in association with other functional capabilities of MPLS enable the implementation of advanced QoS-based routing mechanisms, namely through the definition of explicit routes satisfying certain QoS requirements for each traffic flow of a given FEC.

Having in mind these features and capabilities of MPLS routing a significant number of routing models has been proposed in the literature in recent years. A routing model can be described in terms of various features. A key feature is the *routing optimization framework* which has to do with the scope and nature of the formulation of the routing calculation problem; in this respect we may distinguish network-wide optimization models and flow-oriented optimization models. In the former the objective functions are formulated at network level and depend explicitly on all traffic flows in the network. Examples of these functions are average total traffic carried, total expected revenue, average packet delay or a function which seeks the optimization of the utilization of the arcs of the network in terms of their level of occupancy, as in [13] and [26]. In contrast, flow-oriented optimization models consider the objective functions formulated at the level of each particular node-to-node connection or flow, for example number of arcs of the path, path cost (for a specific link usage path metric) or mean packet delay on the particular traffic stream. Examples of this type of models are the numerous QoS routing models which are based on single-objective constrained shortest path formulations (a review can be seen in [22] and an overview in [4]). Another feature refers to the *nature* of the chosen objective functions and constraints, namely whether the optimization model is single or multiobjective, and the type of functions and constraints (technical, economic, social or other). The *representation* of node-to-node demand requests or traffic offered is also relevant in a telecommunication routing model. We can consider different types of traffic models in terms of the granularity of the representation (for example representation at connection request level or traffic flow level i.e. in terms of a sequence of connections throughout time), or the nature of this representation, namely whether it is deterministic or stochastic.

A recent systematic, in-depth review on main issues, optimization models and algorithms concerning routing methods in communication networks can be seen in [31]. An overview of some applications of MCDA (Multicriteria Decision Analysis) tools in telecommunication strategic planning and negotiation is shown in [16]. A review on applications of MCDA in telecommunications network planning and design, including a section on routing models, is presented in [4]. An overview of a significant number of contributions on multicriteria routing models in telecommunication networks followed by a description of a bi-level hierarchical multicriteria routing model of the flow-oriented optimization type, is put forward in [5].

As discussed in [7], a significant number of routing models for MPLS has been proposed in the literature in recent years which often differ in key instances of the modeling framework. Based on the analysis of the remarkable differences observed in the models proposed in this area, a discussion on key conceptual issues involved in the various modeling approaches and a proposal of a generic hierarchical multiobjective network-wide routing optimization framework or "meta-model" has been presented in [7].

The possibility of applying this modeling framework to a MPLS type network, already outlined in [7], namely by considering two classes of service, QoS traffic (first priority traffic) and Best Effort (BE) traffic (second priority traffic), was a major motivation for this work.

This work presents, in detail, a model for multiobjective routing in MPLS networks formulated within the framework developed by the authors in [7], assuming that there are two classes of services (and different types of traffic flows in each class), namely QoS and BE services. The flows of QoS type (first priority flows), when accepted by the network, have a guaranteed QoS level, related to the required bandwidth, while BE traffic flows, which are treated in the model as second priority flows, are carried by the network in order to obtain the best possible QoS level for the current network routing solution. Another feature of the routing model is the use of alternative routing: when a first choice route assigned to a given micro-flow, belonging to a certain traffic flow (corresponding to a "traffic trunk") is blocked, a second choice route may be attempted. An important feature of this model is the use of hierarchical optimization typically with two optimization levels, including fairness objectives: the first priority objective functions refer to the network level objectives of QoS type flows, namely the total expected revenue and the maximal value of the mean blocking of all types of QoS flows; the second priority objective functions refer to performance metrics for the different types of QoS services and the total expected revenue associated with the BE traffic flows. Another important feature of the model is the use of an approximate stochastic representation of the traffic flows in the network, based on the use of the concept of effective bandwidth for macro-flows and on a generalized Erlang model for estimating the blocking probabilities in the arcs, similar to the one used in [29,26]. After describing in detail the routing model in Sect. 2, including the underlying traffic model, we will present in

Sect. 3 the theoretical foundations of a specialized heuristic strategy for finding "good" compromise solutions to the very complex bi-level routing optimization problem. This theoretical foundation is based on a conjecture concerning the definition of marginal implied costs for QoS flows and BE flows, presented for the first time in this paper, which is an extension and adaptation of earlier definitions of implied cost for single service networks with alternative routing in [18]. The structure of the heuristic procedure for resolving the problem is analogous to the one described in detail in [7,26]. The new version of the heuristic, presented here, is based on a constrained bi-objective shortest path model, the objective functions of which are QoS or BE marginal path implied costs, depending on the class of the routed traffic, and path blocking probabilities. Also, in Sect. 4, a description of a first version of this heuristic will be presented, and some preliminary results for a test network will be revealed in Sect. 5.

2 Description of the Routing Model

The present model can be considered as an application of the multiobjective modeling framework for MPLS networks described by the authors in [7]. This framework (or "meta-model") uses hierarchical optimization with up to three optimization levels: the first priority objective functions refer to the global network level; the second priority objective functions refer to performance metrics for the different types of services supported by the network; the third priority functions are concerned with performance metrics for the micro-flows of packet streams of the same FEC.

It is a network-wide routing optimization approach of a new type, in the form of a hierarchical multiobjective optimization model, which takes into account the nature and relations between the adopted objective functions related to the different types of traffic flows associated with different services. We would like to note that various multiobjective models previously proposed use objective functions chosen to reflect only indirectly network technical-economic objectives. A typical example is the minimization of a utilization cost for all arcs expressed, through empirical functions, in terms of the occupied bandwidth as in [14,13,12,20]. In fact, the pursued objective is to optimize the total traffic carried in the network or the associated expected revenue. One can say that this type of approach is just a rough approximation to the 'hidden' (or implicit) objective function the model seeks to reflect, especially taking into account the random nature of traffic patterns, even in stationary or quasi-stationary network working conditions. Instead, our model considers an explicit representation of the most relevant technical-economic objectives in a network-wide routing optimization, such as the total expected revenue (expressed in terms of the traffic carried of all service types). This aspect of the modeling approach is in line with the school of thought adopted by [18,19,29], in the context of single-objective routing models.

We propose a hierarchy of objective functions by considering in a first approach two levels of optimization with several objective functions in each level. The first level (first priority) includes objective functions formulated at network level for the QoS type traffic and considering the combined effect of all types of traffic flows in the network. The second level refers to average performance metrics of the QoS traffic flows associated with the different types of services supported by the network and the expected revenue of the BE traffic. An important feature of the model is the explicit consideration, as objective functions, of 'fairness' objectives, at the two levels of optimization. These are objectives of min-max type and seek to make the most of the proposed multiobjective formulation. In previous formulations of routing models for these networks, such aims related to fairness are usually not considered explicitly in any form or just represented through constraints on certain performance metrics. Another important feature of the model is the stochastic representation of the traffic flowing in the network as described in [7,29].

We will consider two *classes* of services, namely QoS corresponding to services with certain guaranteed QoS levels, and BE, where the corresponding traffic flows are routed seeking to obtain the best possible quality of service but not at the cost of the QoS of the QoS traffic flows (first priority traffic flows). The *service types* in each class are grouped in the sets $\mathscr{S}_Q$ (for QoS service types) and $\mathscr{S}_B$ (for BE service types), and the traffic flows of each service type $s \in \mathscr{S}_Q$ or $s \in \mathscr{S}_B$ may differ in important attributes, namely the required bandwidth.

The consideration of two (or more) classes of traffic flows in a routing model is a complex issue and different heuristic approaches have been proposed in the literature. Examples of these approaches, typically flow-oriented models, are in [20,23,21]. As for network-wide optimization approaches, [30] describes a bi-objective routing model using lexicographic optimization where a primary objective function is the weighted sum of the carried bandwidth associated with QoS traffic flows and a secondary objective function of the same type is defined for the BE traffic. A heuristic procedure based on a decomposition technique and multicommodity flow programming is developed for obtaining solutions to the problem.

Some definitions relevant to the model are now formally introduced.

Definition 1. A *traffic flow* is a mathematical entity specified by $f_s = (v_i, v_j, \overline{\gamma}_s, \overline{\eta}_s)$ for $s \in \mathscr{S} = \mathscr{S}_Q \cup \mathscr{S}_B$ that corresponds to a stochastic process, in general a marked point process, that describes the arrivals and basic requirements of μ-flows, originated at the MPLS ingress node v_i and destined to the MPLS egress node v_j, using the same LSP and characterized by the vectors of 'attributes' $\overline{\gamma}_s$ and $\overline{\eta}_s$ for service type s.

The vector $\overline{\gamma}_s$ describes the traffic engineering attributes of flows of service type s and the vector $\overline{\eta}_s$ enables the representation of the mechanism(s) of admission control to all links l_k in the network by calls of flow f_s. The set of all traffic flows of type s will be denoted by $\mathscr{F}_s$.

In the teletraffic modeling approach considered here, these attributes include the required effective bandwidth d_s and the mean duration $h(f_s)$ of a μ-flow in f_s. In our model a 'μ-flow' corresponds to a 'call', the term call being used in its broadest sense, that is, as a connection request with certain features. The use of the concept of effective bandwidth [17] in the present context (MPLS networks with explicit routes) was earlier proposed in [29] and used in [25,26]. This enables an upper level network representation in the traffic plane level, through an equivalent multirate loss traffic network.

Consider that we have an approximate teletraffic model that is capable of estimating the node-to-node blocking probabilities $B(f_s)$ for all flows f_s of all service types.

Definition 2. A *routing plan* $\bar{R}$ is a set of loopless paths, for all network services and all flows, which corresponds to a possible routing pattern in the network, assuming that up to $M - 1$ alternative routes may be attempted by any connection request of f_s: $\bar{R} = \cup_{s=1}^{|\mathscr{S}|} R(s)$ for all the network services, where $|\mathscr{S}|$ is the total number of services, $R(s) = \cup_{f_s \in \mathscr{F}_s} R(f_s), s \in \mathscr{S}_Q \cup \mathscr{S}_B$ and $R(f_s) = (r^p(f_s)), p = 1, \cdots, M$.

In the present model (one-stage alternative routing), $M = 2$.

This means that for each flow f_s the first choice route $r^1(f_s)$ will be used unless it is blocked as a result of one of its links l_k not having the required available bandwidth d_s (or as prescribed by a general probabilistic availability function ψ_{ks}). If $r^1(f_s)$ is blocked then the second choice route $r^2(f_s)$ will be attempted by the connection request and the request will be blocked only if $r^2(f_s)$ is also blocked.

Let $A(f_s)$ represent the mean of the traffic offered by traffic flow f_s in Erl and A_s^o represent the total traffic offered by the flows of the service type s, $A_s^o = \sum_{f_s \in \mathscr{F}_s} A(f_s)$ [Erl].

Definition 3. The *average blocking probability for all traffic flows of type s*, for a given routing plan $\bar{R}$, is $B_{ms} = \frac{1}{A_s^o} \sum_{f_s \in \mathscr{F}_s} A(f_s) B(f_s)$.

In particular, $B_{ms|Q} = B_{ms}$ for given $s \in \mathscr{S}_Q$.

Definition 4. The *maximal blocking probability among all traffic flows f_s of QoS class and type s*, is $B_{Ms|Q} = \max_{f_s \in \mathscr{F}_s} \{B(f_s)\}, s \in \mathscr{S}_Q$.

Definition 5. The *maximal average blocking probability among all QoS service types*, is $B_{Mm|Q} = \max_{s \in \mathscr{S}_Q} \{B_{ms}\}$.

Let $w(f_s)$ denote the expected revenue associated with calls of a generic flow f_s and A_s^c be the total carried traffic by traffic flows of type s, $A_s^c = \sum_{f_s \in \mathscr{F}_s} A(f_s)(1 - B(f_s)) = A_s^o(1 - B_{ms})$ [Erl]. Further assume the usual simplification $w(f_s) = w_s, \forall f_s \in \mathscr{F}_s$.

Definition 6. The *total expected revenues associated with QoS(BE) traffic flows*, are given by $W_{Q(B)} = \sum_{s \in \mathscr{S}_{Q(B)}} A_s^c w_s$.

In the framework of the meta-model [7] we may formulate a two-level multiobjective routing optimization problem by separating the total expected revenue in two parts: W_Q for the traffic flows of QoS type and W_B for the traffic flows of BE type, as defined above, and by considering explicitly performance optimization of QoS service types. While W_Q will be a first priority objective function, together with the maximal blocking probability for all QoS service types, $B_{Mm|Q}$, W_B will be a second level objective function. This seeks to guarantee that the routing of BE traffic, in a quasi-stationary situation, will not be made at the cost of a decrease in revenue or of an increase in the blocking probability of QoS traffic flows.

The second level of optimization also concerns QoS service types and includes $2|\mathscr{S}_Q|$ objective functions to be minimized, the mean blocking probability $B_{ms|Q}$ for flows of type $s \in \mathscr{S}_Q$, and the maximal blocking probability $B_{Ms|Q}$, defined above. Note that $B_{Ms|Q}$ represents the fairness objective defined for each service type $s \in \mathscr{S}_Q$.

These considerations led to the following formulation of a *two-level hierarchical optimization problem for two service classes*:

Problem P-M2-S2

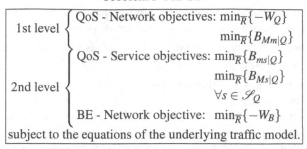

Here the acronym P-M2-S2 stands for 'Problem - Multiobjective with 2 optimization hierarchical levels - with 2 Service classes'.

It is important to note that while QoS and BE traffic flows are treated separately in terms of objective functions in order to take into account their different priority in the optimization model, the interactions among all traffic flows are fully represented in the model. This is guaranteed by the adopted traffic modeling approach underlying the optimization model. This is another major difference in comparison to other routing models proposed for networks with two service classes.

The definition and calculation of the parameters in the expressions are given in [8, Appendix A].

It should be noted that this model is a simplification of the general model for QoS and BE service classes outlined by the authors in [7, Sect.3.3]. In the addressed routing optimization model P-M2-S2 only the macro level representation was considered, having in mind the avoidance of increased complexity in a model, which is by itself very complex, resulting from the inclusion of a third optimization level in the routing model, as well as the corresponding additional computational burden.

The traffic modeling approach is the one used in [7] and earlier in [29,26] for tackling the calculation of blocking probabilities experienced by the traffic flows in network links. It is based on the concept of *effective bandwidth* (see underlying theory in [17]), in association with the definition of MPLS explicit routes. The effective bandwidth can be considered as a stochastic measure of the utilization of network resources enabling the representation (in an approximate manner) of the effects of the variability of the rates of different traffic sources, as well as the effects of statistical multiplexing of different traffic flows in a network. This conceptual tool was used in routing optimization models of multiservice networks of various types as in [29,26]. Hence, the network may be represented in the traffic plane by a multiclass loss traffic network, equivalent to a multirate traffic circuit-switched network.

The basic calculation sub-model enables the blocking probabilities B_{ks}, for connection requests of service type s in link l_k, to be obtained in the form $B_{ks} = \mathscr{L}_s\left(\overline{d_k}, \overline{\rho_k}, C_k\right)$. $\mathscr{L}_s$ represents the basic function (implicit in the analytical model) that gives the marginal blocking probabilities, B_{ks}, in terms of $\overline{d_k} = (d_{k1}, \cdots, d_{k|\mathscr{S}|})$ (vector of equivalent effective bandwidths), $\overline{\rho_k} = (\rho_{k1}, \cdots, \rho_{k|\mathscr{S}|})$ (vector of reduced traffic loads ρ_{ks} offered by flows of type s to l_k) and the link capacity C_k.

This approximation was suggested in [29] in the context of off-line single-objective multiservice routing optimization models and was also used in the multiobjective dynamic alternative routing model [26]. The use of very efficient and robust stochastic approximations is absolutely critical in a routing optimization model of this type, for tractability reasons. The detailed description of the traffic model can be seen in [7,8].

3 Foundations of the Resolution Approach

In the hierarchical multiobjective network-wide optimization routing problem P-M2-S2 we will consider that the routing principle uses alternative routing i.e. the decision variables are the network routing plans $\overline{R}$.

The hierarchical multiobjective alternative routing problem in question is highly 'complex' as a result of two major factors: the strong interdependencies among all objective functions (via the $\{B(f_s)\}$) and the interdependencies between the objective function parameters and the (discrete) decision variables $\overline{R}$ (network route plans). All these interdependencies are defined by the underlying traffic model.

Concerning overall complexity it can be said that the simplest, 'degenerated' single objective version of the problem, corresponding to the single objective function W_Q, one single service and no alternative routing ($M = 1$) is NP-complete in the strong sense, as shown in [11]. Note that our model is a bi-level, multiobjective formulation of this type of problem. This and the interdependencies among the objective functions are a strong indication of extreme intractability of the problem.

Concerning the possible conflict between the objective functions in P-M2-S2, it can be said that in many situations, the maximization of W_Q entails a deterioration on $B(f_s)$, $s \in \mathscr{S}_Q$, for "small" intensity traffic flows $A(f_s)$ which tends to increase $B_{Ms|Q}$ and, as a result, $B_{Mm|Q}$. In single-objective routing models this effect is usually tackled by imposing upper bounds on the values $B(f_s)$. These relations between objective functions of this type have been analyzed in [27]. Note that this is a major factor to justify the interest and potential benefit in using multiobjective approaches when dealing with this type of routing problem.

The resolution (in a multicriteria analysis sense) of the routing model P-M2-S2 will be performed by a heuristic approach, a first version of which is presented in the next section. This heuristic is the extension and adaptation to this problem of the heuristic procedure described in [9] and [26].

The heuristic developed for problem P-M2-S2 is based on the calculation of solutions of a bi-objective shortest path problem. In this problem the path metrics to be minimized will be the marginal implied costs (as defined according to the following analysis) and blocking probabilities.

The implied cost c_{ku} resulting from the acceptance of a call of flow f_u in link l_k is a powerful mathematical concept in routing optimization in circuit-switched networks, and was originally proposed by Kelly [18] for single-rate traffic networks. It was extended to single route (i.e. without alternative routing) multirate traffic networks in [15] and [29]. It can be defined as the expected value of the loss of revenue in all network traffic flows which may use link l_k resulting from the acceptance of a call from f_u associated with the decrease in the capacity of this link. Therefore the implied cost measures the knock-on effects on all network routes (of all traffic flows) resulting from the acceptance of a call from f_u in a link l_k. The authors have adapted in [9] and [26] the definition of c_{ku} to multirate loss networks with alternative routing by extending the corresponding expression given for single-service networks in [18]. This extension implies that the c_{ku} can be calculated from the equations:

$$c_{ku} = \sum_{s \in \mathscr{S}} \frac{1}{1 - B_{ks}} \zeta_{kus} \left[\sum_{f_s : l_k \in r^1(f_s)} \lambda_{r^1(f_s)} \left(s_{r^1(f_s)} + c_{ks} \right) \right.$$

$$\left. + \sum_{f_s : l_k \in r^2(f_s)} \lambda_{r^2(f_s)} \left(s_{r^2(f_s)} + c_{ks} \right) \right] \quad (1)$$

with

$$s_{r^2(f_s)} = w(f_s) - \sum_{l_j \in r^2(f_s)} c_{js}$$

$$s_{r^1(f_s)} = w(f_s) - \sum_{l_j \in r^1(f_s)} c_{js} - (1 - L_{r^2(f_s)}) s_{r^2(f_s)} \quad (2)$$

where $s_{r^p(f_s)}$ is the surplus value of a call on route $r^p(f_s)$, $\lambda_{r^p(f_s)}$ is the marginal traffic carried on $r^p(f_s)$, $L_{r^p(f_s)}$ is the blocking probability for calls of f_s on route $r^p(f_s)$ $(p = 1; 2)$, considering that $r^1(f_s)$ and $r^2(f_s)$ are disjoint paths, and $\zeta_{kus} = \mathscr{L}_s\left(\overline{d_k}, \overline{\rho_k}, C_k - d_{ku}\right) - \mathscr{L}_s\left(\overline{d_k}, \overline{\rho_k}, C_k\right)$ is the increase in the congestion for type s calls on link l_k originated by a decrease in the arc capacity because of the acceptance of a type u call.

The calculation of the implied costs in this form is based on the following conjecture, which is an extension to multirate loss networks with alternative routing of the results in [18, Sect.7] and [28, Sect.3].

Conjecture A: In multirate networks with (one-stage) alternative routing the sensitivity of the revenue W_T with respect to the traffic $A(f_s)$ being offered to a pair of routes (r^1, r^2), when an approximation to the expected revenue is calculated from the solution of the fixed point equations in B_{ks}, can be written

$$\frac{\partial W_T}{\partial A(f_s)} = (1 - L_{r^1(f_s)}) \left(w_{r^1(f_s)} - \sum_{l_k \in r^1(f_s)} c_{ks} \right)$$

$$+ L_{r^1(f_s)}(1 - L_{r^2(f_s)}) \left(w_{r^2(f_s)} - \sum_{l_k \in r^2(f_s)} c_{ks} \right) \quad (3)$$

where the c_{ks} are the implied costs and these satisfy the system of equations (1)-(2).

The validity of this conjecture implies that the implied cost c_{ku} can be interpreted as the exact mathematical measure of the 'knock-on' effects on all network routes of all network flows resulting from the acceptance of a call from f_u in link l_k, in the terms stated above. This in turn is consistent with the calculation of the c_{ks} through the system (1), by applying the theory developed in [18].

It will be further assumed that $w_{r^1(f_s)} = w_{r^2(f_s)} = w(f_s)$. The fixed point equations in this statement result from the traffic model and constitute a system of implicit non-linear equations enabling the calculation of the B_{ks} in terms of link capacities (expressed through matrix $\overline{C} = [C_k]$), the offered traffic matrix $\overline{A} = [A(f_s)]$, and the current network routing plan $\overline{R}$. The calculation of c_{ks} through (1)-(2) also implies the solution of a system of equations.

$$B_{ks} = \beta_{ks}(\overline{B}, \overline{C}, \overline{A}, \overline{R}) \quad \text{and} \quad c_{ks} = \alpha_{ks}(\overline{c}, \overline{B}, \overline{C}, \overline{A}, \overline{R}) \tag{4}$$

with $k = 1, \cdots, |\mathcal{L}|; s = 1, \cdots, |\mathcal{S}|, \overline{B} = [B_{ks}]$ and $\overline{c} = [c_{ks}]$. The numerical resolution of the systems (4) is performed by fixed point iterators, for given $\overline{C}, \overline{A}$ and $\overline{R}$.

In [9] and [26], the authors formulated a bi-level *multiple objective dynamic alternative routing problem for multiservice networks* for a single service class with multiple service types:

$$\text{Problem } \mathcal{P}_G - S$$

$$\begin{vmatrix} \text{Network level: } \min_{\overline{R}}\{-W_T\} \\ \qquad \min_{\overline{R}}\{B_{Mm} = \max_{s \in \mathcal{S}}\{B_{ms}\}\} \\ \text{Service level: } \min_{\overline{R}(s)}\{B_{ms}\}, s = 1, \cdots, |\mathcal{S}| \\ \qquad \min_{\overline{R}(s)}\{B_{Ms} = \max_{f_s \in \mathcal{F}_s}\{B(f_s)\}\}, s = 1, \cdots, |\mathcal{S}| \end{vmatrix}$$

subject to the equations of the teletraffic model enabling the calculation of $\{B(f_s)\}$ in terms of $\{A(f_s)\}$ and $\overline{R}$, where W_T is the total expected network revenue associated with the traffic carried by all service types, B_{ms} is the average blocking probability for all traffic flows of service type s, B_{Ms} is the maximal value of those blocking probabilities, and B_{Mm} is the maximal value among the B_{ms} and is, together with W_T, the first priority objective function. This routing model can be considered as an application of the meta-model associated with P-M3-S2 (Problem - Multiobjective with 3 optimization hierarchical levels - with 2 Service classes) [7] and as a particular case of the addressed model, P-M2-S2, by considering only one service class.

The resolution approach to $\mathcal{P}_G - S$ was based on the calculation of solutions to the bi-objective shortest path problem $\mathcal{P}^{(2)}$, formulated for every end-to-end flow f_s, $\min_{r(f_s) \in \mathcal{D}(f_s)}\{m^n(r(f_s)) = \sum_{l_k \in r(f_s)} m_{ks}^n\}_{n=1;2}$, where $m_{ks}^1 = c_{ks}$ and $m_{ks}^2 = -\log(1 - B_{ks})$, and $\mathcal{D}(f_s)$ is the set of feasible loopless paths for flow f_s, resulting from traffic engineering constraints. The logarithmic function is used to transform the blocking probability into an additive metric.

The use of this constrained bi-objective shortest path problem as a basis for the resolution approach to the network problem $\mathcal{P}_G - S$ relies on the fact that the metric blocking probability tends (at a network level) to minimize the maximal node-to-node blocking probabilities $B(f_s)$ while the metric implied cost tends to maximize the total average revenue W_T (see [10] and [27]). When one states that using the minimization of path implied cost 'tends' to maximize W_T this would be in rigor only valid if the choice of such an 'optimal' path for a given f_s would not change in any form the network working condition, an assumption that is not true, having in mind the interdependencies among $\{c_{ks}\}, \{B_{ks}\}$ and $\overline{R}$ (see (4)). This is the ultimate source of difficulty in devising a heuristic based on this principle, as outlined in the next section.

Nevertheless it was possible to develop a heuristic approach based on this principle that gave very good results when compared with reference routing methods like RTNR (Real Time Network Routing) of AT&T, and DAR (Dynamic Alternate Routing), aimed at maximizing the total expected revenue, as shown in [26].

In order to extend this resolution principle to the problem P-M2-S2 we need to extend the definition of implied costs to a network with two service classes. For this purpose we propose the following definition of *marginal implied costs* associated with QoS (BE) traffic, by extending the original interpretation of implied costs by Kelly [18] to a multirate loss network with two service classes. Hence we will define the *marginal implied cost for QoS traffic*, c_{ku}^Q, associated with the acceptance of a connection (or "call") of traffic f_u of any service type $u \in \mathscr{S}$ on a link l_k, as the expected value of the traffic loss induced on all QoS traffic flows resulting from the capacity decrease in link l_k. In an analogous form one can define the marginal implied cost c_{ku}^B for BE traffic associated with the acceptance of a connection of traffic flow f_u on link l_k.

We will assume, as a conjecture, that the marginal implied costs for QoS (BE) traffic can be estimated by solving a system of equations analogous to (1)-(2), by restraining the summation on the right hand side of (1) to the service types of QoS (BE) class, respectively, by introducing the marginal surplus values $s_{r^i(f_s)}^{Q(B)}$ and the marginal revenues,

$$w^{Q(B)}(f_s) = \alpha^{Q(B)} w(f_s) \text{ with } \alpha^Q + \alpha^B = 1.0 \tag{5}$$

where the coefficients $\alpha^{Q(B)} \in\;]0.0; 1.0[$ satisfy the above normalization condition. This condition and the calculation of the marginal costs through those equations are consistent with the definition of the sensitivity of the marginal revenues associated with QoS and BE traffic, through expressions analogous to (3), as described in the following conjecture.

Conjecture B: In multirate networks with (one-stage) alternative routing and two service classes, the sensitivity of the revenues $W_{Q(B)}$ with respect to the traffic $A(f_s)$ being offered to a pair of routes (r^1, r^2), when an approximation to the expected revenue is calculated from the solution of the fixed point equations in B_{ks}, can be written

$$
\begin{aligned}
\frac{\partial W_{Q(B)}}{\partial A(f_s)} = {} & (1 - L_{r^1(f_s)}) \left(w^{Q(B)}(f_s) - \sum_{l_k \in r^1(f_s)} c_{ks}^{Q(B)} \right) \\
& + L_{r^1(f_s)}(1 - L_{r^2(f_s)}) \left(w^{Q(B)}(f_s) - \sum_{l_k \in r^2(f_s)} c_{ks}^{Q(B)} \right)
\end{aligned}
\tag{6}
$$

In fact, taking into account that $W_T = W_Q + W_B$, (6) and (5), together with the condition $c_{ks} = c_{ks}^Q + c_{ks}^B$, imply (from the additivity property of the derivatives) equation (3) (in conjecture A). The marginal expected revenues per call of f_s, $w^{Q(B)}(f_s)$ (such that $w^Q(f_s) + w^B(f_s) = w(f_s)$) in equation (6) may be interpreted as the part of the expected revenue $w(f_s)$ generated by a connection of f_s that is accepted by the network (for a given choice of the pair of routes $(r^1(f_s), r^2(f_s))$), that is assigned to the calculation of the sensitivity of the revenue from the point of view of traffic losses induced either in the QoS traffic flows or in the BE flows.

In the present model we will consider, as a first approach, $\alpha^Q = \alpha^B = 0.5$ so that no bias is induced in the calculation of the marginal costs through the choice of these factors.

This conjecture plays a central role in the theoretical foundation of the resolution approach, since its validity implies that the marginal implied costs associated with QoS(BE) traffic can be interpreted exactly as stated in the paragraph above. This also implies the consistency of the calculation of the $\{c_{ks}^{Q(B)}\}$ through a system analogous to (1), with the changes associated with the introduction of marginal coefficients $w^{Q(B)}(f_s)$.

The auxiliary bi-objective shortest path problem to be solved will have two possible configurations. The first one will use as link cost coefficients $m_{ks}^1 = c_{ks}^Q$ when one intends to obtain candidate solutions to improve the revenue of the QoS traffic, and the second one will use $m_{ks}^1 = c_{ks}^B$ when one seeks to improve the revenue associated with the BE traffic.

To solve this constrained shortest path problem we will use an adaptation of the previously developed algorithmic approach MMRA-S or Modified Multiobjective Routing Algorithm for multiservice networks, described in [9] and [26].

4 A First Heuristic Approach

Next we describe the main features of a heuristic procedure for solving the model in the context of application to the MPLS network used in [30]. The basic architecture of the heuristic is analogous to the MODR-S (Multiobjective Dynamic Routing for Multiservice networks) heuristic described in [9] and applied in [26], with some relevant adaptations. These adaptations and changes in the heuristic have to do, on the one hand, with the different type of the considered network topology, as shown in Figure 1, that unlike the one for which MODR-S was specified (a fully meshed network) has low connectivity. On the other hand, there are important differences in the objective functions as a result of the existence of two traffic classes, as previously analyzed.

The 'core' of the heuristic is the generation of candidate solutions $(r^1(f_s), r^2(f_s))$ for each f_s, where $r^1(f_s)$ is defined according to the rules described hereafter and $r^2(f_s)$ is typically obtained through a constrained bi-objective shortest path algorithm, devised for problem $\mathscr{P}^{(2)}$, MMRA-S2.

Having in mind the network topology and the need to make a further distinction between real-time QoS services (video and voice services) and non-real time QoS services (such as 'premium data' service), special rules were defined for the selection of candidate first choice routes $r^1(f_s)$. An important parameter defined from these rules is the maximal number of arcs D_s per route for each service type s.

In general, for QoS services, $r^1(f_s)$ is chosen as the direct arc whenever it exists. If no direct path exists, then for real-time QoS services one of the feasible paths with the least number of arcs is chosen. If there is more than one of these paths, the choice is made according to MMRA-S2, by using priority regions defined in the objective function space of $\mathscr{P}^{(2)}$ (see [9,26]). These criteria result from the more stringent constraints on delay and jitter of this type of services, and also having in mind an increase in reliability of the connections. For the remaining QoS services the initial choice of $r^1(f_s)$ is made by using the algorithm MMRA-S2 and the mentioned priority regions.

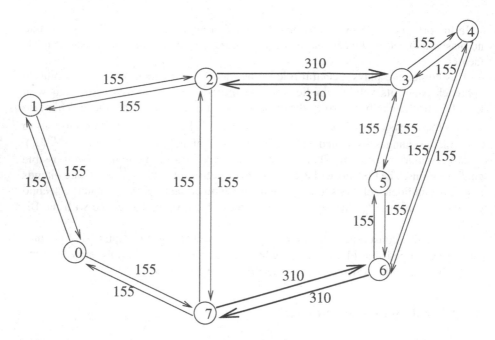

Fig. 1. Test network M1, proposed in [30], with the indication of the bandwidth of each arc C'_k, in Mbps

A similar procedure is applied for obtaining $r^1(f_s)$ for BE traffic flows. Note that for BE traffic flows, the direct path, whenever it exists, is treated exactly as any other path, i.e. no preference is given to the direct path over the other paths. Concerning the calculation of the second choice routes $r^2(f_s)$ for QoS or BE traffic, this is made according to the MMRA-S2 algorithm. These alternative routes may be eliminated through a mechanism designated as Alternative Path Removal (APR) proposed in [26], in order to prevent performance degradation in overload conditions. Therefore, $r_s = r^2(f_s)$ is eliminated whenever $m^1(r_s) > \alpha^{Q(B)} \cdot d_s \cdot z_{APR}$ and $m^2(r_s) > -\log(0.7) \cdot z_{APR}$, where $z_{APR} \in [0.0; 1.0]$ is an empirical parameter which has to be adequately chosen through extensive experimentation with the model. The complete set of rules used in the given network to define candidate routes $(r^1(f_s), r^2(f_s))$ is described in [8, Appendix C].

As for the 'core' algorithm MMRA-S2 it is basically an adaptation to the present model of the bi-objective constrained shortest path algorithm in [26], which is an extension of the algorithm in [10] to a multiservice environment. The main features of MMRA-S are now briefly reviewed. Note that in general there is no feasible solution which minimizes both objective functions of $\mathscr{P}^{(2)}$ simultaneously. Since there is no guarantee of the feasibility of this ideal optimal solution, the resolution of this routing problem aims at finding a best compromise path from the set of non-dominated solutions, according to some relevant criteria previously defined. In this context, since path computation and selection have to be fully automated, such criteria are embedded in the working of the algorithm MMRA [10,27] via preference regions in the objective function space.

The purpose of the MMRA version used in the present context is to calculate solutions to problem $\mathscr{P}^{(2)}$, and is a variant of the algorithm proposed in [1] for a bi-objective shortest path problem of the same type, adapted to the requirements and specifics of the multiobjective dynamic routing model MODR-S [26]. The approach given in [1] (inspired by the one presented in [32] and [6] concerning a procedure to search interactively non-dominated paths) enables the calculation and selection of non-dominated paths in the framework of a routing control mechanism. The procedure satisfies this requirement by integrating the K-shortest paths algorithm [24] and a special concept designated as "soft constraints" (that is constraints not directly incorporated into the mathematical model). The main features of this approach are: i) the representation of QoS requirements through soft constraints corresponding to *requested* and *acceptable thresholds* for each QoS metric; ii) the consideration of this type of thresholds defines priority regions in the objective function space in which non-dominated solutions are searched for; iii) the non-dominated paths are computed by means of an extremely efficient K-shortest path algorithm proposed in [24], designated as MPS algorithm; iv) the adaptation of the preference thresholds to time-varying network working conditions, as required in dynamic routing applications. Further details on the MODR-S teletraffic model and MMRA, in a multiservice network, can be seen in the report [25].

This algorithm was adapted straightforwardly to the present routing model, by incorporating in the definition of the feasible route set $\mathscr{D}(f_s)$ and in the route selection procedure the rules described in detail in [8, Appendix C]. Also the coefficients of the second objective function are the marginal implied costs, either c_{ks}^Q or c_{ks}^B depending on the expected revenue (associated with QoS or BE traffic) we are seeking to improve.

Note that successive application of MMRA-S2 to every traffic flow does not lead to an effective (even less robust) resolution approach to the network routing problem P-M2-S2. The essential reason for this is an instability phenomenon that arises in such a path selection procedure, expressed by the fact that the route sets $\overline{R}$ tend to oscillate between certain solutions, some of which may lead to poor global network performance under the prescribed criteria. This is associated with the complexity and interdependencies in the network problem P-M2-S2, namely the interdependencies between $\{c_{ks}^{Q(B)}\}$ and $\{B_{ks}\}$ and between these two sets of parameters and the current total route set $\overline{R}$.

A general idea of the heuristic is to seek, for each service, a routing solution $\overline{R}(s)$ which may lead to a better performance in terms of W_B, $B_{ms|Q}$ and $B_{Ms|Q}$, $s \in \mathscr{S}_Q$, hence leaving network resources available for traffic flows of other services so that the solutions selected at each step may enable an improvement on the higher level objective functions W_Q and $B_{Mm|Q}$. Therefore the heuristic is constructed in order to seek, firstly for each QoS service and beginning with the higher bandwidth services (considering $s = 1, \cdots, |\mathscr{S}_Q|$) and, secondly, for each BE service and beginning with the higher bandwidth services ($s = |\mathscr{S}_Q| + 1, \cdots, |\mathscr{S}|$), solutions which dominate the initial one, in terms of $B_{ms|Q}$ and $B_{Ms|Q}$ for QoS services and in terms of W_B for BE services, while not worsening any of the network main metrics, W_Q and $B_{Mm|Q}$ (taking into account the optimization priorities in P-M2-S2).

The basis of the heuristic approach (similarly to MODR-S [26]) is to search for the subset of the path set $\overline{R}^a = \cup_{s=1}^{|\mathscr{S}|} \overline{R}^a(s) : \overline{R}^a(s) = \{(r^1(f_s), r^2(f_s)), f_s \in \mathscr{F}_s\}$, the elements of which should be possibly changed in the next route improvement cycle. Detailed analysis and extensive experimentation with MODR-S led us to propose [27,26] a specific criterion for choosing candidate paths for possible routing improvement by increasing order of a function $\xi(f_s)$ of the current $(r^1(f_s), r^2(f_s))$. The criterion depends explicitly on the first choice path $r^1(f_s)$ and on the alternative path $r^2(f_s)$. The adaptation of this criterion to the present model P-M2-S2 considers:

$$\xi(f_s) = F_C^{Q(B)}(f_s) F_L(f_s) \tag{7}$$

if the effect over QoS(BE) traffic is being considered, and where $F_C^{Q(B)}(f_s) = (n_2 - n_1) c_1^{'Q(B)} + c_{r^1(f_s)}^{Q(B)} - c_{r^2(f_s)}^{Q(B)}$, $F_L(f_s) = 1 - L_{r^1(f_s)} L_{r^2(f_s)}$, $c_{r(f_s)}^{Q(B)} = \Sigma_{l_k \in r(f_s)} c_{ks}^{Q(B)}$ and $c_1^{'Q(B)} = \frac{1}{n_1} \Sigma_{l_k \in r^1(f_s)} c_{ks}^{Q(B)} = \frac{1}{n_1} c_{r^1(f_s)}^{Q(B)}$.

The aim of the factor $F_C^{Q(B)}(f_s)$ is to favor (concerning the interest in changing the second choice route when seeking to improve W_Q or W_B) the flows for which the second choice route has a high implied cost and the first choice route a low implied cost. The factor $(n_2 - n_1)$ was introduced for normalizing reasons having in mind that $r^1(f_s)$ has n_1 arcs and $r^2(f_s)$ n_2 arcs, in the considered network. The aim of the second factor $F_L(f_s)$ is to favor the choice of the flows with worse current blocking probability. In the cases where overload conditions led to the elimination of the alternative path (see explanation above), $F_C^{Q(B)}(f_s) = c_{r^1(f_s)}^{Q(B)}$ and $F_L(f_s) = 1 - L_{r^1(f_s)}$.

A second point to be tackled by the heuristic procedure is to specify how many (and which) of the routes with smaller values of $\xi(f_s)$ should possibly be changed by applying MMRA-S2 once again. For this purpose the effect of each candidate route, in terms of the relevant objective functions, is anticipated by solving the corresponding analytical model.

The heuristic is initialized with a set of paths $(r^1(f_s), r^2(f_s))$ that was defined without any previous optimization. The quality of the final solution obtained with the heuristic is dependent on the quality of the initial one. The initial routing plan $\overline{R}_o$ to be used in the heuristic considers that the $r^1(f_s)$ are chosen as follows: the initial path chosen for every flow f_s is the shortest one (that is, the one with minimum number of arcs); if there is more than one shortest path, we choose the one with maximal bottleneck bandwidth (i.e. the minimal capacity of its arcs); if there is more than one shortest path with equal bottleneck bandwidth, the choice is arbitrary. The alternative routes $r^2(f_s)$ are chosen from the candidate paths obtained from MMRA-S2 according to the procedures of solution analysis and selection explained above and in the next paragraphs.

In order to have a "good" initial solution some alternative paths must be eliminated from the initial set of paths. According to the criterion of elimination of the alternative paths proposed in [9] and [26], all paths $r^2(f_s)$ satisfying

$$B(f_s) > \frac{\Sigma_{f_s \in \mathscr{F}_s} B(f_s)}{|\mathscr{F}_s|} \text{ or } B(f_s) > 10\% \tag{8}$$

should be eliminated ($|\mathscr{F}_s|$ is the number of flows of type s). This procedure keeps only 'good' alternative paths in the initial solution, and it seeks to improve both the service performances (especially for the services with higher bandwidth demands) and the network main performance metrics. Note that if the final solution of the heuristic does not dominate the initial one (before the elimination of some alternative paths) in terms of the first level objective functions, then it is this initial solution that should be adopted. However, this situation never occurred in all the tests performed.

The heuristic starts off with a cycle that covers all the services, beginning with the QoS services $s = 1, \cdots, |\mathscr{S}_Q|$ and ending with the BE services $s = |\mathscr{S}_Q| + 1, \cdots, |\mathscr{S}|$. Note that QoS services are treated in the model as first priority traffic, and BE services as second priority traffic. Within each class of service, the algorithm begins with the types of services with higher bandwidth demands. Experience has shown that this ordering of the services generally leads to a better performance of the heuristic. Therefore, the heuristic is set up to find, for each service, and starting with the most demanding services, solutions that dominate the previous one with respect to the first level objective functions W_Q and $B_{Mm|Q}$, if possible without worsening the partial criteria for each service, $B_{ms|Q}$ and $B_{Ms|Q}$ for QoS services and W_B for BE services.

The two main cycles of the heuristic are improvement cycles of the objective functions. Two variables, $nPaths$ and $mPaths$, define the current number of paths, ordered according to $\xi(f_s)$, which are candidates for possible improvements in these two cycles. A specific service protection scheme to prevent excessive network blocking degradation in overload situations, the APR, is used, as described earlier. The parameter z_{APR} varies between 0.0 and 1.0, and its value is defined in the inner cycle of the heuristic.

Concerning the numerical complexity of this heuristic, it can be said that the instructions in the inner cycle of the procedure are executed at least $C_N = |\mathscr{S}|(6|\mathscr{F}_N| - 1)$, where $|\mathscr{F}_N| = \frac{1}{|\mathscr{S}|} \sum_{s \in \mathscr{S}} |\mathscr{F}_s|$ is the average number of traffic flows per service. This figure C_N is an indication of the heuristic numerical complexity just at the level of the 'optimization' procedures. One should note that each calculation of the objective functions and marginal costs (which are used as coefficients in the auxiliary bi-objective shortest path model) involves the numerical resolution of a large system of non-linear equations in $\{B_{ks}\}$ and $\{c_{ks}^{Q(B)}\}$ and such calculations have to be repeated whenever a candidate pair of paths $(r^1(f_s), r^2(f_s))$ is recalculated and analyzed, in terms of its impact on the objective function values.

This heuristic is formalized in the Appendix. A more detailed explanation is in the report [8].

5 Application of the Model

In this section, computational results obtained with the MOR-S2 heuristic in a network case study analogous to the one in [30], are presented. The "quality" of these results concerning W_Q was compared to results obtained with another model proposed in [30] for MPLS networks with two service classes that uses a lexicographic optimization formulation based on a deterministic MCF (*Multicommodity Flow*) model, which gives an upper bound to our objective function W_Q in P-M2-S2. For this purpose the network case study for two service classes in MPLS addressed in [30] was considered.

5.1 Application of the Model to a Network Case Study

In [30], a model is proposed for traffic routing and admission control in multiservice, multipriority networks supporting traffic with different QoS requirements. Having in mind a better understanding of the application case study we will begin with an overview of the relevant features of the model proposed in [30]. Instead of using stochastic traffic models in the calculation of paths, deterministic models are used, in particular mathematical programming models based on MCFs. These models are only a rough approximation in this context, and in fact they tend to under-evaluate the blocking probabilities. As a result, an adaptation of the original model was introduced in [30] in order to obtain more 'correct' models, that is models which give a better approximation in a stochastic traffic environment. The authors of [30] propose a simple technique to adapt the MCF model to a stochastic environment: the requested values of the flows bandwidths in the MCF model are compensated with a factor $\alpha \geq 0.0$, so as to model the effect of the random fluctuations of the traffic that are typical of stochastic traffic models. The higher the variability of the point processes of the stochastic model, the higher is the need for compensation and therefore the higher should α be. A deterministic traffic routing model based on MCFs is specified, where traffic splitting is used. This means that the required bandwidth of each flow may be divided by multiple paths from source to destination, allowing for a better load balancing in the network.

The objective functions of this problem to be maximized are the revenues W_Q and W_B, associated with QoS and BE flows. A bi-criteria lexicographic optimization formulation is considered, concerning the revenues W_Q and W_B, so that the improvements in W_B are to be found under the constraint that the optimal value of W_Q is maintained.

In the deterministic flow-based model [30], a base matrix $T = [T_{ij}]$ with offered bandwidth values from node i to node j [Mbps] is given. A multiplier $m_s \in [0.0; 1.0]$ with $\sum_{s \in \mathcal{S}} m_s = 1.0$ is applied to these matrix values to obtain the offered bandwidth of each flow $f_s = (v_i, v_j, \overline{\gamma}_s, \overline{\eta}(s, \mathcal{L}))$ to the network, $T(f_s) = m_s T_{ij}$. The transformation of this type of matrix into a matrix of traffic intensities, used in our stochastic traffic model, can be carried out as described in Appendix E.1 of the report [8]. The model for calculating the blocking probabilities is described in Appendix B of the same report.

In the application example in [30], results for the QoS flows revenue W_Q are presented for three values of α: $\alpha = 0.0$ corresponds to the deterministic approach; $\alpha = 0.5$ is the compensation factor when calls arrive according to a Poisson process, service times follow an exponential distribution and the network is critically loaded; and $\alpha = 1.0$ for traffic flows with higher 'variability'. The results for the BE revenue W_B are not presented, but for $\alpha = 0.0$ the maximum value is 79.33% of the maximum possible value $W_B^{\max}$, because 20.67% of the traffic is not even admitted to the network due to an admission control scheme. The results for the revenues obtained from the information provided in [30] are $W_{Q|\alpha=0.0}^{[30]} = 65156.00$ and $W_{B|\alpha=0.0}^{[30]} \leq 17462.50$; $W_{Q|\alpha=0.5}^{[30]} = 60829.72$; $W_{Q|\alpha=1.0}^{[30]} = 56338.65$.

Note that the revenue values W_Q in the model [30] should be viewed as upper bounds on the QoS revenue values of the problem P-M2-S2, because of the differences between the two optimization problems and the important differences in the underlying routing control and traffic models, previously referred to. In fact an important feature of the resolution approach of the routing problem in [30] is the admission control of BE traffic flows at the first stage of resolution, so the BE traffic that is actually offered to the network is the fraction of traffic that was not rejected by the admission control. Also note the absence of alternative routing as well as the use of traffic splitting. Therefore, for a specific traffic matrix, the model in [30] tends to obtain smaller values of blocking probability by comparison with a situation without admission control, and this tends to favor higher global revenues. Also the traffic representation, even in the approximated stochastic model [30, Sect.5.4], is a bit rough and tends to under-evaluate the blocking probabilities and to over-estimate the revenues.

5.2   Some Experimental Results

The test network M1 proposed in [30] is displayed in Figure 1. It has $N = 8$ nodes, with 10 pairs of nodes linked by a direct arc. The network has a total of $|\mathscr{L}| = 20$ unidirectional arcs, one in each direction for every pair of adjacent nodes. The bandwidth of each arc C_k' is shown in Figure 1. The number of channels C_k (with basic capacity $u_0 = 16$ kbps) is $C_k = \left\lceil \frac{C_k'}{u_0} \right\rceil$.

There are $|\mathscr{S}| = 4$ service types with the features described in Table 1. The value of $d_s = \frac{d_s'}{u_0}$ [channels] $\forall s \in \mathscr{S}$ presented in the table (where d_s' is the required bandwidth in kbps) is calculated with $u_0 = 16$ kbps. Note that $w_s = d_s, \forall s \in \mathscr{S}$.

Table 1. Service features on the test network M1

Service	Class	d_s' [kbps]	d_s [channels]	w_s	h_s [s]	D_s [arcs]	m_s
1 - video	QoS	640	40	40	600	3	0.1
2 - Premium data	QoS	384	24	24	300	4	0.25
3 - voice	QoS	16	1	1	60	3	0.4
4 - data	BE	384	24	24	300	7	0.25

The traffic flow data information provided by [30] is a base matrix $T = [T_{ij}]$ with offered bandwidth values [Mbps] and a multiplier applied to these matrix values to obtain the offered bandwidth of each flow. Given this information and the variability compensation equations, the values of $A(f_s)$, the average number of offered μ-flows of f_s, during the average service time of a μ-flow can be calculated as shown in [8, Appendix E.1].

In MOR-S2, an initial solution has to be chosen and applied as input data to the heuristic. We chose to consider an initial solution with only one path for each flow, i.e. without a second choice path as we concluded that this is more adequate, and leave it up to the heuristic to find an adequate solution with second choice paths. The initial solution is the same for all the services and the paths are symmetrical.

The objective function values for the initial and final solutions obtained with the MOR-S2 heuristic are in Table 2. The revenue values have 2 decimal places and the blocking probability values have 3 significant figures. The value of the QoS revenue in the final solution is also presented as a percentage of the optimal value obtained in [30].

The MOR-S2 heuristic manages to start off with an initial solution with poor values for the objective functions and still finishes with a solution with significantly better values. The values for all the objective functions, for all values of α, are improved through the heuristic. The QoS revenue of the final solutions are slightly worse than those of the

Table 2. Objective function values for the initial and final solution for MOR-S2 on the test network M1

Objective Functions	$\alpha = 0.0$ Initial	Final	$\alpha = 0.5$ Initial	Final	$\alpha = 1.0$ Initial	Final
W_Q	54803.69	64330.56*	51785.21	60097.78†	49010.41	55978.80‡
$B_{Mm|Q}$	0.413	0.135	0.413	0.0962	0.405	0.0582
$B_{m1|Q}$	0.413	0.135	0.413	0.0962	0.405	0.0582
$B_{m2|Q}$	0.314	0.0159	0.296	0.00811	0.275	0.00279
$B_{m3|Q}$	0.0198	0.00489	0.0174	0.00263	0.0150	0.000436
$B_{M1|Q}$	0.912	0.848	0.882	0.628	0.841	0.440
$B_{M2|Q}$	0.766	0.0427	0.722	0.0305	0.667	0.0111
$B_{M3|Q}$	0.0585	0.0456	0.0517	0.0241	0.0446	0.0143
W_B	15106.57	17391.44	13787.49	17031.62	12445.64	16509.86

*)98.73% of $W_{Q|\alpha=0.0}^{[30]}$; †)98.80% of $W_{Q|\alpha=0.5}^{[30]}$; ‡)99.36% of $W_{Q|\alpha=1.0}^{[30]}$

optimal solution in [30], as expected. However, these QoS revenues can be considered very good as they stand for approximately 99% of the optimal values in [30]. Therefore, we can consider that MOR-S2 has managed to find an adequate "good" compromise routing solution to the routing problem P-M2-S2. In fact these experimental results for three traffic matrices showed that the expected QoS revenue obtained with our heuristic is never less than 98.7% of that upper bound while a substantial improvement on the other objective functions could be obtained with respect to the initial solution, using only shortest path first choice routing, typical of Internet routing conventional algorithms.

6 Conclusions and Further Work

In the emergent MPLS technology for the Internet the implementation of connection-oriented services from origin to destination is possible. This feature in association with other functional capabilities of MPLS enables the implementation of advanced QoS-based routing mechanisms, namely by establishing "explicit routes" (determined at the originating node) for each traffic flow.

Having in mind these features and capabilities of MPLS routing it is possible to explore the multicriteria nature of the routing environment and associated metrics (of technical and economic nature), and devise multicriteria routing models capable of explicitly incorporating various network revenues and performance metrics, including fairness QoS objectives at the level of the services. This enables the formulation of multiobjective network-wide optimization routing models, namely hierarchical multicriteria models, for possible application at the top level of this type of networks.

In this work we described a bi-level multiobjective routing model in MPLS networks, formulated within the general modeling framework developed by the authors in [7], assuming that there are two classes of services (and different types of traffic flows in each class), namely QoS and BE services. The routing model also considers the possibility of using alternative routing when that is advantageous to the first priority objective functions. An important feature of this model is the use of hierarchical optimization with two optimization levels, including fairness objectives. Another feature of the model is the use of an approximate stochastic representation of the traffic flows in the network, based on the use of the concept of effective bandwidth for macro-flows and on a generalized Erlang model for estimating the blocking probabilities in the arcs. Also note that while QoS and BE traffic flows are treated separately in terms of objective functions in order to take into account their different priority in the optimization model, the interactions among all traffic flows are fully represented through the traffic model. This is another advantage in comparison to other routing models proposed for networks with two service classes.

We have also presented the theoretical foundations of a specialized heuristic strategy for finding "good" compromise solutions to the very complex bi-level routing optimization problem. This theoretical foundation was based on a conjecture concerning the definition of marginal implied costs for QoS flows and BE flows, which is an extension and adaptation of earlier definitions of implied costs for single-service networks with alternative routing in [18]. The structure of a first version of a heuristic procedure for resolving the problem was described. The model was applied to a test network previously used in a benchmarking study [30] that uses a lexicographic optimization routing approach, including admission control for BE traffic, based on a deterministic traffic representation, the results of which can be considered as upper bounds with respect to the QoS traffic revenue. These preliminary results seem quite encouraging concerning the potential performance of a multicriteria routing model of this nature.

The major limitation of this type of model is its inherent great complexity and the associated computational burden, which constitute the reverse of its 'ambitious' features, namely, network-wide optimization, multiobjective nature with a significant number of objective functions, use of alternative routing and a stochastic representation of traffic flows of multiple service types. This makes, at present, its potential practical application restrained to networks with a limited number of nodes, such as the core and intermediate (metro-core) level networks of low dimension. The model could also be used as the basis of a periodic type dynamic routing method, similarly to MODR-S [26].

Further work on this model will involve the search for improvements in the heuristic procedure, through sensitivity analysis of the present version, or the possible development of metaheuristics for this very complex network routing problem. Finally a discrete event simulation platform will be developed, which will enable a more exact evaluation of the results of the heuristic in a stochastic dynamic environment closer to real network working conditions.

References

1. Antunes, C.H., Craveirinha, J., Clímaco, J., Barrico, C.: A multiple objective routing approach for integrated communication networks. In: Smith, D., Key, P. (eds.) Proceedings of the 16th International Teletraffic Congress (ITC16) – Teletraffic Engineering in a Competitive World, pp. 1291–1300. Elsevier, Amsterdam (1999)
2. Awduche, D., Chiu, A., Elwalid, A., Widjaja, I., Xiao, X.: Overview and principles of Internet traffic engineering. RFC 3272, Network Working Group (2002)
3. Awduche, D., Malcolm, J., Agogbua, J., O'Dell, M., McManus, J.: Requirements for traffic engineering over MPLS. RFC 2702, Network Working Group (1999)

4. Clímaco, J., Craveirinha, J.: Multicriteria analysis in telecommunication network planning and design – Problems and issues. In: Figueira, J., Greco, S., Ehrgott, M. (eds.) Multiple Criteria Decision Analysis – State of the Art Surveys. International Series in Operations Research & Management Science, vol. 78, pp. 899–951. Springer, Heidelberg (2005)

5. Clímaco, J.C.N., Craveirinha, J.M.F., Pascoal, M.M.B.: Multicriteria routing models in telecommunication networks – Overview and a case study. In: Shi, Y., Olson, D.L., Stam, A. (eds.) Advances in Multiple Criteria Decision Making and Human Systems Management: Knowledge and Wisdom, pp. 17–46. IOS Press, Amsterdam (2007)

6. Clímaco, J.C.N., Martins, E.Q.V.: A bicriterion shortest path algorithm. European Journal of Operational Research 11(4), 399–404 (1982)

7. Craveirinha, J., Girão-Silva, R., Clímaco, J.: A meta-model for multiobjective routing in MPLS networks. Central European Journal of Operations Research 16(1), 79–105 (2008)

8. Craveirinha, J., Girão-Silva, R., Clímaco, J., Martins, L.: A hierarchical multiobjective routing model for MPLS networks with two service classes – Analysis and resolution approach. Research Report 5/2007, INESC-Coimbra (October 2007), http://www.inescc.pt, ISSN 1645-2631

9. Craveirinha, J., Martins, L., Clímaco, J.N.: Dealing with complexity in a multiobjective dynamic routing model for multiservice networks – A heuristic approach. In: Proceedings of the 15th Mini-EURO Conference on Managing Uncertainty in Decision Support Models (MUDSM 2004), Coimbra, Portugal, September 22-24 (2004)

10. Craveirinha, J., Martins, L., Gomes, T., Antunes, C.H., Clímaco, J.N.: A new multiple objective dynamic routing method using implied costs. Journal of Telecommunications and Information Technology 3, 50–59 (2003)

11. Elsayed, H.M., Mahmoud, M.S., Bilal, A.Y., Bernussou, J.: Adaptive alternate-routing in telephone networks: Optimal and equilibrium solutions. Information and Decision Technologies 14, 65–74 (1988)

12. Erbas, S.C.: Utilizing evolutionary algorithms for multiobjective problems in traffic engineering. In: Ben-Ameur, W., Petrowski, A. (eds.) Proceedings of the International Networks Optimization Conference (INOC 2003), Evry/Paris, France, October 27-29, pp. 207–212. Institut National des Télécommunications (2003)

13. Erbas, S.C., Erbas, C.: A multiobjective off-line routing model for MPLS networks. In: Charzinski, J., Lehnert, R., Tran-Gia, P. (eds.) Proceedings of the 18th International Teletraffic Congress (ITC-18), Berlin, Germany, pp. 471–480. Elsevier, Amsterdam (2003)

14. Erbas, S.C., Mathar, R.: An off-line traffic engineering model for MPLS networks. In: Bakshi, B., Stiller, B. (eds.) Proceedings of the 27th Annual IEEE Conference on Local Computer Networks (27th LCN), Tampa, Florida, pp. 166–174. IEEE Computer Society, Los Alamitos (2002)

15. Faragó, A., Blaabjerg, S., Ast, L., Gordos, G., Henk, T.: A new degree of freedom in ATM network dimensioning: Optimizing the logical configuration. IEEE Journal on Selected Areas in Communications 13(7), 1199–1206 (1995)

16. Granat, J., Wierzbicki, A.P.: Multicriteria analysis in telecommunications. In: Proceedings of the 37th Hawaii International Conference on System Sciences, January 5-8 (2004)

17. Kelly, F.P.: Notes on effective bandwidths. In: Kelly, F.P., Zachary, S., Ziedins, I. (eds.) Stochastic Networks: Theory and Applications. Royal Statistical Society Lecture Notes Series, vol. 4, pp. 141–168. Oxford University Press, Oxford (1996)

18. Kelly, F.P.: Routing in circuit-switched networks: Optimization, shadow prices and decentralization. Advances in Applied Probability 20(1), 112–144 (1988)

19. Kelly, F.P.: Routing and capacity allocation in networks with trunk reservation. Mathematics of Operations Research 15(4), 771–793 (1990)

20. Knowles, J., Oates, M., Corne, D.: Advanced multi-objective evolutionary algorithms applied to two problems in telecommunications. BT Technology Journal 18(4), 51–65 (2000)

21. Kochkar, H., Ikenaga, T., Hori, Y., Oie, Y.: Multi-class QoS routing with multiple routing tables. In: Proceedings of IEEE PACRIM 2003, Victoria, B.C., Canada, August 2003, pp. 388–391 (2003)
22. Kuipers, F., Van Mieghem, P., Korkmaz, T., Krunz, M.: An overview of constraint-based path selection algorithms for QoS routing. IEEE Communications Magazine, 50–55 (December 2002)
23. Ma, Q., Steenkiste, P.: Supporting dynamic inter-class resource sharing: A multi-class QoS routing algorithm. In: Proceedings of IEEE Infocom 1999, New York, March 1999, pp. 649–660 (1999)
24. Martins, E.Q.V., Pascoal, M.M.B., Santos, J.L.E.: Deviation algorithms for ranking shortest paths. International Journal of Foundations of Computer Science 10(3), 247–263 (1999)
25. Martins, L., Craveirinha, J., Clímaco, J.: A new multiobjective dynamic routing method for multiservice networks – Modeling, resolution and performance. Research Report 2/2005, INESC-Coimbra (February 2005), http://www.inescc.pt
26. Martins, L., Craveirinha, J., Clímaco, J.: A new multiobjective dynamic routing method for multiservice networks: Modeling and performance. Computational Management Science 3(3), 225–244 (2006)
27. Martins, L., Craveirinha, J., Clímaco, J.N., Gomes, T.: Implementation and performance of a new multiple objective dynamic routing method for multiexchange networks. Journal of Telecommunications and Information Technology 3, 60–66 (2003)
28. Mitra, D., Morrison, J.A., Ramakrishnan, K.G.: ATM network design and optimization: A multirate loss network framework. IEEE/ACM Transactions on Networking 4(4), 531–543 (1996)
29. Mitra, D., Morrison, J.A., Ramakrishnan, K.G.: Optimization and design of network routing using refined asymptotic approximations. Performance Evaluation 36-37, 267–288 (1999)
30. Mitra, D., Ramakrishnan, K.G.: Techniques for traffic engineering of multiservice, multipriority networks. Bell Labs Technical Journal 6(1), 139–151 (2001)
31. Pióro, M., Medhi, D.: Routing, Flow, Capacity Design in Communication and Computer Networks, 1st edn. Morgan Kaufmann Publishers, San Francisco (2004)
32. Rodrigues, J.M., Clímaco, J.C., Current, J.R.: An interactive bi-objective shortest path approach: Searching for unsupported nondominated solutions. Computers & Operations Research 26, 789–798 (1999)
33. Rosen, E., Viswanathan, A., Callon, R.: Multiprotocol label switching architecture. RFC 3031, Network Working Group (January 2001)

Appendix – Formalization of the MOR-S2 Heuristic

I. $\bar{R}_a \leftarrow \bar{R}_o$

II. Compute $\bar{B}$ and $W_Q, B_{Mm|Q}$ for $\bar{R}_a$

III. $W_Q^o \leftarrow W_Q, B_{Mm|Q}^o \leftarrow B_{Mm|Q}$

IV. Eliminate the paths $r^2(f_s)$ in $\bar{R}_a$ that verify (8)

V. $\bar{R}_e \leftarrow \bar{R}_a$

VI. Compute $\bar{B}$ and $W_Q, B_{Mm|Q}$ for $\bar{R}_a$

VII. $\max\{W_Q\} \leftarrow W_Q, \min\{B_{Mm|Q}\} \leftarrow B_{Mm|Q}$

VIII. *For $s = 1$ to $s = |\mathscr{S}|$*

1. $\bar{R}_a(s) \leftarrow \bar{R}_e(s); \bar{R}_*(s) \leftarrow \bar{R}_e(s)$
2. Compute $\bar{B}$ and $B_{ms}, B_{Ms}, s \in \mathscr{S}_Q$ or $W_B, s \in \mathscr{S}_B$ for $\bar{R}_a$
3. $\min\{B_{ms}\}_{ini} \leftarrow B_{ms}, \min\{B_{Ms}\}_{ini} \leftarrow B_{Ms}, s \in \mathscr{S}_Q$ or $\max\{W_B\}_{ini} \leftarrow W_B, s \in \mathscr{S}_B$
4. $mPaths \leftarrow |\mathscr{F}_s|$ (= total number of flows $\leq N(N-1)$), $z_{APR} \leftarrow 1, ape \leftarrow 0$
5. *While ($mPaths \geq |\mathscr{F}_s|-1$) do*
   (a) *$nCycles \leftarrow 2$*
   (b) *$nPaths \leftarrow mPaths$*
   (c) *$\bar{R}_a(s) \leftarrow \bar{R}_e(s)$*
   (d) Compute $\bar{B}$ and $\bar{c}^Q, B_{ms}, s \in \mathscr{S}_Q$ or $\bar{c}^B, W_B, s \in \mathscr{S}_B$ for $\bar{R}_a$
   (e) $\min\{B_{ms}\} \leftarrow B_{ms}, s \in \mathscr{S}_Q$ or $\max\{W_B\} \leftarrow W_B, s \in \mathscr{S}_B$
   (f) *While ($nPaths > 0$) do*
      i. Compute and order the values of the function $\xi(f_s)$ – see (7)
      ii. Find the *$nPaths$* with lower value of $\xi(f_s)$
      iii. Compute with MMRA-S2 new candidate paths for the corresponding O-D pairs and define a new set of first and second choice paths for the service $s, \bar{R}_a(s)$, according to the rules established in [8, Appendix C].
      iv. Compute $\bar{B}$ and $B_{ms}, B_{Ms}, s \in \mathscr{S}_Q$ or $W_B, s \in \mathscr{S}_B$ for $\bar{R}_a$
      v. If $s \in \mathscr{S}_Q$ then
         A. If ($B_{ms} < \min\{B_{ms}\}_{ini}$ and $B_{Ms} < \min\{B_{Ms}\}_{ini}$) then
            – Compute $W_Q, B_{Mm|Q}$
            – If $W_Q > \max\{W_Q\}$ and $B_{Mm|Q} < \min\{B_{Mm|Q}\}$ then
               • $\min\{B_{ms}\}_{ini} \leftarrow B_{ms}, \min\{B_{Ms}\}_{ini} \leftarrow B_{Ms}$
               • $\max\{W_Q\} \leftarrow W_Q, \min\{B_{Mm|Q}\} \leftarrow B_{Mm|Q}$
               • $\bar{R}_*(s) \leftarrow \bar{R}_a(s)$
         B. If ($B_{ms} < \min\{B_{ms}\}$) then
            – $\min\{B_{ms}\} \leftarrow B_{ms}$
         C. Otherwise go to 5.f.vii
      vi. Otherwise ($s \in \mathscr{S}_B$)
         A. If ($W_B > \max\{W_B\}_{ini}$) then
            – Compute $W_Q, B_{Mm|Q}$
            – If $W_Q > \max\{W_Q\}$ and $B_{Mm|Q} < \min\{B_{Mm|Q}\}$ then
               • $\max\{W_B\}_{ini} \leftarrow W_B$
               • $\max\{W_Q\} \leftarrow W_Q, \min\{B_{Mm|Q}\} \leftarrow B_{Mm|Q}$
               • $\bar{R}_*(s) \leftarrow \bar{R}_a(s)$

         B. If $(W_B > \max\{W_B\})$ then
           – $\max\{W_B\} \leftarrow W_B$
         C. Otherwise go to 5.f.vii
    vii. (Update *nPaths*)
         A. $nPaths \leftarrow nPaths - 1$
         B. If $(nPaths = 0$ and $nCycles = 2)$ then
           – $nCycles \leftarrow nCycles - 1$
           – $nPaths \leftarrow |\mathscr{F}_s|$
         C. Compute $\overline{B}$ and $\overline{c}^Q, s \in \mathscr{S}_Q$ or $\overline{c}^B, s \in \mathscr{S}_B$ for $\overline{R}_a$
         D. If $(nPaths \leq 10$ and $ape \geq 1)$ then
           – $z_{APR} \leftarrow nPaths \cdot 0.1$
         E. Otherwise $z_{APR} \leftarrow 1$
    End of the cycle *While (nPaths)*
  (g) $ape \leftarrow ape + 1$
  (h) If $(ape > 1)$ then
    i. $mPaths \leftarrow mPaths - 1$
  End of the cycle *While (mPaths)*
6. $\overline{R}_a(s) \leftarrow \overline{R}_*(s)$

End of the cycle *For (s)*
IX. If $W_Q^o > \max\{W_Q\}$ or $B_{Mm|Q}^o < \min\{B_{Mm|Q}\}$ then

1. The best solution is $\overline{R}_o$

X. Otherwise, the best solution is $\overline{R}_*$
XI. Compute the objective function values for the best solution

Comparison of the Exact and Approximate Algorithms in the Random Shortest Path Problem

Jacek Czekaj[1] and Lesław Socha[2]

[1] University of Silesia, Institute of Physics, 4 Uniwersytecka St., 40-007 Katowice, Poland
jackens@math.us.edu.pl
[2] Cardinal Stefan Wyszyński University in Warsaw, Faculty of Mathematics and Natural Sciences, College of Sciences, 5 Dewajtis St., 01-815 Warszawa, Poland
leslawsocha@poczta.onet.pl

Abstract. The Random Shortest Path Problem with the second moment criterion is discussed in this paper. After the formulation of the problem, exact algorithms, based on general concepts for solving the Multi-objective Shortest Path Problem, are described. Next, several approximate algorithms are proposed. It is shown that the complexity of the exact algorithms is exponential, while the complexity of the approximate algorithms is only polynomial. Computational results for the exact and approximate algorithms, which were performed on large graphs, are shown.

1 Introduction

One of the basic problems in the operational research is the Classic Shortest Path Problem (shortly CSPP) that can be also interpreted as the problem of determination of the path from a given source point to the destination point with minimal transfer time.

In this paper we consider a routing problem to determine the path from a given point to the destination point with minimum transfer time. The simplest solution is to bring the problem to the CSPP.

The CSPP has been studied by many researchers. The first algorithms solving the CSSP were published by Dijkstra [5] for the single source problem and Bellmann [3], Floyd [6] for all pairs problem. There have been many variations and improvements in either analysis or special classes of the given graphs. For example Moffat, Takaoka [11] for the all pairs shortest path problem; Fredman, Tarjan [10] for the single source problem; Frederickson [7] for the single source problem in planar graphs; Johnson [9] for sparse graphs. The detailed study and discussion of the CSPP one can find in monographs, for instance [4] or [1].

Unfortunately, in practice, the travel times are not known exactly. Therefore the travel time from one place to another should be treated as a random variable rather than a single real number. Similar approach can be found in [12] and [13]. This is the main problem discussed in this paper.

In Sect. 2 we present an example showing that in the Random Shortest Path Problem (shortly RSPP) with the second moment criterion the subpath of the shortest path is not necessarily the shortest subpath. We also formally state the Random Shortest Path Problem with the second moment criterion. In Sect. 3 we present two exact algorithms

A. Korytowski et al. (Eds.): System Modeling and Optimization, IFIP AICT 312, pp. 220–238, 2009.
© IFIP International Federation for Information Processing 2009

solving RSPP with the second moment criterion. We show that these exact algorithms have exponential complexity. Therefore in the next section we present approximate algorithms which have only polynomial complexity. In Sect. 5 we show computational results of the presented exact and approximate algorithms, performed on large graphs. In the end of the paper some conclusions are given.

2 Problem Statement

Let $G = (V,E)$ be a directed graph with a finite set of vertices V and a set of edges $E \subseteq V \times V$. We denote by s a source vertex and by t a destination vertex.

In general, in the Shortest Path Problem (shortly SSP) each edge $e \in E$ is associated with some cost (sometimes called weight) c_e. The goal is to find the path, from the source vertex s to the destination vertex t, which minimizes a function of costs related with edges of the path.

In the CSSP this function is a sum, and the costs are real numbers (they usually express the distance between given vertices). There are many algorithms for solving the CSSP, for example, classical Bellman-Ford algorithm or classical Dijkstra algorithm. Since in the next sections we will adopt these algorithms to the Random Shortest Path Problem, therefore we present their pseudocodes below.

Classical-Bellman-Ford-algorithm:
set $c_v = \infty$ for all $v \in V \setminus \{s\}$ and $c_s = 0$
for $i = 1$ **to** $\#V - 1$ **do**
    $relaxation\_state = not\_changed$
    for all $(u,v) \in E$ **do**
        Relax(u,v)
    if $relaxation\_state = not\_changed$ **then**
            break
return path related with c_t

Classical-Dijkstra-algorithm:
set $c_v = \infty$ for all $v \in V \setminus \{s\}$ and $c_s = 0$
$Q = V$
while $Q \neq \emptyset$ **do**
        choose $u \in Q$ such that $c_u = \min\{c_q | q \in Q\}$
        $Q = Q \setminus \{u\}$
        for all $(u,v) \in E$ **do**
            Relax(u,v)
return path related with c_t

Relax(u,v) denotes the following procedure:

Relax(u,v):
if $c_v > c_u + c_{(u,v)}$ **then**
        $c_v = c_u + c_{(u,v)}$
        $\pi_v = u$
        $relaxation\_state = changed$

The detailed description of the CSPP can be found for instance in [4].

In further considerations we will adopt the concept of multi-objective optimization methods. The Multi-objective Shortest Path Problem (shortly MOSPP) has been introduced by Hansen in [8]. In this version of the SSP, each edge of the graph is associated with a vector of real numbers (usually nonnegative). We recall the definition of the standard product order relation in the set $\mathbb{R}^n$:

$$[a_1,\ldots,a_n] < [b_1,\ldots,b_n] \quad \Longleftrightarrow$$
$$\Longleftrightarrow \quad [a_1,\ldots,a_n] \le [b_1,\ldots,b_n] \wedge [a_1,\ldots,a_n] \ne [b_1,\ldots,b_n], \tag{1}$$

where

$$[a_1,\ldots,a_n] \le [b_1,\ldots,b_n] \quad \Longleftrightarrow \quad a_1 \le b_1 \wedge \cdots \wedge a_n \le b_n. \tag{2}$$

Operators $\le$ appearing in the right-hand side of the above equivalence denote the standard order relation in the set $\mathbb{R}$. If (1) holds then we say that the vector $[a_1,\ldots,a_n]$ dominates the vector $[b_1,\ldots,b_n]$. In the MOSPP the goal is to find all paths from the source vertex s to the destination vertex t for which the sum of the vectors related with edges of the path is nondominated by the sum of the vectors related with edges of any other path.

Finally, in the RSPP each edge $e \in E$ is associated with a random variable T_e. The goal is to find the path, from the source vertex s to the destination vertex t, which minimizes a function of random variables related with edges of the path, just like in the CSPP. In practice, the random variable T_e, with $e = (u,v) \in E$, expresses the travel time from the vertex u to the vertex v, or the distance between these vertices. Therefore we assume that every moment of the random variable T_e is nonnegative and finite. We also assume that all the random variables T_e, for $e \in E$, are independent. The notion of the shortest path can be defined in many different ways. It depends only on random variables related with edges of the graph.

For example, if we take as the shortest path, the path with minimal expected value of the sum of random variables related with edges of the path, then the RSPP reduces to the CSPP and can be solved with help of classical Dijkstra or classical Bellman-Ford algorithm. It follows from the equality $\mathscr{E}[X_1 + \cdots + X_n] = \mathscr{E}[X_1] + \cdots + \mathscr{E}[X_n]$, for any random variables $X_1,\ldots,X_n$.

We note also that for any independent random variables $X_1,\ldots,X_n$ holds the equality $\mathscr{V}[X_1 + \cdots + X_n] = \mathscr{V}[X_1] + \cdots + \mathscr{V}[X_n]$, thus if we take as the shortest path, the path with minimal variance of the sum of random variables related with edges of the path, then the RSPP reduces to the CSPP, too.

However, the problem becomes more complicated when as the shortest path we take the path with a minimal sum of the variance and a square of the expected value of the sum of random variables related with edges of the path, i.e. the second moment of the sum of random variables related with edges of the path, because $(\mathscr{E}[X])^2 + \mathscr{V}[X] = \mathscr{E}[X^2]$. Similar criterion has been considered by Murthy and Sarkar in [12]. We show on a simple example that in this case the subpath of the shortest path is not necessarily the shortest subpath.

Example 1. Consider the graph depicted in Fig. 1 and let random variables $T_{(0,1)}$, $T_{(0,2)}$, $T_{(1,3)}$, $T_{(2,3)}$ and $T_{(3,4)}$ have distributions presented in the picture. For any random variable T_e, the notation $\{(x_1,p_1),\ldots,(x_k,p_k)\}$ means that $P(T_e{=}x_1) = p_1,\ldots,P(T_e{=}x_k) = p_k$. Calculating second order moments we obtain

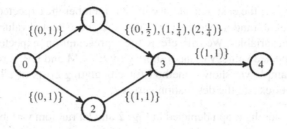

Fig. 1. Example of graph, showing that a subpath of the shortest path is not necessarily the shortest subpath

$$\mathscr{E}\left[(T_{(0,1)}+T_{(1,3)})^2\right] = \tfrac{5}{4} \quad \text{and} \quad \mathscr{E}\left[(T_{(0,2)}+T_{(2,3)})^2\right] = 1, \tag{3}$$

therefore the shortest path between vertices 0 and 3 is the path $\langle 0,2,3 \rangle$, while the shortest path between vertices 0 and 4 is the path $\langle 0,1,3,4 \rangle$. This follows from the equalities

$$\mathscr{E}\left[(T_{(0,1)}+T_{(1,3)}+T_{(3,4)})^2\right] = 3\tfrac{3}{4} \quad \text{and} \quad \mathscr{E}\left[(T_{(0,2)}+T_{(2,3)}+T_{(3,4)})^2\right] = 4. \tag{4}$$

Hence, it follows that the subpath $\langle 0,1,3 \rangle$ of the shortest path $\langle 0,1,3,4 \rangle$ is not the shortest subpath. It means Bellman's "principle of optimality" does not hold!   □

The above example shows that for solving the RSPP with the second moment criterion we cannot directly apply the methods effective for the CSPP.

Now we formulate the RSPP with the second moment criterion. Let

$$\mathscr{P} = \{\langle v_0,v_1,\ldots,v_{n-1},v_n \rangle \mid v_0 = s \wedge v_n = t \wedge (v_0,v_1),\ldots,(v_{n-1},v_n) \in E\} \tag{5}$$

be a set of all paths from the source vertex s to the destination vertex t. The goal is to find a path $\langle v_0,v_1,\ldots,v_{n-1},v_n \rangle \in \mathscr{P}$ such that

$$\mathscr{E}\left[\left(\sum_{i=1}^{n} T_{(v_{i-1},v_i)}\right)^2\right] = \min_{\langle u_0,u_1,\ldots,u_{m-1},u_m \rangle \in \mathscr{P}} \mathscr{E}\left[\left(\sum_{i=1}^{m} T_{(u_{i-1},u_i)}\right)^2\right]. \tag{6}$$

In the next section we present exact algorithms for solving the RSPP.

3   Exact Algorithms

The second moment of the sum of independent random variables $X_1,\ldots,X_n$ can be expressed as follows

$$\mathscr{E}\left[\left(\sum_{i=1}^{n} X_i\right)^2\right] = \sum_{i=1}^{n} \mathscr{E}[X_i^2] + 2\sum_{i=2}^{n} \mathscr{E}[X_i]\sum_{j=1}^{i-1} \mathscr{E}[X_j] \tag{7}$$

or

$$\mathscr{E}\left[\left(\sum_{i=1}^{n} X_i\right)^2\right] = \mathscr{V}\left[\sum_{i=1}^{n} X_i\right] + \left(\mathscr{E}\left[\sum_{i=1}^{n} X_i\right]\right)^2. \tag{8}$$

Therefore, to calculate the cost of a path we must remember the expected values and the second moments of all random variables T_e, for $e \in E$, or expected values and variances of all these random variables. We will refer to the representation [expected value, second moment] and [expected value, variance] shortly by EV-SM and EV-V, respectively.

In the next example we show a method for calculating costs of all possible paths from the source vertex s to the destination vertex t.

Example 2. Consider the graph depicted in Fig. 2 and let random variables related with the edges have vectors of expected value and variance presented in the picture. There is only one edge incoming to the vertex 1, thus $[1,2]$ is the only one cost vector of coming to the vertex 1 (this cost vector is denoted over vertex 1 on the graph). Similarly in the case of the vertices 2 and 3. There are three subpaths to the vertex 4: $\langle 0,1,4 \rangle$, $\langle 0,2,4 \rangle$, $\langle 0,3,4 \rangle$ and the cost vectors of these subpaths are $[2,5]$, $[1,6]$ and $[3,5]$, respectively. Note that the vector $[3,5]$ can by discarded, because if $[\mu, \xi]$ is the cost vector related with some path from the vertex 4 to the destination vertex 8, then $(\mu+3)^2 + (\xi+5)$ is a total cost of the path related with the vector $[3,5]$, while a total cost of the path related with the vector $[2,5]$ is $(\mu+2)^2 + (v+5)$. In general, if $[\mu_1, v_1]$ and $[\mu_2, v_2]$ are cost vectors of two subpaths to the same vertex and $[\mu_1, v_1] < [\mu_2, v_2]$ (conf. (1)), then the vector $[\mu_2, v_2]$ can be discarded, because for any cost vector $[\mu, v]$ related with a path from the given vertex to the destination vertex, the following inequality holds

$$(\mu_1+\mu)^2 + (v_1+v) < (\mu_2+\mu)^2 + (v_2+v). \tag{9}$$

Similar situation appears for the representation EV-SM, because if $[\mu_1, \xi_1]$ and $[\mu_2, \xi_2]$ are cost vectors of two subpaths to the same vertex and $[\mu_1, \xi_1] < [\mu_2, \xi_2]$, then the vector $[\mu_2, \xi_2]$ can be discarded, because for any cost vector $[\mu, \xi]$ expanding the given subpaths to the paths to the destination vertex the following inequality

$$\xi_1 + \xi + 2\mu_1\mu < \xi_2 + \xi + 2\mu_2\mu \tag{10}$$

is satisfied.

Therefore, the RSPP with the second moment criterion reduce to the MOSPP, regardless of the data representation.

According to (7) and (8) the representation EV-V requires less multiplications than the representation EV-SM, which is very important for the time of computations. However, the representation EV-SM has some disadvantages, too. For example, let $[1,2]$ and $[2,1]$ be two vectors in the representation EV-V. These vectors are incomparable (since

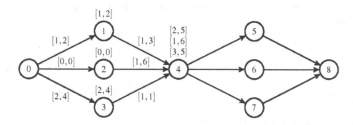

Fig. 2. Illustration of cost vectors propagation

neither $[1,2] < [2,1]$ nor $[1,2] > [2,1]$ holds), thus the both vectors must be remembered. But in the representation EV-SM the corresponding vectors are $[1,2+1^2] = [1,3]$ and $[2,1+2^2] = [2,5]$, thus the vector $[2,5]$ (and so the vector $[2,1]$, too) does not have to be stored! Thus in the representation EV-SM elimination of dominated vectors appears more frequently.

In the next part of this section we discuss the solution of the RSPP with the second moment criterion with help of general methods for solving the MOSPP.

3.1 Extended Bellman-Ford Algorithm

Classical algorithms for the CSPP can be extended for the MOSPP. First, we show an extension of the classical Bellman-Ford algorithm. We associate with each vertex $v \in V$ a list L_v of vectors $[a,b] \in \mathbb{R}^2$. Lists L_v for all $v \in V \setminus \{s\}$ are initially empty and the list L_s related with the source vertex s contains the vector $[0,0]$ at the beginning.

Modifying the procedure of the relaxation of edge we can solve the RSPP with the second moment criterion using the following Extended Bellman-Ford algorithm (shortly EBF):

Extended-Bellman-Ford-algorithm:
set $L_v = \emptyset$ for all $v \in V \setminus \{s\}$ and $L_s = \langle [0,0] \rangle$
for $i = 1$ **to** $\#V - 1$ **do**
    relaxation_state = *not_changed*
    for all $(u,v) \in E$ **do**
        EBF-Relax(u,v)
    if *relaxation_state* = *not_changed* **then**
        break
return path related with minimal value on the list L_t

The procedure **EBF-Relax**(u,v) has the following pseudocode[1]:

EBF-Relax(u,v):
for all $[\mu_u, v_u] \in L_u$ **do**
    set $\mu_v^{new} = \mu_u + \mathscr{E}[T_{(u,v)}]$, $v_v^{new} = v_u + \mathscr{V}[T_{(u,v)}]$ and $\xi_v^{new} = (\mu_v^{new})^2 + v_v^{new}$
    vector_state = *incomparable*
    for all $[\mu_v, v_v] \in L_v$ **do**
        set $\xi_v = \mu_v^2 + v_v$
        if $[\mu_v, \xi_v] > [\mu_v^{new}, \xi_v^{new}]$ **then**
            $L_v = L_v \setminus \langle [\mu_v, v_v] \rangle$
            vector_state = *less*
        if $[\mu_v, \xi_v] \leq [\mu_v^{new}, \xi_v^{new}]$ **then**
            vector_state = *greater_or_equal*
            break
    if *vector_state* = *less* **or** *vector_state* = *incomparable* **or** $L_v = \emptyset$ **then**
        $L_v = L_v \cup \langle [\mu_v^{new}, v_v^{new}] \rangle$
        relaxation_state = *changed*

---

[1] For vectors in representation EV-V. The procedure **EBF-Relax**(u,v) for the representation EV-SM is similar.

Since we have assumed that the expected values and the second moments of all random variables T_e for $e \in E$, are nonnegative, thus there is no cycle with negative cost in the graph.

We will justify that for every vertex $v \in V$, after n iterations of the main loop, the list L_v contains all incomparable vectors related with all possible paths from the source vertex s to the vertex v, where n denotes the number of edges of the most numerous path from the source vertex s to the vertex v. In particular, after $(\#V - 1)$ iterations of the main loop, the list L_t contains all incomparable vectors related with all possible paths from the source vertex s to the destination vertex t (because the number of edges of the most numerous path from the source vertex s to the destination vertex t can have $(\#V - 1)$ edges at most).

Assume first that $v = s$. The list L_s is initialized with $\langle[0,0]\rangle$. There is only one simple path from the source vertex s to the vertex s, i.e., the path $\langle s \rangle$, and the cost vector of that path is $[0,0]$. Every other path from the source vertex s to the vertex s must contain a cycle, thus the cost vector of that path must be greater than or equal to $[0,0]$. Thus, after zero iterations of the main loop, the list L_s contains all incomparable vectors related with all possible paths from the source vertex s to the vertex s.

Suppose now that a vertex $v \neq s$ is achievable from the source s and let $\langle v_0, \ldots, v_n \rangle$ be the most numerous path from the source vertex s to the vertex v (i.e. $v_0 = s$ and $v_n = v$). At first iteration of the main loop, the procedure **EBF-Relax**(v_0, v_1) is executed, hence the list L_{v_1} contains the vector related with the edge (v_0, v_1). At second iteration of the main loop, the procedure **EBF-Relax**(v_1, v_2) is executed, hence the vector related with the path $\langle v_0, v_1, v_2 \rangle$ will be taken into consideration on the list L_{v_2}, i.e. it will be added to the list L_{v_2}, if it is not bigger than any different vector from this list, etc. Finally, after n iterations of the main loop, the procedure **EBF-Relax**(v_{n-1}, v_n) is executed, and the cost vector related with the path $\langle v_0, \ldots, v_n \rangle$ will be taken into consideration on the list L_{v_n}. All different simple paths from the source vertex s to the vertex v have not more than n edges, thus all cost vectors related with these paths will be also taken into consideration on the list L_v after n iterations of the main loop. All paths from the source vertex s to the vertex v having more than n edges must contain a cycle. But the cost vector of every cycle in the graph is nonnegative, thus the cost vector of a path containing a cycle is greater than or equal to the cost vector of the path without cycles. Such a path is simple and as we have justified, the cost vectors related with that path, will be also taken into consideration on the list L_v after n iterations of the main loop.

Let us notice that if the vertex $v \neq s$ is not achievable from the source s, then after zero iterations of the main loop, the list L_v is empty. Since there is no edge (u, v) in the graph, the list L_v will not change in the further iterations of the main loop.

3.2 Generic Label Correcting Algorithm

Another way to solve the RSPP with the second moment criterion is to use a Generic Label Correcting algorithm (shortly GLC) that is a kind of the generalization of the classical Dijkstra algorithm. In this algorithm relaxations of edges are coming in a little bit more natural order than in the EBF algorithm. We present a pseudocode of the GLC algorithm below:

Generic-Label-Correcting-Algorithm:
set $L_v = \emptyset$ for all $v \in V \setminus \{s\}$ and $L_s = \langle [0,0] \rangle$
$Q = \langle s \rangle$
while $Q \neq \emptyset$ **do**
    select $u \in Q$ //typically, with respect to the FIFO principle.
    $Q = Q \setminus \langle u \rangle$
    for all $v \in Successors(u)$ **do**
        relaxation_state $=$ *not_changed*
        EBF-Relax(u,v)
        if *relaxation_state* $=$ *changed* **and** $v \notin Q$ **then**
        $Q = Q \cup \langle v \rangle$
return path related with minimal value on the list L_t

Note that it is possible to move searching of minimal value on the list L_t to the procedure of relaxation of edge what will save quite a lot of memory.

The GLC algorithm starts with $Q = \langle s \rangle$ and $L_s = \langle [0,0] \rangle$. As it has been discussed above, the list L_s contains all incomparable vectors related with all possible paths from the source vertex s to itself and any further relaxation cannot change the contents of this list.

Execution of **EBF-Relax**(s,v) for each $v \in Successors(s)$ can change the contents of the list L_v, and in that case these changes can propagate to the lists related with successors of the vertex v. Therefore, if execution of **EBF-Relax**(s,v) changes the list L_v, then the vertex v must be added to the list Q. Hence cost vectors related with paths going through the vertex v are also updated. The GLC algorithm works as long as the list Q is not empty.

Since there is no cycle with negative cost vector, thus the GLC algorithm must stop, and after its execution, every list L_v contains all incomparable vectors related with all possible paths from the source vertex s to the vertex v.

We show how the GLC algorithm works on a simple example.

Example 3. Consider the graph depicted in Fig. 1. Vectors of expected values and variances related with edges $(0,1)$, $(0,2)$, $(1,3)$, $(2,3)$ and $(3,4)$ are equal to $[0,0]$, $[0,0]$, $[\frac{3}{4}, \frac{11}{16}]$, $[1,0]$ and $[1,0]$, respectively.

- The GLC algorithm starts with $Q = \langle 0 \rangle$ and $L_0 = \langle [0,0] \rangle$, thus at first iteration we obtain $v = 0$, $Q = \langle 0 \rangle \setminus \langle 0 \rangle = \emptyset$ and $Successors(0) = \langle 1,2 \rangle$.
  After execution of **EBF-Relax**$(0,1)$ we obtain $L_1 = \emptyset \cup \langle [0,0] \rangle = \langle [0,0] \rangle$ and $Q = \emptyset \cup \langle 1 \rangle = \langle 1 \rangle$.
  After execution of **EBF-Relax**$(0,2)$ we obtain $L_2 = \emptyset \cup \langle [0,0] \rangle = \langle [0,0] \rangle$ and $Q = \langle 1 \rangle \cup \langle 2 \rangle = \langle 1,2 \rangle$.
- Since $Q = \langle 1,2 \rangle \neq \emptyset$, thus at second iteration we obtain $v = 1$, $Q = \langle 1,2 \rangle \setminus \langle 1 \rangle = \langle 2 \rangle$ and $Successors(1) = \langle 3 \rangle$.
  After execution of **EBF-Relax**$(1,3)$ we obtain:
  $L_3 = \emptyset \cup \langle [\frac{3}{4}, \frac{11}{16}] \rangle = \langle [\frac{3}{4}, \frac{11}{16}] \rangle$ and $Q = \langle 2 \rangle \cup \langle 3 \rangle = \langle 2,3 \rangle$.
- Since $Q = \langle 2,3 \rangle \neq \emptyset$, thus at third iteration we obtain $v = 2$, $Q = \langle 2,3 \rangle \setminus \langle 2 \rangle = \langle 3 \rangle$ and $Successors(2) = \langle 3 \rangle$.
  After execution of **EBF-Relax**$(2,3)$ we obtain:
  $L_3 = \langle [\frac{3}{4}, \frac{11}{16}] \rangle \cup \langle [1,0] \rangle = \langle [\frac{3}{4}, \frac{11}{16}], [1,0] \rangle$ and $Q = \langle 3 \rangle \cup \langle 3 \rangle = \langle 3 \rangle$.

- Since $Q = \langle 3 \rangle \neq \emptyset$, thus at fourth iteration we obtain $v = 3$, $Q = \langle 3 \rangle \setminus \langle 3 \rangle = \emptyset$ and $Successors(3) = \langle 4 \rangle$.
  After execution of **EBF-Relax**$(3,4)$ we obtain:
  $L_4 = \emptyset \cup \langle [1\frac{3}{4}, \frac{11}{16}], [2,0] \rangle = \langle [1\frac{3}{4}, \frac{11}{16}], [2,0] \rangle$ and $Q = \emptyset \cup \langle 4 \rangle = \langle 4 \rangle$
- Since $Q = \langle 4 \rangle \neq \emptyset$, thus at fifth iteration we obtain $v = 4$, $Q = \langle 4 \rangle \setminus \langle 4 \rangle = \emptyset$ and $Successors(4) = \emptyset$.
- Since $Q = \emptyset$, thus we check the list L_4 that is $L_4 = \langle [1\frac{3}{4}, \frac{11}{16}], [2,0] \rangle$. As we can see, $[1\frac{3}{4}, \frac{11}{16}]$ is the cost vector related with the shortest path.

3.3 Complexity

Unfortunately, the time complexity and also the space complexity of the EBF algorithm and the GLC algorithm are exponential. We will show it on a simple example.

Example 4. Consider the graph depicted in Fig. 3(a) and let random variables related with all edges of the graph have vectors in the representation EV-V presented in the picture. Let

$$[0,0], [2^{\alpha_1}, 2^{2n+1-\alpha_1}], [0,0], [2^{\alpha_2}, 2^{2n+1-\alpha_2}], \ldots, [0,0], [2^{\alpha_n}, 2^{2n+1-\alpha_n}] \qquad (11)$$

be vectors which coordinates are the expected values and the variances, related with the edges of a path from the source vertex 0 to the destination vertex n. The sum $\sum_{i=1}^{n} 2^{\alpha_i}$ can be expressed as binary number $(a_{2n}, a_{2n-1}, \ldots, a_2, a_1, 0)_2$. This binary number contains ones at positions $\alpha_1, \ldots, \alpha_n$ and zeros at the other positions, thus all of 2^n possible cost vectors related with all possible paths from the source vertex 0 to the destination vertex n must be different. Furthermore, if we order these cost vectors increasingly by the expected values, then variances of these cost vectors will be ordered decreasing.

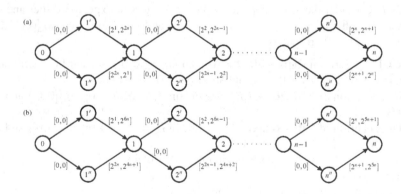

Fig. 3. (a) Graph with $3n + 1$ vertices and $4n$ edges and cost vectors in representation EV-V. (Sums of exponents of powers of coordinates in all non-zero vectors in the graph are equal to $2n + 1$.) All of 2^n paths in this graph are incomparable. (b) Graph with $3n + 1$ vertices and $4n$ edges and cost vectors in representation EV-SM. (Sums of exponents of powers of coordinates in all non-zero vectors in the graph are equal to $6n + 1$.) All of 2^n paths in this graph are incomparable.

Therefore all 2^n possible cost vectors related with all possible paths from the source 0 to the destination n are incomparable.

Now, consider the graph depicted in Fig. 3(b) and let random variables related with all edges of the graph have vectors in the representation EV-SM presented in the picture. Let

$$[0,0], [2^{\alpha_1}, 2^{6n+1-\alpha_1}], [0,0], [2^{\alpha_2}, 2^{6n+1-\alpha_2}], \dots, [0,0], [2^{\alpha_n}, 2^{6n+1-\alpha_n}] \tag{12}$$

be vectors whose coordinates are the expected values and the second moments, related with edges of some path from the source vertex 0 to the destination vertex n. From equation (7) it follows that the vector related with that path is equal to

$$\left[\sum_{i=1}^{n} 2^{\alpha_i}, \ \sum_{i=1}^{n} 2^{6n+1-\alpha_i} + 2 \sum_{i=2}^{n} 2^{\alpha_i} \sum_{j=1}^{i-1} 2^{\alpha_j} \right]. \tag{13}$$

Similarly to the case of graph from Fig. 3(a), we order cost vectors of all 2^n paths increasingly by the expected values, and the factors $\sum_{i=1}^{n} 2^{6n+1-\alpha_i}$ of cost vectors of these paths will be ordered decreasing.

Note that the minimal difference of the sum of the second coordinates of cost vectors of two different paths from the source vertex 0 to the destination vertex n is equal to 2^{5n}. Moreover, the maximal value of the factor $2 \sum_{i=2}^{n} 2^{\alpha_i} \sum_{j=1}^{i-1} 2^{\alpha_j}$ can be estimated as follows:

$$2 \sum_{i=2}^{n} 2^{\alpha_i} \sum_{j=1}^{i-1} 2^{\alpha_j} \leq 2 \sum_{i=2}^{n} 2^{n+i} \sum_{j=1}^{i-1} 2^{n+j} = 2 \sum_{i=2}^{n} 2^{n+i} \cdot 2^{n+i} = 2^{2n+1} \sum_{i=2}^{n} 4^i =$$

$$= 2^{2n+1} \cdot \frac{4^{n+1} - 4^2}{4 - 1} < 2^{2n+1} \cdot \frac{4^{n+1}}{2} = 2^{4n+2} < 2^{5n} \tag{14}$$

for any $n > 2$. It is easy to see that the estimation (14) holds also for $n = 1$ (then $0 < 2^5$) and for $n = 2$ (then $2 \cdot 2^3 \cdot 2^4 < 2^{10}$). Inequality (14) guarantees that order of the second coordinates of cost vectors of all paths will be preserved. Hence it follows that all vectors related with all possible paths from the source vertex 0 to the destination vertex n are incomparable.

Graphs depicted in Fig. 3(a) and Fig. 3(b) show that the pessimistic complexity of the EBF and the GLC algorithm are exponential, regardless of the data representation EV-V or EV-SM.

4 Approximate Algorithms

As we have shown in the previous section, the time complexity and the space complexity of the presented exact algorithms are exponential. Therefore, in this section we consider approximate algorithms, which have polynomial complexity. The first presented algorithm is a kind of modification of the classical Bellman-Ford algorithm. The second one is a simple single criterion approximation, and the last one is a variant of an approximate algorithm for the MOSPP from [15].

4.1 Extended Bellman-Ford Algorithm with Fixed Capacity

The classical Bellman-Ford algorithm fails in the case of the RSPP, because in this algorithm, each vertex is connected with only one currently the best cost related with the shortest path to this vertex. The EBF algorithm solving the MOSPP presented in the previous section can be used in the case of the RSPP, because it assigns to each vertex the list of all nondominated cost vectors related with all paths to this vertex. Unfortunately, as we have shown in the previous section this algorithm has exponential complexity. It is natural to consider a modification of the classical Bellman-Ford algorithm which assigns with each vertex the list of fixed number k of cost vectors. Obviously, such an algorithm is only an approximate algorithm. It is easy to extend the graph shown in Fig. 1, in such a way that the worst subpath will be a subpath of the shortest path.

We present a pseudocode of this modification of the Extended Bellman-Ford algorithm below:

Extended-Bellman-Ford-Algorithm-with-Fixed-Capacity:
set $L_v = \emptyset$ for all $v \in V \setminus \{s\}$ and $L_s = \langle [0,0] \rangle$
for $i = 1$ **to** $\#V - 1$ **do**
    relaxation_state = not_changed
    for all $(u,v) \in E$ **do**
        EBF-FC-Relax(u,v)
    if *relaxation_state = not_changed* **then**
    break
return path related with L_t

The procedure **EBF-FC-Relax**(u,v) has the following pseudocode[2]:

EBF-FC-Relax(u,v):
for all $[\mu_u, \xi_u] \in L_u$ **do**
    set $\mu_v^{new} = \mu_u + \mathcal{E}[T_{(u,v)}]$ and $\xi_v^{new} = \xi_u + \mathcal{E}[T_{(u,v)}^2] + 2\mu_u \mathcal{E}[T_{(u,v)}]$
    if $\#L_v < k$ **then**
        $L_v = L_v \cup \langle [\mu_v^{new}, \xi_v^{new}] \rangle$
        relaxation_state = changed
    else
        find $[\mu_v, \xi_v] \in L_v$ such that $\xi_v = \max\{\xi \,|\, [\mu, \xi] \in L_v\}$.
        if $\xi_v > \xi_v^{new}$ **or** $(\xi_v = \xi_v^{new}$ **and** $\mu_v > \mu_v^{new})$ **then**
            $L_v = L_v \setminus \langle [\mu_v, \xi_v] \rangle \cup \langle [\mu_v^{new}, \xi_v^{new}] \rangle$
            relaxation_state = changed

We show how the Extended Bellman-Ford Algorithm with Fixed Capacity (shortly EBF-FC) works on a simple example.

Example 5. Consider the graph depicted in Fig. 1. Suppose that the edges are ordered as follows: $(0,1)$, $(0,2)$, $(1,3)$, $(2,3)$ and $(3,4)$. Vectors of expected values and the second moments related with these edges are equal to $[0,0]$, $[0,0]$, $[\frac{3}{4}, \frac{5}{4}]$, $[1,0]$ and $[1,0]$ respectively. Assume first that $k = 1$ (conf. the pseudocode).

---

[2] For vectors in representation EV-SM. The procedure **EBF-FC-Relax**(u,v) for the representation EV-V is similar.

- The EBF-FC algorithm starts with $L_0 = \langle [0,0] \rangle$ and $L_1 = L_2 = L_3 = L_4 = \emptyset$.
- After execution of **EBF-FC-Relax**$(0,1)$ we obtain $L_1 = \langle [0+0,0+0+2\cdot 0\cdot 0] \rangle = \langle [0,0] \rangle$.
- After execution of **EBF-FC-Relax**$(0,2)$ we obtain $L_2 = \langle [0+0,0+0+2\cdot 0\cdot 0] \rangle = \langle [0,0] \rangle$.
- After execution of **EBF-FC-Relax**$(1,3)$ we obtain $L_3 = \langle [0+\frac{3}{4},0+\frac{5}{4}+2\cdot 0\cdot \frac{3}{4}] \rangle = \langle [\frac{3}{4},\frac{5}{4}] \rangle$.
- Since $k=1$, $[0+1,0+1+2\cdot 0\cdot 1] = [1,1]$ and $\frac{5}{4} > 1$, thus we have $L_3 = \langle [\frac{3}{4},\frac{5}{4}] \rangle \setminus \langle [\frac{3}{4},\frac{5}{4}] \rangle \cup \langle [1,1] \rangle = \langle [1,1] \rangle$ after execution of **EBF-FC-Relax**$(2,3)$.
- After execution of **EBF-FC-Relax**$(3,4)$ we obtain $L_4 = \langle [1+1,1+1+2\cdot 1\cdot 1] \rangle = \langle [2,4] \rangle$.
- Nothing is changing after the next iteration of the main loop, thus $[2,4]$ is the cost vector related with the returned path. Let us note that we have shown in Example 1 that the cost of the shortest path is not equal to 4, but it is equal to $3\frac{3}{4}$.

Suppose now that $k=2$, i.e. we take into account the Lists L_v for $v \in V$ with the capacity of 2 cost vectors.

- The EBF algorithm starts with $L_0 = \langle [0,0] \rangle$ and $L_1 = L_2 = L_3 = L_4 = \emptyset$.
- After execution of **EBF-FC-Relax**$(0,1)$ we obtain $L_1 = \langle [0+0,0+0+2\cdot 0\cdot 0] \rangle = \langle [0,0] \rangle$.
- After execution of **EBF-FC-Relax**$(0,2)$ we obtain $L_2 = \langle [0+0,0+0+2\cdot 0\cdot 0] \rangle = \langle [0,0] \rangle$.
- After execution of **EBF-FC-Relax**$(1,3)$ we obtain $L_3 = \langle [0+\frac{3}{4},0+\frac{5}{4}+2\cdot 0\cdot \frac{3}{4}] \rangle = \langle [\frac{3}{4},\frac{5}{4}] \rangle$.
- Since $k=2$, thus we obtain $L_3 = \langle [\frac{3}{4},\frac{5}{4}] \rangle \cup \langle [0+1,0+1+2\cdot 0\cdot 1] \rangle = \langle [\frac{3}{4},\frac{5}{4}],[1,1] \rangle$ after execution of **EBF-FC-Relax**$(2,3)$.
- Since $[\frac{3}{4}+1,\frac{5}{4}+1+2\cdot \frac{3}{4}\cdot 1]=[1\frac{3}{4},3\frac{3}{4}]$, $[1+1,1+1+2\cdot 1\cdot 1]=[2,4]$ and $[1\frac{3}{4},3\frac{3}{4}] < [2,4]$, thus after execution of **EBF-FC-Relax**$(3,4)$ we obtain $L_4 = \langle [1\frac{3}{4},3\frac{3}{4}] \rangle$.
- Nothing is changing after the next iteration of the main loop, thus $[1\frac{3}{4},3\frac{3}{4}]$ is the cost vector related with the returned path. As we have shown in Example 1, this is the cost vector of the shortest path.

The above example shows that the EBF-FC algorithm is not an exact algorithm.

Note that the complexity of the EBF-FC algorithm is polynomial. Moreover, it is equal to $O(k\cdot \#V \cdot \#E)$, because the main loop can take at most $\#V \cdot \#E$ iterations and each iteration is a simple execution of the **EBF-FC-Relax** procedure which takes at most k operations (list L_u can have at most k vectors).

4.2 Extended Bellman-Ford Algorithm with Smart Indexing

The second proposed approximate algorithm is a variant of the algorithm presented in [15]. It is another modification of Bellman-Ford algorithm.

Let $\mu_v^{min} = \mu_v^1 < \cdots < \mu_v^r = \mu_v^{max}$ be all possible expected values related with paths from the source vertex s to some vertex $v \in V$. (Note that μ_v^i can be related with many different paths.) Suppose that the currently the best path from the source vertex s to the vertex v has the expected value equal to μ_v^i. If we store the second moment (or variance) of that path at i-th position in array of k elements, then we obtain an exact algorithm. Of course this algorithm is very "expensive" because we have to calculate position i and all possible values $\mu_v^1, \ldots, \mu_v^r$. Since the number r of such values grows exponentially with respect to the distance between the source vertex s and the vertex v, we have to remember exponentially growing lists of values $\mu_v^1, \ldots, \mu_v^r$.

Notice that for each vertex v we can simply estimate (or even calculate) minimal and maximal expected values μ_v^{min} and μ_v^{max}. If we fix the size of lists related with each vertex to k, then we can store second moment (or variance) of that path at the position

$$\left\lceil k \cdot \frac{\mu_v^{new} - \mu_v^{min}}{\mu_v^{max} - \mu_v^{min}} \right\rceil \tag{15}$$

if its expected value is equal to μ_v^{new}. (For small k the variance or the second moment related with different paths will be written in the same position of the array more frequently.)

Extended-Bellman-Ford-Algorith-with-Smart-Indexing:
use classical Bellman-Ford algorithm or classical Dijkstra algorithm to calculate minimal and maximal expected values μ_v^{min} and μ_v^{max} for all $v \in V$
set $\Pi_v^i = \langle \infty, \infty, null \rangle$ for all $v \in V$, $i = 0, \ldots, k$ and $\Pi_s^0 = \langle 0, 0, s \rangle$
for $i = 1$ **to** $\#V - 1$ **do**
    relaxation_state = not_changed
    for all $(u, v) \in E$ **do**
        EBF-SI-Relax(u, v)
    if *relaxation_state = not_changed* **then**
        break
return path related with Π_t

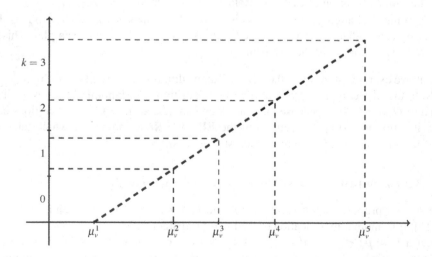

Fig. 4. Illustration of the formula (15)

The procedure **EBF-SI-Relax**(u, v) has the following pseudocode[3]:

EBF-SI-Relax(u, v):
for $i = 0$ **to** k **do**
    let $\langle \mu_u^i, \xi_u^i, \pi_u^i \rangle = \Pi_u^i$
    if $\pi_u^i \neq t$ **then**
        $\mu_v^{new} = \mu_u^i + \mathcal{E}[T_{(u,v)}]$
        $\xi_v^{new} = \xi_u^i + \mathcal{E}[T_{(u,v)}^2] + 2\mu_u^i \mathcal{E}[T_{(u,v)}]$
        if $\mu_v^{min} \neq \mu_v^{max}$ **then**
$$j = \left\lceil k \cdot \frac{\mu_v^{new} - \mu_v^{min}}{\mu_v^{max} - \mu_v^{min}} \right\rceil$$
        else
$$j = 0$$
        let $\langle \mu_v^j, \xi_v^j, \pi_v^j \rangle = \Pi_v^j$
        if $\pi_v^j = null$ **or** $\xi_v^j > \xi_v^{new}$ **then**
            $\Pi_v^j = \langle \mu_v^{new}, \xi_v^{new}, u \rangle$
            relaxation_state = changed

To illustrate the performance of the above algorithm we present a simple example.

Example 6. Consider the graph depicted in Fig. 1. Let $k = 1$, i.e. $k + 1 = 2$ is the array dimension.

- Calculating minimal and maximal expected values we obtain $\mu_0^{min} = \mu_0^{max} = 0$, $\mu_1^{min} = \mu_1^{max} = 0$, $\mu_2^{min} = \mu_2^{max} = 0$, $\mu_3^{min} = \frac{3}{4}$, $\mu_3^{max} = 1$, $\mu_4^{min} = 1\frac{3}{4}$ and $\mu_4^{max} = 2$.
- Since $\mu_0^{min} = \mu_0^{max} = 0$, thus after execution of **EBF-SI-Relax**$(0, 1)$ we obtain $\Pi_1^0 = \langle 0, 0, 0 \rangle$.
- Similarly, after execution of **EBF-SI-Relax**$(0, 2)$ we obtain $\Pi_2^0 = \langle 0, 0, 0 \rangle$.
- After execution of **EBF-SI-Relax**$(1, 3)$ we obtain $\mu_v^{new} = \frac{3}{4}$, $\xi_v^{new} = \frac{5}{4}$ and $j = 0$, thus $\Pi_3^0 = \langle \frac{3}{4}, \frac{5}{4}, 1 \rangle$.
- After execution of **EBF-SI-Relax**$(2, 3)$ we obtain $\mu_v^{new} = 1$, $\xi_v^{new} = 1$ and $j = 1$, thus $\Pi_3^0 = \langle 1, 1, 2 \rangle$.
- After execution of **EBF-SI-Relax**$(3, 4)$, for $i = 0$, we obtain $\mu_v^{new} = 1\frac{3}{4}$, $\xi_v^{new} = 3\frac{3}{4}$ and $j = 0$, thus $\Pi_3^0 = \langle 1\frac{3}{4}, 3\frac{3}{4}, 3 \rangle$. For $i = 1$ we have $\mu_v^{new} = 2$, $\xi_v^{new} = 4$ and $j = 1$, thus $\Pi_3^1 = \langle 2, 4, 3 \rangle$.
- Nothing is changing after the next iteration of the main loop, thus $[1\frac{3}{4}, 3\frac{3}{4}]$ is the cost vector related with the returned path. As we have shown in Example 1, this is the cost vector of the shortest path.

Note that the complexity of the above algorithm is polynomial. Moreover, it is equal to $O(k \cdot \#V \cdot \#E)$, because the main loop can take at most $\#V \cdot \#E$ iterations and each iteration is a simple execution of the **EBF-SI-Relax** procedure which took at most k operations.

---

[3] For vectors in representation EV-SM. The procedure **EBF-SI-Relax**(u, v) for the representation EV-V is similar.

4.3   Extended Bellman-Ford Algorithm with Rounded Values

The next presented approximate algorithm is a simple modification of the EBF algorithm in which we round values of the vectors before the comparison. We call this algorithm Extended Bellman-Ford Algorithm with Rounded Values (shortly EBF-RV). We present a pseudocode of the EBF-RV algorithm below:

Extended-Bellman-Ford-Algorithm-with-Rounded-Values:
use classical Bellman-Ford algorithm or classical Dijkstra algorithm to calculate/ estimate minimal and maximal expected values and variances μ_v^{min}, μ_v^{max}, v_v^{min} and v_v^{max} for all $v \in V$

set $\xi_v^{min} = (\mu_v^{min})^2 + v_v^{min}$ and $\xi_v^{max} = (\mu_v^{max})^2 + v_v^{max}$ for all $v \in V$

set $L_s = \langle [0,0] \rangle$ and $L_v = \emptyset$ for all $v \in V \setminus \{s\}$

for $i = 1$ **to** $\#V - 1$ **do**

    relaxation_state = *not_changed*

    for all $(u,v) \in E$ **do**

        EBF-RV-Relax(u,v)

    if *relaxation_state* = *not_changed* **then**

        break

return path related with minimal value on the list L_t

The procedure **EBF-RV-Relax**(u,v) has the following pseudocode[4]:

EBF-RV-Relax(u,v):

for all $[\mu_u, \xi_u] \in L_u$ **do**

    set $\mu_v^{new} = \mu_u + \mathcal{E}[T_{(u,v)}]$ and $\xi_v^{new} = \xi_u + \mathcal{E}[T_{(u,v)}^2] + 2\mu_u \mathcal{E}[T_{(u,v)}]$

    set $[\widehat{\mu}_v^{new}, \widehat{\xi}_v^{new}] = [[k \cdot \frac{\mu_v^{new} - \mu_v^{min}}{\mu_v^{max} - \mu_v^{min}}], [k \cdot \frac{\xi_v^{new} - \xi_v^{min}}{\xi_v^{max} - \xi_v^{min}}]]$

    vector_state = *incomparable*

    for all $[\mu_v, \xi_v] \in L_v$ **do**

        set $[\widehat{\mu}_v, \widehat{\xi}_v] = [[k \cdot \frac{\mu_v - \mu_v^{min}}{\mu_v^{max} - \mu_v^{min}}], [k \cdot \frac{\xi_v - \xi_v^{min}}{\xi_v^{max} - \xi_v^{min}}]]$

        if $[\widehat{\mu}_v, \widehat{\xi}_v] > [\widehat{\mu}_v^{new}, \widehat{\xi}_v^{new}]$ **then**

            $L_v = L_v \setminus \langle [\mu_v, \xi_v] \rangle$

            vector_state = *less*

        if $[\widehat{\mu}_v, \widehat{\xi}_v] \le [\widehat{\mu}_v^{new}, \widehat{\xi}_v^{new}]$ **then**

            vector_state = *greater_or_equal*

            break

    if *vector_state* = *less* **or** *vector_state* = *incomparable* **or** $L_u = \emptyset$ **then**

        $L_v = L_v \cup \langle [\mu_v^{new}, \xi_v^{new}] \rangle$

        relaxation_state = *changed*

Note that the complexity of the above algorithm is exponential.

---

[4] For vectors in representation EV-SM. The procedure **EBF-RV-Relax**(u,v) for the representation EV-V is similar.

4.4   Single Criterion Approximation Algorithm

As it was mentioned in the Introduction, if we take as the shortest path, the path with the minimal expected value or the variance of the sum of the random variables related with edges of the path, then the RSPP reduces to the CSPP and can be solved with help of the classical Dijkstra algorithm or classical Bellman-Ford algorithm.

This natural observation can be used in the whole series of approximate algorithms.

We can forget for a moment about one of the criteria (expected value or variance) and calculate the shortest path with respect to the other criterion. Next, we can remove from the graph the edge which is the most damaging for the first criterion or indeed cost of the path, and repeat the procedure on the depreciated graph.

If there is no path from the source vertex s to the destination vertex t in the depreciated graph, then the algorithm stops and returns this from the examined paths which has minimal actual cost (second moment).

Single-Criterion-Approximation-algorithm:
while exists a path from the source vertex s to the destination vertex t **do**

use classical Bellman-Ford algorithm or classical Dijkstra algorithm to find the shortest path $p = \langle v_0, \ldots, v_n \rangle \in \mathscr{P}$ with respect to expected value
//or variance, or the second moment
find $(v_{j-1}, v_j) \in p$ such that $\mathscr{V}[T_{(v_{j-1},v_j)}] = \max\limits_{i=1,\ldots,n} \mathscr{V}[T_{(v_{i-1},v_i)}]$

//or $\mathscr{E}[T_{(v_{j-1},v_j)}] = \max\limits_{i=1,\ldots,n} \mathscr{E}[T_{(v_{i-1},v_i)}]$

//or $\mathscr{E}[T^2_{(v_{j-1},v_j)}] + \sum\limits_{\substack{k=1 \\ k \neq j}}^{n} \mathscr{E}[T_{(v_{j-1},v_j)}]\mathscr{E}[T_{(v_{k-1},v_k)}] =$

$\max\limits_{i=1,\ldots,n} \mathscr{E}[T^2_{(v_{i-1},v_i)}] + \sum\limits_{\substack{k=1 \\ k \neq i}}^{n} \mathscr{E}[T_{(v_{i-1},v_i)}]\mathscr{E}[T_{(v_{k-1},v_k)}]$

$E = E \setminus \{(v_{j-1}, v_j)\}$
if the path p has cost lower than the path b **or** b is undefined **then**
$b = p$
return path b

To illustrate the performance of the Single Criterion Approximation algorithm (shortly SCA) we present the case of searching of the shortest path with respect to the variance and removing edge with the greatest expected value.

Example 7. Consider the graph depicted in Fig. 1.

- The shortest path with respect to the variance is the path $p = \langle 0, 2, 3, 4 \rangle$. Since $\mathscr{E}[T_{(0,2)}] = 0$, $\mathscr{E}[T_{(2,3)}] = 1$ and $\mathscr{E}[T_{(3,4)}] = 1$, thus (for example) we have $E = E \setminus \{(2,3)\}$. The path b is undefined, thus $b = p = \langle 0, 2, 3, 4 \rangle$.
- The shortest path with respect of expected value is now the path $p = \langle 0, 1, 2, 4 \rangle$. Since $\mathscr{E}[T_{(0,1)}] = 0$, $\mathscr{E}[T_{(1,3)}] = \frac{3}{4}$ and $\mathscr{E}[T_{(3,4)}] = 1$, thus we have $E = E \setminus \{(3,4)\}$. The cost of the path p is lower than the cost of the path b, thus $b = p = \langle 0, 1, 3, 4 \rangle$.
- There is no path from the source verice 0 to the destination vertex 4, therefore the algorithm returns the path $b = \langle 0, 1, 3, 4 \rangle$. Note that we have shown in Example 1 that b is the shortest path.

Note that the complexity of the SCA algorithm is polynomial. Moreover, it is equal to $O(\#E^2 \cdot \#V)$, because the main loop can take at most $\#E$ iterations (if $E = \emptyset$ then the main loop must finish) and the classical Bellman-Ford algorithm or the classical Dijkstra algorithm have the complexity equal to $O(\#V \cdot \#E)$ and the finding of the edge (v_{j-1}, v_j) can take at most $(\#V - 1)$ operations (path p must be simple, thus it can contain at most $(\#V - 1)$ edges).

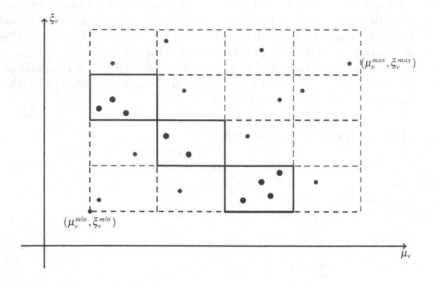

Fig. 5. Illustration of the idea of the EBF-RV algorithm. All the vectors in the highlighted rectangle will be treated as the same vector by the EBF-RV algorithm.

Table 1. Comparison of the time of execution and precision of the presented algorithms. t_1, t_2, t_3 — average factors of times of execution of given algorithms for three different groups of graphs; r_1, r_2, r_3 — average factors of precision of given algorithms for three different groups of graphs; Specification of the i-th group of graphs (for $i = 1, 2, 3$): $\#V = 10000i$, $\#Successors(v) = 10i$ for all vertex $v \in V \setminus \{t\}$. $0 \le \mathscr{E}[T_e] \le 10$ and $0 \le \mathscr{E}[T_e^2] \le 5100$ for all edge $e \in E$.

	EBF	EBF-FC-2	EBF-FC-5	EBF-RV-50	EBF-SI-20	EBF-SI-50	SCA-22	SCA-2E	SCA-2V	SCA-E2	SCA-EE	SCA-EV	SCA-V2	SCA-VE	SCA-VV
t_1	1.000	0.073	0.293	0.191	0.415	*1.144*	0.136	0.082	0.114	0.081	0.101	0.041	0.096	0.069	0.080
t_2	1.000	0.066	0.250	0.109	0.357	*0.911*	0.936	2.086	0.480	2.326	0.431	1.872	0.533	2.012	0.470
t_3	1.000	0.097	0.320	0.102	0.510	*1.081*	—	—	—	—	—	—	—	—	—
r_1	1.000	1.000	1.002	*7.219*	1.000	1.002	2.321	2.321	2.321	2.321	2.398	2.437	2.321	2.321	2.321
r_2	1.000	1.000	1.000	*9.797*	1.000	1.000	2.179	2.179	2.179	2.480	2.845	2.574	2.179	2.179	2.179
r_3	1.000	1.000	1.000	*12.283*	1.000	1.000	—	—	—	—	—	—	—	—	—

In the next section we present computational results of tests of the proposed algorithms.

5 Computational Results

We performed our tests on quite large randomly generated graphs. All these tests were made on Intel Celeron Mobile 1400MHz with 768MB RAM, working under control of Linux operating system with kernel version 2.6.11-6. Table 1 presents factors of times of execution and factors of precision of given algorithms for three different groups of graphs. For example, for the first group of graphs, the EBF-FC-2 algorithm took average 0.073 of time of execution of the exact algorithm (EBF). Every group contained 10 graphs. The i-th group of graphs (for $i = 1, 2, 3$) had $\#V = 10000i$, $\#Successors(v) = 10i$ for all vertex $v \in V \setminus \{t\}$. $0 \le \mathscr{E}[T_e] \le 10$ and $0 \le \mathscr{E}[T_e^2] \le 5100$ for all edge $e \in E$. Results better than the results of algorithm EBF are in bold, and worse results are highlighted in italics.

6 Conclusions

The Random Shortest Path Problem with the second moment criterion was discussed in this paper. It was shown that the complexity of the presented exact algorithms is exponential, while the complexity of the presented approximate algorithms is only polynomial.

Computational results for the exact and approximate algorithms were performed on large graphs. The results show that the approximate algorithms work very well, especially the EBF-FC algorithm which works really fast and produces exact results usually.

References

1. Ahuja, R.K., Magnanti, T.L., Orlin, J.B.: Network Flows: Theory, Algorithms, and Applications. Prentice-Hall, Inc., Englewood Cliffs (1993)
2. Alexopoulos, Ch.: State space partitioning methods for stochastic shortest path problems. Networks 30, 9–21 (1997)
3. Bellman, R.: On a routing problem. Quarterly of Applied Mathematics 16, 87–90 (1958)
4. Cormen, T.H., Leiserson, C.E., Rivest, R.L.: Introduction to Algorithms. MIT Press, Cambridge/ McGraw-Hill Book Company, Boston (1990)
5. Dijkstra, E.W.: A note on two problems in connection with graphs. Numerische Mathematik 1, 269–271 (1959)
6. Floyd, R.W.: Algorithm 97: Shortest path. Comm. ACM 5, 345 (1962)
7. Frederikson, G.N.: Fast algorithms for shortest paths in planar graphs, with applications. SIAM J. Comput. 16, 1004–1022 (1987)
8. Hansen, P.: Bicriterion path problems. In: Proc. 3rd Conf. Multiple Criteria Decision Making – Theory and Applications. Lecture Notes in Economics and Mathematical Systems, vol. 117, pp. 109–127. Springer, Heidelberg (1979)
9. Johnson, D.B.: Efficient algorithms for shortest paths in sparse networks. Journal of the ACM 24, 1–13 (1977)
10. Fredman, M.L., Tarjan, R.E.: Fibonacci heaps and their use in improved network optimization problems. Journal of the ACM 34, 596–615 (1987)

11. Moffat, A., Takaoka, T.: An all pairs shortest path algorithm with expected time $O(n^2 \log n)$. SIAM J. Comput. 16, 1023–1031 (1987)
12. Murthy, I., Sakar, S.: A relaxation-based pruning technique for a class of stochastic shortest path problems. Transportation Science 30, 220–236 (1996)
13. Murthy, I., Sarkar, S.: Exact algorithms for the stochastic shortest path problem with a decreasing deadline utility function. European Journal of Operational Research 103, 209–229 (1997)
14. Skriver, A.J.V., Anderson, K.A.: A label correcting approach for solving bicriterion shortest-path problems. Computers & Operations Research 27, 507–524 (2000)
15. Tsaggouris, G., Zaroliagis, Ch.D.: Improved FPTAS for Multiobjective Shortest Paths with Applications. Technical report no. TR 20050703 (July 2005)
16. Tsaggouris, G., Zaroliagis, Ch.D.: Multiobjective Optization: Improved FPTAS for Shortest Paths and Non-linear Objectives with Applications. Technical report no. TR 2006/03/01 (March 2006)

Free Material Optimization for Plates and Shells*

Stefanie Gaile, Günter Leugering, and Michael Stingl

Institute of Applied Mathematics II, University of Erlangen-Nürnberg,
Martensstr. 3, 91058 Erlangen, Germany
stefanie.gaile@am.uni-erlangen.de

Abstract. In this article, we present the Free Material Optimization (FMO) problem for plates and shells based on Naghdi's shell model. In FMO – a branch of
structural optimization – we search for the ultimately best material properties in a
given design domain loaded by a set of given forces. The optimization variable is
the full material tensor at each point of the design domain. We give a basic formulation of the problem and prove existence of an optimal solution. Lagrange duality
theory allows to identify the basic problem as the dual of an infinite-dimensional
convex nonlinear semidefinite program. After discretization by the finite element
method the latter problem can be solved using a nonlinear SDP code. The article
is concluded by a few numerical studies.

1 Introduction

Structural optimization deals with the problem of finding the stiffest structure subjected
to a set of given loads and boundary conditions, when only a limited amount of material
resources is available. Nowadays, this approach plays an important role in the construction of light-weight structures like airplanes and cars. Large parts of these objects as,
for instance, the fuselage, consist of thin-walled structures like shells and plates. This
is the reason why structural optimization of shells has received a lot of attention in the
design optimization community over the last couple of years. For example, shape optimization techniques have been used to vary the geometry and boundary of a shell with
the goal to stiffen the structure [6]. Various approaches try to identify the optimal topology of a shell in the sense of 0–1–designs. For an overview in the case of plates see [4].
On spherical shells it is possible to calculate the topological derivative and to exploit
this information with the purpose of finding the optimal position of holes [17]. Only
recently, free sizing optimization taking strength and stability constraints into account
has been used to improve the design of shell structures [7].

Another important class of shell design problems is based on material optimization.
Here the design variables reflect not only the distribution of material in the design domain, but also the local properties of the material. The methods used in the area of
material optimization differ in the choice of the admissible set of materials. In [18] a
pseudo density of the material is varied using a SIMP-approach. Rather than cutting the

---

* This work has been supported by the European Commission within the Sixth Framework Programme 2002 – 2006 (FP-6 STREP 30717 PLATO-N).

A. Korytowski et al. (Eds.): System Modeling and Optimization, IFIP AICT 312, pp. 239–250, 2009.

solution space down to 0–1–designs the author proposes to realize the optimal solution using foams that can be produced in manifold densities. In aerospace industry the use of composite materials is very common. In [23] the authors suggest to design composite shells by optimization of the material selection and fibre angles in a laminated shell structure. It is even possible to consider fully anisotropic elasticity tensors as admissible set for the design optimization as shown for Reissner-Mindlin plates in [2]. Finally, there are approaches taking advantage of adaptive methods – either by changing the parametrization of the design space during optimization or by adapting the model via switching between shape and material optimization; see [20].

In this article, we focus on Free Material Optimization, originally introduced for the optimal design of solid bodies by [3]. The design variable used in Free Material Optimization is the full material tensor at each point of the design domain. Therefore it yields not only the optimal material distribution, but also the material properties at each point. Various solution techniques for this problem have been proposed; see, for example, [25]. Due to the high freedom in the design space the resulting material/structure is typically hard to manufacture. Nevertheless it gives valuable information about the optimal material density, symmetry and principal directions, which can be exploited to realize approximations of the optimal design. One possible realization by tapelayering is described in [13]. In the recent years, the formulation of the Free Material Optimization problem has been extended to cover multiple load cases [1], stability control by consideration of global buckling [15] and stress constraints [16]. In this article we propose a formulation of Free Material Optimization based on the linear elastic shell model of Naghdi [21] suited for thin-walled structures like airplanes, cars and pipes.

2   Naghdi's Shell Model

We start with a mathematical description of Naghdi's shell model using the standard notation e.g. described in [21,8,9]. The geometry of a Naghdi shell is described by the midsurface ω – an open bounded two-dimensional set in Euclidean space, which can be parametrized by a sufficiently smooth function $\Phi : \mathbb{R}^2 \to \mathbb{R}^3$, $\Phi \in W^{2,\infty}(\omega)$. This is in contrast to other popular shell models as for example the Kirchhoff-Love model, where one starts from a three-dimensional solid material and makes approximations accounting to the thinness of the shell.

Hence it is advantageous to use curvilinear coordinates denoted by ξ^i (in accordance with common notation in shell theory Latin indices run over 1, 2 and 3, while Greek indices run only over 1 and 2). The covariant basis vectors are then defined by

$$a_\alpha = \frac{\partial \Phi}{\partial \xi^\alpha}, \qquad a_3 = \frac{a_1 \times a_2}{\|a_1 \times a_2\|}. \tag{1}$$

Moreover, the surface covariant derivative of a vector field v is given by

$$v_{\alpha|\mu} = v_{\alpha,\mu} - \Gamma^\lambda_{\alpha\mu} v_\lambda, \tag{2}$$

where $v_{\alpha,\mu}$ is the partial derivative of v_α with respect to ξ^μ and $\Gamma^\lambda_{\alpha\mu}$ is the Christoffel symbol of the midsurface

$$\Gamma^\lambda_{\alpha\mu} = a_{\alpha,\mu} \cdot a^\lambda. \tag{3}$$

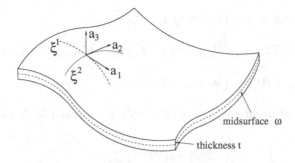

Fig. 1. Curvilinear coordinates on the midsurface

Furthermore the fundamental forms of the midsurface are defined by

- first fundamental form: $a_{\alpha\beta} = a_\alpha \cdot a_\beta$,
- second fundamental form: $b_{\alpha\beta} = -a_{3,\beta} \cdot a_\alpha$,
- third fundamental form: $c_{\alpha\beta} = b_\alpha^\lambda b_{\lambda\beta}$.

It turns out that the midsurface alone contains not enough information to describe bending and shear effects. A remedy is provided by the theory of Cosserat continua: at each point $x \in \omega$ a director vector d is attached to the shell, adding the lacking degrees of freedom [10,22]. These director vectors can be interpreted as material lines along the thickness of the shell. The deformation of the loaded shell can be described by a translation of all points of the midsurface $u \in [H^1(\omega)]^3$ and a rotation of the associated director vectors stemming from the group $SO(2)$. As the rotation of an infinitely-thin straight material line is uniquely defined by a rotation vector normal to that line we introduce $\theta \in [H^1(\omega)]^2$ to represent the rotation by $\theta_\lambda a^\lambda$ [8]. A component on a_3 is not required due to the fact that rotations of the director vectors around their own axis are neglected. Thus we obtain the following displacement formula:

$$U(\xi^1,\xi^2,\xi^3) = u(\xi^1,\xi^2) + \xi^3 \theta_\lambda(\xi^1,\xi^2) a^\lambda(\xi^1,\xi^2). \tag{4}$$

In the remainder of this article, we consider a shell with a Lipschitz boundary $\partial\omega$. The shell is clamped at parts of the boundary. To this end we partition $\partial\omega$ into two sets $\partial\omega_0$ and $\partial\omega_1$ which are open in $\partial\omega$, $\partial\omega = \overline{\partial\omega_0 \cup \partial\omega_1}$ and $\partial\omega_0 \cap \partial\omega_1 = \emptyset$. Then Dirichlet boundary conditions are applied on $\partial\omega_0$ and the shell is subjected to forces and moments on $\partial\omega_1$. Using this, we define the set of admissible displacements to be

$$\mathcal{U} := \{(u,\theta) \in [H^1(\omega)]^5 \mid u = 0 \text{ and } \theta = 0 \text{ on } \partial\omega_0\}. \tag{5}$$

As a consequence we obtain $[H_0^1(\omega)]^5 \subset \mathcal{U} \subset [H^1(\omega)]^5$. It is now possible to deduce formulas for membrane strains $\gamma_{\alpha\beta}$, bending strains $\chi_{\alpha\beta}$ and shear strains ζ_α, respectively:

$$\gamma_{\alpha\beta}(u) = \frac{1}{2}\left(u_{\alpha|\beta} + u_{\beta|\alpha}\right) - b_{\alpha\beta}u_3,$$

$$\chi_{\alpha\beta}(u,\theta) = \frac{1}{2}\left(\theta_{\alpha|\beta} + \theta_{\beta|\alpha} - b_{\beta}^{\lambda}u_{\lambda|\alpha} - b_{\alpha}^{\lambda}u_{\lambda|\beta}\right) + c_{\alpha\beta}u_3, \tag{6}$$

$$\zeta_{\alpha}(u,\theta) = \frac{1}{2}\left(\theta_{\alpha} + u_{3,\alpha} + b_{\alpha}^{\lambda}u_{\lambda}\right).$$

The assumption of linear elasticity in Naghdi's shell model leads to the following Hooke's law:

$$N^{\lambda\mu} = tC^{\lambda\mu\alpha\beta}\gamma_{\alpha\beta},$$

$$M^{\lambda\mu} = \frac{t^3}{12}C^{\lambda\mu\alpha\beta}\chi_{\alpha\beta}, \tag{7}$$

$$m^{\lambda} = tD^{\lambda\alpha}\zeta_{\alpha}.$$

Here $C^{\lambda\mu\alpha\beta}$ and $D^{\lambda\alpha}$ are the elasticity tensors of the shell. $C^{\lambda\mu\alpha\beta}$ is a fourth-order tensor with the following symmetries:

$$C^{\lambda\mu\alpha\beta} = C^{\mu\lambda\alpha\beta}, \quad C^{\lambda\mu\alpha\beta} = C^{\lambda\mu\beta\alpha}, \quad C^{\lambda\mu\alpha\beta} = C^{\alpha\beta\lambda\mu}. \tag{8}$$

$D^{\lambda\alpha}$ is a symmetric second order tensor satisfying $D^{\lambda\alpha} = D^{\alpha\lambda}$. Moreover, the symmetric second order tensors $N^{\lambda\mu}$ and $M^{\lambda\mu}$ are called force resultant and moment resultant, respectively, and m^{λ} is the transverse shear force resultant. Finally t is the thickness of the shell. In the following we assume the thickness of the shell to be constant. Note however that the main results presented in this article remain valid for a thickness profile $t = t(x)$, which remains unchanged during optimization. The symmetry of the tensors allows us to rewrite Hooke's law using the following vectors and matrices:

$$\gamma = \begin{pmatrix} \gamma_{11} \\ \gamma_{22} \\ \sqrt{2}\gamma_{12} \end{pmatrix}, \quad \chi = \begin{pmatrix} \chi_{11} \\ \chi_{22} \\ \sqrt{2}\chi_{12} \end{pmatrix}, \quad \zeta = \begin{pmatrix} \zeta_1 \\ \zeta_2 \end{pmatrix}, \tag{9}$$

$$N = \begin{pmatrix} N_{11} \\ N_{22} \\ \sqrt{2}N_{12} \end{pmatrix}, \quad M = \begin{pmatrix} M_{11} \\ M_{22} \\ \sqrt{2}M_{12} \end{pmatrix}, \quad m = \begin{pmatrix} m_1 \\ m_2 \end{pmatrix}, \tag{10}$$

$$C = \begin{pmatrix} C_{1111} & C_{1122} & \sqrt{2}C_{1112} \\ C_{1122} & C_{2222} & \sqrt{2}C_{2212} \\ \sqrt{2}C_{1112} & \sqrt{2}C_{2212} & 2C_{1212} \end{pmatrix}, \quad D = \begin{pmatrix} D_{11} & D_{12} \\ D_{12} & D_{22} \end{pmatrix}. \tag{11}$$

Then Hooke's law takes the form

$$N(x) = tC(x)\gamma(u(x)),$$

$$M(x) = \frac{t^3}{12}C(x)\chi(u(x),\theta(x)), \tag{12}$$

$$m(x) = tD(x)\zeta(u(x),\theta(x))$$

and the potential energy $\Pi(u,\theta)$ of the Naghdi shell can be written as

$$\Pi(u,\theta) = \frac{1}{2}\int_\omega \left(t\gamma^\top C\gamma + \frac{t^3}{12}\chi^\top C\chi + t\zeta^\top D\zeta\right)dS$$
$$- \int_\omega tf^\top u\,dS - \int_{\partial\omega_1}\left(g_u^\top u + g_\theta^\top\theta\right)dl, \tag{13}$$

where $f \in [L^2(\omega)]^3$ is a given force resultant density and $g_u \in [L^2(\partial\omega_1)]^3$ and $g_\theta \in [L^2(\omega)]^2$ are given traction and moment resultant densities, respectively. The shell is in equilibrium for any $(u,\theta) \in \mathcal{U}$ that minimizes the potential energy

$$\min_{(u,\theta)\in\mathcal{U}} \Pi(u,\theta). \tag{14}$$

It is also possible to treat plates in this context. Assuming a planar midsurface allows to deduce the Reissner-Mindlin plate model from Naghdi's shell model. A planar midsurface has no curvature and thus a constant normal vector a_3. This results in vanishing second and third fundamental forms of the midsurface ω:

$$b_{\alpha\beta} = 0, \quad c_{\alpha\beta} = 0. \tag{15}$$

In this case the formulas for the strains boil down to:

$$\gamma_{\alpha\beta}(u_1,u_2) = \frac{1}{2}\left(u_{\alpha|\beta} + u_{\beta|\alpha}\right),$$
$$\chi_{\alpha\beta}(\theta) = \frac{1}{2}\left(\theta_{\alpha|\beta} + \theta_{\beta|\alpha}\right), \tag{16}$$
$$\zeta_\alpha(u_3,\theta) = \frac{1}{2}\left(\theta_\alpha + u_{3,\alpha}\right).$$

The equilibrium state of the plate is again found by minimizing the potential energy

$$\min_{(u,\theta)\in\mathcal{U}} \Pi(u,\theta) = \frac{1}{2}\int_\omega \left(t\gamma^\top(u_1,u_2)C\gamma(u_1,u_2) + \frac{t^3}{12}\chi^\top(\theta)C\chi(\theta)\right.$$
$$\left. + t\zeta^\top(u_3,\theta)D\zeta(u_3,\theta)\right)dS - \int_\omega tf^\top u\,dS - \int_{\partial\omega_1}\left(g_u^\top u + g_\theta^\top\theta\right)dl. \tag{17}$$

When solving the elasticity problem for a plate this can be separated into the membrane problem

$$\min_{(u_1,u_2)\in\mathcal{U}} \frac{1}{2}\int_\omega t\gamma^\top(u_1,u_2)C\gamma(u_1,u_2)\,dS - \int_\omega t(f_1u_1 + f_2u_2)\,dS$$
$$- \int_{\partial\omega_1}(g_{u_1}u_1 + g_{u_2}u_2)\,dl \tag{18}$$

and the so-called Reissner-Mindlin problem

$$\min_{(u_3,\theta)\in\mathcal{U}} \frac{1}{2}\int_\omega \left(\frac{t^3}{12}\chi^\top(\theta)C\chi(\theta) + t\zeta^\top(u_3,\theta)D\zeta(u_3,\theta)\right)dS$$
$$- \int_\omega tf_3u_3\,dS - \int_{\partial\omega_1}\left(g_{u_3}u_3 + g_\theta^\top\theta\right)dl. \tag{19}$$

3 The Single Load Problem

Up to now we have merely described the physical behavior of the shell. However our overall goal is to find the stiffest structure which is subjected to a given set of loads f, g_u and g_θ. A measure on how much a structure will deform under these loads is given by the compliance

$$\text{comp}(C,D) = -\min_{(u,\theta)\in\mathcal{U}} 2\Pi_{C,D}(u,\theta) = \max_{(u,\theta)\in\mathcal{U}} -2\Pi_{C,D}(u,\theta)$$

$$= \max_{(u,\theta)\in\mathcal{U}} -\int_\omega \left(t\gamma^\top C\gamma + \frac{t^3}{12}\chi^\top C\chi + t\zeta^\top D\zeta \right) dS \qquad (20)$$

$$+2\int_\omega tf^\top u\, dS + 2\int_{\partial\omega_1} \left(g_u^\top u + g_\theta^\top \theta \right) dl.$$

Apparently the compliance is given by twice the negative potential energy in equilibrium. In order to find the stiffest structure possible we now minimize the compliance with respect to the design variables. As we intend to work with Free Material Optimization these variables are the full elasticity tensors C and D. We want to allow for holes and material-no-material situations in the optimal structures, therefore we choose $C \in [L^\infty(\omega)]^{3\times 3}$ and $D \in [L^\infty(\omega)]^{2\times 2}$. As pointed out in Section 2 the matrices have to be symmetric, furthermore they also have to be positive semidefinite as they describe a physical material:

$$C = C^\top \succeq 0, \quad D = D^\top \succeq 0. \qquad (21)$$

As a measure for the amount of material used at a certain point $x \in \omega$ we simply use the summed traces of the matrices $t\left(\text{tr}(C) + \frac{1}{2}\text{tr}(D)\right)$. The factor $\frac{1}{2}$ is necessary to be able to compare the results with the three–dimensional solid case. As we want to limit the material resources, we add the volume constraint

$$\int_\omega t\left(\text{tr}(C) + \frac{1}{2}\text{tr}(D)\right) dS \leq V. \qquad (22)$$

Finally we add box constraints to avoid arbitrarily high material concentrations at single points:

$$0 \leq \rho^- \leq t\left(\text{tr}(C) + \frac{1}{2}\text{tr}(D)\right) \leq \rho^+. \qquad (23)$$

Summarizing (21), (22) and (23) we obtain the set of admissible elasticity tensors

$$\mathscr{C} := \left\{ (C,D) \in [L^\infty(\omega)]^{3\times 3} \times [L^\infty(\omega)]^{2\times 2} \, \middle| \, \begin{array}{l} C = C^\top \succeq 0 \\ D = D^\top \succeq 0 \\ \int_\omega t(\text{tr}(C(x)) + \frac{1}{2}\text{tr}(D(x)))\, dS \leq V \\ 0 \leq \rho^- \leq t(\text{tr}(C) + \frac{1}{2}\text{tr}(D)) \leq \rho^+ \end{array} \right\} \qquad (24)$$

For simplicity of notation we will assume $\rho^- = 0$. But note that all statements presented in this paper are also true for positive ρ^-. We finally are able to formulate the single load problem for shells, in which we seek the design variables C and D which yield the minimal compliance:

$$\min_{(C,D)\in\mathscr{C}}\max_{(u,\theta)\in\mathscr{U}} -\frac{1}{2}\int_\omega \left(t\gamma^\top C\gamma + \frac{t^3}{12}\chi^\top C\chi + t\zeta^\top D\zeta\right)dS$$

$$+\int_\omega tf^\top u\,dS + \int_{\partial\omega_1}\left(g_u^\top u + g_\theta^\top\theta\right)dl. \qquad (25)$$

Introducing the function

$$J((C,D),(u,\theta)):= -\frac{1}{2}\int_\omega \left(t\gamma^\top C\gamma + \frac{t^3}{12}\chi^\top C\chi + t\zeta^\top D\zeta\right)dS$$

$$+\int_\omega tf^\top u\,dS + \int_{\partial\omega_1}\left(g_u^\top u + g_\theta^\top\theta\right)dl \qquad (26)$$

we rewrite the latter optimization problem as

$$\min_{(C,D)\in\mathscr{C}}\max_{(u,\theta)\in\mathscr{U}} J((C,D),(u,\theta)). \qquad (27)$$

In the case of plates we start from the equilibrium problem (17). The uncoupling into the membrane and the Reissner–Mindlin problem is not possible anymore when working with Free Material Optimization, as the material tensor C is one of the optimization variables connecting the membrane and bending terms. Thus the single load problem for Reissner–Mindlin plates takes the form

$$\min_{(C,D)\in\mathscr{C}}\max_{(u,\theta)\in\mathscr{U}} J((C,D),(u,\theta)):= -\frac{1}{2}\int_\omega \left(t\gamma^\top C\gamma + \frac{t^3}{12}\chi^\top C\chi + t\zeta^\top D\zeta\right)dS$$

$$+\int_\omega tf^\top u\,dS + \int_{\partial\omega_1}\left(g_u^\top u + g_\theta^\top\theta\right)dl. \qquad (28)$$

This problem has already been formulated by Bendsøe and Díaz, who propose a solution via analytic derivation of the optimal material properties [2].

We now want to show existence of optimal solutions for problem (27). It can be easily seen that an optimal solution of the single load problem for shells is a saddle-point of the functional $J((C,D),(u,\theta))$. Thus existence of an optimal point follows from a Minimax-Theorem.

Theorem 1. *Problem (27) has an optimal solution* $((C^*,D^*),(u^*,\theta^*)) \in \mathscr{C}\times\mathscr{U}$.

The proof uses the modified Minimax-Theorem presented in [19] that allows for $\mathscr{C}$ to be subset of the dual of a non–reflexive Banach–space – in this case $L^1(\omega)$. The required ellipticity of Naghdi's shell model has been shown in [11,5]. The complete proof can be found in [12].

4   The Primal Problem

In [24] it has been shown that the Free Material Optimization problem for solid material can be transformed into a linear quadratically constrained optimization problem using duality theory. During this section we show that a similar technique can be applied on the Free Material Optimization problem for shells resulting in a convex nonlinear semidefinite program instead of the saddle-point problem given in (27).

Theorem 2. *Problem (27) is equivalent to the Lagrange dual problem of*

$$\max_{\substack{(u,\theta)\in\mathcal{U} \\ \alpha\in\mathbb{R}_0^+ \\ \beta_{u,l}\in L^1(\omega) \\ \beta_{u,l}\geq 0}} \int_\omega tf^\top u\, dS + \int_{\partial\omega_1}(g_u^\top u + g_\theta^\top \theta)dl - \alpha V - \rho^+ \int_\omega \beta_u dS$$

$$\text{subject to}\quad \frac{t}{2}\gamma(u)\gamma(u)^\top + \frac{t^3}{24}\chi(u,\theta)\chi(u,\theta)^\top - t(\alpha + \beta_u - \beta_l)E_3 \preceq 0 \qquad (29)$$

$$\frac{t}{2}\zeta(u,\theta)\zeta(u,\theta)^\top - \frac{t}{2}(\alpha + \beta_u - \beta_l)E_2 \preceq 0$$

where E_n denotes the unit matrix in $\mathbb{R}^n$.

In order to prove this theorem we construct the Lagrangian to problem (29). It can then be shown that this problem is equivalent to the original problem (27). The proof follows the ideas presented in [24, Theorem 3.3.4] and is given in detail in [12].

Problem (29) is a convex nonlinear semidefinite program (SDP). Compared to the original problem formulation (27) problem (29) has several advantages. The matrices C and D are hidden in the problem as Lagrange multipliers. This significantly reduces the number of variables in the discrete problem (compare Section 5). Furthermore problem (29) is convex. As t is the thickness of the shell, it is strictly positive and the matrix constraints of (29) can be simplified to

$$\gamma(u)\gamma(u)^\top + \frac{t^2}{12}\chi(u,\theta)\chi(u,\theta)^\top - 2(\alpha + \beta_u - \beta_l)E_3 \preceq 0, \qquad (30)$$

$$\zeta(u,\theta)\zeta(u,\theta)^\top - (\alpha + \beta_u - \beta_l)E_2 \preceq 0.$$

5   Numerical Treatment

5.1   Discretization

We now intend to solve the infinite-dimensional SDP (29) numerically. For this purpose we discretize the problem by the finite element method [8]. The midsurface ω is partitioned into M elements ω_m. The number of corresponding element nodes is denoted by n. The elasticity matrices $C(x)$ and $D(x)$ are approximated by elementwise constant matrices $(C_1,\dots,C_M)$ and $(D_1,\dots,D_M)$ where $C_m \in \mathbb{R}^{3\times 3}$ and $D_m \in \mathbb{R}^{2\times 2}$ for all $m = 1,\dots,M$. The displacements take the following form

$$U^{3D} = \sum_{i=1}^n \lambda_i(r,s)\left(u^{(i)} + z\frac{t}{2}\theta^{(i)}\right), \qquad (31)$$

where the $\lambda_i(r,s)$ are bilinear 2D Lagrange shape functions. This assures that the Reissner-Mindlin assumption – material lines remain straight and unstretched during deformation – is fulfilled at all nodes of the mesh. Using (31) the discretized membrane strain matrix B_i^γ becomes

$$B_i^\gamma = \begin{pmatrix} \lambda_{i|1} & 0 & -b_{11}\lambda_i & 0\ 0 \\ 0 & \lambda_{i|2} & -b_{22}\lambda_i & 0\ 0 \\ \frac{1}{\sqrt{2}}\lambda_{i|2} & \frac{1}{\sqrt{2}}\lambda_{i|1} & -\sqrt{2}b_{12}\lambda_i & 0\ 0 \end{pmatrix} \qquad (32)$$

The factor $\sqrt{2}$ stems from the vector-matrix-notation introduced in (9). Using this, the discrete counterpart of the dyadic product $\gamma\gamma^{\top}$ reads

$$A_m^{\gamma}(u) = \sum_{i,j \in K} \int_{\omega_m} B_j^{\gamma} U U^{\top} (B_i^{\gamma})^{\top} dx, \tag{33}$$

where K is the index set of nodes associated with the element m. Analogously we derive

$$A_m^{\chi}(u) = \sum_{i,j \in K} \int_{\omega_m} B_j^{\chi} U U^{\top} (B_i^{\chi})^{\top} dx, \tag{34}$$

$$A_m^{\zeta}(u) = \sum_{i,j \in K} \int_{\omega_m} B_j^{\zeta} U U^{\top} (B_i^{\zeta})^{\top} dx. \tag{35}$$

Replacing the forces and moments in problem (29) by their discrete counterparts we get the following discrete single load FMO problem for shells:

$$\max_{\substack{(u,\theta) \in \mathscr{U} \\ \alpha \in \mathbb{R}_0^+ \\ \beta_u, \beta_l \in \mathbb{R}_0^{+M}}} \sum_{i=1}^{n} (t f_i u_i - \rho^+ \beta_{ui}) + \sum_{i \in \partial \omega_1} (g_{ui} u_i + g_{\theta i} \theta_i) dl - V\alpha$$

$$\text{subject to } \frac{t}{2} A_m^{\gamma}(u) + \frac{t^3}{24} A_m^{\chi}(u,\theta) - t(\alpha + \beta_u - \beta_l)E_3 \preceq 0 \tag{36}$$

$$\frac{t}{2} A_m^{\zeta}(u,\theta) - \frac{t}{2}(\alpha + \beta_u - \beta_l)E_2 \preceq 0.$$

Obviously (36) is a finite-dimensional convex nonlinear semidefinite program.

5.2 Examples

Two numerical examples are presented in this section. In order to solve problem (36) we have used the nonlinear SDP code PENNON [14]. Although only the resulting "density" function $t \operatorname{tr}(C) + \frac{1}{2} t \operatorname{tr}(D)$ is depicted, we want to emphasize that the code provides the optimal elasticity matrices C_m and D_m for each element ω_m, $m = 1,\ldots,M$. Thus we gather information about the optimal material symmetry and material directions usable in the manufacturing process.

Example 1. The first example (Fig. 2) serves as a test for the consistency with the two-dimensional solid case. We consider a rectangular plate with in-plane forces. The plate is clamped on one side while forces are applied in the center of the opposite edge and directed in parallel to the boundary. This example known as Michell truss is widely used in topology optimization literature.

The typical material distribution of a Michell truss can be easily recognized in the displayed "density" distribution (Fig. 3). It is also notable that only membrane strains appear as there is no deformation outside the midsurface. Thus there is no material used for the matrix D which accounts to shear effects.

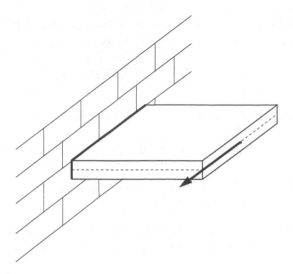

Fig. 2. Michell truss load case

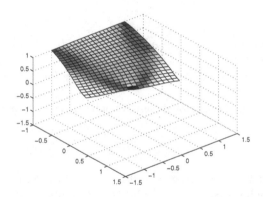

Fig. 3. Michell truss "density distribution"

Example 2. The second example employs all degrees of freedom of the shell. We start with a saddle-shaped midsurface, that is clamped on one side (Fig. 4). A vertical force acts in the center of the opposite edge. This example can be interpreted as optimization of a coat hook fixed to the wall.

The resulting "density distribution" (Fig. 5) shows a firm tip at the location of the load. The shell tries to avoid vertical bending and distributes material over the complete design domain (apart from the corners in the front which are not suited to stiffen this particular structure). No holes can thus be found in contrast to the previous example. This result is not unexpected as stiff triangle-shaped structures appearing in the plane of loading are well known in topology optimization.

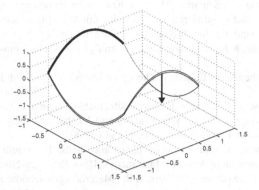

Fig. 4. Saddle-shaped hook load case

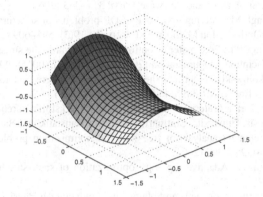

Fig. 5. Saddle-shaped hook "density distribution"

References

1. Ben-Tal, A., Kočvara, M., Nemirovski, A., Zowe, J.: Free material design via semidefinite programming: The multiload case with contact conditions. SIAM J. Optim. 9(4), 813–832 (1999)
2. Bendsøe, M.P., Díaz, A.R.: Optimization of material properties for Mindlin plate design. Structural Optimization 6, 268–270 (1993)
3. Bendsøe, M.P., Guedes, J.M., Haber, R.B., Pedersen, P., Taylor, J.E.: An analytical model to predict optimal material properties in the context of optimal structural design. J. Appl. Mech. Trans. ASME 61, 930–937 (1994)
4. Bendsøe, M.P., Sigmund, O.: Topology Optimization. Theory, Methods and Applications. Springer, Heidelberg (2002)
5. Blouza, M., Le Dret, H.: Naghdi's shell model: Existence, uniqueness and continuous dependence on the midsurface. Journal of Elasticity 64, 199–216 (2001)
6. Camprubí, N., Bischoff, M., Bletzinger, K.-U.: Shape optimization of shells and locking. Computers and Structures 82, 2551–2561 (2004)

7. Cervellera, P., Zhou, M., Schramm, U.: Optimization driven design of shell structures under stiffness, strength and stability requirements. In: 6th World Congresses of Structural and Multidisciplinary Optimization, Rio de Janeiro, Brazil (2005)
8. Chapelle, D., Bathe, K.J.: The Finite Element Analysis of Shells – Fundamentals. Springer, Heidelberg (2003)
9. Ciarlet, P.G.: Mathematical Elasticity. Theory of Shells, vol. 3. North Holland, Amsterdam (2000)
10. Cosserat, E., Cosserat, F.: Théorie des corps déformables. Hermann, Paris (1909)
11. Figueiredo, I., Leal, C.: Ellipticity of Koiter's and Naghdi's models for nonhomogeneous anisotropic shells. Applicable Analysis 70, 75–84 (1998)
12. Gaile, S.: Free material optimization for plates and shells: The single load case. Preprint–Series of the Department of Applied Mathematics, Univ. Erlangen–Nürnberg (2008)
13. Hörnlein, H.R.E.M., Kočvara, M., Werner, R.: Material optimization: Bridging the gap between conceptual and preliminary design. Aerospace Science and Technology 5(8), 541–554 (2001)
14. Kočvara, M., Stingl, M.: PENNON - a code for convex nonlinear and semidefinite programming. Optimization Methods and Software 18(3), 317–333 (2003)
15. Kočvara, M., Stingl, M.: Solving nonconvex SDP problems of structural optimization with stability control. Optimization Methods and Software 19(5), 595–609 (2004)
16. Kočvara, M., Stingl, M.: Free material optimization: Towards the stress constraints. Structural and Multidisciplinary Optimization (2006), doi:10.1007/s00158-007-0095-5
17. Lewiński, T., Sokołowski, J.: Optimal shells formed on a sphere. The topological derivative method. Technical Report RR-3495, INRIA-Lorraine (1998)
18. Lipka, A.: Verbesserter Materialeinsatz innovativer Werkstoffe durch die Topologieoptimierung. PhD thesis, Universität Stuttgart, Institut für Baustatik und Baudynamik (2007)
19. Mach, J.: Finite element analysis of free material optimization problem. Applications of Mathematics 49(4), 285–307 (2004)
20. Maute, K., Ramm, E.: Adaptive topology optimization of shell structures. AIAA 35(11), 1767–1773 (1997)
21. Naghdi, P.M.: The theory of shells and plates. In: Handbuch der Physik VIa/2, pp. 425–640. Springer, Heidelberg (1972)
22. Rubin, M.B.: Cosserat Theories: Shells, Rods and Points. Kluwer Academic Publishers, Dordrecht (2000)
23. Stegmann, J., Lund, E.: Discrete material optimization of general composite shell structures. Int. J. Numer. Meth. Engng. 62, 2009–2027 (2005)
24. Werner, R.: Free Material Optimization - Mathematical Analysis and Numerical Solution. PhD thesis, Friedrich-Alexander-Universität Erlangen-Nürnberg, Institut für Angewandte Mathematik II (2001)
25. Zowe, J., Kočvara, M., Bendsøe, M.P.: Free material optimization via mathematical programming. Mathematical Programming Series B 79, 445–466 (1997)

Elliptic Control by Penalty Techniques
with Control Reduction

Christian Grossmann, Holger Kunz, and Robert Meischner

Dresden University of Technology, Institute of Numerical Mathematics,
D - 01062 Dresden, Germany
Christian.Grossmann@tu-dresden.de

Abstract. The paper deals with the numerical treatment of optimal control problems with bounded distributed controls and elliptic state equations by a wider class of barrier-penalty methods. If the constraints are treated by barrier-penalty techniques then the necessary and sufficient optimality condition forms a coupled system of nonlinear equations which contain not only the usual adjoint and the state equation, but also an approximate projection by means of barrier-penalty terms. Under the made assumptions from the last one the control can be eliminated. This reduced optimality system which does not contain explicitly the controls, but the more regular states and adjoints only, is studied in detail.

1 Problem and Optimality Characterization

In this paper we study the numerical treatment of optimal control problems with bounded distributed controls and elliptic state equations by barrier-penalty methods. Let $\Omega \subset \mathbb{R}^2$ be some bounded convex domain and let be given $q, b, d \in L_\infty(\Omega)$. Further, we abbreviate $U := L_2(\Omega)$. Considered is the following optimal control problem

$$J(y,u) := \frac{1}{2} \int_\Omega (y-q)^2 + \frac{\alpha}{2} \int_\Omega u^2 \to \min!$$

$$\text{s.t. } -\Delta y = u \quad \text{in } \Omega, \tag{1}$$

$$y + \frac{\partial y}{\partial n} = 0 \quad \text{on } \Gamma := \partial\Omega,$$

$$u \in U_{ad},$$

where $\alpha > 0$ denotes a regularization parameter and the set of admissible controls U_{ad} is defined by

$$U_{ad} := \{u \in U : u \le b \quad \text{a.e. in } \Omega\} \tag{2}$$

or in case of additional state constraints by

$$U_{ad} := \{u \in U : u \le b, \, y \le d \quad \text{a.e. in } \Omega\}. \tag{3}$$

The consideration of only one-sided bounds for the controls as well as for the states serves to simplify the presentation, but does not principally restrict the considered class of problems.

A. Korytowski et al. (Eds.): System Modeling and Optimization, IFIP AICT 312, pp. 251–267, 2009.

Throughout this paper let us make the general assumption that $U_{ad} \neq \emptyset$. The numerical treatment of optimal control problems with elliptic state equations and constraints upon controls and partially also upon states have been extensively studied in the literature within the last years. The most popular techniques are semismooth Newton methods (see e.g. [11,12,24]), active set strategies (see [17]) and interior penalty methods (see e.g. [20,25]). A comprehensive discussion about the properties of control problems of the considered type can be found in [23].

In the present paper we widely concentrate upon the convergence properties of the certain barrier-penalty methods for the continuous problem. In particular, rate of convergence estimates for the considered methods are derived. Finite element discretizations are only briefly mentioned. The interaction between parameter selection rules for the embedding and the discretization step size in case of certain barrier methods has been recently analyzed in [16].

The state equations of the given problem (1) are understood in the weak sense of the Sobolev space $V := H^1(\Omega)$. Let the bilinear form $a(\cdot,\cdot) : V \times V \to \mathbb{R}$ be defined by

$$a(y,v) := \int_\Omega \nabla y \cdot \nabla v + \int_\Gamma yv \quad \forall y, v \in V. \tag{4}$$

Then for any $u \in U_{ad}$ the state equation possesses a unique weak solution, i.e. there exists a unique $y \in V$ such that

$$a(y,v) = (u,v) \quad \forall v \in V. \tag{5}$$

Here $(\cdot, \cdot)$ denotes the L_2-inner product. With the continuous embedding $V \hookrightarrow L_2(\Omega)$ by $Su := y$ this also defines a continuous linear mapping $S : L_2(\Omega) \to L_2(\Omega)$ and problem (1) can be reduced to its equivalent form

$$\hat{J}(u) := J(Su, u) \to \min! \quad \text{s.t. } u \in U_{ad}. \tag{6}$$

Theorem 1. *The abstract optimization problem (6) possesses a unique optimal solution $\bar{u}$ and $(S\bar{u}, \bar{u}) \in V \times U_{ad}$ is the related unique optimal solution of (1). The variational inequality*

$$\langle \hat{J}'(\bar{u}), u - \bar{u} \rangle \geq 0 \quad \forall u \in U_{ad} \tag{7}$$

forms a necessary and sufficient condition for $\bar{u} \in U_{ad}$ to solve (1). Further, in case of control constraints only, i.e. if (2) holds, this is equivalent to the coupled system

$$(\bar{y} - q, y) + a(y, \bar{v}) = 0 \quad \forall y \in V,$$

$$a(\bar{y}, v) - (\bar{u}, v) = 0 \quad \forall v \in V, \tag{8}$$

$$\alpha(\bar{u}, u - \bar{u}) - (u - \bar{u}, \bar{v}) \geq 0 \quad \forall u \in U_{ad}.$$

Proof. Since U_{ad} is nonempty, closed and convex and $\hat{J}$ is continuous and strongly convex immediately the existence and uniqueness of the optimal solution follows (c.f. [3], [23]). With the differentiability of $\hat{J}$ and the convexity of U_{ad} we obtain (7). Under the made assumptions the state equation as well as the adjoint equation possess unique solutions which together provide a saddle point of the Lagrangian related to (1), (2) which is characterized by (8). □

We notice that the occurring inequality in (8) is equivalent to

$$\bar{u} = P(\bar{u} - \sigma(\bar{v} + \alpha\bar{u}))$$ (9)

for any $\sigma > 0$, where P denotes the $L_2(\Omega)$-orthogonal-projection onto U_{ad}. With the particular choice $\sigma = 1/\alpha$ this enables to eliminate $\bar{u}$ from the remaining system (see e.g. [11], [15], [24], [23]). This approach has the advantage that only $\bar{y}, \bar{v}$ occur that are much smoother than $\bar{u}$. This fact is important for the discretization and leads to optimal convergence rates as shown in [15].

2   Two Penalties of a General Class Applied to Control Problems

The well know idea of barrier-penalty methods is to augment the objective by some term that penalizes either the closeness to the boundary of U_{ad} in case of interior point methods (for the logarithmic barrier see e.g. Weiser/Gänzler/Schiela [26]) or the violation of constraints that defined U_{ad} in case of pure penalties. Let $\Phi(\cdot, s) : U \to \mathbb{R}$ denote such a parametric barrier-penalty functional, where $s > 0$ is the embedding parameter that has to tend to zero. In our case we apply convex and continuous functionals with the property

$$\lim_{s \to 0+} \Phi(u, s) = 0 \quad \forall u \in U_{ad} \quad \text{and} \quad \lim_{s \to 0+} \Phi(u, s) > 0 \quad \forall u \notin U_{ad}$$ (10)

and obtain for fixed $s > 0$ the augmented control problems

$$\tilde{J}(u, s) := \hat{J}(u) + \Phi(u, s) \to \min! \quad \text{s.t.} \quad u \in U$$ (11)

which are unconstrained. In accordance with the structure of U_{ad} we define Φ via Nemitskij operators either by

$$\Phi(u, s) = \int_{\Omega} \phi(u(x) - b(x), s) \, dx$$
$$\text{or} \Phi(u, s) = \int_{\Omega} \phi(u(x) - b(x), s) \, dx + \int_{\Omega} \phi([Su](x) - d(x), s) \, dx.$$ (12)

Here $\phi : \mathbb{R} \to \mathbb{R}$ denotes some barrier-penalty function that satisfies

$$\frac{\partial}{\partial t} \phi(t, s) = \psi\left(\frac{t}{s}\right) \quad \forall t \in \mathbb{R}$$ (13)

with an appropriate function $\psi : \mathbb{R} \to \mathbb{R}$. For finite dimensional optimization problems a rather general barrier-penalty class has been discussed in Grossmann/Zadlo [8]. In the present paper we apply the concept to the considered infinite dimensional problem, but we restrict us to the specific functions

$$\psi(t) := \max\{0, t\} \quad \text{and} \quad \psi(t) := \delta\left(1 + \frac{t}{\sqrt{t^2 + 1}}\right),$$ (14)

which correspond to

$$\phi(t, s) := \frac{1}{2} s^{-1} \max^2\{0, t\} \quad \text{and} \quad \phi(t, s) := \delta\left(t + \sqrt{s^2 + t^2}\right),$$ (15)

respectively. In the second type the parameter $\delta > 0$ denotes some appropriately chosen constant (compare Theorem 3). While the first type of barrier-penalty functions represents just the standard quadratic loss penalty the second type forms a certain smoothed version of the exact penalty. It has been originally proposed and studied by Kaplan (compare [6]). Another type of a smoothed exact penalty for control problems has been recently studied in [9].

The two types of barrier-penalty techniques considered in our paper show some principle differences in its convergence analysis as will be seen later. Additional properties of the embeddings (14) and (15) are derived in [18] and [19], respectively.

A common property of both barrier-penalty types that follows from the Carathéodory conditions (cf. [27]) is

Lemma 1. *For both of the considered types of barrier-penalty embeddings for any $s > 0$ the related functional $\Phi(\cdot, s)$ is well defined on U.*

For a detailed discussion of the general assumption upon ψ as well as for further types of barrier-penalty functions we refer to Grossmann/Zadlo [8].

Lemma 2. *For any $s > 0$ the penalty problem*

$$\tilde{J}(u,s) \rightarrow min! \quad s.t. \ u \in U \tag{16}$$

possesses a unique solution $u(s)$. The point $u(s) \in U$ forms a solution of the unconstrained problem (16) if and only if

$$\langle \tilde{J}'(u(s),s), u - u(s) \rangle = 0 \quad \forall u \in U \tag{17}$$

holds. In case of control restrictions only, i.e. if U_{ad} is given by (2), the variational equality (17) is equivalent to

$$(y(s) - q, y) - a(y, v(s)) = 0 \quad \forall y \in V,$$

$$-a(y(s), v) + (u(s), v) = 0 \quad \forall v \in V, \tag{18}$$

$$\alpha u(s) + v(s) + \psi((u(s) - b)/s) = 0 \quad a.e. \ in \ \Omega,$$

while in case of control and state constraints, i.e. if U_{ad} is given by (3), the variational equality (17) is equivalent to

$$(y(s) - q, y) + (y, \psi((y(s) - d)/s)) - a(y, v(s)) = 0 \quad \forall y \in V,$$

$$-a(y(s), v) + (u(s), v) = 0 \quad \forall v \in V, \tag{19}$$

$$\alpha u(s) + v(s) + \psi((u(s) - b)/s) = 0 \quad a.e. \ in \ \Omega.$$

Here and in (18) denote $y(s) \in V$ and $v(s) \in V$ the related optimal state and adjoint state, respectively. Further for both cases of constraints holds

$$\limsup_{s \to 0+} \Phi(u(s), s) \leq \min_{u \in U_{ad}} \hat{J}(u). \tag{20}$$

Proof. First we notice that for arbitrary $s > 0$ the augmented objective $\tilde{J}(\cdot, s)$ is well defined for any $u \in U$ since the considered functions $\phi(\cdot, s)$ satisfy the appropriate Carathéodory conditions (c.f. [27]). With the continuity and strong convexity standard arguments imply (c.f. [7], [23]) that a unique minimizer $\bar{u}(s) \in U$ exists. Further, convexity and differentiability of $\tilde{J}(\cdot, s)$ yields the condition (17). Finally, the usual representation of $\tilde{J}(\cdot, s)$ via adjoints leads to the equivalent conditions (18) and (19), respectively.

It remains to prove (20). By assumption $U_{ad} \neq \emptyset$. Let $\tilde{u} \in U_{ad}$ be some arbitrary, but fixed element then with the non-negativity of the functional $\hat{J}$ we obtain

$$\Phi(u(s), s) \leq \hat{J}(u(s)) + \Phi(u(s), s) = \tilde{J}(u(s), s) \leq \tilde{J}(\tilde{u}, s) = \hat{J}(\tilde{u}) + \Phi(\tilde{u}, s) \quad \forall s > 0. \tag{21}$$

Now, property (10) implies that (20) holds. □

Theorem 2. *In case of the quadratic loss penalty functional defined via* $\psi(t) = \max\{0, t\}$ *we have*

$$\lim_{s \to 0+} u(s) = \bar{u}. \tag{22}$$

Further, in case of control constraints only, i.e. U_{ad} defined by (2), holds

$$\|u(s) - \bar{u}\| = O(s^{1/2}) \quad \text{for} \quad s \to 0+. \tag{23}$$

Proof. With

$$\Phi(u, s) = 0 \quad \forall u \in U_{ad}, \quad s > 0 \tag{24}$$

and with the structure of the auxiliary objective $\tilde{J}(\cdot, s)$ and of the reduced objective $\hat{J}$ we obtain

$$\frac{\alpha}{2} \|u(s)\|^2 \leq \hat{J}(u(s)) \leq \tilde{J}(u(s), s) \leq \tilde{J}(\bar{u}, s) = \hat{J}(\bar{u}) \quad \forall s > 0. \tag{25}$$

Hence, $\{u(s)\}_{s>0} \subset U$ is a bounded family in the Hilbert space U. Let $\{s_k\}$ denote an arbitrary sequence with $s_k > 0$, $s_k \to 0+$. Then $\{u^k\}$ defined by $u^k := u(s_k)$, $k = 1, 2, \ldots$ is weakly compact. As a consequence it contains some convergent subsequence. Without loss of generality we may assume that $\{u^k\}$ itself is weakly convergent to some $\tilde{u} \in U$.

Due to Lemma 2 the exists a $c > 0$ with

$$\Phi(u^k, s_k) \leq c \quad \forall k \in \mathbb{N}, \tag{26}$$

$$s_k^{-1} \|[u^k - b]_+\|^2 \leq c \quad \text{and} \quad s_k^{-1} \|[Su^k - d]_+\|^2 \leq c \quad \forall k \in \mathbb{N}. \tag{27}$$

Here and in the sequel as usual $[\cdot]_+$ denotes the positive part completed by zero. Since the functional $u \to \|[u - b]_+\|^2$ and $u \to \|[Su - d]_+\|^2$ are convex and continuous they are also weakly lower semi-continuous. Thus, together with its non-negativity (27) implies

$$\|[\tilde{u} - b]_+\|^2 = 0 \quad \text{and} \quad \|[S\tilde{u} - d]_+\|^2 = 0 \tag{28}$$

which proves

$$\tilde{u} \leq b \quad \text{and} \quad S\tilde{u} \leq d \quad \text{a.e. in } \Omega, \tag{29}$$

i.e. we have $\tilde{u} \in U_{ad}$. Further, the lower semi-continuity of $\hat{J}$ as a consequence of its convexity and continuity together with (25) implies that $\tilde{u}$ is optimal for the original control problem (1). Its solution is unique, namely $\bar{u}$. Because $\{u(s)\}_{s>0}$ is bounded and the selection of any weakly convergent subsequence tends to $\bar{u}$ we have

$$u(s) \rightharpoonup \bar{u} \quad \text{for} \quad s \to 0+. \tag{30}$$

Now, we turn to the strong convergence in $L_2(\Omega)$. From (25) and from the weakly lower semi-continuity of $\hat{J}$ we obtain

$$\hat{J}(\bar{u}) \leq \liminf_{s \to 0+} \hat{J}(u(s)) \leq \limsup_{s \to 0+} \hat{J}(u(s)) \leq \hat{J}(\bar{u}). \tag{31}$$

Hence,

$$\lim_{s \to 0+} \hat{J}(u(s)) = \hat{J}(\bar{u}). \tag{32}$$

Further, since $Su(s) \in V = H^1(\Omega)$ the compact embedding $H^1(\Omega) \hookrightarrow L_2(\Omega)$ the weak convergence of $\{u(s)\}$ in $L_2(\Omega)$ implies the strong convergence of the images $\{Su(s)\}$ in $L_2(\Omega)$. Thus, we obtain

$$\lim_{s \to 0+} \int_\Omega (S(u(s) - q)^2 = \int_\Omega (S(\bar{u} - q)^2. \tag{33}$$

With (32) and with the structure of $\hat{J}$ this results in

$$\lim_{s \to 0+} \|u(s)\| = \|\bar{u}\| \tag{34}$$

Finally, Radon-Riesz Theorem (see e.g. [5, Satz 5.10], [10]) provides the strong convergence, i.e.

$$\lim_{s \to 0+} \|u(s) - \bar{u}\| = 0. \tag{35}$$

Next, we prove the stated order of convergence in case of control constraints only, i.e. if U_{ad} is defined by (2). Under the made assumptions a regular Lagrange multiplier $\bar{\lambda} \in L_\infty(\Omega)$, $\bar{\lambda} \geq 0$ exists such that

$$L(\bar{u}, \lambda) \leq L(\bar{u}, \bar{\lambda}) \leq L(u, \bar{\lambda}) \quad \forall u \in U, \lambda \in U_+^*. \tag{36}$$

Here, due to the occurring regularity the Lagrangian $L(\cdot, \cdot)$ that handles the control bound $u \leq b$ can be represented by

$$L(u, \lambda) = \hat{J}(u) + \langle \lambda, u - b \rangle = \hat{J}(u) + \int_\Omega \lambda(x) (u(x) - b(x)) \, dx. \tag{37}$$

As already used, for the minimizer $u(s)$ of the auxiliary problem holds

$$\tilde{J}(u(s), s) \leq \tilde{J}(\bar{u}, s) = \hat{J}(\bar{u}) \quad \forall s > 0. \tag{38}$$

Thus, together with the right part of the saddle point inequality (36) and with the complementarity $\langle \bar{\lambda}, \bar{u} - b \rangle = 0$ we have

$$\hat{J}(u(s)) + s^{-1} \|[u(s) - b]_+\|^2 \leq \hat{J}(\bar{u}) = L(\bar{u}, \bar{\lambda}) \leq L(u(s), \bar{\lambda}) = \hat{J}(u(s)) + \langle \bar{\lambda}, u(s) - b \rangle. \tag{39}$$

With the right part of the saddle point inequality (36), with $\bar{\lambda} \geq 0$ and with Cauchy's inequality follows

$$\|[u(s) - b]_+\|^2 \leq s \langle \bar{\lambda}, u(s) - b \rangle \leq s \langle \bar{\lambda}, [u(s) - b]_+ \rangle \leq s \|\bar{\lambda}\| \, \|[u(s) - b]_+\|. \tag{40}$$

This leads to

$$\|[u(s) - b]_+\| \leq s \|\bar{\lambda}\| \quad \forall s > 0. \tag{41}$$

Further, we know that with the L_2-projector $\Pi : U \to U_{ad}$ holds

$$[u(s) - b]_+ = u(s) - \Pi u(s). \tag{42}$$

Together with (41) this yields

$$\|u(s) - \Pi u(s)\| \leq s \|\bar{\lambda}\| \quad \forall s > 0. \tag{43}$$

The structure of the objective $\hat{J}$ and the optimality criterion (7) leads to

$$\hat{J}(\Pi u(s)) \geq \hat{J}(\bar{u}) + \langle \hat{J}'(\bar{u}), \Pi u(s) - \bar{u} \rangle + \tfrac{\alpha}{2} \|\Pi u(s) - \bar{u}\|^2$$
$$\geq \hat{J}(\bar{u}) + \tfrac{\alpha}{2} \|\Pi u(s) - \bar{u}\|^2. \tag{44}$$

Thus, with $\hat{J}(u(s)) \leq \hat{J}(\bar{u})$ we have

$$\hat{J}(\Pi u(s)) - \hat{J}(u(s)) \geq \hat{J}(u(s)) - \hat{J}(\bar{u}) + \hat{J}(\Pi u(s)) - \hat{J}(u(s)) \geq \frac{\alpha}{2} \|\Pi u(s) - \bar{u}\|^2. \tag{45}$$

Since $\|u(s)\|$ is bounded for $s \to 0+$ with the local Lipschitz continuity of $\hat{J}$ there is some $c > 0$ such that

$$\|\hat{J}(\Pi u(s)) - \hat{J}(u(s))\| \leq c \|\Pi u(s) - u(s)\|. \tag{46}$$

Now, the estimates (43), (45) result in $\|\Pi u(s) - \bar{u}\| \leq c s^{1/2}$ with some $c > 0$. With the triangle inequality and again with (43) finally we obtain $\|u(s) - \bar{u}\| = O(s^{1/2})$. $\qquad \square$

Remark 1. The second part of Theorem 2 cannot be extended to the case of state constraints due to the lack of regularity of the multipliers. The multipliers related to state constraints as a rule are measures only (compare [1], [2]).

Investigations of penalty type methods for problems with reduced multiplier regularity can be found in [13], [14]. In these papers, in particular, the treatment of state constraints with quadratic loss function methods combined with augmented Lagrangian have been studied. However, the convergence result given in Theorem 2 differs from the mentioned investigations. Especially the proof for the rate of convergence is original.

Barrier methods for state constraints are discussed in [22] and an appropriate discretization of state constraints can be found in [4]. $\qquad \square$

Remark 2. Theorem 2 provides a convergence estimate of order $O(s^{1/2})$ which in practical experiments is exceeded. Here we obtained $\|u_h(s) - \bar{u}_h\| = O(s)$ for the solutions $u_h(s)$ and $\bar{u}_h$ of the auxiliary discrete problem and the discrete problem, respectively. This is quite natural since this has been shown for finite dimensional problems (compare [8]). But this way the discrete estimates would depend upon the dimension. In the computational experiments, however we observed mesh independence of the rate of convergence and related constants. Further studies are on the way to prove this property analytically. $\qquad \square$

Now we turn to the convergence analysis of the smoothed exact penalty type of barrier-penalty embedding.

Theorem 3. *Let $\delta > \frac{1}{2}\|\bar{\lambda}\|_\infty$. Then for the solutions $u(s)$ of the auxiliary problems of the smoothed exact penalty method, i.e. with the barrier-penalty function defined via*

$$\psi(t) = \delta \left(1 + \frac{t}{\sqrt{t^2+1}}\right), \; holds$$

$$\lim_{s \to 0+} \|u(s) - \bar{u}\| = 0. \tag{47}$$

If $\delta > \|\bar{\lambda}\|_\infty$ then some $s_0 > 0$ exists such that

$$u(s) \in U_{ad} \quad \forall s \in (0, s_0] \quad and \quad \|u(s) - \bar{u}\| = O(s^{1/2}) \quad for \quad s \to 0+. \tag{48}$$

Proof. First we notice that the made assumptions $q, b \in L_\infty(\Omega)$ and the convexity of Ω imply $\bar{\lambda} \in L_\infty(\Omega)$. The optimality of $u(s)$ for the auxiliary problem and the properties

$$\phi(t,s) = \delta\left(t + \sqrt{t^2+s^2}\right) \geq \delta\left(t + \sqrt{t^2}\right) = 2\delta\,[t]_+ \quad \forall t \in \mathbb{R}y, \, s > 0 \tag{49}$$

and

$$\phi(t,s) \leq \phi(0,s) = s \quad \forall t \leq 0 \tag{50}$$

yield

$$\hat{f}(u(s)) + 2\delta \int_\Omega [u(s)(x) - b(x)]_+ \, dx \leq \tilde{f}(u(s),s) \leq \tilde{f}(\bar{u},s)$$

$$= \hat{f}(\bar{u}) + \delta \int_\Omega \left(\bar{u}(x) - b(x) + \sqrt{(\bar{u}(x) - b(x))^2 + s^2}\right) dx \tag{51}$$

$$\leq \hat{f}(\bar{u}) + \delta s \int_\Omega dx = \hat{f}(\bar{u}) + \delta \mu(\Omega) s \quad \forall s > 0.$$

Further, from the saddle point inequality we obtain

$$\hat{f}(\bar{u}) \leq \hat{f}(u) + \langle \bar{\lambda}, u - b \rangle \quad \forall u \in U. \tag{52}$$

In particular with $u = u(s)$ and with $\bar{\lambda} \geq 0, \bar{\lambda} \in L_\infty(\Omega)$ follows

$$\hat{f}(\bar{u}) \leq \hat{f}(u(s)) + \int_\Omega \lambda(x)\,(u(s)(x) - b(x))\,dx \leq$$

$$\hat{f}(u(s)) + \|\bar{\lambda}\|_\infty \int_\Omega [u(s)(x) - b(x)]_+\,dx. \tag{53}$$

Thus, we have

$$(2\delta - \|\bar{\lambda}\|_\infty) \int_\Omega [u(s)(x) - b(x)]_+\,dx \leq \delta\mu(\Omega)s \quad \forall s > 0. \tag{54}$$

The convexity and lower semi-continuity of the integral functional at the left hand side, now provides

$$(2\delta - \|\bar{\lambda}\|_\infty) \int_\Omega [\bar{u}(x) - b(x)]_+ dx \le 0 \tag{55}$$

for any weak accumulation point $\tilde{u}$ of the family $\{u(s)\}$ for $s \to 0+$. Hence, as a consequence of the made assumption $\delta > \frac{1}{2}\|\bar{\lambda}\|$ we obtain $\tilde{u} \in U_{ad}$.

Further,

$$\hat{J}(u(s)) \le \tilde{J}(u(s), s) \le \tilde{J}(\bar{u}, s) \le \hat{J}(\bar{u}) + \delta\mu(\Omega)s \quad \forall s > 0 \tag{56}$$

with the weakly lower semi-continuity of $\hat{J}$ yields

$$\hat{J}(\tilde{u}) \le \hat{J}(\bar{u}) \tag{57}$$

for any weak accumulation point $\tilde{u}$ of $\{u(s)\}$ for $s \to 0+$. Since, $\{u(s)\}_{s>0}$ is weakly compact and the optimal solution $\bar{u}$ is unique the whole family converges weakly to $\bar{u}$. Further, (56) leads to (compare the proof of Theorem 2)

$$\lim_{s \to 0+} \hat{J}(u(s)) = \hat{J}(\bar{u}). \tag{58}$$

Now, by the same arguments as in the proof of Theorem 2 with the aid of the Radon-Riesz theorem we finally obtain

$$\lim_{s \to 0+} \|u(s) - \bar{u}\| = 0. \tag{59}$$

Next, we prove the stated order of convergence. To prepare this, first we show that

$$\lim_{s \to 0+} \|\lambda(s) - \bar{\lambda}\|_0 = 0 \tag{60}$$

for $\lambda(s)$ for $s > 0$ defined by

$$\lambda(s) := \psi((u(s) - b)/s). \tag{61}$$

Here $u \to \psi((u-b)/s)$ is understood in the sense of Nemitskij operators. The regularity of $u(s)$ and the Carathéodory properties of ψ guarantee $\lambda(s) \in L_2(\Omega)$. Further, the optimality condition (18) yields

$$\lambda(s) = -v(s) - \alpha u(s). \tag{62}$$

We notice that the state equation and the adjoint equation are stable w.r.t. the input u and y, respectively. With the lifting property that results from the supposed convexity of the domain Ω and with the continuous embedding we obtain

$$\|\bar{y} - y(s)\|_2 \le c\|\bar{u} - u(s)\|_0 \quad \text{and} \quad \|\bar{v} - v(s)\|_2 \le c\|\bar{y} - y(s)\|_0, \tag{63}$$

where $\|\cdot\|_0$ and $\|\cdot\|_2$ denote the L_2-norm and H^2-norm, respectively. Hence, the continuous embedding $H^2(\Omega) \hookrightarrow L_\infty(\Omega)$ and the convergence $\lim_{s \to 0+} \|\bar{u} - u(s)\|_0 = 0$, that has been shown already, imply

$$\lim_{s \to 0+} \|\bar{y} - y(s)\|_\infty = 0 \quad \text{and} \quad \lim_{s \to 0+} \|\bar{v} - v(s)\|_\infty = 0. \tag{64}$$

Now, from (62) follows that $\lambda(s)$ converges for $s \to 0+$ in $L_\infty(\Omega)$ to

$$\tilde{\lambda} := -\bar{v} - \alpha\bar{u}. \tag{65}$$

Next, we show that $\tilde{\lambda} = \bar{\lambda}$, i.e. $\lambda(s)$ converges to the optimal Lagrange multiplier. First, we notice that

$$\lambda(s) \geq 0 \quad \text{a.e. in } \Omega, \quad \forall s > 0. \tag{66}$$

This implies

$$\tilde{\lambda} \geq 0 \quad \text{a.e. in } \Omega. \tag{67}$$

Further, we have

$$\langle \lambda(s), u(s) - b \rangle = \delta \int_\Omega \left(1 + \frac{u(s)(x) - b(x)}{\sqrt{(u(s)(x) - b(x))^2 + s^2}} \right) (u(s)(x) - b(x)) \, dx \tag{68}$$

$$= \delta \int_\Omega \left((u(s)(x) - b(x)) + \frac{u(s)(x) - b(x)^2}{\sqrt{(u(s)(x) - b(x))^2 + s^2}} \right) dx.$$

This leads to

$$|\langle \lambda(s), u(s) - b \rangle| \leq \delta \int_\Omega \left| u(s)(x) - b(x) + |u(s)(x) - b(x)| \right| dx$$

$$= 2\delta \int_\Omega [u(s)(x) - b(x)]_+ \, dx \leq 2\delta \sqrt{\mu(\Omega)} \, \|[u(s)(x) - b(x)]_+\|_0. \tag{69}$$

With $\lim_{s \to 0+} \|u(s) - \bar{u}\|_0 = 0$ and $\|[\bar{u} - b]_+\|_0 = 0$, now we obtain

$$\lim_{s \to 0+} |\langle \lambda(s), u(s) - b \rangle| = 0. \tag{70}$$

Hence, taking the already shown convergence of $\{\lambda(s)\}$ and $\{u(s)\}$ into account this yields

$$\langle \tilde{\lambda}, \bar{u} - b \rangle = 0. \tag{71}$$

Thus, $\tilde{\lambda}$ forms an optimal multiplier for the original problem related to the control constraint $u \leq b$. The structure of the constraints imply the uniqueness of the Lagrangian multiplier. This leads to $\tilde{\lambda} = \bar{\lambda}$.

Now, we turn to the proof of finite feasibility of the method. By definition we have

$$\lambda(s)(x) = \delta \left(1 + \frac{u(s)(x) - b(x)}{\sqrt{(u(s)(x) - b(x))^2 + s^2}} \right) \tag{72}$$

and consequently

$$\frac{\|\lambda(s)\|_\infty}{\delta} = \operatorname*{ess\,sup}_{x \in \Omega} \left| 1 + \frac{u(s)(x) - b(x)}{\sqrt{(u(s)(x) - b(x))^2 + s^2}} \right| \quad \forall s > 0. \tag{73}$$

With the shown convergence $\lim_{s \to 0+} \|\lambda(s) - \bar{\lambda}\|_\infty = 0$ and with the assumption $\delta > \|\bar{\lambda}\|_\infty$ this guarantees that some $s_0 > 0$ exists with

$$\operatorname*{ess\,sup}_{x \in \Omega} \left| 1 + \frac{u(s)(x) - b(x)}{\sqrt{(u(s)(x) - b(x))^2 + s^2}} \right| < 1 \quad \forall s \in (0, s_0]. \tag{74}$$

Thus, we have

$$u(s) < b \quad \text{a.e. in } \Omega, \quad \forall s \in (0, s_0] \tag{75}$$

Since only upper bounds for the constraints are considered this proves the stated property $u(s) \in U_{ad}$ for any $s \in (0, s_0]$.

Finally, we show the rate of convergence. With the optimality of $u(s)$ for the auxiliary problems we obtain

$$\hat{J}(\bar{u}) + \langle \hat{J}'(\bar{u}), u(s) - \bar{u} \rangle + \frac{\alpha}{2} \|u(s) - \bar{u}\|^2 \le$$
$$\tilde{J}(u(s), s) \le \tilde{J}(\bar{u}, s) \le \hat{J}(\bar{u}) + \delta \mu(\Omega) s \quad \forall s > 0. \tag{76}$$

Since $u(s) \in U_{ad}$ for $s \in (0, s_0]$ the optimality of $\bar{u}$ for the original problem guarantees

$$\langle \hat{J}'(\bar{u}), u(s) - \bar{u} \rangle \ge 0 \quad \forall s \in (0, s_0]. \tag{77}$$

Thus, (76) leads to

$$\frac{\alpha}{2} \|u(s) - \bar{u}\|^2 \le \delta \mu(\Omega) s \quad \forall s \in (0, s_0]. \tag{78}$$

This completes the proof of the theorem. $\square$

Remark 3. The method of smoothed exact penalties has been proposed originally by A.A.Kaplan (compare [6]). Theorem 3 provides for this technique a convergence estimate of order $O(s^{1/2})$ which in practical experiments is exceeded. Here as earlier for the quadratic loss penalty we obtained again $\|u_h(s) - \bar{u}_h\| = O(s)$ for the solutions $u_h(s)$ and $\bar{u}_h$ of the auxiliary problem and the discrete problem, respectively. In case of discretized problems this is a consequence of convergence results for finite dimensional problems (compare [8]). But, as already mentioned, the discrete estimates would depend upon the dimension. In the computational experiments, however we also observed mesh independence of the rate of convergence and related constants. Further studies are on the way to prove this property analytically.

The smoothed exact penalty method requires quite regular Lagrangian multipliers. This restricts its application to the treatment of control constraints. For continuous problems with state constraints it can be combined with other penalty types like the quadratic loss. In the discrete case the smoothed exact penalty can be applied, but then instead of a fixed parameter δ as described above some function should be used to adapt the method better to the local behavior of the multipliers.

An essential advantage of the smoothed exact penalty is that the auxiliary objective is defined on the whole $L_2(\Omega)$, but still guarantees feasibility already for certain positive s. So it really combines properties of interior point methods like logarithmic barriers with penalty methods. $\qquad\square$

3 Control Reduction and Discretization

The treatment of the restrictions by the augmented problem (11) leads to the necessary and sufficient optimality conditions (18) and (19) in case of bound on controls only and in case of additional restrictions upon states, respectively. For any $s > 0$ these system possesses unique solutions $(\bar{y}(s), \bar{v}(s), \bar{u}(s)) \in V \times V \times U$. The structure of the considered functions ψ guarantee that from the last equation, i.e. from

$$\alpha u(s) + v(s) + \psi((u(s) - b)/s) = 0 \quad \text{a.e. in } \Omega,$$

in both cases the optimal control $\bar{u}$ can be determined in dependence of $v(s)$. Due to the regularity $v(s) \in H^2(\Omega) \hookrightarrow C(\bar{\Omega})$ of the adjoint this can be done by pointwise elimination. Let us denote this by $u(s) = g(v(s), s)$. Thus, (18) and (19) leads to the parametric control reduced optimality system

$$(y(s) - q, y) - a(y, v(s)) = 0 \quad \forall y \in V,$$
$$-a(y(s), v) + (g(v(s), s), v) = 0 \quad \forall v \in V \tag{79}$$

and

$$(y(s) - q, y) + \psi((y(s) - d)/s) - a(y, v(s)) = 0 \quad \forall y \in V,$$
$$-a(y(s), v) + (g(v(s), s), v) = 0 \quad \forall v \in V, \tag{80}$$

respectively. Both conditions form a coupled system of weakly nonlinear partial differential equations. There holds

Theorem 4. *For any $s > 0$ each of the systems (79) and (80) possesses a unique solution $(y(s), v(s)) \in V \times V$ and $u(s) := g(v(s), s)$ forms the optimal solution of the related parametric barrier-penalty problem (16).*

Since the optimal state $y(s)$ as well as the optimal adjoint state $v(s)$ possess a higher regularity than the optimal control $u(s)$ problem (79) as well as (80) allows a more efficient treatment by discretization techniques, e.g. by finite elements, than the full system. However, it has to be noticed that in case of state constraints the limit properties of barrier-penalty functions asymptotically lead to ill-conditioned problems for $s \to 0+$.

The control reduction via the mapping g requires its efficient evaluation. As already mentioned, due to the regularity of the adjoint states $v(s)$ this can be done pointwise. In the case of the quadratic loss penalty the piecewise linear structure of the function Ψ allows an explicit evaluation of g while in the case of the smoothed exact penalty an additional iteration process to evaluate g is needed. For this purpose Newton's method has been applied with an appropriate choice of the initial iterates.

Conforming finite element discretizations $V_h \subset V$ (cf. [7]) can be applied to the control reduced systems (79) and (80). This leads to the finite dimensional system of nonlinear equations

$$(y_h(s) - q, y_h) - a(y_h, v_h(s)) = 0 \quad \forall y_h \in V_h,$$

$$-a(y_h(s), v_h) + (g(v_h(s), s), v_h) = 0 \quad \forall v_h \in V_h. \tag{81}$$

and

$$(y_h(s) - q, y_h) + (\psi((y_h(s) - d)/s, y_h) - a(y_h, v_h(s)) = 0 \quad \forall y_h \in V_h,$$

$$-a(y_h(s), v_h) + (g(v_h(s), s), v_h) = 0 \quad \forall v_h \in V_h. \tag{82}$$

in case of (2) and (3), respectively. Like in the continuous case system (79) defines uniquely the solution $(y_h(s), v_h(s)) \in V_h \times V_h$. Further, we obtain $u_h(s) = g(v_h(s), s)$ which unlike in full discretization does not use an a-priori discretization of the space U. The convergence theory for control reduced finite element discretizations as developed in [15], [20], [21], [26] can be carried over to the system (81) and partially also to (82).

4 Numerical Examples

Finally we report on some numerical experiments that show the applicability of the proposed barrier-penalty embeddings. As already noticed, the experimental rate of convergence exceeds the estimate $O(s^{1/2})$ and further research is on the way to prove this also by a sharper analysis.

We consider piecewise linear conforming finite elements $V_h \subset V$ with a criss-cross triangulation applied to the control reduced systems (79) and (80), respectively.

Example 1

$$J(y, u) := \frac{1}{2}\|y - q\|_0^2 + \frac{\alpha}{2}\|u\|_0^2 \rightarrow \min!$$

s.t.

$$-\Delta y = u \quad \text{in } \Omega = [0, 1]^2,$$

$$y + \frac{\partial y}{\partial n} = 0 \quad \text{on } \Gamma := \partial\Omega,$$

$$u \in U_{ad} := \{u \in U : -4 \le u \le 12 \quad \text{a.e. in } \Omega\} \tag{83}$$

with $q(x_1, x_2) = x_1 + x_2$. The graphs in Fig. 1 and Fig. 2 show the discrete solution obtained with the quadratic loss penalty for $s = 10^{-10}$ over a grid with $N = 900$ grid points.

Similar results are obtained by means of the smoothed exact penalty generated by

$$\psi(t) = \delta\left(1 + \frac{t}{\sqrt{1 + t^2}}\right). \tag{84}$$

Unlike in the quadratic loss penalty case here feasibility is obtained for sufficiently small $s > 0$ (see Fig. 3).

Further, we have applied the long step path following concept to the proposed barrier-penalty techniques. In this technique only one Newton iteration is performed at each parameter level s. We reduced the embedding parameter by the linear reduction rule $s_{k+1} = \rho \cdot s_k$ with some $\rho \in (0, 1)$. The obtained experimental order of convergence of

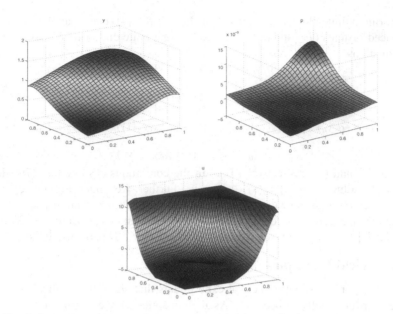

Fig. 1. Optimal (left) and adjoint (middle) states, and optimal control (upper right)

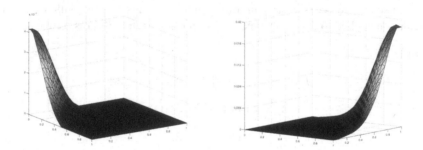

Fig. 2. Multiplier for the lower (left) and upper bound (right)

this long-step path following method for $\rho = 0.5$ applied to the considered Example 1 is given below.

s	$EOC_0(y)$	$EOC_1(y)$	$EOC_0(u)$
1	1.0097	0.9419	1.3334
2^{-4}	1.00	0.9955	1.0279
2^{-8}	1.00	1.00	1.0026
2^{-12}	1.00	1.00	1.00
2^{-16}	1.00	1.00	1.00
2^{-20}	1.00	1.00	1.00

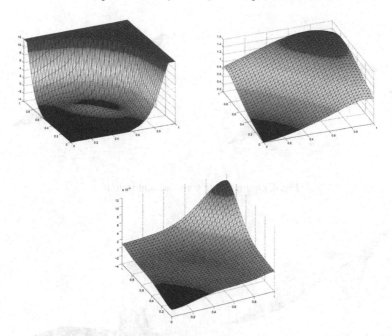

Fig. 3. Optimal control $\bar{u}$ (left), optimal state $\bar{y}$ (right), and adjoint state $\bar{v}$ (down)

Here ECO_0 and ECO_1 denote the experimental order of convergence in the L_2-norm and in the H^1-norm, respectively. These orders in case of the states are defined via the obtained solutions $y(s)$ and $y(\tilde{s})$ for different parameters $s > 0$ and $\tilde{s} > 0$, respectively, by

$$EOC_j(y) := (\ln(\|y(s) - y_{ref}\|_j) - \ln(\|y(\tilde{s}) - y_{ref}\|_j)) / (\ln(s) - \ln(\tilde{s})), \qquad (85)$$

where y_{ref} denotes the reference solution obtained as limit for $s \to 0$ and $\|\cdot\|_j$ are the considered norms. The $EOC_0(u)$ is analogously defined.

Example 2. In the next numerical experiment we modified the above considered Example 1 by the additional state constraint

$$y(x) \leq 1.2 \quad \text{in} \quad \Omega, \qquad (86)$$

but with no bounds on controls, i.e. we consider the control problem

$$J(y,u) := \frac{1}{2}\|y - q\|_0^2 + \frac{\alpha}{2}\|u\|_0^2 \to \min!$$

s.t.
$$-\Delta y = u \quad \text{in } \Omega = [0,1]^2,$$
$$y + \frac{\partial y}{\partial n} = 0 \quad \text{on } \Gamma := \partial\Omega, \qquad (87)$$
$$u \in U_{ad} := \{u \in U : Su \leq 1.2 \quad \text{a.e. in } \Omega\}$$

where $q(x_1,x_2) = x_1 + x_2$ and S denotes the solution operator of the elliptic boundary value problem that defines the states $y \in V$ for given controls $u \in U$. Numerical results

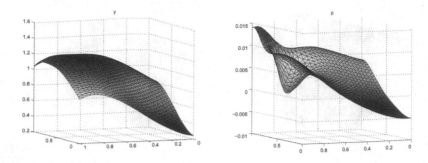

Fig. 4. Optimal state y and adjoint state v

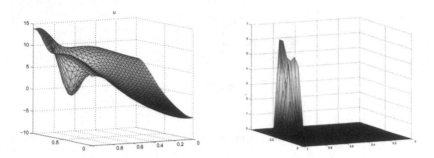

Fig. 5. Optimal control u and the multiplier for the state constraint

for $N = 256$ grid points and the embedding parameter $s = 10^{-3}$ are given in Fig. 4 and Fig. 5.

Acknowledgements. We thank the anonymous referees for their critical comments and advice.

References

1. Casas, E., Mateos, M.: Uniform convergence of the FEM applications to state constrained control problems. Comput. Appl. Math. 21, 67–100 (2002)
2. Casas, E., Mateos, M., Tröltzsch, F.: Error estimates for the numerical approximation of boundary semilinear elliptic control problems. Comput. Optim. Appl. 31, 193–219 (2005)
3. Céa, J.: Optimisation. Théorie et algorithmes, Paris, Dunod. IX (1971)
4. Deckelnick, K., Hinze, M.: A finite element approximation to elliptic control problems in presence of control and state constraints. Hamburger Beiträge zur Angewandten Mathematik 1 (2007)
5. Elstrodt, J.: Maß- und Integrationstheorie, 2nd edn. Springer, Berlin (1999)
6. Grossmann, C., Kaplan, A.A.: Strafmethoden und modifizierte Lagrange Funktionen in der nichtlinearen Optimierung. Teubner, Leipzig (1979)
7. Grossmann, C., Roos, H.-G., Stynes, M.: Numerical Treatment of Partial Differential Equations. Springer, Berlin (2007)

8. Grossmann, C., Zadlo, M.: A general class of penalty/barrier path-following Newton methods for nonlinear programming. Optimization 54, 161–190 (2005)
9. Gugat, M., Herty, M.: The smoothed-penalty algorithm for state constrained optimal control problems for partial differential equations (submitted for publication) (2008)
10. Hewitt, E., Stromberg, K.: Real and Abstract Analysis. A Modern Treatment of the Theory of Functions of a Real Variable, 3rd printing. Springer, Berlin (1975)
11. Hintermüller, M., Ito, M.K., Kunisch, K.: The primal-dual active set strategy as a semismooth Newton method. SIAM J. Optim. 13, 865–888 (2003)
12. Hintermüller, M., Kovtunenko, V., Kunisch, K.: Semismooth Newton methods for a class of unilaterally constrained variational problems. Adv. Math. Sci. Appl. 14, 513–535 (2004)
13. Hintermüller, M., Kunisch, K.: Feasible and noninterior path-following in constrained minimization with low multiplier regularity. SIAM J. Control Optim. 45, 1198–1221 (2006)
14. Hintermüller, M., Kunisch, K.: Path-following methods for a class of constrained minimization problems in function space. SIAM J. Optim. 17, 159–187 (2006)
15. Hinze, M.: A variational discretization concept in control constrained optimization: The linear-quadratic case. Comput. Optim. Appl. 30, 45–61 (2005)
16. Hinze, M., Schiela, A.: Discretization of interior point methods for state constrained elliptic optimal control problems: Optimal error estimates and parameter adjustment. Priority program 1253, preprint SPP1253-03-03 (2007)
17. Ito, K., Kunisch, K.: The primal-dual active set method for nonlinear optimal control problems with bilateral constraints. SIAM J. Control Optim. 43, 357–376 (2004)
18. Kunz, H.: Eine kontrollreduzierte Strafmethode für elliptische Steuerprobleme. Diploma thesis, TU Dresden (2007)
19. Meischner, R.: Geglättete exakte Strafen und kontrollreduzierte elliptische Steuerung. Diploma thesis, TU Dresden (2007)
20. Schiela, A.: The control reduced interior point method: A functional space oriented algorithmic approach. Ph.D. thesis, FU Berlin (2006)
21. Schiela, A.: Convergence of the control reduced interior point method for PDE constrained optimal control with state constraints. ZIB Report 06-16, Zuse-Zentrum Berlin (2006)
22. Schiela, A.: Barrier methods for control problems with state constraints. ZIB Report 07-07, Zuse-Zentrum Berlin (2007)
23. Tröltzsch, F.: Optimale Steuerung partieller Differentialgleichungen. Vieweg, Wiesbaden (2005)
24. Ulbrich, M.: Semismooth Newton methods for operator equations in function spaces. SIAM J. Optim. 13, 805–841 (2003)
25. Ulbrich, M., Ulbrich, S.: Superlinear convergence of affine-scaling interior-point Newton methods for infinite-dimensional nonlinear problems with pointwise bounds. SIAM J. Control Optim. 38, 1938–1984 (2000)
26. Weiser, M., Gänzler, T., Schiela, A.: A control reduced primal interior point method for PDE constrained optimization. ZIB Report 04-38, Zuse-Zentrum Berlin (2004)
27. Zeidler, E.: Nonlinear Functional Analysis and Its Applications. II: Nonlinear Monotone Operators. Springer, New York (1990)

Applications of Topological Derivatives and Neural Networks for Inverse Problems

Marta Grzanek[1] and Katarzyna Szulc[2]

[1] University of Łódź, Faculty of Mathematics and Computer Science,
Banacha 22, 90-238 Łódź, Poland
marta@math.uni.lodz.pl
[2] Université Henri Poincaré Nancy 1, Institut Elie Cartan, B.P. 239,
54506 Vandœvre-Lès-Nancy Cedex, France
katarzyna.szulc@iecn.u-nancy.fr

Abstract. Numerical method of identification for small circular openings in the domain of integration of an elliptic equation is presented. The method combines the asymptotic analysis of PDE's with an application of neural networks. The asymptotic analysis is performed in singularly perturbed geometrical domains with the imperfections in form of small voids and results in the form of the so-called topological derivatives of observation functionals for the inverse problem under study. Neural networks are used in order to find the mapping which associates to the observation shape functionals the conditional expectation of the size and location of the imperfections. The observation is given by a finite number of shape functionals. The approximation of the shape functionals by using the topological derivatives is used to prepare the training data for the learning process of an artificial neural network. Numerical results of the computations are presented and the probabilistic error analysis of such an identification method of the holes by neural network is performed.

1 Introduction

The paper describes a new numerical method which can be used for numerical identification of the conditional expectation of *small imperfections* in geometrical domains. The numerical method is implemented in neural networks, therefore, it requires the learning data set. Such a set can be constructed in general with a high cost of the computational effort. We propose the specific method of construction which reduces the cost by careful asymptotic analysis of the mathematical model which describes the observation of the real object. Such an observation is given by a finite number of functionals with their values depending on the imperfections to be identified. The simplest applied problem could be:

Identify a microcrack in the elastic body Ω from the finite number of elastic modes $\Lambda = (\lambda_1, \ldots, \lambda_M)$.

The microcrack should be characterized by a finite number of parameters, say $\Phi = (\ell_1, \ldots, \ell_N)$, and we should have in hand the relation between the elastic modes Λ and the form of microcrack Φ. This relation can be formally denoted by

$$\mathscr{F} : \Phi \mapsto \Lambda$$

A. Korytowski et al. (Eds.): System Modeling and Optimization, IFIP AICT 312, pp. 268–281, 2009.

and the learning set for neural network includes the set of the pointwise values $(\Phi_k, \Lambda_k), k = 1, \dots$ obtained from mathematical model of such a relation in order to model the inverse relation $(\Lambda_k, \Phi_k), k = 1, \dots$. The inverse relation is approximated by a neural network after the appropriate training procedure. Then, this approximation in hand, we can try to answer the question:

Given real observation of elastic modes of an elastic body, determine the size, the shape and the location of a finite number of microcracks in the body if the set Λ of elastic modes is measured for the specific body.

If we want to find one microcrack, the formal description of our problem is just

$$\widetilde{\Phi} = \mathscr{F}^{-1}(\widetilde{\Lambda}) . \tag{1}$$

This problem is an inverse problem which is, it seems, quite important for applications. There are difficult questions associated with such a problem:

- what is the meaning of the inverse mapping
- how for given value of observation Λ the required value $\widetilde{\Phi}$ of parameters which characterize the unknown defect or imperfection in the form of a microcrack can be computed in the way that the proposed method is robust
- if the method is convergent and in what sense

The inverse problem is a subject of the research, in the paper we restrict ourselves to a simplified variant which shows that the proposed methodology is promising and can be possibly developed for some real life inverse problems. The framework of our method includes the PDE model of our object and its asymptotic analysis in singularly perturbed geometrical domains, finite number of parameters which model the imperfections to be identified, finite number of observation functionals which can be evaluated from the PDE model and measured from the real life object, and finally the neural network which model the inverse mapping according to the standard rules in such approach. We refer to [3] for an introduction to the methodology of numerical solutions to inverse problems with the asymptotic analysis and neural networks. For voids of arbitrary shape, the topological derivatives are determined in function of the so-called polarization tensors and virtual mass tensors, we refer the reader to [6] for the detailed description of the results for spectral problems and the energy functional for Laplacian.

In [3], a particular problem is considered that guarantees existence of the inverse mapping. In this paper we don't know whether the inverse mapping exists. Based on probability theory [12] we obtain only the conditional expectation of the inverse mapping. So in our meaning **the "inverse mapping" is the conditional expectation of multifunction that for the vector of observation calculates the location and the size of imperfections.**

2 Preliminaries

We propose numerical method for identification of a finite number of imperfections in geometrical domain $\Omega_0 \subset \mathbb{R}^2$. The imperfections $\mathscr{B}_j(y_j)$, $j = 1, \dots, k$ are small voids or holes of radii ρ_j, $j = 1, \dots, k$, included in Ω_0, so the geometrical domain with the given number k of imperfections is denoted by Ω_k (see Section 3.1 for details). We point out that k holes $\mathscr{B}_j(y_j)$, $j = 1, \ldots, k$ are included in the domain Ω_k.

The identification procedure is based on the knowledge of N shape functionals $\mathscr{J}_i(\Omega_k)$, $i = 1, \ldots, N$, which are evaluated for the solutions $u(\Omega_k)$ of a PDE defined in Ω_k. Since $\Omega_k = \Omega_0 \setminus \bigcup_{j=1}^{k} \overline{\mathscr{B}_j(y_j)}$, and $\mathscr{B}_j(y_j) = \{x : | x - y_j | < \rho_j\}$, it follows that $\mathscr{J}_i(\Omega_k)$ depends on $3k$ parameters, i.e., $y_j \in \mathbb{R}^2$ and $\rho_j > 0$, $j = 1, \ldots, k$.

The numerical procedure means, e.g., the *minimization* with respect to $3k$ parameters of the gap between the values of observation functionals which are given for the real object, and the values which are obtained from the mathematical model. The values from the mathematical model are determined in the form of the function

$$\mathscr{G}_k : \mathbb{R}^{3k} \ni (y_1, \rho_1, \ldots, y_k, \rho_k) \mapsto (\mathscr{J}_1(\Omega_k), \ldots, \mathscr{J}_N(\Omega_k)) \in \mathbb{R}^N \tag{2}$$

Therefore, the inverse problem can be defined by e.g., the minimization of the following goal functional

$$\min_{(Y,\rho)} \operatorname{dist}(\mathscr{G}_k(Y,\rho), \mathscr{J}(\Omega_k)) \tag{3}$$

where $(Y,\rho) = (y_1, \rho_1, \ldots, y_k, \rho_k)$ and $\mathscr{J}(\Omega_k) = (\mathscr{J}_1, \ldots, \mathscr{J}_N) \in \mathbb{R}^N$ are the values determined for the real object.

In order to simplify the numerical procedure, the mapping $\mathscr{G}_k(Y,\rho)$ is replaced in (3) by its approximation given in terms of the so-called topological derivatives. In Section 3 the inverse problem is introduced. The observation shape functionals (4) are defined in terms of solutions by (5). The proposed approximation of the observation operator (6) is given by formula (7).

In Section 3 some observation shape functionals are introduced in (4) and the form of the topological derivatives for such functionals is presented in Theorem 1. In Subsection 3.2 the observation shape functionals, used in Section 4, depend on the solution to the Laplace equation defined in Ω_0 and in each domain Ω_k, $k = 3$, with the homogeneous Neumann boundary conditions on the boundaries γ_j of the holes $\mathscr{B}_j$, respectively.

In Subsection 4.3 the numerical realization of the identification procedure based on the application of artificial neural networks is described in detail. The computations are performed for one hole, the results of computations are given in the last part of the paper.

The proposed numerical procedure can be characterized by the following features. The learning data sets for the neural networks are constructed using the asymptotic analysis of the observation shape functionals. Such a method of construction is very useful from the numerical point of view but unfortunately restricts the validity of the proposed approach only for small radii of imperfections. We assume also that the given data for the inverse problem are exact in the sense that there are some unknown imperfections in the form of circular holes which furnish the prescribed values of the observation shape functionals. Therefore, the inverse mapping $\mathscr{G}_k^{-1}(\mathscr{J}(\Omega_k)) = (Y,\rho)$ can be defined and it is given by the neural network. In fact, the situation is more complicated, since we can only model the conditional expectation, we refer to Section 4 for details. So instead of values of inverse mapping we calculate only conditional expectation of these values. The positive conclusion for our approach is the "probabilistic" convergence of the numerical procedure which, to the best of our knowledge, is an original contribution to the field of inverse problems.

3 Shape Optimization

In this paper we present the identification method for finite number of holes in a given geometrical domain $\Omega_0 \subset \mathbb{R}^2$. Actually, we want to identify the locations $y_j \in \Omega_0$ and the radii ρ_j, $j = 1, \ldots, k$ of k small holes $\mathscr{B}_j(y_j) = \{x : |x - y_j| < \rho_j\}$ included in Ω_0.

The procedure of identification is based on the values of specific shape functionals $\mathscr{J}_i(\Omega_k)$, $i = 1, \ldots, N$, which depend on the solutions $u(x)$, $x \in \Omega_k$, for the Laplace equation in the domain Ω_k. Here we denote by Ω_0, the domain without hole, $\Omega_1 = \Omega_0 \setminus \overline{\mathscr{B}_1(y_1)}$, for $k = 1$, is the domain with one hole and for $k \geq 2$, Ω_k is the domain with k holes. For our numerical examples we use $k = 1$ or $k = 3$. By the proposed identification procedure we can compute the coordinates $y_j \in \Omega_0$ of centers of the holes and the radii ρ_j such that given values of observation from physical model denoted by $\mathscr{J}_1, \ldots, \mathscr{J}_N$, $j = 1, \ldots, k$ coincide approximatively with the values $\mathscr{J}_i(\Omega_k)$, $i = 1, \ldots, N$, evaluated from the mathematical model (see Fig. 1).

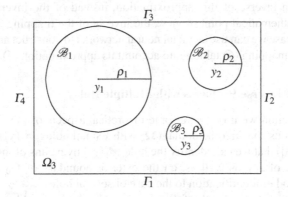

Fig. 1. Domain Ω_3 with imperfections $\mathscr{B}_1$, $\mathscr{B}_2$, $\mathscr{B}_3$

Let us consider the following inverse problem. We assume that the shape functionals take the form

$$\mathscr{J}_i(\Omega_k) = \int_{\Omega_k} \mathscr{F}_i(x, u(x), \nabla u(x)) dx \quad i = 1, \ldots, N, \tag{4}$$

where $\Omega_k = \Omega_0 \setminus \bigcup_{j=1}^{k} \overline{\mathscr{B}_j(y_j)}$, $\Omega_k \subset \mathbb{R}^2$, are given domains. We assume that $\mathscr{F}_i(\cdot, \cdot, \cdot)$ are given smooth functions. The partial differential equations for $u \in H^1(\Omega_k)$ are defined as follows:

$$\begin{cases} \Delta u = f \text{ in } \Omega_k, \\ \quad u = g \text{ on } \Gamma_1 \cup \Gamma_3, \\ \quad \frac{\partial u}{\partial n} = h \text{ on } \Gamma_2 \cup \Gamma_4, \\ \quad \frac{\partial u}{\partial n} = 0 \text{ on } \bigcup_{j=1}^{k} \gamma_j, \end{cases} \tag{5}$$

where f, g, h are given continuous functions in $\overline{\Omega_0}$ and $H^1(\Omega)$ stands for the standard Sobolev space of functions in $L^2(\Omega)$ along with the first order derivatives. The identification procedure for the location and size of several holes by using neural networks is presented in the present paper. Therefore, we are interested from numerical point of view, in the generalized inverse of the mapping

$$\mathcal{G}_k : \mathbb{R}^{3k} \ni (y_j, \rho_j, \dots, y_k, \rho_k) \mapsto (\mathcal{J}_1(\Omega_k), \dots, \mathcal{J}_N(\Omega_k)) \in \mathbb{R}^N \tag{6}$$

where (y_j, ρ_j) represents the center and radius of the hole $\mathcal{B}_j$. To this end we construct its approximation in the form

$$\mathcal{G}_{ki} \cong \mathcal{J}_i(\Omega_k) + \sum_{j=1}^{k} \frac{\rho_j^2}{2} \mathcal{T}_{\Omega_0} \mathcal{J}_i(y_j) \tag{7}$$

and determine an inverse of this approximation, instead of the inverses of mapping $\mathcal{G}_k$. From the mathematical point of view, the inverse of the mapping $\mathcal{G}_k$ is difficult to evaluate. In this case we can use artificial neural networks to construct an approximation of the inverse of mapping (6) taking into account its approximation (7).

3.1 Analysis of Inverse Problems with Multiple Holes

In the present section we investigate some theoretical aspects of the inverse problem. To begin with, let us consider the domain Ω_k with k small holes $\mathcal{B}_j(y_j)$, $j = 1, \dots, k$. It is well known ([3]) that we can identify the hole $\mathcal{B}_1(y_1)$ by means of the measurements performed inside of Ω_0, as well as, on the exterior boundary of Ω_0. We propose to extend this method of identification to the case of several holes.

In such a case we consider the following boundary value problem defined in Ω_k. Find $u \in H^1(\Omega_k)$ such that

$$\begin{cases} \Delta u = 0 \text{ in } \Omega_k, \\ \quad u = 1 \text{ on } \Gamma_1, \\ \quad \frac{\partial u}{\partial n} = 0 \text{ on } \Gamma_2 \cup \Gamma_4 \cup \bigcup_{j=1}^{k} \gamma_j, \\ \quad u = 0 \text{ on } \Gamma_3, \end{cases} \tag{8}$$

where γ_j denotes the boundary of the hole $\mathcal{B}_j(y_j)$. We denote by $u = u(\Omega_k)$ the solution to (8) in Ω_k.

The asymptotic expansion of the shape functional (3) is given by

$$\mathcal{J}_i(\Omega_k) = \mathcal{J}_i(\Omega_0) + \sum_{j=1}^{k} \frac{\rho_j^2}{2} (\mathcal{T}_{\Omega_0} \mathcal{J}_i)(y_j) + o(\rho^2), \quad \text{with } \rho = (\rho_1, \dots, \rho_k). \tag{9}$$

where $\mathscr{T}_{\Omega_0}\mathscr{J}_i$ is the function defined in Ω_0, it is the topological derivative of the shape functional $\mathscr{J}_i(\Omega)$ evaluated at y_j in the domain Ω_0. For simplicity, we assume that

$$\mathscr{F}_i(x,u,q) = F_i(u) + G_i(q). \tag{10}$$

Hence, the shape functional $\mathscr{J}_i(\Omega_k)$ takes the form

$$\mathscr{J}_i(\Omega_k) = \int_{\Omega_k} [F_i(u(\Omega_k)) + G_i(\nabla u(\Omega_k))] \, dx \tag{11}$$

The formula for topological derivative of the shape functional [10,11] defined by (11) is given in the following theorem [3].

Theorem 1. *The topological derivative of functional (11) at a point $y \in \Omega_0$ is given by the following formula*

$$\mathscr{T}_{\Omega_0}\mathscr{J}_i(y) = -\frac{1}{2\pi} [2\pi F_i(u(y)) + g_i(\nabla u(y)) + 2\pi f(y)v(y) + 4\pi \nabla u(y) \cdot \nabla v(y)] \tag{12}$$

where $\nabla u(y) = (a,b)^\top$ and the function g_i depending on the gradient of solution at the point $y \in \Omega_0$ takes the form

$$g_i(\nabla u(y)) = \frac{1}{2\pi} \int_0^{2\pi} G_i(a\sin^2\vartheta - b\sin\vartheta\cos\vartheta, -a\sin\vartheta\cos\vartheta + b\cos^2\vartheta)d\vartheta. \tag{13}$$

The adjoint state $v \in H^1_{\Gamma_1}(\Omega_0) = \{\phi \in H^1(\Omega_0) \mid \phi = 0 \text{ on } \Gamma_1\}$ solves the boundary value problem

$$-\int_{\Omega_0} \nabla v \cdot \nabla \phi \, dx = -\int_{\Omega_0} [F_i'(u)\phi + G_{iq}(\nabla u) \cdot \nabla \phi] dx, \quad \forall \phi \in H^1_{\Gamma_1}(\Omega_0). \tag{14}$$

3.2 Numerical Example of Observation Shape Functionals

Let Ω_0 be the square $(0,1) \times (0,1)$. We consider three cases, the domain Ω_1 with one hole, and the domains Ω_2, Ω_3 with two and three holes, respectively. In each case the asymptotic approximations of solutions in Ω_1, Ω_2, Ω_3 can be computed in the fixed domain Ω_0. Therefore, we consider four boundary value problems defined in Ω_0 for harmonic functions $\Delta u = 0$ in Ω_0 with the different boundary conditions.

For the first boundary value problem the following boundary conditions are prescribed for the solution $u = u_1$ with $i = 1$,

$$\begin{aligned} u_1 &= 1 - x_1 \text{ on } (0,1) \times \{0\}, \\ u_1 &= 0 \quad\quad \text{ on } \{1\} \times (0,1), \\ u_1 &= 0 \quad\quad \text{ on } (0,1) \times \{1\}, \\ u_1 &= 1 - x_2 \text{ on } \{0\} \times (0,1). \end{aligned} \tag{15}$$

The other three cases $i = 2,3,4$ of boundary conditions are obtained from the above conditions by applying the successive rotation by the angle $\frac{\pi}{2}$. The observation shape functionals $\mathscr{J}_i = \mathscr{J}_i(\Omega_0) = \int_{\Omega_0} [F_i(u) + G_i(\nabla u)] \, dx$ are defined as follows: for $i = 1,2,3,4$

$$\mathscr{J}_{[1+3(i-1)]} = \int_\Omega u_i^2 dx, \quad \mathscr{J}_{[2+3(i-1)]} = \int_\Omega \left(\frac{\partial u_i}{\partial x_1}\right)^2 dx,$$

$$\mathscr{J}_{[3+3(i-1)]} = \int_\Omega \left(\frac{\partial u_i}{\partial x_2}\right)^2 dx \tag{16}$$

In the perforated domains $\Omega_k = \Omega_0 \setminus \bigcup_{j=1}^{k} \overline{\mathscr{B}_j(y_j)}$, $y_j = (y_j^1, y_j^2)$, we prescribe for the solution $u = u(\Omega_k)$ the homogenous Neumann boundary conditions on the boundaries γ_j of the balls $\mathscr{B}_j(y_j)$.

Topological derivatives of the shape functionals are obtained from Theorem 1 by direct computation of the function g_i. For the functionals in the form $\mathscr{J}_{[1+3(i-1)]}$ the topological derivative is given by

$$\mathscr{T}_{\Omega_0} \mathscr{J}_1(y_j) = -[u(y_j)]^2 + 2\nabla u(y_j) \nabla v_1(y_j), \tag{17}$$

where $v = v_i$ is the adjoint state, defined for $i = 1,2,3$ by a solution to the boundary value problem

$$\begin{cases} \Delta v = -F_i'(u) + \mathrm{div}(G_{iq}(\nabla u)) & \text{in } \Omega_0, \\ v = 0 & \text{on } \Gamma_1 \cup \Gamma_3, \\ \frac{\partial v}{\partial n} = 0 & \text{on } \Gamma_2 \cup \Gamma_4, \end{cases} \tag{18}$$

For the functionals in the form $\mathscr{J}_{[2+3(i-1)]}$ and $\mathscr{J}_{[3+3(i-1)]}$, we observe that

$$\frac{\partial u}{\partial \tau} = \cos \vartheta \cdot \frac{\partial u}{\partial x_1} + \sin \vartheta \cdot \frac{\partial u}{\partial x_2} \tag{19}$$

hence we can use Theorem 1 with

$$G(\nabla u) = (\sin \vartheta \frac{\partial u}{\partial \tau})^2 = (-\sin^2 \vartheta \frac{\partial u}{\partial x_1} + \sin \vartheta \cos \vartheta \frac{\partial u}{\partial x_2})^2 \tag{20}$$

As a result we obtain

$$\mathscr{T}_{\Omega_0} \mathscr{J}_2(y_j) = -\pi \left[\frac{3}{2} \left(\frac{\partial u}{\partial x_1} \right)^2 + \frac{1}{2} \left(\frac{\partial u}{\partial x_2} \right)^2 + 4(\nabla u \cdot \nabla v_2) \right](y_j) \tag{21}$$

$$\mathscr{T}_{\Omega_0} \mathscr{J}_3(y_j) = -\pi \left[\frac{1}{2} \left(\frac{\partial u}{\partial x_1} \right)^2 + \frac{3}{2} \left(\frac{\partial u}{\partial x_2} \right)^2 + 4(\nabla u \cdot \nabla v_3) \right](y_j) \tag{22}$$

where u is the harmonic function in Ω_0 with the boundary conditions determined for $i = 1$ by (15), v_1 is the adjoint state for $\mathscr{J}_1(\Omega_0) = \int_\Omega u_1^2 dx$, v_2 is the adjoint state for $\mathscr{J}_2(\Omega_0) = \int_{\Omega_0} (\frac{\partial u_1}{\partial x_1})^2 dx$ and v_3 is the adjoint state for $J_3(\Omega_0) = \int_{\Omega_0} (\frac{\partial u_1}{\partial x_2})^2 dx$. The remaining formulae for topological derivatives are obtained in the same way from (17) and (14), (18).

4   Neural Networks

We are going to present the method which is successfully tested for numerical solution of the inverse problem under considerations. We assume that the distance between the observations from real object and from the mathematical model equals zero, it means that there is (Y, ρ) such that

$$\mathscr{G}_k(Y, \rho) = \mathscr{J}(\Omega_k) \in \mathbb{R}^N. \tag{23}$$

Thus, it makes sense to consider the inverse mapping $\mathscr{G}_k^{-1}$. Basing on probability theory [12] we obtain only the conditional expectation of the inverse mapping. So in our meaning the "inverse mapping" is the conditional expectation of multifunction that for the vector of observation calculates the location and the size of imperfections. It is general case for the problem that is considered in [3]. Moreover, Theorem 4.1 in [3] shows, under appropriate conditions, the existence of the inverse mapping. Therefore, the neural network is constructed to approximate the inverse mapping $\mathscr{G}_k^{-1}$. In practice, using probability theory once more, we can only approximate the conditional expectation of the location and radii of k holes instead of the inverse mapping $\mathscr{G}_k^{-1}$. Our procedure uses formula (7) to construct data set for learning of the network. We use this asymptotic formula to calculate the shape functionals $\mathscr{J}_j(\Omega_k)$, $i = 1, \ldots, N$. k is a fixed integer, and $k = 3$ for this part of the paper. In other words we are going to determine the size $\rho_j > 0$ and the location $y_j \in \mathbb{R}^2$ of centers of three holes. The numerical result, however, is the conditional expectation of unknown values.

For the vector of actual observations $\mathscr{J}(\Omega_3) = (\mathscr{J}_1(\Omega_3), \ldots, \mathscr{J}_{12}(\Omega_3)) \in \mathbb{R}^{12}$ which describes the unknown properties of imperfections, we want to find $(Y, \rho) = (y_1, \rho_1, y_2, \rho_2, y_3, \rho_3) \in \mathbb{R}^9$ such that the vector $\mathscr{G}_k(Y, \rho) \in \mathbb{R}^{12}$ evaluated from the mathematical model coincides with the given vector of actual observations.

In the learning process [2], for random distribution [1] of (Y, ρ) the observation vectors $\mathscr{G}_k(Y, \rho)$ are computed from the mathematical model of the body $\Omega_3 \subset \mathbb{R}^2$ by an application of the asymptotics obtained for $\rho_j \to 0$ with $j = 1, 2, 3$. Therefore, a systematic error is introduced to the learning procedure by taking values of topological derivatives in order to compute the approximate values of $\mathscr{G}_k(Y, \rho)$.

We describe briefly the neural networks. Fundamental element of artificial neural networks is an artificial neuron. An artificial neuron has many inputs and a single output. Each of inputs has weight. Output signal is calculated inside neuron based on input information. Each artificial neuron has activation function. Additionally, every neuron has one extra input named bias. This input is always equal to one and has its own weight. Let n be the number of neuron's inputs, and denote input vector by $x = [x_0, x_1, \ldots, x_n]$, where $x_0 = 1$, weight vector by $w = [w_0, w_1, \ldots, w_n]$, as well as activation function by f. The model of this neuron is described in Fig. 2.

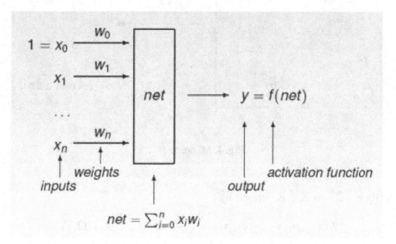

Fig. 2. Model of artificial neuron

A multilayer feedforward neural network is a set of neurons that are divided in separate groups named layers. Input vector represents input vector for all neurons in the first layer. Each connection between input and neuron has its own weight. Each neuron of the first layer generates output signal. Vector of output signals from the first layer represents input vector to the second layer. An example of typical multilayer feedforward network is shown in Fig. 3.

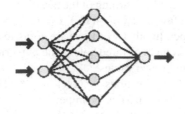

Fig. 3. Model of multilayer feedforward neural network

4.1 Inverse Problem

We consider a particular case of the mapping $\mathcal{G}$ for $k = 3$ and $N = 12$ (see Fig. 4). It means that we consider the domain Ω_3 as a square $(0,1) \times (0,1)$ with three holes $\mathcal{B}_1$, $\mathcal{B}_2$, $\mathcal{B}_3$.

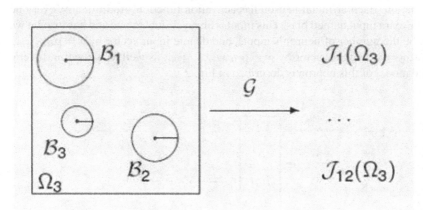

Fig. 4. Mapping $\mathcal{G}$

Mapping $\mathcal{G} : \mathbb{R}^9 \to \mathbb{R}^{12}$ is defined by

$$\mathcal{G} : (y_1, \rho_1, y_2, \rho_2, y_3, \rho_3) \to (\mathcal{J}_1(\Omega_3), \dots, \mathcal{J}_{12}(\Omega_3)), \tag{24}$$

where
(y_1, ρ_1) − center and radius of first hole,
(y_2, ρ_2) − center and radius of second hole,
(y_3, ρ_3) − center and radius of third hole,
$\mathscr{J}_1(\Omega_3), \ldots, \mathscr{J}_{12}(\Omega_3)$ − shape functionals.

For domain with three holes mapping (24) calculates vector of shape functionals (Fig. 4).

Let $\mathscr{G}^{-1} : \mathbb{R}^{12} \to \mathbb{R}^9$ be **the inverse mapping** defined by

$$\mathscr{G}^{-1} : (\mathscr{J}_1(\Omega_3), \ldots, \mathscr{J}_{12}(\Omega_3)) \to (y_1, \rho_1, y_2, \rho_2, y_3, \rho_3). \tag{25}$$

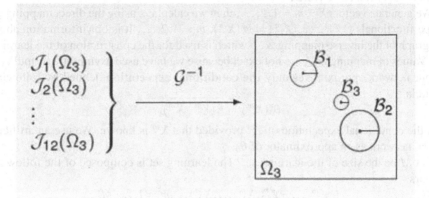

Fig. 5. Mapping $\mathscr{G}^{-1}$

The inverse mapping $\mathscr{G}^{-1}$ for vector of shape functionals calculates location and size of three holes in our domain (Fig. 5). Any analytical formula for mapping $\mathscr{G}^{-1}$ is unknown. Our aim is construction of an approximation of this mapping using artificial neural networks.

Multilayer feedforward neural network is capable of arbitrarily accurate approximations to arbitrary mapping [12]. To approximate mapping $\mathscr{G}^{-1}$ by neural network we have to construct network and learning set. Appropriate neural network consists of 4 hidden layers. Let q_1, q_2, q_3 and q_4 denote the numbers of neurons in hidden layers. ϕ_1, ϕ_2, ϕ_3 and ϕ_4 denote sigmoidal activation functions for each of hidden layers. Furthermore, our network has one output layer with 9 neurons and linear activation function. Input vector has 12 components and represents the vector of shape functionals. Numbers of neurons in hidden layers are not fixed, and increase to infinity when the size of learning set increases to infinity. Output vector represents vector of data for three holes. The first and the second components represent the center of the first hole, the third entry represents the radius of the first hole. In the same way we denote all entries.

To construct learning set we use probability theory. We consider a sequence $\{Z^m\}$, $m = 1, 2, \ldots$ of vectors, where $Z^m = (Y^m, X^m)$. $X^m = (X_1^m, \ldots, X_9^m)$ describes the three holes. Its coordinates denote:

- (X_1^m, X_2^m)– the center of first hole,
- X_3^m– the radius of first hole,
- (X_4^m, X_5^m)– the center of second hole,
- X_6^m– the radius of second hole,
- (X_7^m, X_8^m)– the center of third hole,
- X_9^m– the radius of third hole.

We generate vector X^m randomly by choosing three holes in a domain Ω_0. $Y^m = (Y_1^k, \ldots, Y_9^m)$– describes unknown value of function $\mathscr{G}^{-1}$. Using notation of previous section we have that $(Y_1^m, Y_2^m) = y_1$, $Y_3^m = \rho_1$, $(Y_4^m, Y_5^m) = y_2$, $Y_6^m = \rho_2$, $(Y_7^m, Y_8^m) = y_3$, $Y_9^m = \rho_3$.

We generate vectors X^m, $m = 1, 2, \ldots$, in a random way. For $m \to \infty$ the whole square $(0, 1) \times (0, 1)$ can be "covered".

We generate vectors X^m, $m = 1, 2, \ldots$, then we calculate, using the direct mapping $\mathscr{G}$, shape functionals $(\mathscr{J}_1^m, \ldots, \mathscr{J}_{12}^m) = \mathscr{G}(X^m)$, $m = 1, 2, \ldots$. It is our information about the graph of the inverse mapping $\mathscr{G}^{-1}$ which is used for the construction of the learning set. Values of mapping $\mathscr{G}^{-1}$ are not exact because we have used asymptotic formula (7) so our network approximates only **the conditional expectation** defined by following formula

$$\theta_o(X^m) = E(Y^m \mid X^m). \tag{26}$$

It is the conditional expectation of Y^m provided that X^m is known. We use an artificial neural network as an approximator of θ_o.

Let M be the size of the learning set. The learning set is composed of the following vectors

$$(\mathscr{J}_1^m, \ldots, \mathscr{J}_{12}^m, X_1^m, \ldots, X_9^m), \quad m = 1, \ldots, M$$

where $(\mathscr{J}_1^m, \ldots, \mathscr{J}_{12}^m)$ is the input vector for neural network, and $(X_1^m, \ldots, X_9^m)$ is the output vector (required).

Furthermore, we have dependence

$$(\mathscr{J}_1^m, \ldots, \mathscr{J}_{12}^m) = \mathscr{G}(X_1^m, \ldots, X_9^m), \quad m = 1, \ldots, M.$$

To solve our problem we use artificial neural networks. Each of networks can be represented by some mapping $f^q : \mathbb{R}^{12} \to \mathbb{R}^6$. Parameter q describes number of all neurons in hidden layers. This value depends on the numbers of neurons in all hidden layers. We construct a sequence of networks as a sequence of approximators to θ_o. We let networks where q_1, q_2, q_3, q_4 and M grows [12]. For a given M the learning network provides an approximation to the unknown regression function θ_o.

Set Θ as a function space containing θ. $\theta(\cdot) := f^q(\cdot, \delta^q)$ where δ^q is a set parameters of networks. The function space Θ contains θ_o. We construct a sequence of "sieves" $\{\Theta_M\}$ for $M = 1, 2, \ldots$ where Θ_M is a function space containing networks learned by $M - elements$ of the learning sets.

The "connectionist sieve estimator" $\tilde{\theta}_M$ is defined as a solution to the least squares problem (appropriate for learning $E(Y^M \mid X^M)$)

$$\min_{\theta \in \Theta_M} M^{-1} \sum_{m=1}^{M} [Y^m - \theta(X^m)]^2, \tag{27}$$

for $M = 1, 2, \ldots$.

We can apply the following result of White [12] to obtain a convergence result for our method.

Theorem 2. *Under some technical assumptions there exists sieve estimator $\tilde{\theta}_M$ such that*

$$\frac{1}{M} \sum_{m=1}^{M} [Y^m - \tilde{\theta}_M(X^m)]^2 = \min_{\theta \in \Theta_M} \frac{1}{M} \sum_{m=1}^{M} [Y^m - \theta(X^m)]^2, \qquad (28)$$

for $M = 1, 2, \ldots$. Furthermore, $d(\tilde{\theta}_M, \theta_0) \xrightarrow{P} 0$ (i.e., for all $\varepsilon > 0$ $P[\omega \in domain : d(\tilde{\theta}_M(\omega), \theta_0) > \varepsilon] \to 0$ as $M \to \infty$), where d measures the distance between functions and the convergence is in measure.

4.2 Conclusions

In this paper we use artificial neural networks as an approximator of unknown mapping which, for a given vector of shape functionals, calculates locations and radii of three holes. In [3] the authors consider the same problem for one hole. Here, we consider more complicated problem with some additional features. Our learning data set is not exact because asymptotic formula (7) is used to calculate the set. Therefore the obtained network is an approximator of the conditional expectation θ_0.

Our aim is to determine a solution of problem (27). This problem is defined for exact unknown values Y^m however network is learned based on inexact data. In Theorem 2 we present a result on the existence of a solution $\tilde{\theta}_M$ to problem (27) as well as on the convergence of the method. It follows that $d(\tilde{\theta}_M, \theta_0) \xrightarrow{P} 0$. It means that distance between: $\tilde{\theta}_M$, the solution of problem (27) and θ_0, the conditional expectation tends to zero in probability measure.

The numerical results presented below show that the method is very efficient and robust, however there are still many open problems in the mathematical analysis of the proposed method.

4.3 Numerical Example for One Hole

We consider an example with one hole instead of three holes. The example with one hole needs less complicated network and less time to perform the learning process as well as it has the same rules during the creation and learning process. We used Matlab 6.1 to generate learning data, testing data and to create and train neural network. Our network has two hidden layers: first with 24 neurons and second with 12 neurons. In both layers we use sigmoidal activation function. Output layer is composed of three neurons. First and second output are the conditional expectation of location of hole and third output is the radius of hole. Input vector has 12 components and describes 12-element vector of shape functionals. The number of shape functionals is fixed for numerical example (more details [3]). Based on vector of shape functionals network calculates output of network.

We prepared two (different) sets, a learning set and a testing set. The learning set contains 1000 elements, and the testing set contains 10 elements. The number of elements in both sets is fixed for numerical example. The method of constructing learning set and testing set is the same. In the domain Ω_0 we generate randomly distributed hole. We calculate using mapping $\mathscr{G}$ the vector of shape functionals in domain with holes. So we have pair of vectors: the first vector describes holes and the second vector includes the of shape functionals. We repeat this procedure for a finite fixed number of iterations. To train the network we use vectors from the testing set. The vectors of shape functionals are the inputs for network, the vectors which describe holes are the required output from the network.

We assume tolerated learning error: 10^{-5} and number of learning epochs: 700. Consequently, we get following results:

- error on learning data is 0.0083 ($< 1\%$),
- error on testing data is 0.0081 ($< 1\%$).

In the following figure (Fig. 6) we present the result that we obtained for the testing set. We prepared, in the same way as the learning set, the testing set. The testing set contains 10 pairs of vectors: describes holes and vectors of topological derivatives. We put 10 holes in the figure, drawn in bold. These are the real holes from the testing set. Each of the vectors of topological derivatives was put to the neural network. On the output we obtain 10 vectors. Each of them describes one hole. It is the answer of network for the vector of observations. In the figure, 10 holes drawn in solid lines are the outputs of the network. Finally, in the figure we present differences between results for 10 pairs of holes: from testing set (bold) and outputs of network (solid).

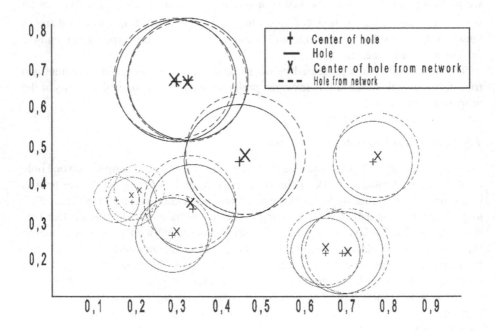

Fig. 6. Numerical result

References

1. Baron, A.R.: Universal approximation bounds for superpositions of a sigmoidal function. IEEE Transactions on Information Theory 39, 930–945 (1993)
2. Hagan, M., Menhaj, M.: Training feedforward networks with the Marquardt algorithm. IEEE Trans. on Neural Networks 5(6), 989–993 (1994)
3. Jackowska-Strumillo, L., Sokołowski, J., Żochowski, A., Henrot, A.: On Numerical Solution of Shape Inverse Problems. Computational Optimization and Applications 23, 231–255 (2002)
4. Mazja, W.G., Nazarov, S.A., Plamenevskii, B.A.: Asymptotische Theorie elliptischer Randwertaufgaben in singulär gestörten Gebieten, vol. 1, p. 432. Akademie Verlag, Berlin (1991); English transl.: Asymptotic Theory of Elliptic Boundary Value Problems in Singularly Perturbed Domains, vol. 1, p. 435. Birkhäuser Verlag, Basel (2000)
5. Nazarov, S.A., Plamenevsky, B.A.: Elliptic Problems in Domains with Piecewise Smooth Boundaries. De Gruyter Exposition in Mathematics 13 (1994), Walter de Gruyter
6. Nazarov, S.A., Sokołowski, J.: Spectral problems in the shape optimisation. Singular boundary perturbations. Asymptotic Analysis 56(3-4), 159–204 (2008)
7. Roche, J.R., Sokołowski, J.: Numerical methods for shape identification problems. Special issue of Control and Cybernetics: Shape Optimization and Scientific Computations 5, 867–894 (1996)
8. Shumacher, A.: Topologieoptimierung von Bauteilstrukturen unter Verwendung von Lochpositionierungskriterien. Ph.D. Thesis, Universität-Gesamthochschule-Siegen, Siegen (1995)
9. Sokołowski, J., Zolesio, J.-P.: Introduction to Shape Optimization. Shape Sensitivity Analysis. Springer, Berlin (1992)
10. Sokołowski, J., Żochowski, A.: Modeling of topological derivatives for contact problems. Numerische Mathematik 102(1), 145–179 (2005)
11. Sokołowski, J., Żochowski, A.: On topological derivative in shape optimization. SIAM Journal on Control and Optimization 37(4), 1251–1272 (1999)
12. White, H.: Connectionist Nonparametric Regression: Multilayer Feedforward Networks Can Learn Arbitrary Mappings. Neural Networks 3, 535–549 (1990)

Factorization by Invariant Embedding of a Boundary Value Problem for the Laplace Operator

Jacques Henry[1], Bento Louro[2], and Maria C. Soares[2]

[1] INRIA Bordeaux Sud-Ouest, IMB Université Bordeaux 1 351, cours de la libération,
33405 Talence, France
jacques.henry@inria.fr
[2] Departamento de Matemática, Faculdade de Ciências e Tecnologia,
Universidade Nova de Lisboa, Monte de Caparica, 2829-516 Caparica, Portugal
bjl@fct.unl.pt, mcs@fct.unl.pt

Abstract. This work concerns the factorization of a second order elliptic boundary value problem defined in a star-shaped bounded regular domain, in a system of uncoupled first order initial value problems, using the technique of invariant embedding. The family of domains is defined by a homothety. The method yields an equivalent formulation to the initial boundary value problem by a system of two uncoupled Cauchy problems. The singularity at the origin of the homothety is studied.

1 Introduction

The invariant embedding technique consists in embedding the initial problem in a family of similar problems depending on a parameter, which are solved recursively. It has been used by Bellman [1] and Lions [6] (in the infinite dimensional case) to derive the optimal feedback law in linear-quadratic optimal control problems. It yields a factorization of the optimality system. In our approach, the invariant embedding is used spatially. Each problem is defined over a subdomain limited by a mobile boundary (see Fig. 1), depending on the parameter. Defining an operator relating the value of the solution, or its derivative, with the mobile boundary condition (Dirichlet-Neumann or Neumann-Dirichlet, for example), we find a family of operators on functions defined on the mobile boundary satisfying a Riccati equation. The method applied to cylindrical domains has been presented in [4,5]. Here we particularize the study to the two dimensional Poisson equation with a Dirichlet boundary condition: $-\Delta u = f$, in Ω, $u_{|\Gamma_a} = 0$. The assumption on the shape of the domain is less restrictive than in [4,5] and the invariant embedding is realized by a homothety. The family of curves which limits the subdomains defined by the invariant embedding are homothetic to one another, and we consider the moving boundary starting on the outside boundary of the domain and shrinking to a point. We show some results dealing with the singularity that will appear at that point. The factorization of the boundary value problem can be viewed as an infinite dimensional generalization of the block Gauss LU factorization.

A. Korytowski et al. (Eds.): System Modeling and Optimization, IFIP AICT 312, pp. 282–292, 2009.
© IFIP International Federation for Information Processing 2009

Fig. 1. The star-shaped domain

2 Definition of the Problem and Regularization

Let $\Omega \subset \mathbb{R}^2$ be an open bounded regular domain containing the origin O, star-shaped with respect to O, with boundary $\Gamma_a = \partial\Omega$. We consider the Poisson problem with Dirichlet data

$$(\mathscr{P}) \begin{cases} -\Delta u = f, \text{ in } \Omega \\ u|_{\Gamma_a} = 0, \end{cases} \tag{1}$$

where $f \in L^2(\Omega)$. In spite of the particularization to the Laplacian operator in this definition, we believe that the same procedure could be applied to any strongly elliptic self-adjoint problem.

Applying the (spatial) invariant embedding method to this problem, we must start defining a family of subdomains limited by a boundary sweeping over the initial domain Ω.

We start dealing with the case where the family of curves which limits the subdomains, starts on the boundary of the domain and shrinks homothetically to a point. Since the mobile boundary reduces to a point, a singularity will necessary appear at that point. We must make, as a consequence, a regularization around this point and a possible way to do it, is to define an auxiliary domain, where we introduce a fictitious boundary around the singular point. In this case, however, we introduce a perturbation of the solution so, naturally, we must choose the new boundary condition, in a way that we can obtain the convergence of this auxiliary problem to the initial one. With this purpose, we will consider the following auxiliary problem:

$$(\mathscr{P}_\varepsilon) \begin{cases} -\Delta u_\varepsilon = f, \text{ in } \Omega \setminus \Omega_\varepsilon \\ u_\varepsilon|_{\Gamma_a} = 0 \\ \int_{\Gamma_\varepsilon} \dfrac{\partial u_\varepsilon}{\partial n}\, d\Gamma = 0, \ u_\varepsilon|_{\Gamma_\varepsilon} \text{ is constant.} \end{cases} \tag{2}$$

Here, Ω_ε is an open regular domain verifying $\overline{\Omega}_\varepsilon \subset \Omega$ and Γ_ε, which is homothetical to Γ_a with ratio $\varepsilon < 1$, is the boundary of Ω_ε. This problem is well posed. We can justify the choice of the boundary conditions on Γ_ε with the fact that the condition

$$\int_{\Gamma_\varepsilon} \frac{\partial u_\varepsilon}{\partial n}\, d\Gamma_\varepsilon = 0 \tag{3}$$

corresponds to a null total flux.

It's easy to see that $U_\varepsilon = \{u_\varepsilon \in H^1(\Omega \setminus \Omega_\varepsilon) : u_{\varepsilon|_{\Gamma_a}} = 0 \wedge u_{\varepsilon|_{\Gamma_\varepsilon}}$ is constant$\}$ is a Hilbert space and that the variational formulation of problem $(\mathscr{P}_\varepsilon)$ is

$$\begin{cases} u_\varepsilon \in U_\varepsilon \\ \displaystyle\int_{\Omega \setminus \Omega_\varepsilon} \nabla u_\varepsilon \nabla v_\varepsilon \, d\Omega = \int_{\Omega \setminus \Omega_\varepsilon} f \, v_\varepsilon \, d\Omega, \forall v_\varepsilon \in U_\varepsilon. \end{cases} \tag{4}$$

We prove that when $\varepsilon \to 0$, problem $(\mathscr{P}_\varepsilon)$ reduces to problem $(\mathscr{P})$, that is, u_ε, the solution of problem $(\mathscr{P}_\varepsilon)$, converges to u, the solution of problem $(\mathscr{P})$, by means of the next theorem, which proof can be found in [3]:

Theorem 1. *Let u_ε (respectively, u) be the solution of ($\mathscr{P}_\varepsilon$) (respectively, ($\mathscr{P}$)) and*

$$\tilde{u}_\varepsilon = \begin{cases} u_\varepsilon, & in \ \Omega \setminus \Omega_\varepsilon \\ u_\varepsilon = u_\varepsilon|_{\Gamma_\varepsilon}, & in \ \Omega_\varepsilon. \end{cases} \tag{5}$$

Then,

$$\tilde{u}_\varepsilon \underset{\varepsilon \to 0}{\longrightarrow} u, \, H^1(\Omega) - strong. \tag{6}$$

3   Invariant Embedding in a Star-Shaped Domain

We start defining polar coordinates by means of $x = \rho \cos(\theta), y = \rho \sin(\theta), 0 < \rho \le \varphi(\theta)$, where $\rho = \varphi(\theta)$ defines the boundary Γ_a. Here, $\varphi(\theta) \in \mathscr{C}^1([0, 2\pi])$ is such that $\varphi(2\pi) = \varphi(0), \varphi'(2\pi) = \varphi'(0)$ and $0 < k_0 < \varphi(\theta) < k_1$. In the coordinates (τ, θ), where $\tau = \rho/\varphi(\theta)$, the Laplace equation becomes

$$\left(\frac{1}{\varphi^2(\theta)} + \frac{(\varphi'(\theta))^2}{\varphi^4(\theta)} \right) \frac{\partial^2 u}{\partial \tau^2} + \left(-2 \frac{\varphi'(\theta)}{\varphi^3(\theta)} \right) \frac{1}{\tau} \frac{\partial^2 u}{\partial \tau \partial \theta}$$
$$+ \left(-\frac{\varphi''(\theta)}{\varphi^3(\theta)} + 2 \frac{(\varphi'(\theta))^2}{\varphi^4(\theta)} + \frac{1}{\varphi^2(\theta)} \right) \frac{1}{\tau} \frac{\partial u}{\partial \tau} + \frac{1}{\varphi^2(\theta)} \frac{1}{\tau^2} \frac{\partial^2 u}{\partial \theta^2} = -f \tag{7}$$

Now, let α be the angle $(\mathbf{OM}, \mathbf{n})$ where M is a point on Γ_a and $\mathbf{n}$ is the outward normal to Γ_a at M. We assume that $-\pi/2 < \alpha_0 \le \alpha \le \alpha_1 < \pi/2$. We consider the homothety of center O and ratio $0 < \tau < 1$, which transforms Ω to Ω_τ with boundary Γ_τ, and the following system of curvilinear coordinates: for $M \in \Omega$, (τ, t) are such that M', the image of M by a $1/\tau$ homothety, belongs to Γ_a and $t, 0 \le t < t_0$, is the curvilinear abscissa of M' on Γ_a (where t_0 is the length of Γ_a). This new system of coordinates and the one defined previously are related through the equalities $\cos(\alpha) \, dt = \varphi \, d\theta$ and $\tan(\alpha) = \frac{\varphi'}{\varphi}$. In these coordinates, the exterior normal to Γ_τ can be written as $\frac{\partial}{\partial n} = -\frac{1}{\varphi \cos(\alpha)} \frac{\partial}{\partial \tau} + \frac{\tan(\alpha)}{\tau} \frac{\partial}{\partial t}$, and the Laplace equation takes the form

$$-\frac{\partial}{\partial \tau} \left(\frac{\tau}{\varphi \cos(\alpha)} \frac{\partial u_\tau}{\partial \tau} - \tan(\alpha) \frac{\partial u_\tau}{\partial t} \right)$$
$$-\frac{\partial}{\partial t} \left(-\tan(\alpha) \frac{\partial u_\tau}{\partial \tau} + \frac{\varphi}{\tau \cos(\alpha)} \frac{\partial u_\tau}{\partial t} \right) = \tau f \varphi \cos(\alpha). \tag{8}$$

Using the technique of invariant embedding, we embed problem ($\mathscr{P}_\varepsilon$) in a family of similar problems defined on $\Omega \setminus \Omega_\tau = \{(s,t) \in]\tau, 1[\times]0, t_0[\}$, for every $\tau \in [\varepsilon, 1)$. For each problem we impose the Neumann boundary condition $\frac{\partial u_\tau}{\partial n}|_{\Gamma_\tau} = h$, where Γ_τ is the moving boundary:

$$
(\mathscr{P}_{\tau,h})
\begin{cases}
-\dfrac{\partial}{\partial \tau}\left(\dfrac{\tau}{\varphi \cos(\alpha)} \dfrac{\partial u_\tau}{\partial \tau} - \tan(\alpha)\dfrac{\partial u_\tau}{\partial t} \right) \\[2mm]
\quad - \dfrac{\partial}{\partial t}\left(-\tan(\alpha)\dfrac{\partial u_\tau}{\partial \tau} + \dfrac{\varphi}{\tau \cos(\alpha)}\dfrac{\partial u_\tau}{\partial t} \right) = \tau f \varphi \cos(\alpha), \text{ in } \Omega \setminus \Omega_\tau \\[2mm]
u_\tau|_{\Gamma_a} = 0 \\[2mm]
u_{\tau|_{t=0}} = u_{\tau|_{t=t_0}}, \quad \dfrac{\partial u_\tau}{\partial t}|_{t=0} = \dfrac{\partial u_\tau}{\partial t}|_{t=t_0} \\[2mm]
\dfrac{\partial u_\tau}{\partial n}|_{\Gamma_\tau} = -\dfrac{1}{\varphi \cos(\alpha)}\dfrac{\partial u_\tau}{\partial \tau} + \dfrac{\tan(\alpha)}{\tau}\dfrac{\partial u_\tau}{\partial t} = h.
\end{cases}
\tag{9}
$$

In order to apply a method similar to the one used by Lions [6] for decoupling the optimality conditions associated to an optimal control problem of a parabolic equation, we define $P(\tau)h = \gamma_{\tau|_{\Gamma_\tau}}$, where γ_τ is the solution of

$$
\begin{cases}
\dfrac{\partial}{\partial \tau}\left(\dfrac{\tau}{\varphi \cos(\alpha)} \dfrac{\partial \gamma_\tau}{\partial \tau} - \tan(\alpha)\dfrac{\partial \gamma_\tau}{\partial t} \right) \\[2mm]
\quad + \dfrac{\partial}{\partial t}\left(-\tan(\alpha)\dfrac{\partial \gamma_\tau}{\partial \tau} + \dfrac{\varphi}{\tau \cos(\alpha)}\dfrac{\partial \gamma_\tau}{\partial t} \right) = 0, \text{ in } \Omega \setminus \Omega_\tau \\[2mm]
\gamma_{\tau|_{\Gamma_a}} = 0 \\[2mm]
\dfrac{\partial \gamma_\tau}{\partial n}|_{\Gamma_\tau} = h \\[2mm]
\gamma_{\tau|_{t=0}} = \gamma_{\tau|_{t=t_0}}, \quad \dfrac{\partial \gamma_\tau}{\partial t}|_{t=0} = \dfrac{\partial \gamma_\tau}{\partial t}|_{t=t_0}
\end{cases}
\tag{10}
$$

and $r(\tau) = \beta_{\tau|_{\Gamma_\tau}}$, where β_τ is the solution of

$$
\begin{cases}
-\dfrac{\partial}{\partial \tau}\left(\dfrac{\tau}{\varphi \cos(\alpha)} \dfrac{\partial \beta_\tau}{\partial \tau} - \tan(\alpha)\dfrac{\partial \beta_\tau}{\partial t} \right) \\[2mm]
\quad - \dfrac{\partial}{\partial t}\left(-\tan(\alpha)\dfrac{\partial \beta_\tau}{\partial \tau} + \dfrac{\varphi}{\tau \cos(\alpha)}\dfrac{\partial \beta_\tau}{\partial t} \right) = \tau f \varphi \cos(\alpha), \text{ in } \Omega \setminus \Omega_\tau \\[2mm]
\beta_{\tau|_{\Gamma_a}} = 0 \\[2mm]
\dfrac{\partial \beta_\tau}{\partial n}|_{\Gamma_\tau} = 0 \\[2mm]
\beta_{\tau|_{t=0}} = \beta_{\tau|_{t=t_0}}, \quad \dfrac{\partial \beta_\tau}{\partial t}|_{t=0} = \dfrac{\partial \beta_\tau}{\partial t}|_{t=t_0}.
\end{cases}
\tag{11}
$$

By linearity of $(\mathcal{P}_{\tau,h})$, the following relation holds true

$$u_\tau(\tau) = P(\tau)h + r(\tau). \tag{12}$$

Then taking $h = \dfrac{\partial u}{\partial n}(\tau)$ on Γ_τ, it is clear that $u_\tau(\tau') = u(\tau')$ for $\tau \le \tau' \le 1$. Then

$$u(\tau) = P(\tau)\frac{\partial u}{\partial n}(\tau) + r(\tau). \tag{13}$$

Taking the derivative of the previous equality, in a formal way, with respect to τ, we can derive (cf Sect. 4) the following system of uncoupled, first order in τ, equations:

$$\begin{cases}
\dfrac{\partial P}{\partial \tau} - \dfrac{\varphi \sin(\alpha)}{\tau}\dfrac{\partial}{\partial t}P + \dfrac{P}{\tau}\dfrac{\partial}{\partial t}\left(\varphi \sin(\alpha)\right) + \dfrac{P}{\tau^2}\dfrac{\partial}{\partial t}\left(\varphi \cos(\alpha)\dfrac{\partial}{\partial t}P\right) \\
\quad - \dfrac{P}{\tau} = -\varphi \cos(\alpha)\, \mathrm{I} \\[2mm]
\dfrac{\partial r}{\partial \tau} - \dfrac{\varphi \sin(\alpha)}{\tau}\dfrac{\partial r}{\partial t} + \dfrac{P}{\tau^2}\dfrac{\partial}{\partial t}\left(\varphi \cos(\alpha)\dfrac{\partial r}{\partial t}\right) = -Pf\varphi \cos(\alpha) \\[2mm]
P\left(-\dfrac{1}{\varphi \cos(\alpha)}\dfrac{\partial u}{\partial \tau} + \dfrac{\tan(\alpha)}{\tau}\dfrac{\partial u}{\partial t}\right) + r = u.
\end{cases} \tag{14}$$

Again from (13), and considering the initial conditions on Γ_a, we also obtain $P(1) = 0$ and $r(1) = 0$, which corresponds to the initial conditions of the first two equations above. We will define the initial condition for the last equation in Sect. 5. The Riccati equation for P which depends only on the operator and the shape of the domain in problem $\mathcal{P}$, can be solved for decreasing τ once for all. For each data (f), the problem is now solved by integrating two Cauchy problems: the one on r for τ decreasing from 1 to 0 and the one on u backwards in τ.

4   Arriving to the Uncoupled System

Considering u_ε, solution of (4) and v_ε, solution of the homogeneous equation $\Delta v_\varepsilon = 0$ with arbitrary boundary condition on Γ_ε and using the Green formula, we obtain in the (τ,t) coordinates

$$\int_\varepsilon^1 \int_0^{t_0} \frac{\tau}{\varphi \cos(\alpha)}\frac{\partial v_\varepsilon}{\partial \tau}\frac{\partial u_\varepsilon}{\partial \tau} - \tan(\alpha)\left(\frac{\partial v_\varepsilon}{\partial t}\frac{\partial u_\varepsilon}{\partial \tau} + \frac{\partial v_\varepsilon}{\partial \tau}\frac{\partial u_\varepsilon}{\partial t}\right)$$

$$+ \frac{\varphi}{\tau \cos(\alpha)}\frac{\partial v_\varepsilon}{\partial t}\frac{\partial u_\varepsilon}{\partial t}\, dt\, d\tau = -\varepsilon \int_0^{t_0} \left(\frac{1}{\varphi \cos(\alpha)}\frac{\partial v_\varepsilon}{\partial \tau} - \frac{\tan(\alpha)}{\varepsilon}\frac{\partial v_\varepsilon}{\partial t}\right)\Bigg|_{\tau=\varepsilon} u_\varepsilon|_{\tau=\varepsilon}\, dt. \tag{15}$$

A similar formula holds for $\Omega \setminus \Omega_\tau = \{(s,t) \in\,]\tau, 1[\times]0, t_0[\}$.

Deriving the resulting equality with respect to the variable τ, we obtain

$$
\frac{\partial}{\partial \tau}\left(\int_\tau^1\int_0^{t_0}\left(\frac{s}{\varphi\cos(\alpha)}\frac{\partial v_\tau}{\partial \tau}\frac{\partial u_\tau}{\partial \tau} - \tan(\alpha)\left(\frac{\partial v_\tau}{\partial t}\frac{\partial u_\tau}{\partial \tau} + \frac{\partial v_\tau}{\partial \tau}\frac{\partial u_\tau}{\partial t}\right)\right.\right.
$$
$$
\left.\left. + \frac{\varphi}{s\cos(\alpha)}\frac{\partial v_\tau}{\partial t}\frac{\partial u_\tau}{\partial t}\right)dt\,ds\right) = -\frac{\partial}{\partial \tau}\left(\int_0^{t_0}\left(\frac{\tau}{\varphi\cos(\alpha)}\frac{\partial v_\tau}{\partial \tau} - \tan(\alpha)\frac{\partial v_\tau}{\partial t}\right)u_\tau\,dt\right) \Rightarrow
$$
$$
\int_0^{t_0}\left(\frac{\tau}{\varphi\cos(\alpha)}\frac{\partial v_\tau}{\partial \tau}\frac{\partial u_\tau}{\partial \tau} - \tan(\alpha)\left(\frac{\partial v_\tau}{\partial t}\frac{\partial u_\tau}{\partial \tau} + \frac{\partial v_\tau}{\partial \tau}\frac{\partial u_\tau}{\partial t}\right) + \frac{\varphi}{\tau\cos(\alpha)}\frac{\partial v_\tau}{\partial t}\frac{\partial u_\tau}{\partial t}\right)dt
$$
$$
= \int_0^{t_0}\frac{\partial}{\partial \tau}\left(\left(\frac{\tau}{\varphi\cos(\alpha)}\frac{\partial v_\tau}{\partial \tau} - \tan(\alpha)\frac{\partial v_\tau}{\partial t}\right)u_\tau\right)dt = -\int_0^{t_0}\frac{\partial}{\partial \tau}\left(\tau\frac{\partial v_\tau}{\partial n}u_\tau\right)dt.
$$

$$(16)$$

Then, using (13) and the Laplace equation (8) we have successively,

$$
\int_0^{t_0}\left(\frac{\tau}{\varphi\cos(\alpha)}\frac{\partial v_\tau}{\partial \tau}\frac{\partial u_\tau}{\partial \tau} - \tan(\alpha)\left(\frac{\partial v_\tau}{\partial t}\frac{\partial u_\tau}{\partial \tau} + \frac{\partial v_\tau}{\partial \tau}\frac{\partial u_\tau}{\partial t}\right) + \frac{\varphi}{\tau\cos(\alpha)}\frac{\partial v_\tau}{\partial t}\frac{\partial u_\tau}{\partial t}\right)dt
$$
$$
= -\int_0^{t_0}\left(\frac{\partial}{\partial \tau}\left(\tau\frac{\partial v_\tau}{\partial n}\right)P\frac{\partial u_\tau}{\partial n} + \tau\frac{\partial v_\tau}{\partial n}\frac{\partial P}{\partial \tau}\frac{\partial u_\tau}{\partial n} + \tau\frac{\partial v_\tau}{\partial n}P\frac{\partial}{\partial \tau}\left(\frac{\partial u_\tau}{\partial n}\right)\right.
$$
$$
\left. + \frac{\partial}{\partial \tau}\left(\tau\frac{\partial v_\tau}{\partial n}\right)r + \tau\frac{\partial v_\tau}{\partial n}\frac{\partial r}{\partial \tau}\right)dt
$$
$$
= \int_0^{t_0}\left(\frac{\partial}{\partial t}\left(\tan(\alpha)\frac{\partial v_\tau}{\partial \tau} - \frac{\varphi}{\tau\cos(\alpha)}\frac{\partial v_\tau}{\partial t}\right)P\frac{\partial u_\tau}{\partial n} - \tau\frac{\partial v_\tau}{\partial n}\frac{\partial P}{\partial \tau}\frac{\partial u_\tau}{\partial n} + \frac{\partial v_\tau}{\partial n}P\frac{\partial u_\tau}{\partial n}\right.
$$
$$
+ \frac{\partial v_\tau}{\partial n}P\left[\frac{\partial}{\partial t}\left(\tan(\alpha)\frac{\partial u_\tau}{\partial \tau} - \frac{\varphi}{\tau\cos(\alpha)}\frac{\partial u_\tau}{\partial t}\right) - f\tau\varphi\cos(\alpha)\right]
$$
$$
\left. + \frac{\partial}{\partial t}\left(\tan(\alpha)\frac{\partial v_\tau}{\partial \tau} - \frac{\varphi}{\tau\cos(\alpha)}\frac{\partial v_\tau}{\partial t}\right)r - \tau\frac{\partial v_\tau}{\partial n}\frac{\partial r}{\partial \tau}\right)dt.
$$

$$(17)$$

Therefore, using once again (13) and integrating by parts in t right hand side, the equality becomes

$$
\int_0^{t_0}\left(\frac{\tau}{\varphi\cos(\alpha)}\frac{\partial v_\tau}{\partial \tau}\frac{\partial u_\tau}{\partial \tau} - \tan(\alpha)\frac{\partial v_\tau}{\partial t}\frac{\partial u_\tau}{\partial \tau} - \tan(\alpha)\frac{\partial v_\tau}{\partial \tau}\frac{\partial u_\tau}{\partial t}\right.
$$
$$
\left. + \frac{\varphi}{\tau\cos(\alpha)}\frac{\partial v_\tau}{\partial t}\frac{\partial u_\tau}{\partial t}\right)dt
$$
$$
= \int_0^{t_0}\left(-\tan(\alpha)\frac{\partial v_\tau}{\partial \tau}\frac{\partial u_\tau}{\partial t} + \frac{\varphi}{\tau\cos(\alpha)}\frac{\partial v_\tau}{\partial t}\frac{\partial u_\tau}{\partial t}\right.
$$
$$
- \tau\frac{\partial v_\tau}{\partial n}\frac{\partial P}{\partial \tau}\frac{\partial u_\tau}{\partial n} + \frac{\partial v_\tau}{\partial n}P\frac{\partial u_\tau}{\partial n} - \frac{\partial v_\tau}{\partial t}\tan(\alpha)\frac{\partial u_\tau}{\partial \tau} + \frac{\partial v_\tau}{\partial t}\frac{\varphi}{\tau\cos(\alpha)}\frac{\partial u_\tau}{\partial t}
$$
$$
\left. - \frac{\partial v_\tau}{\partial n}Pf\tau\varphi\cos(\alpha) - \tau\frac{\partial v_\tau}{\partial n}\frac{\partial r}{\partial \tau}\right)dt.
$$

$$(18)$$

After simplification, we obtain

$$
\int_0^{t_0} \frac{\tau}{\varphi \cos(\alpha)} \frac{\partial v_\tau}{\partial \tau} \frac{\partial u_\tau}{\partial \tau}\, dt
$$

$$
= \int_0^{t_0} \left(-\tau \frac{\partial v_\tau}{\partial n} \frac{\partial P}{\partial \tau} \frac{\partial u_\tau}{\partial n} + \frac{\partial v_\tau}{\partial n} P \frac{\partial u_\tau}{\partial n} + \frac{\partial v_\tau}{\partial t} \frac{\varphi}{\tau \cos(\alpha)} \frac{\partial u_\tau}{\partial t} \right.
\tag{19}
$$

$$
\left. - \frac{\partial v_\tau}{\partial n} P f \tau \varphi \cos(\alpha) - \tau \frac{\partial v_\tau}{\partial n} \frac{\partial r}{\partial \tau} \right) dt.
$$

Expressing the t and τ derivatives in terms of normal derivatives, we get

$$
\int_0^{t_0} \left(\tau \varphi \cos(\alpha) \frac{\partial v_\tau}{\partial n} \frac{\partial u_\tau}{\partial n} - \varphi \sin(\alpha) \frac{\partial v_\tau}{\partial n} \frac{\partial}{\partial t} \left(P \frac{\partial u_\tau}{\partial n} \right) \right.
$$

$$
\left. - \varphi \sin(\alpha) \frac{\partial v_\tau}{\partial n} \frac{\partial r}{\partial t} - \varphi \sin(\alpha) \frac{\partial}{\partial t} \left(P \frac{\partial v_\tau}{\partial n} \right) \frac{\partial u_\tau}{\partial n} \right) dt
$$

$$
= \int_0^{t_0} \left(-\tau \frac{\partial v_\tau}{\partial n} \frac{\partial P}{\partial \tau} \frac{\partial u_\tau}{\partial n} + \frac{\partial v_\tau}{\partial n} P \frac{\partial u_\tau}{\partial n} - \tau \frac{\partial v_\tau}{\partial n} \frac{\partial r}{\partial \tau} \right.
\tag{20}
$$

$$
+ \frac{\varphi \cos(\alpha)}{\tau} \frac{\partial}{\partial t} \left(P \frac{\partial v_\tau}{\partial n} \right) \frac{\partial}{\partial t} \left(P \frac{\partial u_\tau}{\partial n} \right)
$$

$$
\left. + \frac{\varphi \cos(\alpha)}{\tau} \frac{\partial}{\partial t} \left(P \frac{\partial v_\tau}{\partial n} \right) \frac{\partial r}{\partial t} - \frac{\partial v_\tau}{\partial n} P f \tau \varphi \cos(\alpha) \right) dt.
$$

From the principle of invariant embedding, $\dfrac{\partial u_\tau}{\partial n}$ on Γ_τ is arbitrary so that we can separate the parts depending and independent of this quantity, obtaining

$$
\left(\tau \varphi \cos(\alpha) \frac{\partial v_\tau}{\partial n}, \frac{\partial u_\tau}{\partial n} \right) - \left(\varphi \sin(\alpha) \frac{\partial}{\partial t} \circ P \frac{\partial v_\tau}{\partial n}, \frac{\partial u_\tau}{\partial n} \right)
$$

$$
- \left(\varphi \sin(\alpha) \frac{\partial v_\tau}{\partial n}, \frac{\partial}{\partial t} \circ P \frac{\partial u_\tau}{\partial n} \right) = \left(-\tau \frac{\partial v_\tau}{\partial n}, \frac{\partial P}{\partial \tau} \frac{\partial u_\tau}{\partial n} \right)
\tag{21}
$$

$$
+ \left(\frac{\partial v_\tau}{\partial n}, P \frac{\partial u_\tau}{\partial n} \right) + \left(\frac{\varphi \cos(\alpha)}{\tau} \frac{\partial}{\partial t} \circ P \frac{\partial v_\tau}{\partial n}, \frac{\partial}{\partial t} \circ P \frac{\partial u_\tau}{\partial n} \right)
$$

and

$$
\left(\varphi \sin(\alpha) \frac{\partial v_\tau}{\partial n}, \frac{\partial r}{\partial t} \right) = \left(P \frac{\partial v_\tau}{\partial n}, f \tau \varphi \cos(\alpha) \right) + \left(\tau \frac{\partial v_\tau}{\partial n}, \frac{\partial r}{\partial \tau} \right)
$$

$$
- \left(\frac{\varphi \cos(\alpha)}{\tau} \frac{\partial}{\partial t} \circ P \frac{\partial v_\tau}{\partial n}, \frac{\partial r}{\partial t} \right),
\tag{22}
$$

where $\dfrac{\partial v_\tau}{\partial n}$ is an arbitrary test function. This corresponds to (14).

5   Defining $u(0)$

In this section we study the limit of problem $(\mathscr{P}_\varepsilon)$ when ε goes to zero, that is when the hole shrinks to the origin. This is useful to define an initial condition for the equation for u in the factorized form.

Theorem 2. *Considering u_ε the solution of problem ($\mathscr{P}_\varepsilon$), $u_\varepsilon|_{\Gamma_\varepsilon}$ is bounded by a constant not depending on ε.*

Proof. The first part of the proof consists on showing that we have

$$\inf_{\Gamma_\varepsilon} w_\varepsilon \leq u_\varepsilon|_{\Gamma_\varepsilon} \leq \sup_{\Gamma_\varepsilon} w_\varepsilon, \tag{23}$$

where $w_\varepsilon \in H_0^1(\Omega)$ is the solution of the problem

$$-\Delta w_\varepsilon = \tilde{f}_\varepsilon = \begin{cases} f, \Omega \setminus \Omega_\varepsilon \\ 0, \Omega_\varepsilon. \end{cases} \tag{24}$$

From $-\Delta w_\varepsilon = \tilde{f}_\varepsilon$, in $H_0^1(\Omega)$, we find

$$\int_\Omega -\Delta w_\varepsilon = \int_\Omega \tilde{f}_\varepsilon = \int_{\Omega \setminus \Omega_\varepsilon} f = -\int_{\Gamma_a} \frac{\partial w_\varepsilon}{\partial n}. \tag{25}$$

On the other hand, from the formulation of problem ($\mathscr{P}_\varepsilon$) and choosing a test function equal to one, we find

$$\int_{\Omega \setminus \Omega_\varepsilon} -\Delta u_\varepsilon = \int_{\Omega \setminus \Omega_\varepsilon} f = -\int_{\Gamma_\varepsilon} \frac{\partial u_\varepsilon}{\partial n} - \int_{\Gamma_a} \frac{\partial u_\varepsilon}{\partial n} = -\int_{\Gamma_a} \frac{\partial u_\varepsilon}{\partial n}. \tag{26}$$

Therefore, we have the equality $\int_{\Gamma_a} \dfrac{\partial u_\varepsilon}{\partial n} = \int_{\Gamma_a} \dfrac{\partial w_\varepsilon}{\partial n}$.

Let us now suppose that $u_\varepsilon|_{\Gamma_\varepsilon} = c_\varepsilon < \inf_{\Gamma_\varepsilon} w_\varepsilon$. Then, $u_\varepsilon - w_\varepsilon$ satisfies:

$$\begin{cases} -\Delta(u_\varepsilon - w_\varepsilon) = 0, \text{ in } \Omega \setminus \Omega_\varepsilon \\ (u_\varepsilon - w_\varepsilon)|_{\Gamma_a} = 0 \\ (u_\varepsilon - w_\varepsilon)|_{\Gamma_\varepsilon} < 0. \end{cases} \tag{27}$$

From (27), and using the maximum principle, we can also conclude that $u_\varepsilon - w_\varepsilon \leq 0$, in $\Omega \setminus \Omega_\varepsilon$ and, in fact, $u_\varepsilon - w_\varepsilon < 0$, in $\Omega \setminus \Omega_\varepsilon$. As a consequence, using the definition of directional derivative, we find that $\dfrac{\partial u_\varepsilon}{\partial n}|_{\Gamma_a} \geq \dfrac{\partial w_\varepsilon}{\partial n}|_{\Gamma_a}$.

From $\dfrac{\partial(u_\varepsilon - w_\varepsilon)}{\partial n}|_{\Gamma_a} \geq 0$ and $\int_{\Gamma_a} \dfrac{\partial(u_\varepsilon - w_\varepsilon)}{\partial n} = 0$ we conclude that $\dfrac{\partial(u_\varepsilon - w_\varepsilon)}{\partial n}|_{\Gamma_a} = 0$. Therefore, we have $u_\varepsilon - w_\varepsilon < 0$, in $\Omega \setminus \Omega_\varepsilon$, and $(u_\varepsilon - w_\varepsilon) = 0$, in Γ_a. Using Lemma 3.4 of [2], for each point of Γ_a, we find $\dfrac{\partial(u_\varepsilon - w_\varepsilon)}{\partial n} > 0$ a.e. on Γ_a and we reach a contradiction. So, we must have $\inf_{\Gamma_\varepsilon} w_\varepsilon \leq c_\varepsilon$.

Analogously, one can show that $c_\varepsilon \leq \sup_{\Gamma_\varepsilon} w_\varepsilon$.

For the second part of the proof, using [2] (Theorem 8.15, page 189, with $q = 4$), we can show that $\|w_\varepsilon\|_{L^\infty(\Omega)}$ is bounded by a constant not depending on ε (it only depends on constants concerning $\|f\|_{L^2(\Omega)}$ and the size of Ω) and the result follows. $\square$

Now we are able to establish the value of u on the origin.

Theorem 3. *Let $f \in \mathscr{C}^{0,\alpha}(\Omega)$ Then, when ε converges to 0, $u_{\varepsilon|_{\Gamma_\varepsilon}}$ converges to $u(0)$.*

Proof. Considering u the solution of problem $(\mathscr{P})$, since $f \in \mathscr{C}^{0,\alpha}(\Omega)$, we have $u \in \mathscr{C}^{2,\alpha}(\Omega)$. Let, as previously, $-\Delta w_\varepsilon = \tilde{f}_\varepsilon$, $w_\varepsilon \in H_0^1(\Omega)$. Therefore, $v_\varepsilon = w_\varepsilon - u$ satisfies $-\Delta(v_\varepsilon) = \tilde{g}_\varepsilon$, where $\tilde{g}_\varepsilon = \begin{cases} -f, \Omega_\varepsilon \\ 0, \Omega \setminus \Omega_\varepsilon. \end{cases}$ Using again [2] we can show that $\|v_\varepsilon\|_{L^\infty(\Omega)} \le k(\|v_\varepsilon\|_{L^2(\Omega)} + \|\tilde{g}_\varepsilon\|_{L^2(\Omega)})$, where k is a constant not depending on ε. When $\varepsilon \to 0$ we have $\|v_\varepsilon\|_{L^2(\Omega)} \to 0$ and $\|\tilde{g}_\varepsilon\|_{L^2(\Omega)} \to 0$, then $\|v_\varepsilon\|_{L^\infty(\Omega)} \to 0$. So, for $\delta > 0$ there exists $\varepsilon > 0$ such that $|v_\varepsilon(x)| \le \frac{\delta}{2}$ and $|u(x) - u(0)| \le \frac{\delta}{2}, \forall x \in \Omega_\varepsilon \cup \Gamma_\varepsilon$. Then, for $x \in \Gamma_\varepsilon$, $|w_\varepsilon(x) - u(0)| = |v_\varepsilon(x) + u(x) - u(0)| \le \delta$ and consequently, $-\delta \le \inf_{\Gamma_\varepsilon}(w_\varepsilon(x)) - u(0) = \inf_{\Gamma_\varepsilon}(w_\varepsilon(x) - u(0)) \le \sup_{\Gamma_\varepsilon}(w_\varepsilon(x) - u(0)) = \sup_{\Gamma_\varepsilon}(w_\varepsilon(x)) - u(0) \le \delta$. Using (23), we find $-\delta \le \inf_{\Gamma_\varepsilon} w_\varepsilon - u(0) \le u_{\varepsilon|_{\Gamma_\varepsilon}} - u(0) \le \sup_{\Gamma_\varepsilon} w_\varepsilon - u(0) \le \delta$, which implies that $u_{\varepsilon|_{\Gamma_\varepsilon}} \to u(0)$, when $\varepsilon \to 0$. $\qquad\square$

6   Conclusion

Considering $H_{\tau,p}^1(\mathscr{I})$, where $\mathscr{I}$ denotes the interval $(0, t_0)$, to be the space of functions v verifying $v \in L^2(\mathscr{I})$, $\frac{1}{\cos(\alpha)}\frac{\partial v}{\partial t} \in L^2(\mathscr{I})$ and such that v has periodic boundary conditions $v(0) = v(t_0)$, we can define $H_{\tau,p}^{1/2}(\mathscr{I})$ as the $1/2$ interpolate between $H_{\tau,p}^1(\mathscr{I})$ and $L^2(\mathscr{I})$, and $\left(H_{\tau,p}^{1/2}(\mathscr{I})\right)'$ as the $1/2$ interpolate between $(H_{\tau,p}^1(\mathscr{I}))'$ and $L^2(\mathscr{I})$. The final result is synthesized as follows - denoting by $(.,.)$ the scalar product in $L^2(\mathscr{I})$, then P, r and u_τ satisfy:

1. The operator

$$P \in \mathscr{L}\left(L^2(\mathscr{I}), H_{\tau,p}^1(\mathscr{I})\right) \cap \mathscr{L}\left(\left(H_{\tau,p}^{1/2}(\mathscr{I})\right)', H_{\tau,p}^{1/2}(\mathscr{I})\right)$$
$$\cap \mathscr{L}\left((H_{\tau,p}^1(\mathscr{I}))', L^2(\mathscr{I})\right), \tag{28}$$

bounded as a function of τ, satisfies, for every $h, \bar{h}$ in $L^2(\mathscr{I})$, the Riccati equation

$$\left(\frac{dP}{d\tau}h, \bar{h}\right) - \left(\frac{\varphi \sin\alpha}{\tau}h, \frac{\partial}{\partial t}\circ P\bar{h}\right) - \left(\frac{\partial}{\partial t}\circ Ph, \frac{\varphi \sin\alpha}{\tau}\bar{h}\right)$$
$$- \left(\frac{\varphi \cos\alpha}{\tau^2}\frac{\partial}{\partial t}\circ Ph, \frac{\partial}{\partial t}\circ P\bar{h}\right) - \left(\frac{1}{\tau}h, P\bar{h}\right) = -(\varphi \cos\alpha\, h, \bar{h}) \tag{29}$$

in $\mathscr{D}'(0,1)$, with the initial condition $P(1) = 0$.

2. For every h in $H_{\tau,p}^{1/2}(\mathscr{I})$, r satisfies the equation

$$\left(\frac{\partial r}{\partial \tau}, h\right) - \left(\frac{\varphi \sin\alpha}{\tau}\frac{\partial r}{\partial t}, h\right) - \left(\frac{\varphi \cos\alpha}{\tau^2}\frac{\partial r}{\partial t}, \frac{\partial}{\partial t}\circ Ph\right) = -(\varphi \cos\alpha f, Ph) \tag{30}$$

in $\mathscr{D}'(0,1)$, with the initial condition $r(1) = 0$.

3. For every h in $\left(H_{\tau,p}^{1/2}(\mathcal{I})\right)'$, u satisfies the equation

$$\left(\frac{1}{\varphi\cos\alpha}\frac{\partial u}{\partial\tau}, Ph\right) - \left(\frac{\tan\alpha}{\tau}\frac{\partial u}{\partial t}, Ph\right) + \langle u,h\rangle_{H_{\tau,p}^{1/2}(\mathcal{I}),\left(H_{\tau,p}^{1/2}(\mathcal{I})\right)'}$$
$$= \langle r,h\rangle_{H_{\tau,p}^{1/2}(\mathcal{I}),\left(H_{\tau,p}^{1/2}(\mathcal{I})\right)'} \tag{31}$$

in $\mathscr{D}'(0,1)$, with the initial condition given by Theorem 3.

Moreover, taking $Q = \dfrac{P}{\tau}$ as unknown and $\mu = \log(\tau)$ as variable, the above Riccati equation can also take the simpler form

$$\left(\frac{dQ}{d\mu}h,\bar{h}\right) - \left(\varphi\sin\alpha\, h, \frac{\partial}{\partial t}\circ Q\bar{h}\right) - \left(\frac{\partial}{\partial t}\circ Qh, \varphi\sin\alpha\,\bar{h}\right)$$
$$- \left(\varphi\cos\alpha\frac{\partial}{\partial t}\circ Qh, \frac{\partial}{\partial t}\circ Q\bar{h}\right) = -\left(\varphi\cos\alpha\, h,\bar{h}\right), \tag{32}$$

and, identically, the equation for the residue r becomes

$$\left(\frac{\partial r}{\partial\mu},h\right) - \left(\varphi\sin\alpha\frac{\partial r}{\partial t},h\right) - \left(\varphi\cos\alpha\frac{\partial r}{\partial t}, \frac{\partial}{\partial t}\circ Qh\right) = -\left(e^{2\mu}\varphi\cos\alpha f, Qh\right), \tag{33}$$

with initial conditions, respectively, $Q(0) = 0$ and $r(0) = 0$. Then u satisfies

$$\left(\frac{1}{\varphi\cos\alpha}\frac{\partial u}{\partial\mu}, Qh\right) - \left(\tan\alpha\frac{\partial u}{\partial t}, Qh\right) + \langle u,h\rangle_{H_{\tau,p}^{1/2}(\mathcal{I}),\left(H_{\tau,p}^{1/2}(\mathcal{I})\right)'}$$
$$= \langle r,h\rangle_{H_{\tau,p}^{1/2}(\mathcal{I}),\left(H_{\tau,p}^{1/2}(\mathcal{I})\right)'} \tag{34}$$

The initial condition on u given by Theorem 3 is now valid at $\mu = -\infty$. These equations allow us to seek an explicit formula for the solution of $(\mathscr{P})$, through homographic transformation, as the Riccati equation has constant coefficient in μ.

From the numerical point of view, one can consider a spatial discretization of the problem adapted to the system of coordinates (t,τ) (or (t,μ)), which leads to a linear system of equations. Then there exists a particular discretization of the system (32), (33), (34) through which we can recover the Gauss block LU factorization of this linear system. That is why we claim that the proposed factorization is an infinite dimensional generalization of the Gauss factorization. But other discretizations exist that give new directly computable discretizations of the original Poisson boundary value problem. They will be presented elsewhere.

References

1. Bellman, R.: Dynamic Programming. Princeton University Press, Princeton (1957)
2. Gilbarg, D., Trudinger, N.: Elliptic Partial Differential Equations of Second Order. Springer, Berlin (1983)
3. Henry, J., Louro, B., Soares, M.C.: A factorization method for elliptic problems in a circular domain. C. R. Math. Acad. Sci. 339(3), 175–180 (2004)

4. Henry, J., Ramos, A.: Factorization of second order elliptic boundary value problems by dynamic programming. Nonlinear Analysis 59, 629–647 (2004)
5. Henry, J., Ramos, A.: Study of the initial value problems appearing in a method of factorization of second-order elliptic boundary value problems. Nonlinear Analysis 68, 2984–3008 (2008)
6. Lions, J.L.: Optimal Control of Systems Governed by Partial Differential Equations. Springer, New York (1971)
7. Soares, M.C.: Factorization by Invariant Embedding of Elliptic Problems: Circular and Star-shaped Domains, PhD Thesis, Faculdade de Ciências e Tecnologia, Universidade Nova de Lisboa (2006)

Nonlinear Cross Gramians

Tudor C. Ionescu and Jacquelien M.A. Scherpen

University of Groningen, Faculty of Mathematics and Natural Sciences ITM,
Nijenborgh 4, 9747AG, Groningen, The Netherlands
t.c.ionescu@rug.nl, j.m.a.scherpen@rug.nl

Abstract. We study the notion of cross Gramians for nonlinear gradient systems, using the characterization in terms of prolongation and gradient extension associated to the system. The cross Gramian is given for the variational system associated to the original nonlinear gradient system. We obtain linearization results that correspond to the notion of a cross Gramian for symmetric linear systems. Furthermore, first steps towards relations with the singular value functions of the nonlinear Hankel operator are studied and yield promising results.

1 Introduction

In this paper, we give an extension of the cross Gramian notion for nonlinear gradient systems. The gradient systems are an important class of nonlinear systems, endowed with a pseudo-Riemannian metric on the state-space manifold, such that the drift is a gradient vectorfield with respect to this metric and a potential function and the input vectorfields are gradient with respect to the same metric and output, see e.g. [3,15] and references therein. Examples of gradient systems include nonlinear electrical circuits and certain dissipative systems. The linear counterpart is a symmetric system. With respect to model reduction, for linear systems it is showed in [1,4,14] that exploiting the symmetry, model reduction becomes more efficient. This is based on the notion of cross Gramian, that is the solution of a Sylvester equation, which can be solved in an efficient way. The cross Gramian for a symmetric system contains information about both controllability and observability at the same time and moreover the squared cross Gramian is the product of the controllability and observability Gramians. Then the Hankel singular values are the eigenvalues of the cross Gramian. Moreover, the cross Gramian can be obtained using only one of the Gramians of the system and the metric.

For nonlinear systems the problem is more complicated and not yet tackled in the literature. The notion of symmetry for a nonlinear system is now best studied by considering nonlinear gradient systems. We use the associated prolongation and gradient extension and the results in [3]. A nonlinear system is gradient if the two latter systems have the same input-output behavior. Using this property and its consequences, we give the definition of the cross Gramian for the variational system (which is a gradient system, too) as the nonlinear, non-trivial extension of the concept of the cross Gramian for linear systems. Furthermore, we give a nonlinear counterpart of the Sylvester equation. Using the cross Gramian and the theory of Hankel singular values as in [6,11], first steps towards proving that the squared eigenvalues of the nonlinear cross Gramian are directly related to the Hankel singular values of the system, are set. In this case, instead

A. Korytowski et al. (Eds.): System Modeling and Optimization, IFIP AICT 312, pp. 293–306, 2009.

of balancing, only solving a nonlinear Sylvester equation, a metric and an eigenvalue decomposition suffice for obtaining the Hankel singular values of the gradient system.

The paper is outlined as follows. In Section 2 we give an overview of the cross Gramian technique for linear systems. To show the line of thinking in the nonlinear case, in Section 3, we give a review of the definitions of the prolongation and gradient extension and the property of a nonlinear system being gradient itself, this being a natural extension of the linear symmetric system notion, to the nonlinear case. In Section 4, we analyze some linearization results which motivate the reasoning in Section 5, where the definition of the nonlinear Gramian is presented and the conjecture about the relation for singular value functions is stated. Finally an example is given in Section 6 and in Section 7 some conclusions end this paper.

A nonlinear system is defined here as:

$$\begin{cases} \dot{x} = f(x) + g(x)u \\ y = h(x) \end{cases}, \tag{1}$$

where $x \in \mathcal{M}$ is the state vector, $u \in \mathbb{R}^m$ is the vector of inputs and $y \in \mathbb{R}^p$ is the output. $\mathcal{M}$ is a smooth manifold, of dimension n. We make the following assumptions:

Assumption 1. $f(x), g(x), h(x)$ are smooth;

Assumption 2. The system is square, i.e. $m = p$;

Assumption 3. x_0 is an asymptotically stable equilibrium point of the system and $h(x_0) = 0$;

Assumption 4. System (1) is asymptotically reachable from x_0 (i.e. for any x, there exists an input u and $t \geq 0$, such that $x = \phi(t, 0, x_0, u)$, with ϕ being the trajectory obtained by integrating the first equation in (1)).

Assumption 5. System (1) is zero-state observable (i.e. if $u(t) = 0$, $y(t) = 0$ then $x(t) = 0$).

Assumptions 4 and 5 are related to the minimality of the system, see [12].

Notation: Let $\mathcal{M}$ be a smooth manifold and $V(x)$ a smooth vectorfield, $x \in \mathcal{M}$. Then we denote by $\text{grad}_G V$ the gradient of $V(x)$ on the manifold $\mathcal{M}$ endowed with the pseudo-Riemannian metric G. In local coordinates $\text{grad}_G V = -G^{-1}(x)\frac{\partial V(x)}{\partial x}$ (see [15] for details). $\frac{\partial V(x)}{\partial x}$ means the row vector $\left[\frac{\partial V(x)}{\partial x_1} \ ... \ \frac{\partial V(x)}{\partial x_n} \right]$. $\mathbb{R}$ is the set of real numbers.

2   Linear Systems Case

If the system (1) is linear, then it can be written as:

$$\begin{cases} \dot{x} = Ax + Bu \\ y = Cx \end{cases}, \tag{2}$$

where $A \in \mathbb{R}^{n \times n}, B \in \mathbb{R}^{n \times m}, C \in \mathbb{R}^{p \times n}$ are constant matrices. In this case, Assumption 1 is automatically satisfied. We consider system (2) satisfying Assumptions 2-5. Assumptions 4 and 5 are equivalent to the minimality of the system (see e.g. [16] for more details). A linear system has a corresponding unique dual system defined as:

$$\begin{cases} \dot{z} = A^T z + C^T u_d \\ y_d = B^T z \end{cases}.$$

(3)

Because (1) is controllable and observable, and these properties are dual to each other (i.e. if the pair (A, B) is controllable, then (B^T, A^T) is observable), it follows immediately that the dual system (3) is controllable and observable, i.e. minimal, too.

The definition of the cross Gramian for a linear square system is:

Definition 1. [14] Let (2) be a square system. Then the cross Gramian X is defined as the solution of the Sylvester equation:

$$AX + XA + BC = 0.$$

(4)

If the system is asymptotically stable, then the cross Gramian can be equivalently defined as: $X = \int_0^\infty e^{At} BC e^{At} dt$.

Another important definition is the one of the Hankel operator associated to the linear system (2):

$$\mathcal{H}(u) = \int_{-\infty}^0 H(t - \tau) u(-\tau) d\tau$$

(5)

where $t > 0$ and H is the impulse response of the system (2). The singular values of the Hankel operator are fundamental for the balanced truncation model order reduction. Each singular value represents a measure for the importance of each state component in the output response of system (2) to a certain input (see e.g. [2] for more details). The cross Gramian possesses some interesting properties being related to the above defined Hankel operator and the Hankel singular values of a linear square system.

Theorem 1. *[14] For square linear systems the non-zero eigenvalues of the cross Gramian X are the non-zero eigenvalues of the Hankel operator associated to the system.*

However, the singular value problem is different, that is the singular values of the cross Gramian are not the Hankel singular values of the system. Still, there is a relation of majorization between the two as shown below.

Theorem 2. *[14] For a square linear system, the following relations hold:* $\sum_{i=1}^k \sigma_i \geq \sum_{i=1}^k \pi_i$ *and* $\sum_{i=k+1}^n \sigma_i \leq \sum_{i=k+1}^n \pi_i$, *where σ_i are the Hankel singular values, π_i are the singular values of X, and k is the index for which σ_k is much larger than σ_{k+1}.*

For symmetric systems, the cross Gramian X has more attractive properties, useful for model reduction.

First we give the definition of a symmetric linear system:

Definition 2. [1,4,14] A square linear, system $G(s) = C(sI - A)^{-1}B$, with the state-space realization (2) is called symmetric if $G(s) = G^T(s)$.

Proposition 1. *Assume that system (2) satisfies assumptions 2-5. Then system (2) is symmetric if and only if there exists an invertible symmetric matrix T such that $A^T T = TA$, $C^T = TB$, i.e. the system and its dual are input-output (externally) equivalent.*

In, for instance [1,14], model reduction based on the balancing procedure, for this type of systems is considered. The symmetry property is exploited, making the procedure more efficient. Basically, the Sylvester equation from Definition 1 is solved and the cross Gramian is obtained. It will directly provide the Hankel singular values of the system. We refer to the results presented in [14,1,4], which are summarized in the sequel.

Defining the controllability Gramian as W and the observability Gramian as M, they are the solutions of the following Lyapunov equations, respectively:

$$AW + WA^T + BB^T = 0 \tag{6}$$

$$A^T M + MA + C^T C = 0. \tag{7}$$

The following theorem summarizes the properties of X in relation with W and M.

Theorem 3. *[14,4] Let (2) be a square asymptotically stable symmetric system in the sense of Definition 2. If X is the solution of (4) then the following relations are equivalent:*

1. *$X^2 = WM > 0$;*
2. *If T is the symmetry transformation as in Proposition 1, then $X = WT = T^{-1}M$;*
3. *The Hankel singular values of (2) are the absolute values of the eigenvalues of X.*

For symmetric systems, when compared to the classical balancing procedure, there are two advantages: the first is that instead of solving two Lyapunov equations, whose computational complexity is known to be a drawback, only one Sylvester equation is solved. The second advantage consists of avoiding in this way the balancing procedure. Since the Hankel singular values satisfy $\sigma_i = \sqrt{\lambda_i}$, $\lambda_i \in \lambda(WM)$, $i = 1, ..., n$, the problem of finding them turns into an eigenvalue problem of the cross Gramian X.

Remark 1. *There exists a relation between the controllability and observability operators, and the cross Gramian. Define by $x = \mathscr{C}(u) = \int_0^\infty e^{At} Bu(t)dt$, the controllability operator and by $y = \mathscr{O}(x) = Ce^{At}x$ the observability operator of the system (A,B,C). Then, by the definition of the cross Gramian, we have: $Xx = \mathscr{C}\mathscr{O}(x)$. It can be proven that, under minimality and symmetry assumptions as in the definitions presented here, the eigenvalues of the $\mathscr{C}\mathscr{O}\mathscr{C}\mathscr{O}$ operator are the squared Hankel singular values of the system, i.e. the eigenvalues of $\mathscr{H}^*\mathscr{H}$.*

3 Review of Gradient Systems

The nonlinear extension of the notion of symmetric systems is the gradient systems. The property of a system being gradient is described in terms of necessary and sufficient conditions satisfied by the prolongation (variational) system and the gradient extension associated with (1). We will give a brief overview of the results in [3,15].

Definition 3. [3,15] A nonlinear system (1) is called a gradient system if:

1. There exists a pseudo-Riemannian metric G, on the manifold $\mathcal{M}$, given as $\sum\limits_{i,j=1}^{m} g_{ij}(x)dx_i \otimes dx_j$, with $g_{ij}(x) = g_{ji}(x)$ smooth functions of x, and $G(x) = [g_{ij}(x)]_{i,j=1...n}$ invertible, for all x.
2. There exists a smooth potential function $V : \mathcal{M} \to \mathbb{R}$,

such that the system (1) can be written as:

$$\begin{cases} \dot{x} = \text{grad}_G V(x) - \sum\limits_{i=1}^{m} u_i \text{grad}_G h_i(x), \ \ x \in \mathbb{R}^n \\ y_i = h_i(x), \ i = 1,...,m \end{cases} \tag{8}$$

In local coordinates $x = [x_1 \ x_2 \ ... \ x_n]^T \in \mathcal{M}$, the system can be written as:

$$\begin{cases} \dot{x} = -G^{-1}(x)\dfrac{\partial^T V}{\partial x}(x) + G^{-1}(x)\dfrac{\partial^T h}{\partial x}(x)u \\ y = h(x) \end{cases} \tag{9}$$

Next, we present the definition of the prolonged system associated with (1).

Definition 4. [3] The prolongation Σ_p of (1) is defined by:

$$\begin{cases} \dot{x} = f(x) + g(x)u \\ \dot{v} = \dfrac{\partial f(x)}{\partial x}v + \sum\limits_{j=1}^{m} u_j \dfrac{\partial g_j(x)}{\partial x}v + g(x)u_p \\ y = h(x), \ y_p = \dfrac{\partial h(x)}{\partial x}v \end{cases} \tag{10}$$

where $v \in T\mathcal{M}$, the tangent bundle of the manifold $\mathcal{M}$.

3.1 The Riemannian Metric on $T^*\mathcal{M}$

Since a canonical pseudo-Riemannian structure on the cotangent bundle $T^*\mathcal{M}$ of the manifold $\mathcal{M}$ does not exist, a pseudo-Riemannian metric cannot be defined directly. In this case a torsion-free affine connection defined on the manifold $\mathcal{M}$ and its Riemannian extension G^C to $T^*\mathcal{M}$ are used.

Definition 5. An affine connection on a manifold $\mathcal{M}$ is defined as an assignment ∇ : $(X,Y) \to \nabla_X Y$, where X, Y and $\nabla_X Y$ are vectorfields on $\mathcal{M}$, satisfying the following properties: it is $\mathbb{R}$-bilinear, $\nabla_{fX} Y = f\nabla_X Y$ and $\nabla_X(fY) = f\nabla_X Y + X(f)Y$, for every $f \in C^\infty(\mathcal{M})$.

Let X and Y be any two vectorfields on $\mathcal{M}$. Their symmetric product is given as: $< X :$ $Y >= \nabla_X Y + \nabla_Y X$. We introduce the construction that associates to each vectorfield X a function V^X on $T^*\mathcal{M}$, given by $V^X(x,p) =< p, X(x) >$, $x \in \mathcal{M}$, $p \in T^*\mathcal{M}$.

If ∇ is a torsion-free affine connection (see [3] and references therein for more details) then it defines a pseudo-Riemannian metric G^C as a unique (0,2)-tensor on $T^*\mathcal{M}$ which satisfies:

$$G^C(X,Y) = -V^{<X:Y>} \tag{11}$$

Now the gradient vectorfield associated with the function $V^X \in C^\infty(T^*\mathcal{M})$, X vectorfield on $\mathcal{M}$, can be expressed locally as:

$$\text{grad}_{G^C} V^X = X_i \frac{\partial}{\partial x_i} + p_i \left(\frac{\partial X_i}{\partial x_j} + 2\Gamma^a_{jk} X_k \right) \frac{\partial}{\partial p_j}, \tag{12}$$

where X is a vectorfield on $\mathcal{M}$, $i,j,k = 1,,n$ and Γ^i_{jk} represent the Christoffel symbols of the affine connection ∇ (relation (2.8) in [3]).

For our purpose, we assume that G^C is properly defined ([3]) and we will use the local expression from (12) to express the gradient extension of (1), comprising all the terms $2p_i\Gamma^a_{jk}X_k\frac{\partial}{\partial p_j}$ in a function $\mathscr{F}$.

3.2   The Gradient Extension of a Nonlinear System

Definition 6. The gradient extension of (1) is defined by:

$$\begin{cases} \dot{x} = f(x) + g(x)u \\ \dot{p} = \dfrac{\partial^T (f(x) + g(x)u)}{\partial x} p + \mathscr{F}\left(g_{ij}(x), \dfrac{\partial g_{ij}(x)}{\partial x_k}, f_k(x), u, g(x), p\right) + \dfrac{\partial h(x)}{\partial x} u_g, \\ y = h(x), \; y_g = g^T(x)p, \; i,j,k = 1,...,n. \end{cases} \tag{13}$$

Remark 2. *Notice that for the linear system (2) the prolongation is the system itself written twice and the gradient extension contains the system itself and the dual of the prolonged variable part, yielding, respectively:*

$$\begin{cases} \dot{x} = Ax + Bu \\ \dot{v} = Av + Bu_p \\ y = Cx, \; y_p = Cv \end{cases}, \begin{cases} \dot{x} = Ax + Bu \\ \dot{p} = A^T p + C^T u_g \\ y = Cx, \; y_g = B^T p \end{cases}. \tag{14}$$

Remark 3. *According to [3, Corollaries 3.3, 3.6] (1) is zero-state observable if and only the prolonged system is zero-state observable and the zero-state observability of (1) implies the zero-state observability of the gradient extension.*

The main result, useful for our purpose, is:

Theorem 4. *[3, Theorem 5.4, Corollary 4.4] Let (1) be as in Assumption 4. Assume that there exists a torsion-free affine connection on $\mathcal{M}$ with which the system is compatible, and that the system is observable with its observability distribution having constant dimension. Then, system (1) is a gradient control system, as in Definition 3, if and only if the prolonged system Σ_p and the gradient extension Σ_g have the same input-output behavior.*

Remark 4. *In the linear systems case, this result becomes a property between the system itself and its dual counterpart, which immediately leads to the definition of symmetric systems. The metric is given by the matrix T, showing that a linear symmetric system is a particular case (linear version) of the gradient system.*

Lemma 1. *[3, Lemmas 5.5, 5.6] If (1) is a gradient control system, then there exists a diffeomorphism $\phi(x,v) = (x, G(x)v)$, such that $(x,p) = \phi(x, G(x)v)$, where v and p satisfy (14), and $G(x)$ is the matrix associated to the metric.*

Remark 5. *For linear systems this means, indeed that $p = Tv$.*

4 Linearization Results

For (1) satisfying Assumptions 1 and 3 we define the observability function ([9])

$$L_o(x) = \frac{1}{2} \int_0^\infty \|y(t)\|_{L_2}^2 dt, \ x(0) = x, \ x(\infty) = x_0 \tag{15}$$

and the controllability function ([9])

$$L_c(x) = \min_{u \in L_2^-, x(0)=x, \, x(-\infty)=x_0} \frac{1}{2} \int_{-\infty}^0 \|u(t)\|_{L_2}^2 dt \tag{16}$$

If the system satisfies Assumption 4 as well, then $L_c(x)$ exists, is finite, $L_c(x) > 0$, $L_c(x_0) = 0$ and satisfies the Hamilton-Jacobi equation ([9]):

$$\frac{\partial L_c}{\partial x} f(x) + \frac{1}{2}\frac{\partial L_c}{\partial x} g(x) g^T(x) \frac{\partial^T L_c}{\partial x} = 0 \tag{17}$$

such that $-\left(f(x) + g(x)g^T(x)\dfrac{\partial L_c(x)}{\partial x}\right)$ is asymptotically stable. If the system also satisfies Assumption 5, then $L_o(x)$ exists, is finite, $L_o(x) > 0$, $L_o(x_0) = 0$, and satisfies the nonlinear Lyapunov equation ([9]):

$$\frac{\partial L_o}{\partial x} f(x) + \frac{1}{2} h^T(x) h(x) = 0. \tag{18}$$

Suppose $x_0, u = 0$ is an equilibrium point and assume that $h(x_0) = 0$. Then $-G^{-1}(x_0)\frac{\partial^T V}{\partial x}(x_0) = 0$. Taking Taylor series expansion in system (8), we can write:

$$\dot{x} = G^{-1}(x_0)\frac{\partial^2 V}{\partial x^2}(x_0)(x - x_0) + \left[\sum_{i,j=1}^n \frac{\partial g_{ij}}{\partial x_i}(x_0)\frac{\partial V}{\partial x_j}(x_0)\right]_{i,j=1\ldots n}(x - x_0) + \ldots \tag{19}$$

Since $\frac{\partial V}{\partial x_j}(x_0) = 0$, $j = 1,...,n$, then the linearization of the gradient system (8) yields:

$$\begin{cases} \dot{\bar{x}} = -G^{-1}(x_0)\frac{\partial^2 V}{\partial x^2}(x_0)\bar{x} + G^{-1}(x_0)\frac{\partial^T h}{\partial x}(x_0)u \\ \bar{y} = \frac{\partial h}{\partial x}(x_0)\bar{x} \end{cases} . \tag{20}$$

Lemma 2. *The system (20) is a gradient (symmetric) system with the metric $T = G(x_0)$.*

Proof. Denote $G = G(x_0)$, $Q = \frac{\partial^2 V}{\partial x^2}(x_0)$. Since V is smooth, Q is symmetric. G, by definition is symmetric and invertible. Then:

$$\begin{aligned} H(s) &= C(sI + G^{-1}Q)^{-1}G^{-1}C^T = C(sG^{-1}G + G^{-1}Q)^{-1}G^{-1}C^T \\ &= C[G^{-1}(sI + QG^{-1})G]^{-1}G^{-1}C = CG^{-1}(sI + QG^{-1})C^T = H^T(s). \end{aligned} \tag{21}$$

$\square$

Let W and M be the controllability and the observability Gramians of (20), respectively, and assume $W > 0$, $M > 0$, i.e. (20) is controllable and observable. Then:

$$M = \frac{\partial^2 L_o}{\partial x^2}(x_0), \quad W^{-1} = \frac{\partial^2 L_c}{\partial x^2}(x_0). \tag{22}$$

The asymptotic reachability of the nonlinear system implies its accessibility and this implies the controllability of the linear system, see [12]. Since the linearized system is assumed symmetric, controllability implies observability, and this implies the local zero-state observability of the nonlinear system. So, locally there exists a duality of the controllability and observability property, which motivates the search for a cross Gramian for the nonlinear gradient system.

The linearized system is gradient and then, according to Theorem 3, statement 2, we have that near x_0:

$$\left(\frac{\partial^2 L_o}{\partial x^2}(x)\right)^{-1}G(x) = G^{-1}(x)\frac{\partial^2 L_c}{\partial x^2}(x). \tag{23}$$

Remark 6. *Given a system (1), the linearization of the prolonged system Σ_p around $x_0, v = 0, u = u_p = 0$ and of the gradient extension Σ_g around $x_0, p = 0, u = u_g = 0$, respectively, we obtain the linear systems (14). If the system is symmetric then $p = Tv$, $G(x_0) = T$.*

Since the duality in properties takes place between the v part and the p part of the two systems, we are going to extract these parts from the nonlinear system and study them.

4.1   The Isomorphic Case

Another case related to linearization is that when the system is equivalent to a linear system, as treated in [15]. This means that there exists a coordinate transformation $x' = \eta(x)$, such that in the new coordinates, the system is described by a linear state-space realization. If the equivalent linear system is a gradient system, as well, and the transformation η is an isometry (see e.g. [8]), then the gradient system is said to be isomorphic to the linear symmetric system. Then, the linear idea of cross Gramian can be extended to the nonlinear gradient system via the diffeomorphism η and the isometry relation, as follows.

Denote by $\bar{L}_o(x') = \frac{1}{2}x'^T \bar{M} x'$ the observability function and by $\bar{L}_c(x') = \frac{1}{2}x'^T \bar{W}^{-1} x'$ the controllability function of the linear system, where

$$\bar{M} = \frac{\partial^2 \bar{L}_o}{\partial x'^2}(x'), \ \bar{W}^{-1} = \frac{\partial^2 \bar{L}_c}{\partial x'^2}(x') \tag{24}$$

are the constant Gramians of the linear system. Moreover

$$\bar{L}_o(x') = L_o(\eta^{-1}(x)), \ \bar{L}_c(x') = L_c(\eta^{-1}(x)).$$

This leads to the following relation:

$$\frac{\partial \bar{L}_o}{\partial x'}(x') = \frac{\partial L_o}{\partial x}(\eta^{-1}(x')) \left(\frac{\partial \eta}{\partial x}(\eta^{-1}(x')) \right)^{-1} \tag{25}$$

Let T be the matrix associated with the metric for the symmetric linear system. Then, according to Theorem 3, statement 2, $\bar{W}T = T^{-1}\bar{M}$ that can be rewritten as $\bar{M}^{-1}T = T^{-1}\bar{W}^{-1}$. Postmultiplying with x' we get: $(\partial^2 \bar{L}_o(x')/\partial x'^2)^{-1} Tx' = T^{-1}(\partial^T \bar{L}_c(x')/\partial x')$. Using relation (25) and $x' = \eta(x)$ we can write:

$$\frac{\partial^T L_c}{\partial x}(x) = \frac{\partial^T \eta}{\partial x}(x) \cdot T \cdot \left(\frac{\partial^2 \bar{L}_o}{\partial x^2}(\eta(x)) \right)^{-1} \cdot T \cdot \eta(x) \tag{26}$$

which shows that the observability function, the metric, and the isomorphism between the systems give the controllability function of the gradient system. This can be called a nonlinear version of the cross Gramian idea for this particular case, and it motivates the search for the nonlinear cross Gramian in the general case.

5 Nonlinear Cross Gramian

In this section, we will make an analysis of the variational part of the prolonged system. Denote by:

$$\Sigma'_p : \begin{cases} \dot{v} = \dfrac{\partial (f(x) + g(x)u)}{\partial x} v + g(x)u_p \\ y_p = \dfrac{\partial h(x)}{\partial x} v \end{cases}, \tag{27}$$

where x is considered a parameter varying according to (1).

Since the system is asymptotically stable, by the definition of its variational associated system, the latter is also asymptotically stable. By Theorem 4, Σ'_p has the same input-output behavior as the system Σ'_g, given by:

$$\begin{cases} \dot{p} = \dfrac{\partial^T (f(x) + g(x)u)}{\partial x} p + \mathcal{F}(g_{ij}(x), \frac{\partial g_{ij}(x)}{\partial x_k}, f_k(x), u, g(x), p) + \dfrac{\partial^T h(x)}{\partial x} u_g \\ y_g = g^T(x)p \end{cases}, \tag{28}$$

where x again is a parameter varying as in (1). According to Lemma 1, there exists a coordinate transformation such that $p = \psi(x,v)$, given by $\psi(x,v) = G(x)v$, where $G(x)$ is symmetric and invertible (as in the definition of (8)) and is given by the pseudo-Riemannian metric. Applying the coordinate transformation on Σ'_p, we get:

$$G(x)g(x) = \frac{\partial^T h(x)}{\partial x} \quad \text{and} \quad \frac{\partial h(x)}{\partial x} G^{-1}(x)p = g^T(x)p. \tag{29}$$

Remark 7. *In the linear systems case, everything fits with the definition and characterization of the property of symmetry. Moreover, the linearization of Σ'_p and Σ'_g around an equilibrium point $(x_0,0,0,0)$ yields the v part and p part of (14), with $p = Tv$, with T invertible and symmetric.*

Based on the local existence of the cross Gramian, we make an analysis of the observability function of Σ'_p. In this case, $u = 0$, $u_p = 0$ and Σ'_p becomes:

$$\begin{cases} \dot{v} = \dfrac{\partial f(x)}{\partial x} v \\[2mm] y_p = \dfrac{\partial h(x)}{\partial x} v \end{cases} . \tag{30}$$

Assuming the zero-state observability combined with the asymptotic stability of Σ'_p implies the existence of the observability function $L^p_o(x,v) > 0$, $L^p_o(x_0,0) = 0$, defined as:

$$L^p_o(x,v) = \frac{1}{2} \int_t^\infty y_p^T(\tau)y_p(\tau)d\tau \tag{31}$$

and satisfying the nonlinear Lyapunov equation:

$$\frac{\partial L^p_o(x,v)}{\partial v} \frac{\partial f(x)}{\partial x} v + \frac{1}{2}v^T \frac{\partial^T h(x)}{\partial x} \frac{\partial h(x)}{\partial x} v = -\frac{\partial L^p_o(x,v)}{\partial x} f(x). \tag{32}$$

Since the system is linear in v, without loss of generality, we can write $L^p_o(x,v)$ as:

$$L^p_o(x,v) = \frac{1}{2}v^T \mathcal{L}(x)v \tag{33}$$

with $\mathcal{L}(x)$ symmetric, positive definite and with smooth elements.

In the sequel, we determine the nonlinear counterpart of the Sylvester equation which in the linear case gives the cross Gramian. Taking the derivative with respect to v and using (29), we get:

$$\frac{\partial^2 L^p_o(x,v)}{\partial v^2} \frac{\partial f(x)}{\partial x} v + \frac{\partial^T f(x)}{\partial x} \frac{\partial^T L^p_o(x,v)}{\partial v} + G(x)g(x)\frac{\partial h(x)}{\partial x} v = -\frac{\partial^2 L^p_o(x,v)}{\partial v \partial x} f(x). \tag{34}$$

Applying the coordinate transformation, $p = G(x)v$, on (30) we get:

$$\frac{\partial f}{\partial x} v = G^{-1}(x)\frac{\partial^T f}{\partial x} G(x)v + \overline{\mathscr{F}}(g_{ij}(x), \tfrac{\partial g_{ij}(x)}{\partial x_k}, f_k(x), u, g(x), p), \tag{35}$$

where $\overline{\mathscr{F}} = \mathscr{F} - G^{-1}(x)\dot{G}(x)v$.

Premultiplying the equation with v^T and using (35) we obtain:

$$p^T G^{-1}(x)\mathcal{L}(x)\frac{\partial f(x)}{\partial x}v + p^T \frac{\partial f(x)}{\partial x}G^{-1}(x)\mathcal{L}(x)v + p^T g(x)\frac{\partial h}{\partial x}v =$$
$$-v^T \frac{\partial^2 L_o^p(x,v)}{\partial v\partial x}f(x) - \overline{\mathcal{F}}^T \mathcal{L}(x)v. \tag{36}$$

Using the coordinate transformation (29) and equation (32) we get:

$$p^T G^{-1}(x)\mathcal{L}(x)\frac{\partial f}{\partial x}v + \frac{1}{2}p^T g(x)\frac{\partial h}{\partial x}v =$$
$$-v^T \frac{\partial^2 L_o^p(x,v)}{\partial v\partial x}f(x) + \frac{\partial L_o^p(x,v)}{\partial x}f(x) - \overline{\mathcal{F}}^T \mathcal{L}(x)v. \tag{37}$$

Remark 8. *In the linear systems case, (34) becomes:* $v^T M Av + v^T A^T Mv + p^T BCv$
$= 0$. *Since* $v = T^{-1}p$, *we get:*

$$p^T T^{-1}MAv + p^T AT^{-1}Mv + p^T BCv = 0. \tag{38}$$

Using the symmetry property, this immediately leads to the Sylvester equation (4). Moreover, the relation $X = T^{-1}M$ *is satisfied as in Theorem 3. Equation (38) becomes*

$$XA + \frac{1}{2}BC = 0.$$

We are now ready to define the cross Gramian for a nonlinear gradient system.

Definition 7. We call

$$\mathcal{X}(x) = G^{-1}(x)\mathcal{L}(x) \tag{39}$$

the *cross-Gramian* matrix associated to Σ_p' and it satisfies (37).

This is an extension of statement 2 in Theorem 3, i.e. the cross Gramian is given by the gradient metric and the observability Gramian. In order to explain the cross Gramian and its importance we present in a nutshell the study of Hankel singular values for a nonlinear system (1) as in [6,11]. Suppose that (1) is asymptotically reachable from $x(0)$, then the controllability function $L_c(x)$ exists and is positive definite, with $L_c(x_0) = 0$.

If $\mathcal{H}(u)$ is the Hankel operator of the system then for finding out the Hankel singular values of the system the differential problem is solved: $(d\mathcal{H}(u))^*\mathcal{H}(u) = \lambda u$, where $(d\mathcal{H}(u))^*$ represents the adjoint of the $(d\mathcal{H}(u))$ operator (see [6] for further details). A solution for this problem is given by the following result:

Lemma 3. *[6] If there exists* $\lambda \neq 0$ *such that*

$$\frac{\partial L_o}{\partial x}(x(0)) = \lambda \frac{\partial L_c}{\partial x}(x(0)), \tag{40}$$

then λ *is an eigenvalue of the operator* $(d\mathcal{H}(u))^*\mathcal{H}(u)$, *with the corresponding eigenvector* $u = \mathcal{C}^\dagger(x(0))$, *where* $\mathcal{C}(u)$ *is the controllability operator associated to (1).*

Remark 9. *In the linear case, this problem becomes:* $Mx(0) = \lambda W^{-1}x(0)$. *Since* $W > 0$, *we can write* $WMx(0) = \lambda x(0)$ *and if, moreover, the system is gradient, then, according to Theorem 6 we have:* $X^2x(0) = \lambda x(0)$, X *being the cross Gramian. This means that* λ *is the squared Hankel singular value* σ, *which for a symmetric system is an eigenvalue of* X.

Still, in order to make the connection between λ's and the Hankel singular values of (1) the Hankel norm is involved. The following results give the relation:

Theorem 5. *[6] Suppose that the linearization of (1) has non-zero distinct Hankel singular values. Then, there exists a neighborhood* $U \subset \mathbb{R}$ *of* 0 *and* $\rho_i(s) > 0$, $i = 1, ...n$ *such that:* $\min\{\rho_i(s), \rho_i(-s)\} \geq \max\{\rho_{i+1}(s), \rho_{i+1}(-s)\}$ *holds for all* $s \in U$, $i = 1, ..., n-1$. *Moreover, there exist* $\xi_i(s)$, *satisfying the following:*

$$L_c(\xi_i(s)) = s^2/2, L_o(\xi_i(s)) = \rho_i(s)s^2/2, \frac{\partial L_o}{\partial x}(\xi_i(s)) = \lambda_i(s)\frac{\partial L_c}{\partial x}(\xi_i(s)),$$

$$\lambda_i(s) = \rho_i^2(s) + \frac{s}{2}\frac{d\rho_i^2(s)}{ds}. \tag{41}$$

Even more, if $U = \mathbb{R}$, *the Hankel norm of the system is* $\sup_s \rho_1(s)$.

The $\rho_i(s)$ are a clear extension of the Hankel singular values for a nonlinear system and they can be obtained from the Hankel singular value functions of the nonlinear system, as defined in [9]. The following result establishes this link:

Theorem 6. *[11] If (1) is in input-normal, output-diagonal form, i.e.* $L_c(x) = x^T x/2$, $L_o(x) = x^T \text{diag}(\tau_1(x), ..., \tau_n(x))x/2$, *then*

$$\rho_i^2(x_j) = \tau_i(0, ..., x_j, ..., 0), \quad i \neq j,$$

$$\rho_j^2(x_j) = \tau_j(0, ..., x_j, ..., 0) + \frac{1}{2}\frac{\partial \tau_j}{\partial x_j}(0, ..., x_j, ..., 0)x_j. \tag{42}$$

Returning to our case, we state the following

Conjecture 1. *Let (1) be a nonlinear gradient system with the associated variational system* Σ'_p. *If* λ_i, $i = 1, ..., n$, *satisfy Theorem 5, then they are the squared eigenvalues of* $\mathcal{X}(x)$.

We aim at proving this conjecture by finding the meaning of the gradient extension in the context of the balancing procedure (following the reasoning in e.g. [6]), in order to be able to obtain an equivalent of equation (40) written in terms of the cross Gramian. In this way, the λ's in (40) associated to Σ_p, are related to the eigenvalues of the cross Gramian and thus, the Hankel singular value functions can be obtained from solving an eigenvalue problem for the cross Gramian.

Remark 10. *For linear systems this falls into place with the theory for symmetric systems, see Remark 9.*

Then using Theorem 5, the Hankel singular values of the original system are obtained, avoiding the balancing procedure.

6 Example

Given a double mass double spring system (see Figure 1), we compute the cross Gramian of the gradient system associated to it.

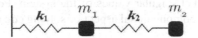

Fig. 1. Double mass double spring system

The system is given by:

$$\begin{cases} m_1\ddot{x}_1 + k_1(x_1) + k_2(x_1,x_2) = 0 \\ m_2\ddot{x}_2 - k_2(x_1,x_2) + u = 0 \end{cases}, \tag{43}$$

where x_1,x_2 are the displacements, $m_1,m_2 > 0$ are the masses and $k_1(x_1),k_2(x_1,x_2)$ are the corresponding elastic forces, with the initial conditions $x_1(0) = 1, x_2(0) = 0$. The potential energy of the system is given by $V(x)$, smooth, such that $\frac{\partial V(x)}{\partial x_1} = k_1(x_1)$, $\frac{\partial V(x)}{\partial x_2} = k_2(x_1,x_2)$. We choose $k_1(x_1) = -x_1^3$ and $k_2(x_1,x_2) = x_1 - x_2$ (elastic coefficients constant and equal to 1). We take $m_1 = m_2 = 1$. The Hamiltonian of the system is $H(x) = \frac{1}{2}\dot{x}^T M^{-1}(x)\dot{x} + V(x)$, with $M(x) = I_2$. Since $M > 0$, $G(x) = M^{-1}(x) = I_2$ can define a Riemannian metric on $\mathbb{R}^n$ (e.g. see [10, Chapter 6, Section 6.1]). The associated gradient system, of the form (8), is:

$$\dot{x} = \begin{bmatrix} -x_1^3 \\ x_1 - x_2 \end{bmatrix} + \begin{bmatrix} 2 \\ 1 \end{bmatrix} u, \quad y = x_1. \tag{44}$$

Denote $\mathscr{L}(x(t)) = [l_{ij}(x(t))]_{i,j=1,2} = [\mathbf{l}_{ij}(t)]_{i,j=1,2}$. Solving equation (37) associated to (44), for all $v \in T\mathscr{M}$, yields the following parameter-varying system to be solved:

$$\begin{cases} \dfrac{d\mathbf{l}_{11}(t)}{dt} = 3x_1^2(t)\mathbf{l}_{11}(t) - \mathbf{l}_{12}(t) - 1 \\ \dfrac{d\mathbf{l}_{12}(t)}{dt} = \left(\frac{3}{2}x_1^2(t) + 1\right)\mathbf{l}_{12}(t) + \mathbf{l}_{22}(t) \cdot \\ \dfrac{d\mathbf{l}_{22}(t)}{dt} = \mathbf{l}_{22}(t) \end{cases} \tag{45}$$

Solving system (44) for $u(t) = 0$, $t > 0, x_1(0) = 1$ we get $x_1(t) = \dfrac{1}{\sqrt{2t+1}}$. Substituting in (45) we obtain a time varying system. We solve it using approximation of 3rd order and obtain:

$$\mathscr{L}(t) = \mathscr{X}(t) = \begin{bmatrix} 3 + 10t + 9t^2 + 2t^3 & -t - \frac{3}{2}t^2 - \frac{1}{6}t^3 \\ -t - \frac{3}{2}t^2 - \frac{1}{6}t^3 & 1 + t + \frac{1}{2}t^3 + \frac{1}{6}t^3 \end{bmatrix} \tag{46}$$

The eigenvalue functions of the cross Gramian are given as:

$$\begin{aligned} \lambda_1(t) &= 3 + 10t + 10t^2 - 3t^3 + O(t^4) \\ \lambda_2(t) &= 1 + t + 0.9t^3 + O(t^4) \end{aligned} \tag{47}$$

7   Conclusions and Future Work

We present here the nonlinear counterpart of the cross Gramian for gradient systems. We do this in terms of the variational system. The reason is that in the next step we want to prove that the eigenvalues obtained from the cross Gramian are related in a direct manner to the Hankel singular values of the system. For later concern we will also take into account the computational aspect of solving equation (37).

Acknowledgements. The authors gratefully acknowledge the contribution of National Research Organization and reviewers' comments.

References

1. Aldaheri, R.W.: Model order reduction via real Schur-form decomposition. Int. J. Control 53(3), 709–716 (1991)
2. Antoulas, A.C.: Approximation of Large-scale Systems. SIAM, Philadelphia (2005)
3. Cortes, J., van der Schaft, A.J., Crouch, P.E.: Characterization of gradient control systems. SIAM J. Contr. & Opt. 44, 1192–1214 (2005)
4. Fernando, K.V., Nicholson, H.: On the structure of balanced and other principal representations of SISO systems. IEEE Trans. Automat. Contr. 28(2), 228–231 (1983)
5. Fujimoto, K., Scherpen, J.M.A.: Singular Value Analysis of Hankel Operators for General Nonlinear Systems. In: Proceedings of ECC (2003)
6. Fujimoto, K., Scherpen, J.M.A.: Nonlinear Input-Normal Realizations Based on the Differential Eigenstructure of Hankel Operators. IEEE Trans. on Aut. Control 50 (2005)
7. Gray, W.S., Mesko, J.P.: Observability functions for linear and nonlinear systems. System and Control Letters 33, 99–113 (1999)
8. Marsden, J.E., Ratiu, T., Abraham, R.: Manifolds, Tensor Analysis and Applications. Springer, New York (2001)
9. Scherpen, J.M.A.: Balancing for nonlinear systems. System & Control Letters 21, 143–153 (1993)
10. Scherpen, J.M.A.: Balancing for Nonlinear Systems. PhD Thesis, Univ. of Twente, Netherlands (1994)
11. Scherpen, J.M.A., Fujimoto, K., Gray, W.S.: Hamiltonian extensions, Hilbert adjoints and singular value functions for nonlinear systems. In: Proc. of the 39th IEEE CDC 2000, vol. 5, pp. 5102–5107 (2000)
12. Scherpen, J.M.A., Gray, W.S.: Minimality and local state decompositions of a nonlinear state space realization using energy functions. IEEE Trans. on Aut. Contr. 45(11), 2079–2086 (2000)
13. Scherpen, J.M.A.: Duality and singular value functions of the nonlinear normalized right and left coprime factorizations. In: Proceedings of the 44th IEEE CDC-ECC, pp. 2254–2259 (2005)
14. Sorensen, D.C., Antoulas, A.C.: The Sylvester equation and approximate balanced truncation. Linear Algebra and Its Applications 351-352, 671–700 (2002)
15. van der Schaft, A.J.: Linearization of Hamiltonian and gradient systems. IMA J. Math. Contr. & Inf. 1, 185–198 (1984)
16. Zhou, K., Doyle, J.C., Glover, K.: Robust Optimal Control. Prentice-Hall, Englewood Cliffs (1996)

Estimation of Regularization Parameters in Elliptic Optimal Control Problems by POD Model Reduction

Martin Kahlbacher and Stefan Volkwein

Institute for Mathematics and Scientific Computing,
University of Graz, Heinrichstrasse 36, 8010 Graz, Austria
martin.kahlbacher@uni-graz.at, stefan.volkwein@uni-graz.at

Abstract. In this article parameter estimation problems for a nonlinear elliptic problem are considered. Using Tikhonov regularization techniques the identification problems are formulated in terms of optimal control problems which are solved numerically by an augmented Lagrangian method combined with a globalized sequential quadratic programming algorithm. For the discretization of the partial differential equations a Galerkin scheme based on proper orthogonal decomposition (POD) is utilized, which leads to a fast optimization solver. This method is utilized in a bilevel optimization problem to determine the parameters for the Tikhonov regularization. Numerical examples illustrate the efficiency of the proposed approach.

1 Introduction

Parameter estimation problems for partial differential equations are very important in application areas. Using Tikhonov regularization techniques (see, e.g., [20]) these problems can often be expressed in terms of constrained optimal control problems so that numerical optimization can be applied to solve the parameter identification problems numerically. Here, we apply an augmented Lagrangian method (see, e.g., [2,3]) combined with a globalized sequential quadratic programming (SQP) algorithm as described in [6]. In this article we continue our successful development of solution methods for parameter estimation problems for nonlinear elliptic partial differential equations (PDEs); see [12,13,22]. The goal is to derive efficient, robust and fast solvers where the PDEs are discretized by a Galerkin scheme based on proper orthogonal decomposition. POD is a powerful method to derive low-dimensional models for nonlinear systems. It is based on projecting the system onto subspaces consisting of basis elements that contain characteristics of the expected solution. This is in contrast to, e.g., finite element techniques, where the elements of the subspaces are uncorrelated to the physical properties of the system that they approximate. It is successfully used in different fields including signal analysis and pattern recognition (see, e.g., [5]), fluid dynamics and coherent structures (see, e.g., [7,15]) and more recently in control theory (see, e.g., [10]). The relationship between POD and balancing is considered in [9,19,23]. In contrast to POD approximations, reduced-basis element methods for parameter dependent elliptic systems are investigated in [1,8,16,18], for instance.

A. Korytowski et al. (Eds.): System Modeling and Optimization, IFIP AICT 312, pp. 307–318, 2009.

In the present paper we determine numerically parameters in a Tikhonov regularization. This regularization technique is used to formulate the identification problem in terms of an optimal control problem. For any admissible parameter $p \in P_{ad} \subset \mathbb{R}^N$ let $u(p)$ denote the solution to the underlying semilinear elliptic PDE. The identification problem is to find a parameter $p^* \in P_{ad}$ so that for a given (measurement) data u_d (e.g., on the boundary or on a part of the domain) the quantity $\|u^* - u_d\|$ is minimal, where $u^* = u^*(p^*)$. For a precise introduction we refer to Sect. 2. For the Tikhonov regularization we take a $\kappa > 0$ and solve the optimal control problem

$$\min_{(p,u)} \frac{1}{2}\|u - u_d\|^2 + \frac{\kappa}{2}\|p\|^2 \quad \text{subject to (s.t.)} \quad u \text{ solves PDE for } p \in P_{ad}. \quad (1)$$

By (u^κ, p^κ) we denote a (local) optimal solution to (1). Then we introduce the following bilevel optimization problem:

$$\min_{\kappa} \|u^\kappa - u_d\|^2 \quad \text{s.t.} \quad (p^\kappa, u^\kappa) \text{ solves (1) for } \kappa \geq \kappa_a \quad (2)$$

with $\kappa_a > 0$. To solve (2) numerically we apply the MATLAB routine fmincon, where the solution pair (p^κ, u^κ) to (1) is computed by a fast optimization solver based on a POD Galerkin projection.

Note that the inner optimization problem (1) is non-convex, thus there might exist more than one local minimum. By varying the Tikhonov parameter κ we search an optimal κ^* so that (1) for $\kappa = \kappa^*$ yields a solution (p^*, u^*) for which the error in a given norm between the state u^* and the *noisy* measuring data u_d is minimal.

A similar approach compared to the method of solving the bilevel problem above is to fix κ, but start the inner optimization loop with varying starting values (p^0, u^0). In this work we only deal with the previous case (bilevel problem with varying κ), though. In both methods we exploit the fact that – using the POD approximation – one optimization loop takes very little time. Thus, it is no matter of temporal cost to solve an optimization problem like (1) many times successively.

The paper is organized in the following manner. In Sect. 2 we introduce the underlying parameter estimation problem. The POD method is briefly reviewed in Sect. 3. The POD basis is used to derive a POD Galerkin projection for the optimal control problem. Finally, numerical examples are carried out in the last section. In particular, we apply the reduced-basis method to obtain appropriate snapshots for the POD basis computation in one of the numerical tests.

2   The Identification Problem

Let $\Omega \subset \mathbb{R}^d$, $d = 2, 3$, be an open, bounded and connected set with Lipschitz-continuous boundary $\Gamma = \partial\Omega$. Let $q > d/2 + 1$ and $r > d + 1$. For given $f \in L^q(\Omega)$, $g \in L^r(\Gamma)$, $c, q \in L^\infty(\Omega)$ with $c \geq c_a > 0$ in Ω almost everywhere (a.e.) and $q \geq q_a \geq 0$ in Ω a.e., $\sigma \geq 0$ we consider the nonlinear problem

$$-c\Delta u + qu + e^u = f \quad \text{in } \Omega,$$
$$c\frac{\partial u}{\partial n} + \sigma u = g \quad \text{in } \Gamma. \quad (3)$$

There exists a unique weak solution $u \in H_b^1(\Omega) = H^1(\Omega) \cap L^\infty(\Omega)$ satisfying

$$\int_\Omega c\nabla u \cdot \nabla \varphi + (qu + e^u)\varphi \, dx + \int_\Gamma \sigma u\varphi \, ds = \int_\Omega f\varphi \, dx + \int_\Gamma g\varphi \, ds \qquad (4)$$

for all $\varphi \in H^1(\Omega)$, where the Banach space $H_b^1(\Omega)$ is endowed with the common norm $\|u\|_{H_b^1(\Omega)} = \|u\|_{H^1(\Omega)} + \|u\|_{L^\infty(\Omega)}$ for $u \in H_b^1(\Omega)$. Moreover, this solution belongs to $C(\overline{\Omega})$. For a proof we refer the reader to [4], for instance.

2.1 Estimation of Diffusion and Potential Parameter

The goal of the first estimation problem is to identify the parameter pair

$$p = (c, q) \in P_{ad}^1 = \left\{ \tilde{p} = (\tilde{c}, \tilde{q}) \in \mathbb{R}^2 \,\middle|\, \tilde{c} \geq c_a \text{ and } \tilde{q} \geq q_a \right\}$$

from measurements for the weak solution $u \in H_b^1(\Omega)$ to (3) on the boundary Γ and on a subset Ω_m of the domain Ω. Let α_1, α_2 denote nonnegative weights, κ_c, κ_q be positive regularization parameters and $c_d, q_d \in \mathbb{R}$ stand for nominal parameters. Introducing the quadratic cost functional

$$J_1(p, u) = \frac{\alpha_1}{2} \int_\Gamma |u - u_\Gamma|^2 \, ds + \frac{\alpha_2}{2} \int_{\Omega_m} |u - u_\Omega|^2 \, dx + \frac{\kappa_c}{2} |c - c_d|^2 + \frac{\kappa_q}{2} |q - q_d|^2$$

for $p = (c, q) \in \mathbb{R}^2$ and $u \in H^1(\Omega)$ we express the identification problem as the following constrained optimal control problem

$$\min J_1(p, u) \quad \text{s.t.} \quad p = (c, q) \in P_{ad}^1 \text{ and } u \in H_b^1(\Omega) \text{ satisfy (4)}. \qquad (5)$$

Throughout the paper we suppose that (5) admits at least one local solution $x^* = (p^*, u^*)$ with $p^* = (c^*, q^*) \in P_{ad}^1$.

2.2 Estimation of Varying Diffusion Parameter

In the second example we suppose that Ω is split into two measurable disjunct subsets Ω_i, $i = 1, 2$, and that c is constant on Ω_i, i.e., $c \equiv c_i$ on Ω_i for $i = 1, 2$. Hence, we introduce the set of admissible parameters by

$$P_{ad}^2 = \left\{ \tilde{p} = (\tilde{c}_1, \tilde{c}_2) \in \mathbb{R}^2 \,\middle|\, \tilde{c}_i \geq c_a \text{ for } i = 1, 2 \right\}. \qquad (6)$$

The goal is to identify c from given measurements for the weak solution $u \in H_b^1(\Omega)$ to (3) on the boundary Γ. Let α_1 denote a nonnegative weight, κ_1, κ_2 be positive regularization parameters and $c_{1,d}, c_{2,d} \in \mathbb{R}$ stand for nominal potential parameters. Introducing the cost functional

$$J_2(p, u) = \frac{\alpha_1}{2} \int_\Gamma |u - u_\Gamma|^2 \, ds + \frac{\kappa_1}{2} |c_1 - c_{1,d}|^2 + \frac{\kappa_2}{2} |c - c_{2,d}|^2 \qquad (7)$$

for $p = (c_1, c_2) \in \mathbb{R}^2$ and $u \in H^1(\Omega)$ we express the identification problem as the following constrained optimal control problem

$$\min J_2(p, u) \quad \text{s.t.} \quad p = (c_1, c_2) \in P_{ad}^2 \text{ and } u \in H_b^1(\Omega) \text{ satisfies (4)}. \qquad (8)$$

We assume that (8) admits at least one local solution $x^* = (p^*, u^*)$ with $p^* = (c_1^*, c_2^*)$.

3 The POD Method

In this section we introduce briefly the POD method. Suppose that for points $p_j \in P_{ad}^i$, $j = 1, \ldots, n$ and $i = 1, 2$, we know (at least approximately) the solution u_j to (3), e.g., by utilizing a finite element or finite difference discretization. We set

$$\mathcal{V} = \text{span} \{ u_1, \ldots, u_n \} \subset H_b^1(\Omega) \subset H^1(\Omega) \tag{9}$$

with $d = \dim \mathcal{V} \leq n$. Then the *POD basis of rank* $\ell \leq d$ is given by the solution to

$$\min_{\psi_1, \ldots, \psi_\ell} \sum_{j=1}^n \beta_j \left\| u_j - \sum_{i=1}^\ell \langle u_j, \psi_i \rangle_{H^1(\Omega)} \psi_i \right\|_{H^1(\Omega)}^2 \quad \text{s.t.} \quad \langle \psi_i, \psi_j \rangle_{H^1(\Omega)} = \delta_{ij} \tag{10}$$

with nonnegative weights $\{\beta_j\}_{j=1}^n$. For the choice of the β_j's we refer to [11,14].

The solution to (10) is characterized by the eigenvalue problem

$$\mathcal{R} \psi_i = \lambda_i \psi_i, \quad 1 \leq i \leq \ell, \tag{11}$$

where $\lambda_1 \geq \lambda_2 \geq \ldots \geq \lambda_\ell \geq \ldots \geq \lambda_d > 0$ denote the eigenvalues of the linear, bounded, self-adjoint, and nonnegative operator $\mathcal{R} : H^1(\Omega) \to \mathcal{V}$ defined by

$$\mathcal{R}z = \sum_{j=1}^n \beta_j \langle u_j, z \rangle_{H^1(\Omega)} u_j \quad \text{for } z \in H^1(\Omega); \tag{12}$$

see [7,14,21]. Suppose that we have determined a POD basis $\{\psi_i\}_{i=1}^\ell$. We set

$$V^\ell = \text{span} \{ \psi_1, \ldots, \psi_\ell \} \subset \mathcal{V} \subset H^1(\Omega). \tag{13}$$

Then the following relation holds

$$\sum_{j=1}^n \beta_j \left\| u_j - \sum_{i=1}^\ell \langle u_j, \psi_i \rangle_{H^1(\Omega)} \psi_i \right\|_{H^1(\Omega)}^2 = \sum_{i=\ell+1}^d \lambda_i, \tag{14}$$

i.e., a rapid decay of the eigenvalues λ_i indicates that the vectors $u_1, \ldots, u_n$ can be well approximated by taking only a few ansatz functions $\{\psi_i\}_{i=1}^\ell$ with $\ell \ll d$.

Now we introduce the *POD Galerkin scheme for* (4) as follows: the function $u^\ell = \sum_{i=1}^\ell u_i^\ell \psi_i \in V^\ell$ solves

$$\int_\Omega c \nabla u^\ell \cdot \nabla \psi \, dx + \int_\Omega \left(q u^\ell + e^{u^\ell} \right) \psi \, dx + \int_\Gamma \sigma u^\ell \psi \, ds$$
$$= \int_\Omega f \psi \, dx + \int_\Gamma g \psi \, ds \quad \text{for all } \psi \in V^\ell. \tag{15}$$

Problem (15) is a nonlinear system for the ℓ unknown modal coefficients $u_1^\ell, \ldots, u_\ell^\ell \in \mathbb{R}$. If

$$\mathcal{E}(\ell) = \frac{\sum_{i=1}^\ell \lambda_i}{\sum_{i=1}^d \lambda_i} \approx 1 \quad \text{for } \ell \ll d, \tag{16}$$

holds, (15) is called a *low-dimensional model* for (4).

4 Numerical Experiments

In this section we present numerical examples for the identification problem. The numerical tests are executed on a standard 3.0 GHz desktop PC. We are using the MATLAB 7.1 package together with FEMLAB 3.1.

Run 1 (*Problem* (5)). Suppose that the domain Ω is given by

$$\Omega = \left\{ \mathbf{x} = (x_1, x_2) \mid \frac{x_1^2}{1.2^2} + x_2^2 < 1 \right\} \subset \mathbb{R}^2; \tag{17}$$

see Fig. 1. In (3) we choose $f = 5$, $\sigma = 3/2$, and $g = -1$. For $c_{ex} = 1.2$ and $q_{ex} = 11$ we calculate a finite element (FE) solution $u_{ex}^h = u^h(c_{ex}, q_{ex})$ with 1275 degrees of freedom. The parameter $p_{ex} = (c_{ex}, q_{ex})$ is our reference parameter.

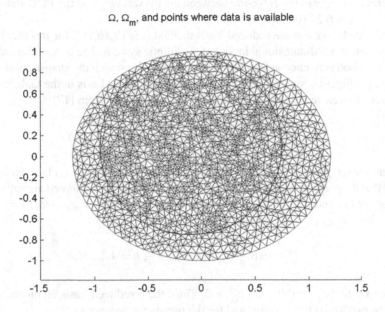

Fig. 1. Run 1: Domain Ω and the interior points for which we have measurements

Basis computation. We distinguish three different techniques in deriving a basis for the Galerkin projection.

1) First we compute 20 snapshots by varying the parameters c and q simultaneously. We define the equidistant grid

$$(c, q) \in \{0.2, 0.8, 1.4, 2\} \times \{1, 8, 15, 22, 29\} \tag{18}$$

and calculate a POD model with $\ell = 6$ basis functions. In (10) we choose trapezoidal weights. Thus, we consider

$$\min \sum_{i=1}^{4} \sum_{j=1}^{5} \beta_i \tilde{\beta}_j \left\| u^h(c_i, q_j) - \sum_{k=1}^{\ell} \langle u^h(c_i, q_j), \psi_k \rangle_{H^1(\Omega)} \psi_k \right\|_{H^1(\Omega)}^2 \tag{19}$$

where

$$\beta_1 = \frac{c_2 - c_1}{2}, \quad \beta_i = \frac{c_{i+1} - c_{i-1}}{2} \text{ for } i = 2, 3, \quad \beta_4 = \frac{c_4 - c_3}{2} \tag{20}$$

and

$$\tilde{\beta}_1 = \frac{q_2 - q_1}{2}, \quad \tilde{\beta}_j = \frac{q_{j+1} - q_{j-1}}{2} \text{ for } j = 2, 3, 4, \quad \tilde{\beta}_5 = \frac{q_5 - q_4}{2}. \tag{21}$$

The relative error in the H^1-norm between the FE state u_{ex}^h and the POD state $u_{ex}^\ell = u^\ell(c_{ex}, q_{ex})$ is $6.2 \cdot 10^{-4}$.

2) Alternatively, we use the reduced-basis method (see [8,16,18], for instance) in order to obtain a 6-dimensional model of the elliptic system. The idea of the reduced-basis method is to choose the parameter instances for which the snapshots are computed intelligently and to use these snapshots directly as basis in the Galerkin projection. Therefore we apply the simplified formula taken from [17]:

$$q_k^{rb} = \exp(-\ln \gamma + k \cdot \delta^q) - \frac{1}{\gamma} \quad \text{for } k = 1, \dots, N, \tag{22}$$

where we set $\gamma = 0.02$, $q_{max} = 29$, $N = 3$, and $\delta^q = \ln(\gamma \cdot q_{max} + 1)/N$. Hence, we find that the parameters for which the snapshots should be computed are: $q_1^{rb} = 4.3$, $q_2^{rb} = 12.68$, and $q_3^{rb} = 29$. Analogously we set $\gamma = 0.02$, $c_{max} = 2$, $M = 2$, and $\delta^c = \ln(\gamma \cdot c_{max} + 1)/M$ and choose

$$c_k^{rb} = \exp(-\ln \gamma + k \cdot \delta^c) - \frac{1}{\gamma}, \quad k = 1, \dots, M, \tag{23}$$

hence we find $c_1^{rb} = 0.91$ and $c_2^{rb} = 2$. Thus, the 6 reduced-basis elements are the solutions $u^h(c, q)$ to (3) computed for the parameter instances

$$(c, q) \in \{0.91, 2\} \times \{4.3, 12.68, 29\} \tag{24}$$

The relative error in the H^1-norm between the FE state u_{ex}^h and the reduced order model $u_{ex}^{rb} = u^{rb}(c_{ex}, q_{ex})$ is $1.7 \cdot 10^{-4}$.

3) The best approximation of the FE state can be obtained by combining both methods (POD and reduced-basis). Therefore we compute 20 snapshots at the parameter instances calculated by the reduced-basis ansatz (i.e., we set $N = 5$ and $M = 4$ and use the formula from above again). We find that the snapshots should be computed at the 20 snapshot pairings

$$(c, q) \in \{0.43, 0.91, 1.43, 2\} \times \{2.23, 5.57, 10.53, 17.95, 29\}. \tag{25}$$

Then we construct a 6-dimensional POD basis. The relative error in the H^1-norm between the FE state u_{ex}^h and this reduced order model $u_{ex}^{\ell,rb} = u^{\ell,rb}(c_{ex}, q_{ex})$ is now about 10^{-4}.

We proceed by using this POD basis for the reduced-order modeling. The computation of the POD solution takes 437 seconds (411 seconds thereof are for the computation of the 20 FE snapshots whereas one solve of the nonlinear POD model only takes 0.06 seconds). From Table 1 it can be observed that the relative error between the FE state and the POD state decreases as the number of POD basis functions increases.

Table 1. Run 1: Relative errors between the FE state and the POD state for increasing number of POD basis functions

	$\ell = 4$	$\ell = 5$	$\ell = 6$	$\ell = 7$
$\dfrac{\|u^h - u^{\ell,rb}\|_{H^1(\Omega)}}{\|u^h\|_{H^1(\Omega)}}$	1.2e-3	5.3e-4	1.0e-4	1.1e-5

Identification problem. Now turn to the identification problem. Let $c_a = q_a = 0.01$ to ensure that both parameters are positive. Moreover, we choose $c_d = q_d = 0$, i.e., no a-priori knowledge on the parameters is available. We add a random noise of 8% to the FE state u_{ex}^h. For the weights in the cost functional we take $\alpha_1 = \alpha_2 = 1000$, and we choose

$$\Omega_m = \left\{ \mathbf{x} = (x_1, x_2) \in \Omega \mid (x_1 + 0.1)^2 + (x_2 - 0.1)^2 < 0.85^2 \right\} \tag{26}$$

for the partial measurement. Furthermore, we suppose that measurements are not given on the whole subdomain Ω_m, but only on 381 points (of totally 762 grid points) in Ω_m. The points for which we have measurements (besides the points on the boundary) are indicated by the circles in Fig. 1. Now we consider the bilevel optimization problem (compare (2))

$$\min_{\kappa = (\kappa_c, \kappa_q)} \int_{\Gamma} |u^{\kappa} - u_{\Gamma}|^2 \, ds \quad \text{s.t.} \quad (c^{\kappa}, q^{\kappa}, u^{\kappa}) \text{ solves (5) for } \kappa_c, \kappa_q \geq 10^{-16} \tag{27}$$

By using the MATLAB function fmincon we determine – after 56.2 seconds – the optimal weighting parameters $\kappa_c^* = 0.1691$ and $\kappa_q^* = 10^{-16}$. For these optimal weights we solve the reduced order model by means of an augmented Lagrange-SQP algorithm and use the POD Galerkin projection. Altogether 50 SQP iterations are required and we find numerically an optimal solution (c^*, q^*, u^*) to (27); in particular, $c^* = 1.1972$ and $q^* = 10.9827$. Thus,

$$\frac{\|p_{ex} - p^*\|_2}{\|p_{ex}\|_2} \approx 0.16\% \quad \text{with } p_{ex} = (c_{ex}, q_{ex}) \text{ and } p^* = (c^*, q^*). \tag{28}$$

The relative errors in the state variable to the exact (unnoisy) data and to the noisy data are stated for 3 different norms in Table 2. The CPU time for the optimization is small compared to the POD computation time. The POD optimization algorithm for (5)

Table 2. Run 1: Relative errors of the suboptimal state u^* compared to the exact data u^h_{ex} and to the noisy data u_Γ for the optimal $(\kappa^*_c, \kappa^*_q) = (0.1691, 10^{-16})$ and for $(\kappa^{(j)}_c, \kappa^{(j)}_q)$, $j = 1, 2, 3$

	$\dfrac{\|u^*-u\|_{L^2(\Gamma)}}{\|u\|_{L^2(\Gamma)}}$	$\dfrac{\|u^{(1)}-u\|_{L^2(\Gamma)}}{\|u\|_{L^2(\Gamma)}}$	$\dfrac{\|u^{(2)}-u\|_{L^2(\Gamma)}}{\|u\|_{L^2(\Gamma)}}$	$\dfrac{\|u^{(3)}-u\|_{L^2(\Gamma)}}{\|u\|_{L^2(\Gamma)}}$
$u = u^h_{ex}$	0.004592	0.091625	0.013806	0.018451
$u = u_\Gamma$	0.037749	0.095162	0.042008	0.044252

only takes 1.7 seconds. For comparison, when we use the FE discretized model in the augmented SQP-Lagrange algorithm, it takes about 290 seconds to obtain a solution. Note that for the choice $\kappa^{(1)}_c = 5 \cdot \kappa^*_c$ and $\kappa^{(1)}_q = \kappa^*_q$, we find the solution $c^{(1)} = 1.1746$ and $q^{(1)} = 10.9273$, which gives

$$\frac{\|p_{ex} - p^{(1)}\|_2}{\|p_{ex}\|_2} \approx 0.7\% \quad \text{with } p^{(1)} = (c^{(1)}, q^{(1)}) \tag{29}$$

and the relative errors are as stated in Table 2. The same can be done with $\kappa^{(2)}_c = 0.2 \cdot \kappa^*_c$ and $\kappa^{(2)}_q = \kappa^*_q$. We find $c^{(2)} = 1.2021$ and $q^{(2)} = 10.9947$. Thus,

$$\frac{\|p_{ex} - p^{(2)}\|_2}{\|p_{ex}\|_2} \approx 0.05\% \quad \text{with } p^{(2)} = (c^{(2)}, q^{(2)}). \tag{30}$$

Finally, we choose $\kappa^{(3)}_c = \kappa^{(3)}_q = 10^{-16}$. The resulting parameters are $c^{(3)} = 1.2034$ and $q^{(3)} = 10.9978$, which gives

$$\frac{\|p_{ex} - p^{(3)}\|_2}{\|p_{ex}\|_2} \approx 0.04\% \quad \text{with } p^{(3)} = (c^{(3)}, q^{(3)}) \tag{31}$$

We observe that the relative error in the coefficients is smaller for both $p^{(2)}$ and $p^{(3)}$ compared to p^*. However, we observe from Table 2 that the relative errors of the PDE solution u^* on the boundary Γ are the smallest ones. Note that in (27) the term $\|u - u_\Gamma\|^2$ is minimized. For the absolute errors we refer to Table 3. Also the absolute errors are for κ^* the smallest ones, in particular also the error of $u^* - u^h_{ex}$.

Table 3. Run 1: Absolute errors of the suboptimal state u^* compared to the exact data u^h_{ex} and to the noisy data u_Γ for the optimal $(\kappa^*_c, \kappa^*_q) = (0.1691, 10^{-16})$ and for $(\kappa^{(j)}_c, \kappa^{(j)}_q)$, $j = 1, 2, 3$

	$\|u^* - u\|_{L^2(\Gamma)}$	$\|u^{(1)} - u\|_{L^2(\Gamma)}$	$\|u^{(2)} - u\|_{L^2(\Gamma)}$	$\|u^{(3)} - u\|_{L^2(\Gamma)}$
$u = u^h_{ex}$	0.000166	0.003320	0.000500	0.000667
$u = u_\Gamma$	0.001363	0.003437	0.001517	0.001598

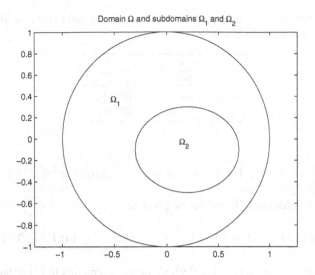

Domain Ω and subdomains Ω_1 and Ω_2

Fig. 2. Run 2: Domain Ω and subdomains Ω_1, Ω_2

Run 2 (*Problem* (8)). Now let $\Omega = \{\mathbf{x} = (x_1, x_2) \,|\, x_1^2 + x_2^2 < 1\}$ be the open unit circle in $\mathbb{R}^2$ and the subdomains Ω_1, Ω_2 be given as

$$\Omega_1 = \Omega \setminus \Omega_2, \quad \Omega_2 = \left\{ \mathbf{x} = (x_1, x_2) \in \Omega \,\Big|\, \frac{(x_1 - 0.2)^2}{a^2} + \frac{(x_2 + 0.1)^2}{b^2} < 1 \right\} \quad (32)$$

with $a = 0.5$ and $b = 0.4$; see Fig. 2. In (3) we choose $q \equiv 20$, $f \equiv 4$, $\sigma = 2$, and $g(\mathbf{x}) = 10 + \cos(\pi x_1 / 2) \cdot \cos(\pi x_2 / 2)$. For $p_{ex} = (c_{1,ex}, c_{2,ex}) = (0.8, 1.3)$ we compute the FE solution with 1070 degrees of freedom. To derive a POD basis we choose the diffusion values $p_j = (\eta_k, \eta_l) \in \mathbb{R}_+^2$, $1 \leq j \leq n$, with

$$j = 5(k-1) + l \text{ for } 1 \leq k, l \leq 5, \quad \eta_k = 0.5 + \frac{k-1}{4} \text{ for } k = 1, \dots, 5 \quad (33)$$

and compute the corresponding FE solutions $u_j^h = u^h(p_j) \in H^1(\Omega)$ to (3), i.e., we have $n = 25$ snapshots $\{u_j^h\}_{j=1}^n$. The computation of the snapshots requires 307 seconds. Next we compute the POD basis of rank $\ell = 7$ as described in Sect. 3 and construct the POD model $u^\ell(\bar{c})$ which has a relative error to the FE state u_{ex}^h of $1.38 \cdot 10^{-4}$. Now (2) has the form

$$\min_{\kappa = (\kappa_1, \kappa_2)} \int_\Gamma |u^\kappa - u_\Gamma|^2 \, ds \quad \text{s.t.} \quad (c_1^\kappa, c_2^\kappa, u^\kappa) \text{ solves (8) for } \kappa_1, \kappa_2 \geq 10^{-16}. \quad (34)$$

In the optimization algorithm for noisy data (3%) we choose $\alpha_1 = 100$ and find the optimal weight $\kappa^* = (\kappa_1^*, \kappa_2^*) = (0.7534, 0.0023)$. The corresponding optimal coefficient is $p^* = (0.7873, 1.3247)$. Moreover, the relative and absolute errors in the state variable are stated in Table 4. If we take $\kappa^{(1)} = (\kappa_1^{(1)}, \kappa_2^{(1)}) = (10^{-16}, 10^{-16})$ instead of κ^*,

Table 4. Run 2: Relative errors of the suboptimal state u^* compared to the exact data u_{ex}^h and to the noisy data u_Γ for $\kappa_1 = 0.7534$ and $\kappa_2 = 0.0023$

	$\dfrac{\|u^*-u\|_{L^2(\Gamma)}}{\|u\|_{L^2(\Gamma)}}$	$\|u^* - u\|_{L^2(\Gamma)}$
$u = u_{ex}^h$	0.004276	0.016811
$u = u_\Gamma$	0.012713	0.050184

the result is $p^{(1)} = (0.7902, 1.4185)$ solves (8). Then, $\|p_{ex} - p^{(1)}\|_2/\|p_{ex}\|_2 \approx 8\%$, but $\|p_{ex} - p^*\|_2/\|p_{ex}\|_2 \approx 2\%$.

Now, let the subdomains Ω_1 and Ω_2 be given as

$$\Omega_1 = \Omega \setminus \Omega_2, \quad \Omega_2 = \left\{ \mathbf{x} = (x_1, x_2) \in \Omega \,|\, x_1^2 + (x_2 + 0.1)^2 < 0.75^2 \right\}. \quad (35)$$

We choose $p_{ex} = (c_{1,ex}, c_{2,ex}) = (1.2, 0.9)$, all other parameters in (3) remain the same. Moreover, the measuring data u_d is much more noisy (15%) than before. In this case we observe that – due to the bigger noise – both components of the ideal κ^* are far away from zero (see Fig. 3). The cost funtional in (2) for $\kappa^{(1)} = (10^{-16}, 10^{-16})$ has a value of 0.2757, while for $\kappa^* = (0.3465, 0.6675)$ the cost is only 0.2745. However, the relative error in the parameter $p = (p_1, p_2)$ is much smaller for the solution using $\kappa^{(1)}$ rather than κ^*. We observe $\|p_{ex} - p^{(1)}\|_2/\|p_{ex}\|_2 \approx 0.8\%$, but $\|p_{ex} - p^*\|_2/\|p_{ex}\|_2 \approx 14\%$.

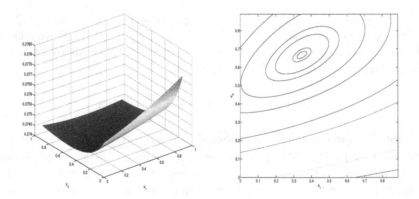

Fig. 3. Run 2: Cost functional in (2) for a grid of different $\kappa = (\kappa_1, \kappa_2)$ (left plot) and contour plot of the cost functional. The absolute minimum is approximately at $\kappa^* = (0.35, 0.67)$ (right plot).

Acknowledgements. The authors gratefully acknowledge partial support by the Austrian Science Fund FWF under grant no. P19588-N18 and by the SFB Research Center "Mathematical Optimization in Biomedical Sciences" (SFB F32).

References

1. Barrault, M., Maday, Y., Nguyen, N.C., Patera, A.T.: An empirical interpolation method: application to efficient reduced-basis discretization of partial differential equations. Comptes Rendus de'l Académie des Sciences Paris I(339), 667–672 (2004)
2. Bertsekas, D.P.: Constrained Optimization and Lagrange Multipliers. Academic Press, New York (1982)
3. Bertsekas, D.P.: Nonlinear Programming, 2nd edn. Athena Scientific, Belmont (1999)
4. Casas, E., Raymond, J.-P., Zidani, H.: Optimal control problem governed by semilinear elliptic equations with integral control constraints and pointwise state constraint. In: Desch, W., et al. (eds.) Control and estimation of distributed parameter systems. International conference in Vorau, Austria, July 14-20. Birkhauser, Basel (1996); ISNM, Int. Ser. Numer. Math. 126, 89–102 (1998)
5. Fukuda, K.: Introduction to Statistical Recognition. Academic Press, New York (1990)
6. Hintermüller, M.: A primal-dual active set algorithm for bilaterally control constrainted optimal control problems. Quarterly of Applied Mathematics 61, 131–160 (2003)
7. Holmes, P., Lumley, J.L., Berkooz, G.: Turbulence, Coherent Structures, Dynamical Systems and Symmetry. In: Cambridge Monographs on Mechanics. Cambridge University Press, Cambridge (1996)
8. Ito, K., Ravindran, S.S.: Reduced basis method for unsteady viscous flows. Int. J. of Comp. Fluid Dynamics 15, 97–113 (2001)
9. Lall, S., Marsden, J.E., Glavaski, S.: Empirical model reduction of controlled nonlinear systems. In: Proceedings of the IFAC Congress, vol. F, pp. 473–478 (1999)
10. Ly, H.V., Tran, H.T.: Modelling and control of physical processes using proper orthogonal decomposition. Mathematical and Computer Modeling 33, 223–236 (2001)
11. Kahlbacher, M., Volkwein, S.: Galerkin proper orthogonal decomposition methods for parameter dependent elliptic systems. Discussiones Mathematicae: Differential Inclusions, Control and Optimization 27, 95–117 (2007)
12. Kahlbacher, M., Volkwein, S.: Model reduction by proper orthogonal decomposition for estimation of scalar parameters in elliptic PDEs. In: Wesseling, P., Onate, E., Periaux, J. (eds.) Proceedings of ECCOMAS CFD, Egmont aan Zee (2006)
13. Kahlbacher, M., Volkwein, S.: Estimation of diffusion coefficients in a scalar Ginzburg-Landau equation by using model reduction. Submitted (2007), http://www.uni-graz.at/imawww/reports/archive-2007/IMA05-07.pdf
14. Kunisch, K., Volkwein, S.: Galerkin proper orthogonal decomposition methods for a general equation in fluid dynamics. SIAM J. Numer. Anal. 40, 492–515 (2002)
15. Sirovich, L.: Turbulence and the dynamics of coherent structures, parts I-III. Quarterly of Applied Mathematics XLV, 561–590 (1987)
16. Machiels, L., Maday, Y., Patera, A.T.: Output bounds for reduced-order approximations of elliptic partial differential equations. Computer Methods in Applied Mechanics and Engineering 190, 3413–3426 (2001)
17. Maday, Y., Patera, A.T., Turinici, G.: Global a priori convergence theory for reduced-basis approximations of single-parameter symmetric coercive elliptic partial differential equations. Comptes Rendus de'l Académie des Sciences Paris I(335), 289–294 (2002)
18. Maday, Y., Rønquist, E.M.: A reduced-basis element method. Journal of Scientific Computing 17, 1–4 (2002)
19. Rowley, C.W.: Model reduction for fluids, using balanced proper orthogonal decomposition. International Journal of Bifurcation and Chaos 15, 997–1013 (2005)

20. Vogel, C.R.: Computational Methods for Inverse Problems, Philadlphia. SIAM Frontiers in Applied Mathematics (2002)
21. Volkwein, S.: Model Reduction using Proper Orthogonal Decomposition. Lecture Notes, Institute of Mathematics and Scientific Computing, University of Graz, http://www.uni-graz.at/imawww/volkwein/POD.pdf
22. Volkwein, S., Hepberger, A.: Impedance Identification by POD Model Reduction Techniques (2008) (submitted), http://www.uni-graz.at/imawww/reports/archive-2008/IMA01-08.pdf
23. Willcox, K., Peraire, J.: Balanced model reduction via the proper orthogonal decomposition. American Institute of Aeronautics and Astronautics (AIAA) 40, 2323–2330 (2002)

Identification of Material Models
of Nanocoatings System Using
the Metamodeling Approach

Magdalena Kopernik, Andrzej Stanisławczyk, Jan Kusiak, and Maciej Pietrzyk

AGH University of Science and Technology, Mickiewicza 30, 30-059 Kraków, Poland
magdalenakopernik@interia.pl, astan@agh.edu.pl,
kusiak@agh.edu.pl, pietrzyk@agh.edu.pl

Abstract. Hard systems of nanocoatings deposited using PVD (physical vapor deposition) are used in the artificial heart prosthesis. Correct determination of nanomaterial parameters is crucial for accuracy of simulation. The objective of this work is identification of material parameters of nanocoatings in hard system using the inverse analysis based on the artificial neural network metamodeling. The inverse analysis was preceded by the development of the Finite Element Method (FEM) model dedicated to the nanoindentation test of the hard nanocoatings system. The performed sensitivity analysis is focused on determination of parameters, having the highest influence on FEM model response. The obtained, reliable FEM model was used next in the inverse analysis. The objective of that analysis was evaluation of the parameters of the individual layers of the nanocoating system. In order to decrease the computation time connected with the inverse analysis, the metamodeling approach was proposed. The used metamodel was based on the artificial neural network technique. The obtained results confirm the usefulness of the presented method in the identification of the material properties of the complex, nanocoating systems.

1 Introduction

Thin hard nanocoating systems exhibit interesting tribological and functional properties, which are difficult to achieve in conventional, homogenous materials. On the other hand, due to very small scale and contrasting physical properties in adjacent, very thin layers, physical and numerical modeling of these systems face essential difficulties. Hard nanocoatings and their systems are usually investigated in experimental nanoindentation tests, because other, standard experimental methods performed in macro and micro scale are not suitable for such case [12]. Analytical methods for nanoindentation tests, which lead to evaluation of mechanical properties, were developed by Oliver and Pharr [12]. However, all these solutions are dedicated to monolayer materials. Therefore, the authors of the present work have undertaken some attempts towards the numerical FEM modeling of multilayer system [7]. FEM modeling of nanoindentation test appears difficult, because of the small thickness of layers, which involves necessity of mesh regeneration. Accuracy of the FEM simulation of the layered, multimaterial system depends on adequate evaluation of the properties of every single layer, which is crucial in modeling of nanocoatings. As the result of mentioned above difficulties, the

A. Korytowski et al. (Eds.): System Modeling and Optimization, IFIP AICT 312, pp. 319–330, 2009.
© IFIP International Federation for Information Processing 2009

direct numerical model for nanoindentation test is computationally expensive. Therefore, development of the alternative, computationally effective method, based on the metamodel principle, is the main objective of the present work.

The first part of work describes the nanoindentation test of hard nanocoatings and explains how hardness in nanoscale is measured. The next part of the paper is dedicated to the development of the efficient and robust FEM model of nanoindentation test. FEM modeling was preceded by the sensitivity analysis oriented towards the determination of material model parameters and nanotest settings, which have the greatest influence on a response of generated FEM model of nanocoatings system.

The main objective of the present work is the inverse analysis, which allows the identification of material parameters of inner nanocoating in system of hard nanocoatings composed of various nanomaterial layers. The metamodel approach [9] based on the FEM modeling and artificial neural network techniques [6,10] is proposed in the paper.

2   Nanoindentation Tests

The first objective of the work was investigation of properties of tribological hard nanocoatings system, which is composed of TiAlN [3] and TiN [2]. These materials are deposited on the elastic substrate like carbide using PVD technique. Titanium nitride is used for some particular and the most demanding applications, because it increases the biocompatibility of the material. An artificial left blood chamber and its constructional element, which is an aortic valve, are the good examples of biotechnological application [11] of these materials, especially of TiN. The properties (hardness and Young's modulus) of a specimen are examined in the experimental nanoindentation tests.

2.1   Examined Material

The specimen (technical material) of titanium nitride basis and thin mixed hard elastic-plastic nanolayers deposited on elastic substrate was investigated. The material system of eleven PVD, thin material layers on carbide (infinite thickness) is shown in Fig. 1. Two different coatings are deposited periodically. Coating 2 (TiN, an elastic material) is 40 nm thick and is repeated three times. Coating 1 (TiAlN, an elasticplastic material) is 400 nm thick and is repeated four times.

2.2   Experiment

The objective of the nanoindentation test is to evaluate the mechanical properties of indented material like hardness and Young's modulus. The experimental nanoindentation test is performed in load or depth controlled mode using a Nano Test System [3,2]. Diamond ($E = 1141$ MPa, $v = 0.07$), Berkovich pyramid (tip radius $R = 150$ nm, pyramid angle $\alpha = 70.32°$) penetrates into the specimen. The schematic illustration of experiment and the top view of Berkovich indent is shown in Fig. 2.

The multistage process of deformation in nanoindentation test is performed in the case of testing the multicoating material. This procedure is necessary for specimen composed of nanocoatings to eliminate the effect of scatter in results and to create a

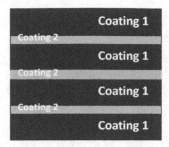

Fig. 1. Analyzed system of nanocoatings

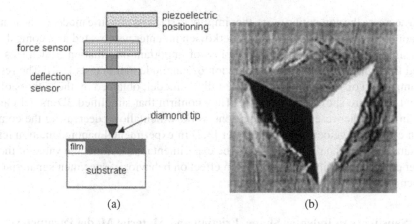

Fig. 2. (a) Schematic illustration of nanoindentation test [5], (b) The real view of Berkovich indent [5]

possibility to achieve the response of bottom layers during long term deformation process. The indentation test supplies force versus indentation depth (tool displacement) data. The load is the main output from the experiment and the Martens hardness for deformed material is calculated on the basis of force/displacement or depth data [12,3,2].

3   FEM Model

The earlier research of the authors [7,9,10,8] is focused mainly on overcoming numerical difficulties occurring in FEM simulation of deformation process of hard nanocoatings, caused by the nanothickness of layers, necessity of remeshing and scaling operations, as well as the multimaterial, multistage character of simulation and efforts to decrease the computing costs.

The objective of the present work was the development of FEM model of nanoindentation test accounting for different control parameters of the test, like indenter shape and friction conditions, as well as sensitivity of the response of the specimen with respect to material model parameters. The selected results, which are crucial for the development of efficient FEM model of nanoindentation, are presented below.

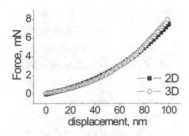

Fig. 3. Force versus depth results for 2D FEM axisymmetric and 3D FEM models [8]

To decrease the computing costs, the simplified 2D axisymmetric model of the nanoindentation test was considered and the Berkovich indenter was treated as a conical one. Such simplified model does not cause a loss of important information, which was validated through the full 3D FEM simulation of nanoindentation tests [1,4]. The results of comparison of 2D simulation and the full 3D model, obtained by the authors of the present work are shown in Fig. 3 [8]. They confirm that simplified 2D model can be used in the further research of the present work, which allows decreasing the computation costs. The velocity of the indenter [3,2] in experimental nanoindentation test is constant and very small. According to the experimental procedure, the value of the indenter constant velocity does not have an effect on behavior of specimen's material and in each simulation is equal to 1 nm/s.

3.1 Sensitivity to Indenter Shape, Friction and Material Model Parameters

The design of conditions of an effective nanoindentation test, as well as adequate choice of the FEM model parameters, were preceded by the sensitivity analysis. This method allows the estimation of the influence of the individual process parameters on the value of the analyzed one. The considered parameter was the total load (force) of the nanoindentation test, therefore, its sensitivity with respect to the process parameters was determined. The performed sensitivity analysis was based on the finite difference approximation. The sensitivity coefficients φ_{p_j} were defined as:

$$\varphi_{p_j}\Big|_{\mathbf{p}^*} := \frac{p_j^*}{F_{av}(\mathbf{p}^*)} \frac{\partial F_{av}}{\partial p_j}\Big|_{\mathbf{p}^*} \cong \frac{p_j^*}{F_{av}(\mathbf{p}^*)} \frac{F_{av}(\mathbf{p}^* + \Delta p_j \mathbf{e_j}) - F_{av}(\mathbf{p}^*)}{\Delta p_j} \tag{1}$$

where: $\mathbf{p}^* = (R, \alpha, \mu, E, K, n)$ - vector composed of considered parameters, $\mathbf{e_j}$ - vector of the canonical basis, Δp - variation of the parameter p, F_{av} - average value of the total load, calculated as follows:

$$F_{av} = \frac{1}{t} \int_0^t F(\tau) d\tau \tag{2}$$

where: $F(\tau)$ - the load at the time τ, t - total time of the process.

The sensitivities of the total load of the nanoindentation test with respect to the indenter shape parameters (R, α), friction coefficient (μ) and specimen's material parameters

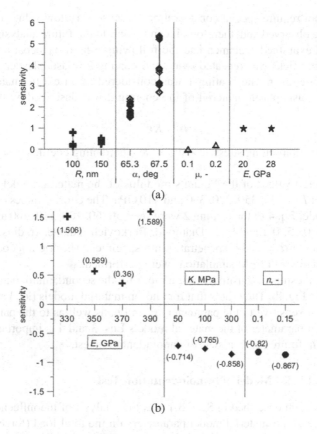

Fig. 4. (a) Sensitivity of the load with respect to geometrical indenter tip parameters: R, α and friction coefficient μ for monocoating specimen. (b) Sensitivity of the load with respect to the material model parameters E, K and n for the specimen composed of 3 hard nanocoatings.

(E, K, n) were analyzed. Two different specimens were examined: an elastic monocoating specimen - 400 nm thick and a specimen composed of 3 hard nanocoatings.

For the first, monocoating specimen, twelve Berkovich indenters with four tip radii equal to 100, 110, 150 and 160 nm, as well as three tip vertex angles: 65.3, 67.5 and 70.32°, all with round tip were investigated. The Coulomb friction law was assumed with the following Coulomb friction coefficient values: $\mu = 0.1$, $\mu = 0.15$, $\mu = 0.2$ and $\mu = 0.25$. The Poisson ratio $\nu = 0.177$ and four elastic moduli: $E_1 = 20$, $E_2 = 22$, $E_3 = 28$, and $E_4 = 30$ GPa were used in the material model defined by:

$$\sigma = E\varepsilon \tag{3}$$

where: σ - work-hardening stress, ε - strain, E - Young's modulus.

196 FEM simulations were performed. The obtained results of the sensitivity analysis presented in Fig. 4a indicate that the nanoindentation test is the most sensitive to the geometrical parameters of the tip, especially to the higher values of the tip angle. The

friction does not require special consideration, because very low values of sensitivity coefficients are observed and, therefore, it can be omitted in future analysis.

The second examined specimen had the following 3 hard nanocoatings: coating 1 (elastic, 400 nm thick) was repeated twice and coating 2 (elasticplastic, 40 nm thick) was a single interlayer. The coating 1 was considered as an elastic material defined by Eq. 3. The elasticplastic material of the coating 2 was described by the following relationship:

$$\sigma = K\varepsilon^n \tag{4}$$

where: σ - work-hardening stress, ε - strain, K - hardening coefficient, n - hardening exponent.

The considered values of the Young's modulus of the material model Eq. 3 of the coating 1 were: $E = 330, 350, 370, 390$ and 410 GPa. The chosen values of parameters in material model Eq. 4 of the coating 2 were: $K = 50, 60, 100, 110, 300$ and 310 MPa, while $n = 0.1, 0.15, 0.2$ and 0.25. Diamond, Berkovich indenter (radius $R = 150$ nm and pyramid angle $\alpha = 70.32°$) penetrates into specimen. The friction coefficient μ is assumed 0. Finally, 144 FEM simulations were performed.

The obtained results of sensitivity calculations for the second, multicoating specimen are presented in Fig.4b. They show that for chosen material models Eq. 3 and Eq. 4, the load is the most sensitive to the parameters E and n, as well as to the parameter K. It means that each parameter of the material models Eqs. 3 and 4 is important and has to be considered in future FEM models of nanoindentation test.

3.2   The Final FEM Model of Nanoindentation Test

The aim of research described in Sect. 3.1 was the analysis of the influence of the deformation process and material model parameters on the total load (force), as well as on the evaluation of optimal conditions and input settings for FEM model of nanoindentation test of the hard nanocoatings system. The defined process and material model parameters used in the developed FEM model of nanoindentation test are:

– the angle of indenter $\alpha = 70.32°$ and tip radius $R = 150$ nm,
– the indenter velocity $v = 1$ nm/s and final displacement $d = 100$ nm,
– the parameters E, K, n in used material models Eqs. 3 and 4, which are specified in the last section of Sect. 3.1 and presented in Fig. 4b,
– the specimen has three coatings with material models are described by Eqs. 3 and 4, respectively for elastic coating 1 and elasticplastic coating 2,
– the frictionless conditions between indenter and specimen ($\mu = 0$).

The final, used in further calculations of the present work, FEM mesh has 7000 nodes and 13 000 elements (Fig. 5a).

The described FEM model was implemented into the FORGE 2 code. The example of equivalent strain distribution is plotted in Fig. 5b. It can be seen that the maximum of strain is located in the inner coating.

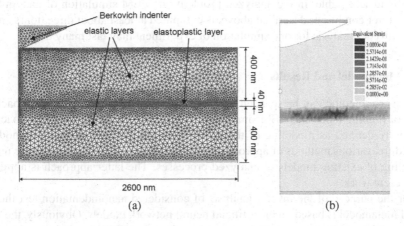

Fig. 5. (a) The developed FEM model (specimen and indenter) of the nanoindentation test. (b) Distributions of the equivalent strain of 3 hard nanocoatings.

4 Inverse Analysis with Metamodel

Generally, the main goal of the inverse analysis is evaluation of the real parameters of the model for the tested material. The aim of the present work is evaluation of these parameters on the basis of the nanoindentation test for multi-nanocoating systems. The known and described widely in [12] analytical methods used in experimental nanoindentation test lead to evaluation of mechanical properties (hardness, Young's modulus) and they produce desired results, but only for monolayer specimen. It is impossible to extrapolate these solutions to the multilayered nanocoatings.

Therefore, the objective of this research is to evaluate the properties of the inner layers of the multinanocoating system. The inverse analysis was suggested by the authors to solve this problem. The load measured for the whole nanocoatings system is the main output from the nanoindentation experiment, which is indirectly used in the goal function of the classical inverse approach. Since the inverse analysis of such complex nanomaterial system is very time-consuming procedure, the classical inverse analysis was coupled with the artificial neural network (ANN). The ANN approach allows significant reduction of the computational costs.

4.1 Classical Inverse Approach

The identification of material model can be done using the classical inverse approach. The objective of the inverse analysis is to find, using the optimization procedure, the material model parameters, which give the best matching between results of the FEM simulation and the experiment. The discrepancy between these values is the optimization goal function, which has to be minimized.

Unfortunately, in many cases, the evaluation of goal function requires numerous time-consuming FEM simulations. It makes the computation time of the whole inverse analysis unacceptable. In the analyzed problem one FEM simulation of nanoindentation test, for conditions described above, is computed at least about three hours and the whole inverse analysis for one simulation of experiment may last many days.

4.2 Metamodel and Results

The inverse method can be speeded up by using the fast metamodeling approach, instead of running thousands FEM simulations. The idea of the metamodel approach can be briefly defined as modeling of the existing model. Usually, in the metamodeling procedure, various methods of approximation or artificial intelligence tools are used to modeling of existing models of analyzed processes. The latter approach is applied in the present work.

For the purpose of the inverse analysis of considered nanoindentation test the proposed metamodel is based on the artificial neural network models. Obviously, the ANN metamodel creation demands numerous time-consuming FEM simulations as the input data. But this is done once, and later on, the whole inverse procedure may be performed fast for many simulations of experiment. The idea of the metamodel creation and its application in the inverse analysis are shown in Fig. 6.

As it was mentioned, the metamodel of the FEM output data of the nanoindentation test, obtained in the FORGE 2 simulations, is based on the artificial neural network approach. The Multi Layer Perceptron (MLP) of the 4-2-1 architecture is used (logistic transfer functions in the first and second layers; linear activation function in the output layer - see Fig. 7).

The ANN input data are the parameters E, K and n of the material models Eqs. 3 and 4 of the analyzed multinanocoating specimen, as well as the indenter displacement d. The ANN output data corresponds to force F.

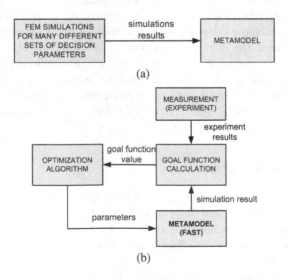

Fig. 6. The metamodel: (a) creation, (b) application in the inverse analysis

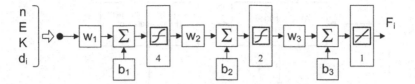

Fig. 7. The metamodel based on the ANN

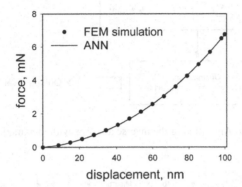

Fig. 8. Results of the artificial neural network test for work-hardening curves of multi-nanocoating specimen

144 data sets for various n, K and E were used. Each set was composed of 25 values of force versus displacement data. 142 sets of data were training data, and two sets were used as for ANN test. The network was trained using Levenberg-Marquardt algorithm [6]. The network was tested for $n = 0.15$, $E = 370\,\text{GPa}$, $K = 100\,\text{MPa}$ and the results are shown in Fig. 8. Root mean square error for the two test sets is equal to $20\,\mu\text{N}$, which confirms good predictive capability of the network.

The trained network was used next as the metamodel in the inverse analysis. The analyzed goal function of the inverse problem was the root mean square error between experimental data and the output of the network:

$$\phi(n, E, K) = \sqrt{\frac{1}{N} \sum_{i=1}^{N} \left(F_{EXP}(i) - F_{ANN}(n, E, K, d_i)\right)^2} \qquad (5)$$

where: F_{EXP} - force vs displacements simulated by FEM, F_{ANN} - ANN predicted values of the force, d_i - displacements, N - number of computing steps.

To find the minimum value of the goal function (Eq. 5) the hybrid optimization procedure was applied. The genetic algorithm was used in the first phase to the localization of the minimum, while the Quasi-Newton algorithm was used in the final search. The whole algorithm of the inverse method with metamodel is shown in Fig. 9.

The experimental data was generated by FEM simulation for the set of material model parameters: $n = 0.175$, $E = 400\,\text{GPa}$ and $K = 270\,\text{MPa}$. The results for examined case are presented in Fig. 10. Evaluated minimum of the goal function (Eq. 5) is

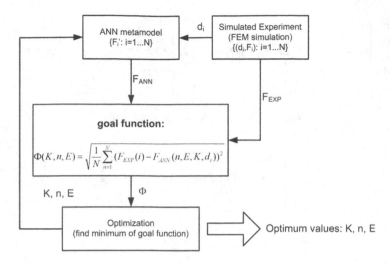

Fig. 9. Algorithm of the inverse analysis with metamodel

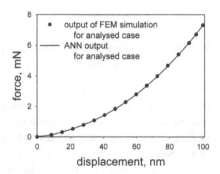

Fig. 10. Results of the inverse analysis for the analyzed case

found at $n = 0.17$, $E = 397$ GPa and $K = 331$ MPa for analyzed case. The goal function value is $\phi = 21\,\mu$N.

It is shown in Fig. 10 that the experimental (simulated) points match very well the found solution. Unfortunately, the problem is irreversible. It means that for one set of force versus displacement many different solutions can be found. This statement is also confirmed by the second plot, which is shown in Fig. 11. This plot presents logarithm of goal function (Eq. 5) for the examined case. The chosen goal function for simulated experimental data set takes minimal values in some area located around experimental parameters. The minimum of goal function is shallow and therefore, the optimization result depends on the starting point. The ambiguity problem will be greater for the real experimental data (not simulated), because there is a big scatter in experimental results. Thus, the future form of used material model (Eq. 4) should be modified by adding more parameters or chosen the more complex material model.

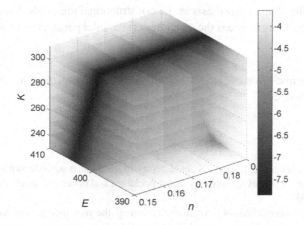

Fig. 11. Plot of the logarithm of the goal function for the experimental data set

5 Conclusions

The presented research and reached results lead to the following conclusions:

- Presented metamodeling approach is useful for considered optimization problem when the evaluation of the goal function is time-consuming. The proposed algorithm allows radical decrease of the number of long-term FEM calculations. Despite of the initial computational efforts connected with the ANN training, the final use of the network as the metamodel in the evaluation of material model parameters lasts only a few seconds. So, the reached time-profit is very high.
- The quality of results is very good. Trained ANN gives good compatibility with the test set.
- There is a certain disadvantage of presented approach - a weak ambiguity, because similar output curves can be obtained for different combination of chosen input material model parameters. This observation is proved by the plot of the used goal function (Eq. 5), which is shown in Fig. 11. The goal function takes minimal values in some area located around chosen experimental parameters. The minimum of goal function is shallow and vast. Therefore, the analyzed problem is irreversible. The disadvantage of such approach appears, because it is impossible to find out precisely, which combination of parameters in material model is the best solution of the inverse problem. Therefore, the future research of the authors will be focused on solving the uniqueness problem by modification of the form of material model, by adding more parameters or choosing a more complex material model.
- The ANN based metamodel can be used together with another optimization procedure, for example the one defined by heuristic algorithms.
- The key aspects of all the prospects and conclusions are important from the point of view of the authors, because the examined system of nanocoatings will be used for demanding biomedical application. These coatings will be also deposited on polyurethane by PLD (Pulsed Laser Deposition) technique and are supposed to be

used for artificial heart prosthesis as the constructional materials. Numerical model of such artificial organ needs the exact material model parameters of all its material layers.

Acknowledgements. Financial assistance of the MNiSzW, project no. N507 136 32/ 3962, is acknowledged.

References

1. Albrecht, H.J., Hannach, T., Hase, S., et al.: Nanoindentation: a suitable tool to determine local mechanical properties in microelectronic packages and materials? Arch. Appl. Mech. 74, 728–738 (2005)
2. Beake, B.D., Ranganathan, N.: An investigation of the nanoindentation and nano/microtribological behaviour of monolayer, bilayer and trilayer coatings on cemented carbide. Mat. Sci. Eng. A23, 46–51 (2006)
3. Beake, B.D., Smith, J.F., Gray, A., et al.: Investigating the correlation between nano-impact fracture resistance and hardness/modulus ratio from nanoindentation at 25-500 C and the fracture resistance and lifetime of cutting tools with Ti1-xAlxN (x=0.5 and 0.67) PVD coatings in milling operations. Surface and Coatings 201, 4585–4593 (2007)
4. Chollacoop, N., Dao, M., Suresh, S.: Depth-sensing instrumented indentation with dual sharp indenters. Acta Mater 51, 3713–3729 (2003)
5. Fischer-Cripps, A.C.: Nanoindentation. Springer, New York (2004)
6. Koker, R., Altincock, N., Demir, A.: Neural network based prediction of mechanical properties of particulate reinforced metal matrix composites using various training algorithms. Materials and Design 28, 616–627 (2007)
7. Kopernik, M., Pietrzyk, M.: 2D numerical simulation of elasto-plastic deformation of thin hard coating systems in deep nanoindentation test with sharp indenter. Archives of Metallurgy and Materials 52, 299–310 (2007)
8. Kopernik, M., Szeliga, D.: Modelling of nanomaterials – sensitivity analysis to determine the nanoindentation test parameters. Computer Methods in Materials Science 7, 255–261 (2007)
9. Kopernik, M., Stanisławczyk, A., Kusiak, J., et al.: Identification of material models in hard system of nanocoatings using metamodel. In: Korytowski, A., Mitkowski, W., Szymkat, M. (eds.) Abstr., 23rd IFIP TC7 Conference on System Modelling and Optimization, Kraków (2007)
10. Kopernik, M., Stanisławczyk, A., Szeliga, D.: Problems of material models of hard nanocoatings. In: Dems, K. (ed.) Proc. 17th Int. CMM Conf., Łódź-Spała (2007)
11. Kustosz, R., Major, R., Wierzchoń, T., et al.: Designing a new heart. Academia 3, 14–17 (2004)
12. Oliver, C., Pharr, G.M.: An improved technique for determining hardness and elastic modulus using load and displacement sensing indentation experiment. J. Mater. Res. 7, 1564–1583 (1992)

Using Self-adjoint Extensions in Shape Optimization

Antoine Laurain[1] and Katarzyna Szulc[2]

[1] University of Graz, Department of Mathematics and Scientific Computing, Graz, Austria
antoine.laurain@uni-graz.at
[2] Institut Élie Cartan, University Henri Poincaré, Nancy1, France
katarzyna.szulc@iecn.u-nancy.fr

Abstract. Self-adjoint extensions of elliptic operators are used to model the solution of a partial differential equation defined in a singularly perturbed domain. The asymptotic expansion of the solution of a Laplacian with respect to a small parameter ε is first performed in a domain perturbed by the creation of a small hole. The resulting singular perturbation is approximated by choosing an appropriate self-adjoint extension of the Laplacian, according to the previous asymptotic analysis. The sensitivity with respect to the position of the center of the small hole is then studied for a class of functionals depending on the domain. A numerical application for solving an inverse problem is presented. Error estimates are provided and a link to the notion of topological derivative is established.

1   Introduction

The standard approach in shape optimization consists in performing smooth perturbations of the boundary of a domain Ω in the normal direction. This technique does not allow topological changes in the domain. From a numerical point of view, topological changes can be obtained using levelset methods with this technique, but these changes are restricted and have no theoretical background.

In order to overcome this difficulty, several techniques have been introduced. We refer to [2] for recent developments, and to [1] for the method of homogenization in topology optimization. Other techniques rely on the simplified framework of the asymptotic analysis of the problems. In particular, the internal topology variations are introduced in [16], the necessary optimality conditions for simultaneous topology and shape optimization are derived in [17], the Steklov-Poincaré operators for modeling of small holes are used in [4].

The use of self-adjoint extensions of elliptic operators for modeling the solution in singularly perturbed domains was introduced in [7,10,11], and an alternative approach for simultaneous topology and shape optimization using self-adjoint extensions was presented in [12,13].

In this paper we develop a numerical method based on the application of self-adjoint extensions of elliptic operators in shape optimization. The singular perturbation of the geometrical domain Ω in $\mathbb{R}^2$ is defined by a small opening ω_ε^h of diameter $O(\varepsilon)$ and of center h. The main idea of self-adjoint extensions is to model such a small defect ω_ε^h by a concentrated *action*, the so-called potential of zero-radii. In this way the solution $u_h(x,\varepsilon)$ which has a singular behavior as $\varepsilon \to 0$ is replaced by a function with the

A. Korytowski et al. (Eds.): System Modeling and Optimization, IFIP AICT 312, pp. 331–349, 2009.
ⓒ IFIP International Federation for Information Processing 2009

singularity at the center h of the defect. Such an approach is well-known in modeling of physical processes in material with defects, we refer the reader e.g. to [3,8,15]. The interesting feature of using self-adjoint extensions is that it can be extended to spectral problems and evolution boundary value problems. Also, from a numerical point of view, the singularity created by the small ε-perturbation is costly because the mesh has to be refined in the neighborhood of this small hole and the geometry of the hole has to be parameterized. To circumvent this problem, we use the concept of self-adjoint extensions of elliptic operators to define an approximation of u_h which is defined on the fixed domain Ω.

In the first chapter, the asymptotic analysis of the singularly perturbed Dirichlet problem is performed using the method of compound asymptotic expansions. The solution $u_h(x, \varepsilon)$ of the perturbed problem is approximated by a sequence of limit problems. The 2-dimensional case considered in this paper leads to a specific asymptotic analysis, due to the nature of the fundamental solution in 2D, which is of logarithm type.

In the second chapter, the self-adjoint extension of the Laplace operator with Dirichlet boundary conditions is introduced. The approximation of the solution $u_h(x, \varepsilon)$ is then the solution $\mathbf{v}_h$ of a differential equation involving the self-adjoint extension. Error estimates for this approximation with respect to ε are given. Then, for the numerical application, the sensitivity with respect to the position h of the hole is studied, and the continuity with respect to h is proven for certain functionals in L_p spaces.

In the third chapter, the numerical problem is considered. We want to minimize the L_2-distance between the approximation $\mathbf{v}_h$ and a data z measured on a subset Ω_2 of the domain Ω. The first-order derivative with respect to h of the functional is computed for use in the conjugate gradient method used in the numerical algorithm.

In the fourth chapter, a link is established with the so-called topological derivative. The topological derivative can be recovered using the self-adjoint extension model. Usually, the topological derivative can be used to solve the problem under consideration, however we show here that our algorithm is more precise than the topological derivative because it involves additional terms of the expansion of the perturbed solution. Actually, in numerical tests, our algorithm always converges to the true solution as the space step goes to zero, while the topological derivative can be quite far from the true solution.

In the fifth chapter, the numerical algorithm is presented. We use a Fletcher-Reeves conjugate gradient algorithm associated with a line search to minimize the functional. Finite differences are used to discretize the problems. Finally, in the sixth chapter, numerical results are presented.

2   Problem Formulation

Let Ω and ω, with $0 \in \omega$, $0 \in \Omega$ be two open subsets of $\mathbb{R}^2$ with smooth boundaries. Let $\varepsilon > 0$ be a small parameter and $h \in \mathbb{R}^2$. We define the perturbed domains Ω_ε^h and ω_ε^h in the following way: $\omega_\varepsilon = \{x \in \mathbb{R}^2, \, x = \varepsilon \xi, \, \xi \in \omega\}$ and $\Omega_\varepsilon = \Omega \setminus \omega_\varepsilon$, $\omega_\varepsilon^h = \{x = y + h, \, y \in \omega_\varepsilon\}$ and $\Omega_\varepsilon^h = \Omega \setminus \omega_\varepsilon^h$. We consider the following perturbed problem in $\mathbb{R}^2$, with f in $L^2(\Omega)$:

$$-\Delta u_h(x,\varepsilon) = f(x) \quad \text{in } \Omega_\varepsilon^h, \tag{1}$$
$$u_h(x,\varepsilon) = 0 \quad \text{on } \partial\Omega, \tag{2}$$
$$u_h(x,\varepsilon) = 0 \quad \text{on } \partial\omega_\varepsilon^h. \tag{3}$$

In order to approximate the solution of (1)-(3), we use the technique of *compound asymptotic expansions*. The main idea of this technique is to look for an approximation in the form of a series with respect to the power of ε, with the coefficients given by a sequence of limit problems defined either in the unperturbed domain Ω or in $\mathbb{R}^2 \setminus \omega$. The limit problems defined on $\mathbb{R}^2 \setminus \omega$ are called *boundary layers* because they correspond to solutions concentrated on the boundary of ω_ε^h and vanishing at finite distance of ω_ε^h. The boundary conditions verified by a problem are determined by the discrepancy left by the higher-order limit problem. Due to the nature of the fundamental solution in dimension 2, i.e. a logarithm, a specific procedure needs to be used, which leads to an expansion containing powers of $\ln\varepsilon$. Even if the full expansion can be obtained in the case of the Dirichlet equation we are looking at, we restrict ourselves to the first term of the expansion, which is the only term of interest for our purposes.

2.1 First Limit Problem

The first approximation v^0 solves:

$$-\Delta v^0(x) = f(x) \quad \text{in } \Omega, \tag{4}$$
$$v^0(x) = 0 \quad \text{on } \partial\Omega. \tag{5}$$

Since f is in $L^2(\Omega)$ and Ω is smooth, v^0 is in $H^2(\Omega)$. This approximation is satisfying outside a neighborhood of the boundary of the hole ω_ε^h. Due to the Dirichlet conditions on the boundary of the hole ω_ε^h, $u_h(x,\varepsilon)$ will be better approximated by

$$-\Delta v_h(x) = f(x) + \beta_h\delta(x-h) \quad \text{in } \Omega. \tag{6}$$
$$v_h(x) = 0 \quad \text{on } \partial\Omega. \tag{7}$$

We then have

$$v_h(x) = v^0(x) + \beta_h G(x,h),$$

where $G(x,y)$ is the generalized Green function defined by

$$-\Delta_x G(x,y) = \delta(x-y) \quad \text{in } \Omega, \tag{8}$$
$$G(x,y) = 0 \quad \text{on } \partial\Omega, \tag{9}$$

and $\delta(x-y)$ is the Dirac mass at y. The function $u_h(x,\varepsilon)$ is then approximated outside a neighborhood of ω_ε^h by

$$u_h(x,\varepsilon) \simeq v^0(x) + \beta_h G(x,h).$$

The function G admits the following representation:

$$G(x,h) = -\left\{ (2\pi)^{-1}\log|x-h| + \mathscr{G}(x,h) \right\}, \tag{10}$$

where $|\cdot|$ stands for the euclidean norm in $\mathbb{R}^2$. The function $\mathscr{G}$ is the regular part of the Green function solution of

$$-\Delta_x \mathscr{G}(x,y) = 0 \quad \text{in } \Omega, \tag{11}$$

$$\mathscr{G}(x,y) = -(2\pi)^{-1} \log|x-y| \quad \text{on } \partial\Omega, \tag{12}$$

For $x \in \partial\Omega$ and as $h \to 0$, we can use Taylor's formula to expand $-(2\pi)^{-1}\log|x-h|$ in (12) with respect to h and obtain

$$-(2\pi)^{-1}\log|x-h| = -(2\pi)^{-1}\log|x| + (2\pi)^{-1}\langle h, \frac{x}{|x|^2}\rangle + r_h,$$

with $\|r_h\|_{L^\infty(\partial\Omega)} = O(|h|^2)$. Thus $\mathscr{G}(x,h)$ admits the expansion

$$\mathscr{G}(x,h) = \mathscr{G}(x,0) + \mathscr{S}_h(x) + \mathscr{R}_h(x), \tag{13}$$

with $\mathscr{S}_h(x)$ and $\mathscr{R}_h(x)$ solutions of

$$-\Delta\mathscr{S}_h(x) = 0 \quad \text{in } \Omega, \tag{14}$$

$$\mathscr{S}_h(x) = (2\pi)^{-1}\langle h, \frac{x}{|x|^2}\rangle \quad \text{on } \partial\Omega. \tag{15}$$

$$-\Delta\mathscr{R}_h(x) = 0 \quad \text{in } \Omega, \tag{16}$$

$$\mathscr{R}_h(x) = r_h \quad \text{on } \partial\Omega. \tag{17}$$

Finally, $\mathscr{G}(h,h)$ can be decomposed into

$$\mathscr{G}(h,h) = \mathscr{G}(0,0) + \langle h, \nabla\mathscr{G}(0,0)\rangle + \mathscr{S}_h(0) + \mathscr{R}_h(h) + O(|h|^2). \tag{18}$$

Since $\|r_h\|_{L^\infty(\partial\Omega)} = O(|h|^2)$ we get $\|\mathscr{R}_h\|_{L^\infty(\Omega)} = O(|h|^2)$. We also have $\|\mathscr{S}_h\|_{L^\infty(\Omega)} = O(|h|)$. The approximation $u_h(x,\varepsilon) \simeq v^0(x) + \beta_h G(x,h)$ does not verify the boundary condition (3) on the hole. Consequently, a boundary layer $w_h^0(\xi_h,\varepsilon)$ must be added, which depends on the fast variable ξ_h defined as $\xi_h = \varepsilon^{-1}(x-h)$, in order to compensate for the induced discrepancy. Expanding $v^0(x) + \beta_h G(x,h)$ when $x \to h$ we get

$$v^0(x) + \beta_h G(x,h) = v^0(x) - \beta_h\{(2\pi)^{-1}\log|x-h| + \mathscr{G}(x,h)\}$$
$$= v^0(h) - \beta_h\{(2\pi)^{-1}\log|\varepsilon\xi_h| + \mathscr{G}(h,h)\} + z_\varepsilon^h(x). \tag{19}$$

The estimates on the rest $z_\varepsilon^h(x)$ will be addressed later in Section 3.2. In view of (19), we introduce the boundary layer $w_h^0(\xi,\varepsilon)$ solution of the following system

$$-\Delta_{\xi_h} w_h^0(\xi_h,\varepsilon) = 0 \quad \text{in } \mathbb{R}^2 \setminus \omega_h, \tag{20}$$

$$w_h^0(\xi_h,\varepsilon) = -v^0(h) + \beta_h\{(2\pi)^{-1}\log|\varepsilon\xi_h| + \mathscr{G}(h,h)\} \quad \text{on } \partial\omega_h. \tag{21}$$

The solution of (20)-(21) is

$$w_h^0(\xi_h,\varepsilon) = -v^0(h) + \beta_h\{(2\pi)^{-1}\log\varepsilon + \mathscr{G}(h,h)\} + \beta_h\mathscr{E}_h^0(\xi_h), \tag{22}$$

with $\mathscr{E}_h^0(\xi_h)$ solution of

$$-\Delta_{\xi_h}\mathscr{E}_h^0(\xi_h) = 0 \quad \text{in } \mathbb{R}^2 \setminus \omega_h, \tag{23}$$

$$\mathscr{E}_h^0(\xi_h) = (2\pi)^{-1}\log|\xi_h| \quad \text{on } \partial\omega_h. \tag{24}$$

The function $\mathscr{E}_h^0(\xi_h)$ admits the following expansion w.r.t. ξ_h

$$\mathscr{E}_h^0(\xi_h) = (2\pi)^{-1}L + O(|\xi_h|^{-1}), \tag{25}$$

where L is a constant depending only on the shape of ω. The quantity $\exp(L)$ is called the logarithmic capacity of ω. Thus, $w_h^0(\xi_h,\varepsilon)$ admits the expansion

$$w_h^0(\xi_h,\varepsilon) = -v^0(h) + \beta_h\left\{(2\pi)^{-1}\log\varepsilon + \mathscr{G}(h,h)\right\} + \beta_h(2\pi)^{-1}L + O(|\xi_h|^{-1}).$$

In order to have $w_h^0(\xi_h,\varepsilon) \to 0$ as $|\xi_h| \to \infty$, a condition is imposed on β_h:

$$\beta_h = \left\{(2\pi)^{-1}(\log\varepsilon + L) + \mathscr{G}(h,h)\right\}^{-1}v^0(h), \tag{26}$$

so that

$$w_h^0(\xi_h,\varepsilon) = \beta_h\mathscr{E}_h^0(\xi_h).$$

As a consequence, the solution $u_h(x,\varepsilon)$ of (1)-(3) can be represented by

$$u_h(x,\varepsilon) = v^0(x) + \beta_h G(x,h) + \tilde{u}_h^0(x,\varepsilon), \tag{27}$$

where the function $\tilde{u}_h^0(x,\varepsilon)$ is solution of the following problem

$$-\Delta\tilde{u}_h^0(x,\varepsilon) = 0 \quad \text{in } \Omega_\varepsilon^h \tag{28}$$

$$\tilde{u}_h^0(x,\varepsilon) = 0 \quad \text{on } \partial\Omega \tag{29}$$

$$\begin{aligned}\tilde{u}_h^0(x,\varepsilon) = &-(v^0(x) - v^0(h))\\ &+ \beta_h\left\{(2\pi)^{-1}(\log|\xi_h| - L)\right\}\\ &+ \beta_h\{\mathscr{G}(x,h) - \mathscr{G}(h,h)\} \quad \text{on } \partial\omega_\varepsilon^h.\end{aligned} \tag{30}$$

3   Self-adjoint Extension of the Laplacian with Dirichlet Conditions

3.1   Self-adjoint Extension

For the sake of simplicity, we assume that $h = 0$ in what follows (without loss of generality) and we will return to the general case in the next section. In what follows, we use the notation β instead of β_0. The self-adjoint extension of the Laplace operator with Dirichlet boundary conditions is defined as follows: let $\mathscr{A}_0$ be the Laplacian operator $-\Delta_x$ in $L_2(\Omega)$ with the domain of definition

$$\mathscr{D}(\mathscr{A}_0) = \left\{v \in C_0^\infty(\overline{\Omega} \setminus \{0\})\right\} \tag{31}$$

The inclusion $v \in \mathscr{D}(\mathscr{A}_0)$ indicates that v satisfies the boundary conditions (3) and is equal to zero in the neighborhood of the center 0 of ω_ε, this last condition mimicking the Dirichlet condition (3).

Introduce the cut-off function $\chi_\delta(x) = \chi(\delta x)$ where χ is such that $\chi \in C^\infty(\mathbb{R}^2)$ and

$$\chi(x) = 1 \text{ for } |x| < 1, \tag{32}$$
$$\chi(x) = 0 \text{ for } |x| > 2. \tag{33}$$

We assume that δ is chosen such that χ_δ has compact support in Ω. The closure $\overline{\mathscr{A}_0}$ and the adjoint $\mathscr{A}_0^*$ of the operator $\mathscr{A}_0$ are given by the following lemma:

Lemma 1. *The closure $\overline{\mathscr{A}_0}$ and the adjoint $\mathscr{A}_0^*$ of the operator $\mathscr{A}_0$ are given by the differential expression $-\Delta_x$, with the respective domain of definition:*

$$\mathscr{D}(\overline{\mathscr{A}_0}) = \left\{ v \in H^2(\Omega), \, v(0) = 0, \, v = 0 \text{ on } \partial\Omega \right\} \tag{34}$$

and

$$\mathscr{D}(\mathscr{A}_0^*) = \left\{ v : v(x) = \chi_\delta(x)\left(-\frac{a}{2\pi}\log r + b\right) + \bar{v}(x), \, \bar{v} \in \mathscr{D}(\overline{\mathscr{A}_0}), \, a, b \in \mathbb{R} \right\} \tag{35}$$

Note that in (35), it can be shown that the domain $\mathscr{D}(\mathscr{A}_0^*)$ does not depend on the cut-off function χ_δ. Since the domain of definition of the initial operator $\mathscr{A}_0$ is restricted, the domain of definition of the adjoint is large, and the two operators $\overline{\mathscr{A}_0}$ and $\mathscr{A}_0^*$ are not self-adjoints. However, there exists a family of self-adjoint operators $\mathscr{A}$, such that $\mathscr{A}_0 \subset \mathscr{A} \subset \mathscr{A}_0^*$ and the domain of definition $\mathscr{D}(\mathscr{A})$ contains all the required singular solutions for the Dirichlet problem in Ω.

The family of self-adjoint extensions of the operator $\mathscr{A}_0$ is built by restricting the domain of the operator $\mathscr{A}_0^*$. The abstract boundary condition $b = Sa$ is added in the definition of $\mathscr{D}(\mathscr{A}_0^*)$ with a given coefficient S. In our case, we will obtain S depending on the asymptotic expansion of v_h w.r.t. ε. With such an S, the influence of the small hole can be modeled. Therefore, the following theorem can be proved.

Theorem 1. *Let* **A** *be the restriction of the operator $\mathscr{A}_0^*$ to the vector space*

$$\mathscr{D}(\mathbf{A}) = \{v \in \mathscr{D}(\mathscr{A}_0^*) : b = Sa\} \tag{36}$$

where $S = S(\varepsilon) = (2\pi)^{-1}(\log\varepsilon + L)$, L is a constant which depends on the shape of ω. Then **A** *is a self-adjoint operator.*

The following equation

$$\mathbf{A}v = f \in L_2(\Omega) \tag{37}$$

admits a unique solution $\mathbf{v} \in \mathscr{D}(\mathbf{A})$ *and* **v** *is given by*

$$\mathbf{v}(x) = v^0(x) + \beta G(x, 0) \qquad \forall x \in \Omega.$$

Proof. 1) It is enough to prove the following: if for $\mathbf{v}, f \in L_2(\Omega)$ the following equality is true

$$(\mathbf{v}, \mathbf{A}z)_\Omega = (f, z)_\Omega \qquad \forall z \in \mathscr{D}(\mathbf{A}), \tag{38}$$

then $\mathbf{v} \in \mathscr{D}(\mathbf{A})$ and $f = \mathbf{A}\mathbf{v}$. Since $\mathscr{A}_0 \subset \mathbf{A}$, we can see that $\mathbf{v} \in \mathscr{D}(\mathscr{A}_0^*)$ and $\mathscr{A}_0^*\mathbf{v} = f$. Thus, it is only necessary to show that $\mathbf{v} \in \mathscr{D}(\mathbf{A})$. In view of (38) we can write the Green's formula:

$$0 = (\mathbf{v}, \mathbf{A}\mathbf{z})_\Omega - (\mathscr{A}_0^*\mathbf{v}, \mathbf{z})_\Omega \tag{39}$$

$$= \lim_{\delta \to 0} \int_{\Omega \backslash \mathbb{B}_\delta} (\mathbf{z}\Delta_x\mathbf{v} - \mathbf{v}\Delta_x\mathbf{z}) \, dx \tag{40}$$

$$= \lim_{\delta \to 0} \int_{\partial\mathbb{B}_\delta} \mathbf{v}\partial_n\mathbf{z} - \mathbf{z}\partial_n\mathbf{v} \, ds_x + \int_{\partial\Omega} \mathbf{v}\partial_n\mathbf{z} - \mathbf{z}\partial_n\mathbf{v} \, ds_x \tag{41}$$

$$= \lim_{\delta \to 0} \int_{\partial\mathbb{B}_\delta} \mathbf{v}\partial_n\mathbf{z} - \mathbf{z}\partial_n\mathbf{v} \, ds_x. \tag{42}$$

Since $\mathbf{v} \in \mathscr{D}(\mathscr{A}_0^*)$, $v = 0$ on $\partial\Omega$ and as a consequence $\int_{\partial\Omega} \mathbf{v}\partial_n\mathbf{z} - \mathbf{z}\partial_n\mathbf{v} \, ds_x = 0$. In what follows we introduce the notation $r = |x|$. Replacing $\mathbf{v}$ and $\mathbf{z}$ by the asymptotic expansions given in the definition of $\mathscr{D}(\mathscr{A}_0^*)$, with the coefficients denoted respectively a, b and p, q, we get

$$\begin{aligned} 0 &= \lim_{\delta \to 0} \delta \int_0^{2\pi} (b - a\frac{1}{2\pi}\log r)\frac{\partial}{\partial r}(q - p\frac{1}{2\pi}\log r)|_{r=\delta} \\ &\quad -(q - p\frac{1}{2\pi}\log r)\frac{\partial}{\partial r}(b - a\frac{1}{2\pi}\log r)|_{r=\delta} \, d\phi \\ &= \lim_{\delta \to 0} a(q - p\frac{1}{2\pi}\log\delta) - p(b - a\frac{1}{2\pi}\log\delta) \\ &= aq - bp \\ &= (Sa - b)p, \end{aligned} \tag{43}$$

and the conclusion is $b = Sa$ which means that $\mathbf{v} \in \mathscr{D}(\mathbf{A})$. Here we have used the relation $q = Sp$ since $\mathbf{z} \in \mathscr{D}(\mathbf{A})$.

2) First of all, the unicity of the solution is proved. Let $\mathbf{v}_1$ and $\mathbf{v}_2$ be two functions in $\mathscr{D}(\mathbf{A})$. Then the difference $\mathbf{v} = \mathbf{v}_1 - \mathbf{v}_2$ verifies $\mathbf{A}\mathbf{v} = 0$ in Ω and $\mathbf{v} = 0$ on $\partial\Omega$. Thus $\mathbf{v}$ is the fundamental solution of the Laplacian

$$\mathbf{v} = \mu G(x,0) = -\mu \left((2\pi)^{-1}\log r + \mathscr{G}(x,0) \right) \tag{44}$$

where G and $\mathscr{G}$ are defined in (8)-(9) and (11)-(12). The asymptotic representation (44) gives coefficients $a = \mu$ and $b = -\mathscr{G}(0,0)\mu$. From the definition of $\mathscr{D}(\mathbf{A})$ we get $b = Sa$, thus we obtain

$$-\mathscr{G}(0,0)\mu = ((2\pi)^{-1}\log\varepsilon + L)\mu$$

which implies $\mu = 0$. Thus $\mathbf{v} \equiv 0$ and we have proved unicity of the solution.

Now it remains to show that

$$\mathbf{v}(x) = v^0(x) + \beta G(x,0) \tag{45}$$

is solution of $\mathbf{A}\mathbf{v} = f$. In view of definitions (8)-(9) and (4)-(5) of $G(x,0)$ and v^0, respectively, we clearly have $-\Delta_x u(x,\varepsilon) = f(x)$ in $\Omega \backslash 0$ and $u(x,\varepsilon) = 0$ on $\partial\Omega$. We also have $a = \beta$ and $b = v^0(0) - \beta\mathscr{G}(0,0)$, so that the relation $b = Sa$ is satisfied. As a consequence, we get $\mathbf{v} \in \mathscr{D}(\mathbf{A})$. $\qquad\square$

3.2 Estimates

From now on, we will write $\mathbf{v}_h$ instead of $\mathbf{v}$ to stress the dependence of $\mathbf{v}$ on h. According to (37), $\mathbf{v}_h$ corresponds to the first-order terms in the expansion (27). Therefore we will now give an estimate for the L^2-norm of $\tilde{u}_h^0 = u_h - \mathbf{v}_h$. Define $\tilde{u}_h^1$, $\tilde{u}_h^2$ and $\tilde{u}_h^3$ harmonic functions on Ω_ε such that

$$\tilde{u}_h^0 = \tilde{u}_h^1 + \tilde{u}_h^2 + \tilde{u}_h^3$$

and for all $x \in \partial\omega_\varepsilon^h$ we have in view of (28)-(30)

$$\tilde{u}_h^1(x,\varepsilon) = -(v^0(x) - v^0(h)), \tag{46}$$
$$\tilde{u}_h^2(x,\varepsilon) = \beta_h\left\{(2\pi)^{-1}(\log|\xi_h| - L)\right\}, \tag{47}$$
$$\tilde{u}_h^3(x,\varepsilon) = \beta_h\left\{\mathscr{G}(x,h) - \mathscr{G}(h,h)\right\}. \tag{48}$$

Since $f \in L^2(\Omega)$, $v^0 \in H^2(\Omega)$ and by the Sobolev-Rellich theorem, we have

$$v^0 \in C^0(\overline{\Omega}) \text{ and } \nabla v^0 \in L^\infty(\Omega).$$

Therefore we get

$$\sup_{x\in\partial\omega_\varepsilon^h} |v^0(x) - v^0(h)| \le M_1\varepsilon \sup_{x\in\omega_\varepsilon^h} |\nabla v^0(x)|,$$

where M_1 depends only on the shape of ω; $M_1 = 1$ if $\omega = B(0,1)$. By the maximum principle we get

$$\|\tilde{u}_h^1\|_{L^\infty(\Omega_\varepsilon)} \le M_1\varepsilon \sup_{x\in\omega_\varepsilon^h} |\nabla v^0(x)|$$

and

$$\|\tilde{u}_h^1\|_{L^2(\Omega_\varepsilon)} \le M_1\varepsilon \sup_{x\in\omega_\varepsilon^h} |\nabla v^0(x)|,$$

with M_1 depending only on the shape of Ω and ω. In a similar way, since $\mathscr{G}(\cdot,h) \in C^\infty(\overline{\Omega})$, we have

$$\sup_{x\in\partial\omega_\varepsilon^h} |\mathscr{G}(x,h) - \mathscr{G}(h,h)| \le M_3\varepsilon \sup_{x\in\omega_\varepsilon^h} |\nabla\mathscr{G}(x,h)|,$$

and

$$\|\tilde{u}_h^3\|_{L^\infty(\Omega_\varepsilon)} \le M_3\beta_h\varepsilon \sup_{x\in\omega_\varepsilon^h} |\nabla\mathscr{G}(x,h)|,$$

$$\|\tilde{u}_h^3\|_{L^2(\Omega_\varepsilon)} \le M_3\beta_h\varepsilon \sup_{x\in\omega_\varepsilon^h} |\nabla\mathscr{G}(x,h)|,$$

with M_3 depending only on the shape of Ω and ω. Now consider the case of $\tilde{u}_h^2$. If $\omega = B(0,1)$, we get $\tilde{u}_h^2 \equiv 0$. In the more general case of any shape for ω, we get according to (25)

$$\tilde{u}_h^2 = O(|\xi_h|^{-1}),$$

and since

$$\int_{\Omega_\varepsilon^h} |\xi_h|^{-2}dx = \int_{\Omega_\varepsilon^h} \varepsilon^2|x-h|^{-2}dx \le \tilde{M}_2\varepsilon^2|\log\varepsilon|,$$

where $\tilde{M}_2$ is some constant independent of ε and h. In view of the expression of β_h we get

$$\|\tilde{u}_h^2\|_{L^2(\Omega_\varepsilon)} \leq \tilde{M}_2|\beta_h|\varepsilon|\log\varepsilon|^{\frac{1}{2}} \leq M_2|v^0(h)|\varepsilon,$$

and M_2 depends only on the shape of Ω and ω. Gathering the estimates for $\tilde{u}_h^1, \tilde{u}_h^2$ and $\tilde{u}_h^3$, we obtain

$$\|\tilde{u}_h^0\|_{L^2(\Omega_\varepsilon)} \leq M\varepsilon, \tag{49}$$

with M depending only on the shape of Ω and ω.

3.3 Derivative with Respect to the Position of the Hole

Recall that the function $u_h(x,\varepsilon)$ of (1)-(3) can be represented by

$$u_h(x,\varepsilon) = v^0(x) + \beta_h G(x,h) + \tilde{u}_h^0(x,\varepsilon). \tag{50}$$

It can also be represented in a form derived from (35)

$$u_h(x,\varepsilon) = -\frac{a_h}{2\pi}\log r_h + b_h + \bar{u}_h(x,\varepsilon) \tag{51}$$

where

$$a_h = \beta_h = \left\{(2\pi)^{-1}(\log\varepsilon + L) + \mathscr{G}(h,h)\right\}^{-1}v^0(h), \tag{52}$$

$$b_h = v^0(h) - \beta_h\mathscr{G}(h,h) = Sa_h \tag{53}$$

with $S = (2\pi)^{-1}(\log\varepsilon + L)$, and (r_h, θ_h) stand for the polar coordinates of center h. The coefficient S depends on ε but does not depend on the position of the hole h. The function $\bar{u}_h$ belongs to the set $\mathscr{D}(\overline{\mathscr{A}_0})$.

3.4 Energy Functionals in L_p

We consider functionals of the form:

$$\mathscr{F}(u,\varepsilon) = \int_{\Omega_\varepsilon} F(x,u)\,dx \tag{54}$$

with $u \in \mathscr{D}(A)$. We make an assumption on the functional (54), sufficient for further asymptotic analysis. Namely, the following inequality holds for some $p \in [1,\infty[$ and for all $u,v \in L_p(\Omega_\varepsilon)$

$$|\mathscr{F}(u,\varepsilon) - \mathscr{F}(v,\varepsilon)| \leq c\|u-v\|_{L_p(\Omega_\varepsilon)}\left(\|u\|_{L_p(\Omega_\varepsilon)}^{p-1} + \|v\|_{L_p(\Omega_\varepsilon)}^{p-1}\right), \tag{55}$$

where the constant c depends on Ω, but it is independent of the parameter ε and of the functions u,v. We assume also that the same inequality holds in unperturbed domain Ω,

$$|\mathscr{F}(u,0) - \mathscr{F}(v,0)| \leq c\|u-v\|_p\left(\|u\|_p^{p-1} + \|v\|_p^{p-1}\right), \tag{56}$$

where $\|\cdot\|_p$ denotes the norm in $L_p(\Omega)$. Let $u_h(x,\varepsilon)$ be the solution of equation (1)-(3). The function u_h is extended by zero over the opening ω_ε^h, and the extended function is still denoted u_h. Since $u_h \in H_0^1(\Omega_\varepsilon)$ we also have $u_h \in H_0^1(\Omega)$; see [6, Prop. 3.1.4, p. 78]. Then, thanks to the imbedding $H^1(\Omega) \subset L_p(\Omega)$, $p \in [1,\infty[$ and to inequality (56) we have

$$|\mathscr{F}(u,0) - \mathscr{F}(u_h,0)| \leq c\|u - u_h\|_p \left(\|u\|_p^{p-1} + \|u_h\|_p^{p-1}\right). \tag{57}$$

According to (51) we can write

$$\|u - u_h\|_p = \left\| -\frac{a_0}{2\pi}\log r + \frac{a_0}{2\pi}\log r_h + S(a_0 - a_h) + \bar{u} - \bar{u}_h \right\|_p. \tag{58}$$

In view of the expansion (18) of $\mathscr{G}(h,h)$ and the smoothness of v^0 solution of (4)-(5) we get

$$\|a_0 - a_h\|_p = \|\beta - \beta_h\|_p \leq c|h| \tag{59}$$

where c is a constant depending only on Ω, for ε small enough, according to the expression (52) of β_h. For b_h we obtain a similar result because of the relation $b_h = Sa_h$

$$\|b_0 - b_h\|_p = \|Sa - Sa_h\|_p \leq Sc|h| \tag{60}$$

and $S = (2\pi)^{-1}(\log \varepsilon + L)$. Since $\bar{u}$ and $\bar{u}_h$ belong to $\mathscr{D}(\bar{\mathscr{A}}_0)$, $\bar{u}$ and $\bar{u}_h$ are in $H^2(\Omega)$, and we obtain the same inequality for $\|\bar{u} - \bar{u}_h\|_p$

$$\|\bar{u} - \bar{u}_h\|_p \leq c|h|. \tag{61}$$

The only term that remains to estimate in (58) is $-\frac{a_0}{2\pi}\log r + \frac{a_0}{2\pi}\log r_h$, since we have proven the continuity of $\|a_0 - a_h\|_p$ in (59), we only have to estimate $\|\log r - \log r_h\|_p$. Possibly changing the coordinates, we may suppose that $h = (|h|, 0)$. Then we can split the following integral into two parts

$$\int_\Omega |\log r_h - \log r|^p \, dx = I_0^h + I_1^h$$

with

$$\begin{aligned}
I_0^h &= \int_{|x|<2|h|} |\log|x + |h|e_1| - \log|x||^p \, dx \\
&= |h|^2 \int_{|\xi|<2} |\log|\xi + e_1| - \log|\xi||^p \, d\xi \\
&\leq c|h|^2.
\end{aligned} \tag{62}$$

We have used the change of variables $x = |h|\xi$. We also have $e_1 = (1,0)$. Further,

$$\begin{aligned}
I_1^h &= \int_{\Omega \setminus |x|<2|h|} \left| \log \frac{|x|^2 - 2|h||x_1 + |h|^2}{|x|^2} \right|^p \, dx \\
&\leq C_\alpha \int_{\Omega \setminus |x|<2|h|} \left(\left(\frac{|h|}{r}\right)^{\alpha p} + \left(\frac{|h|^2}{r^2}\right)^{\alpha p} \right) \\
&\leq C_\alpha \left(|h|^{\alpha p} \int_{2|h|}^D r^{-\alpha p + 1} \, dr + |h|^{2\alpha p} \int_{2|h|}^D r^{-2\alpha p + 1} \, dr \right),
\end{aligned} \tag{63}$$

with $\alpha \in]0,1]$ and C_α is a constant depending only on α. Thus, if $\alpha p < 1$, then

$$I_1^h \leq C_\alpha \left(|h|^{\alpha p} |h|^{-\alpha p+2} + |h|^{2\alpha p} |h|^{-2\alpha p+2} \right) = C_\alpha |h|^2.$$

Finally we obtain

$$\| \log r - \log r_h \|_p \leq C_\alpha |h|^{2/p}. \tag{64}$$

Therefore, choosing $p \in]1,2]$ and combining (59), (60), (61) and (64) we obtain

$$|\mathscr{F}(u,0) - \mathscr{F}(u_h,0)| \leq M|h|, \tag{65}$$

where M is a constant which depends only on the shape of Ω.

4   Least Squares Functional

In this section, the domain Ω is split into two disjoint open sets Ω_1 and Ω_2 so that $\overline{\Omega} = \overline{\Omega_1} \cup \overline{\Omega_2}$, and we introduce a least squares functional $\mathscr{J}_\varepsilon$ which measures the L_2 distance between some data z and an approximation $\mathbf{v}_h$ of (1)-(3) on Ω_2 (see Fig. 1). The data z corresponds to the solution in a domain with a hole ω_ε^h whose position in Ω_1 is unknown. In what follows, we assume that ε is known. By minimizing $\mathscr{J}_\varepsilon$ w.r.t. h, we are able to find the position of the hole ω_ε^h.

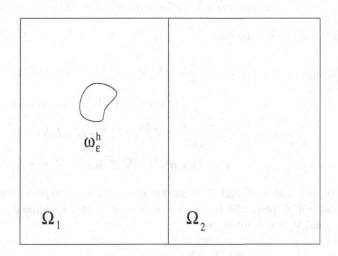

Fig. 1. The domain ω

From now on, the domain ω is assumed to be a ball of radius 1. The general case is easily deduced from this particular case. We consider a cost functional defined as follows:

$$\mathscr{J}_\varepsilon(h) := \frac{1}{2} \int_{\Omega_2} (\mathbf{v}_h(x) - z(x))^2 dx \tag{66}$$

where z is a given observation in $L^2(\Omega_2)$ and $\mathbf{v}_h$ is given by:

$$\mathbf{v}_h(x) = v^0(x) + \frac{G(x,h)}{\frac{\log \varepsilon + L}{2\pi} + \mathscr{G}(h,h)} v^0(h), \quad \forall x \in \Omega. \tag{67}$$

In the case where ω_ε is the ball $B(x_0, \varepsilon)$, we get $|\xi_h| = 1$ on $\partial \omega_h$ and therefore $L = 0$. We want to solve the minimization problem

$$\min_{h \in \Omega_1} \mathscr{J}_\varepsilon(h). \tag{68}$$

To this end we use a Fletcher-Reeves algorithm with a line search procedure, therefore we must first compute the gradient of $\mathscr{J}_\varepsilon$ w.r.t. h.

$$\nabla \mathscr{J}_\varepsilon(h) = \int_{\Omega_2} (\mathbf{v}_h(x) - z(x)) \nabla_h \mathbf{v}_h(x) dx \tag{69}$$

where ∇_h denotes the gradient with respect to h. To compute $\nabla_h \mathbf{v}_h(x)$ let us simplify (67) by introducing:

$$\lambda(h) = \left(\frac{\log \varepsilon + L}{2\pi} + \mathscr{G}(h,h) \right)^{-1}. \tag{70}$$

Thus (67) can be written as

$$\mathbf{v}_h(x) = v^0(x) + \lambda(h) G(x,h) v^0(h) \quad \forall x \in \Omega. \tag{71}$$

The gradient $\nabla_h \mathbf{v}_h(x)$ takes the form:

$$\begin{aligned}
\nabla_h \mathbf{v}_h(x) &= \lambda(h) \left[v^0(h) \left(\frac{x-h}{2\pi r_h^2} - \nabla_y \mathscr{G}(x,h) \right) + G(x,h) \nabla v^0(h) \right] \\
&\quad - \lambda(h)^2 G(x,h) v^0(h) \nabla_h[\mathscr{G}(h,h)], \\
&= \lambda(h) \left[v^0(h) \left(\frac{x-h}{2\pi r_h^2} - \nabla_y \mathscr{G}(x,h) \right) + G(x,h) \nabla v^0(h) \right] \\
&\quad - \lambda(h)^2 G(x,h) v^0(h) [\nabla_x \mathscr{G}(h,h) + \nabla_y \mathscr{G}(h,h)].
\end{aligned} \tag{72}$$

where $r_h = |x - h|$, and $\nabla_x \mathscr{G}$ and $\nabla_y \mathscr{G}$ are the gradients with respect to the first and second variables of $\mathscr{G}$, respectively. The value of $\nabla_x \mathscr{G}$ is clearly defined according to (11) and (12), and $\nabla_y \mathscr{G}$ is solution of

$$\begin{aligned}
-\Delta_x[\nabla_y \mathscr{G}](x,y) &= 0 \quad \text{in } \Omega \\
\nabla_y \mathscr{G}(x,y) &= (2\pi)^{-1} \frac{x-y}{\|x-y\|^2} \quad \text{on } \partial\Omega.
\end{aligned} \tag{73}$$

5   Topological Derivative

A new idea was introduced first by Schumacher in 1994 with the so-called "bubble method", where the parameterized setting is kept and holes are created in the domain according to a certain criterion. This idea was later developed by Sokołowski and Żochowski [16], and Guillaume and Masmoudi [4], with the introduction of the *topological derivative*. The topological derivative measures the variation of a cost functional depending on the shape of a domain, when a small change in the topology of this domain is performed, for instance with the creation of a small hole of any shape.

Let Ω and ω_ε be two open sets in $\mathbb{R}^N$, and $\omega_\varepsilon \subset B(h,\varepsilon)$ where $B(h,\varepsilon)$ is a ball of radius $\varepsilon > 0$ centered at $h \in \Omega$. Denote $\Omega_\varepsilon = \Omega \setminus \overline{\omega}_\varepsilon$, if the cost functional $J(\Omega)$ is differentiable with respect to the creation of this small hole ω_ε, then we can write the expansion

$$J(\Omega_\varepsilon) = J(\Omega) + \rho(\varepsilon)\mathscr{T}(h) + o(\rho(\varepsilon)), \tag{74}$$

with $\rho(\varepsilon) \to 0$ as $\varepsilon \to 0$, $\rho(\varepsilon) > 0$, and $\mathscr{T}(h)$ is the so-called topological derivative of J. First note that the topological derivative is a pointwise expression defined at every point of the domain, therefore it is usually easy to compute and gives an efficient criteria for a descent direction in a gradient method.

There is a link between the self-adjoint extension introduced in Section 3 and the notion of topological derivative. Indeed, both are obtained through the asymptotic expansion w.r.t. ε of the solution of (1)-(3). Actually, it is possible to recover the usual topological derivative from the self-adjoint extension model. In order to do so, we write the expansion with respect to ε of the functional $\mathscr{J}_\varepsilon$:

$$\mathscr{J}_\varepsilon(h) = \frac{1}{2}\int_{\Omega_2} (\mathbf{v}_h(x) - z(x))^2 dx$$

$$= \frac{1}{2}\int_\Omega (\mathbf{v}_h(x) - z(x))^2 \mathbb{1}_{\Omega_2}(x) dx$$

$$= \frac{1}{2}\int_\Omega (v^0(x) - z(x))^2 \mathbb{1}_{\Omega_2}(x) dx$$

$$+ \int_\Omega (v^0(x) - z(x))(\lambda(h)G(x,h)v^0(h))\mathbb{1}_{\Omega_2}(x) dx + \mathscr{R}_1(\varepsilon,h), \tag{75}$$

with

$$\mathscr{R}_1(\varepsilon,h) = (\lambda(h)v^0(h))^2 \int_{\Omega_2} G(x,h)^2 dx \leq M(\log\varepsilon)^{-2}, \tag{76}$$

and M depends only on the shape of Ω and Ω_2. Then we introduce the following adjoint state p

$$-\Delta p(x) = (v^0(x) - z(x))\mathbb{1}_{\Omega_2}(x) \quad \text{in } \Omega, \tag{77}$$

$$p(x) = 0 \quad \text{on } \partial\Omega, \tag{78}$$

and after an integration by parts we obtain

$$\mathscr{J}_\varepsilon(h) = \frac{1}{2}\int_{\Omega_2} (v^0(x) - z(x))^2 dx + \lambda(h)v^0(h)p(h) + \mathscr{R}_1(\varepsilon,h), \tag{79}$$

Since $\lambda(h)$ admits the expansion

$$\lambda(h) = 2\pi(\log\varepsilon)^{-1} + O((\log\varepsilon)^{-2}) \quad \text{as} \quad \varepsilon \to 0,$$

we can expand (79) further:

$$\mathscr{J}_\varepsilon(h) = \frac{1}{2} \int_{\Omega_2} (v^0(x) - z(x))^2 dx - 2\pi |\log\varepsilon|^{-1} v^0(h)p(h) + \mathscr{R}_2(\varepsilon,h), \qquad (80)$$

with

$$\mathscr{R}_2(\varepsilon,h) = O((\log\varepsilon)^{-2}) \quad \text{as} \quad \varepsilon \to 0. \qquad (81)$$

Therefore we have obtained expansion (74) with $\rho(\varepsilon) = 2\pi |\log\varepsilon|^{-1}$ and $\mathscr{T}(h) = -v^0(h)p(h)$. We will use this result later to initialize our algorithm.

Remark 1. *As expected, the topological derivative $\mathscr{T}(h)$ is easy to compute because it requires only the solution of two Laplacian equations on Ω. The drawback is that the rest $\mathscr{R}_2(\varepsilon,h)$ is of order $(\log\varepsilon)^{-2}$, which is not negligible for numerical calculations compared to the main term in $(\log\varepsilon)^{-1}$. Therefore, there is a lack of precision when using the topological derivative to localize the point, and we will see in our numerical tests that this lack of precision can lead to inaccurate results.*

6   Algorithm

The observation z in the functional $\mathscr{J}_\varepsilon$ corresponds to measured data. For the numerical tests, the position of the hole is known beforehand, and the observation is computed accordingly.

One should note that if h^* is the real position of the center of the hole, used to compute the function z, we cannot expect h^* to be the solution of the minimization problem (68) because the corresponding solution $\mathbf{v}_{h^*}$ is only an approximation of z. One can be more precise by looking at $\mathscr{J}_\varepsilon(h^*)$:

$$\mathscr{J}_\varepsilon(h^*) = \frac{1}{2} \int_{\Omega_2} (\mathbf{v}_{h^*}(x) - z(x))^2 dx.$$

According to (49), there exists M such that

$$\|v_{h^*} - z\|_{L^2(\Omega_\varepsilon)} \le M\varepsilon,$$

and as a consequence

$$\mathscr{J}_\varepsilon(h^*) \le M^2 \varepsilon^2,$$

which means that h^* is not necessarily the optimal solution.

Usually, the topological derivative can be used to find the position of the hole. A possible way to proceed is to compute the topological derivative $\mathscr{T}(h)$ at every point $h \in \Omega_1$, and look for the minimum of $\mathscr{T}(h)$ which should give an approximate position of the unknown hole. Unfortunately, due to the lack of information since z is known only on Ω_2, and according to Remark 1, the numerical tests show that this is not enough to find the exact position of the hole. It is possible to go further in the expansion of the topological derivative to have a better approximation, but then we obtain a function which is difficult to evaluate at every point of the domain because for every h, one needs to solve a Laplace equation (to find the function $\mathscr{G}$ and compute $\mathscr{G}(h,h)$).

Therefore, the idea of the algorithm is to initialize the position of the hole by using the topological derivative, and then to use a Fletcher-Reeves conjugate gradient. This can be related to the speed method in shape optimization, since once the hole is created, moving the position h is equivalent to moving the boundary of the small hole with a uniform speed (the shape of the hole remains constant). However, the formulas for the derivatives of $\mathscr{J}_\varepsilon$ have been derived easily and the numerical application is also less involved than for the usual speed method setting, where the boundary of the hole needs to be parameterized.

6.1 The Discrete Algorithm

The usual Fletcher-Reeves algorithm is as follows. At the step $k+1$ the point h_k becomes

$$h_{k+1} = h_k + t_k d_k, \tag{82}$$

where d_k is the direction of descent given by

$$d_k = -\nabla_h \mathscr{J}(h_k) + \frac{\nabla_h \mathscr{J}(h_k)^T \nabla_h \mathscr{J}(h_k)}{\nabla_h \mathscr{J}(h_{k-1})^T \nabla_h \mathscr{J}(h_{k-1})} d_{k-1}, \tag{83}$$

and the time step t_k is given by a line-search procedure.

Algorithm: In the subsequent algorithm subscript l, for all quantities refers to the discrete counterpart of the respective continuous variable, while the superscript k refers to the k-th iteration of the algorithm. The discretization is based on finite differences. We assume that the discretized domain is given by a uniform grid with mesh size l. We denote the grid points by $\mathbf{x}_i, i = 1,..,N$. For the discretization of the Laplace operator we use the standard five points stencil. The grid functions $v_l^0, G_l,...$ are defined on the grid points.

Step 1: Set $k = 0$. Compute $v_l^{0,k}$ from (4)-(5) and the topological derivative $\mathscr{T}_l^k(h)$ at every $h \in \Omega_1$. Deduce a starting point h_0 by taking $h_0 = \mathrm{argmin}_{h \in \Omega_1} \mathscr{T}_l^k(h)$. Set a tolerance γ.

Step 2: Compute $G_l^k, \mathscr{G}_l^k$ from the discrete relaxed system corresponding to (8)-(9) and (11)-(12) and deduce $\lambda_l^k(h)$ and $\mathbf{v}_l^k$. Evaluate the cost functional $J_{\varepsilon,l}^k(h^k)$.

Step 3: If the direction d^k verifies $\|d^k\| < \gamma l (1 + d^0)$ and $\|h^k - h^{k-1}\| < \gamma l$ then stop; otherwise continue with step 4.

Step 4: Update h^k. Put $k := k + 1$. Go to step 2.

7 Numerical Example

Several examples are presented here, with different source terms f. The domain Ω is taken as the square $\Omega = [0,1] \times [0,1]$. For each example, the number of iterations are given, the initial value for h (given by the topological derivative) and the final value of h are compared to the real position of the hole. We also give a plot for the grid 512×512 of the convergence of the functional $\mathcal{J}_\varepsilon(h)$, of the measured data z and of the reconstructed solution $\mathbf{v}_{h_f}$.

The observation z is artificial, which means that we know *ad hoc* the location of the hole h^* but we start the procedure from another value of h. In order to compute this observation z precisely, we use finite differences with a Shortley-Weller approximation [5] to discretize the Laplacian on the boundary of Ω_ε. The position h^* denotes the real center of the hole. The set ω is chosen as a ball of radius 1. The notation h_0 and h_f denote the initial and final value of h, respectively. In every example we take $\Omega_1 = B(h^*, 0.4)$, but h^* takes different values. In what follows, (x_1, x_2) denotes the Cartesian coordinates in $\mathbb{R}^2$.

Example 1. In the first example, the data is $f \equiv 70$ and $h^* = (0.5, 0.5)$. Due to the very simple and symmetric situation, the topological derivative is enough to find the position of the hole and no further iterations are needed after the initialization. Results are presented in Table 1.

Table 1. Example 1

l	iterations	h_0	h_f	h^*
$1/128$	1	$(0.5,0.5)$	$(0.5,0.5)$	$(0.5,0.5)$
$1/256$	1	$(0.5,0.5)$	$(0.5,0.5)$	$(0.5,0.5)$
$1/512$	1	$(0.5,0.5)$	$(0.5,0.5)$	$(0.5,0.5)$

Example 2. In the second example, the data is $f(x_1, x_2) = 100x_1^2 x_2 + 10$ and $h^* = (0.4, 0.4)$. One can see in this example that the topological derivative (which gives h_0) was far from the optimal solution while our algorithm converges to the true solution as $l \to 0$. The corresponding figure is Fig. 2. Results are presented in Table 2.

Table 2. Example 2

l	iterations	h_0	h_f	h^*
$1/128$	11	$(0.6953, 0.6640)$	$(0.3750, 0.3570)$	$(0.4, 0.4)$
$1/256$	12	$(0.6953, 0.6679)$	$(0.3913, 0.3828)$	$(0.4, 0.4)$
$1/512$	14	$(0.6933, 0.6718)$	$(0.3965, 0.3959)$	$(0.4, 0.4)$

Example 3. In the third example, the data is $f(x_1, x_2) = 100\sin(4\pi x_1)\sin(4\pi x_2) + 10$ and $h^* = (0.6, 0.6)$. The topological derivative gives again an initialization far from the optimal solution while our algorithm converges to the true solution as $l \to 0$. The corresponding figure is Fig. 3. Results are presented in Table 2.

Example 4. In the fourth example, the data is

$$f(x_1, x_2) = -(60\cos(4\pi r) + 30)\mathbb{1}_{\{r < 0.4 \ \& \ r > 0.1\}}$$

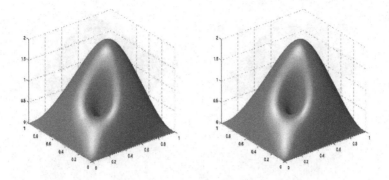

Fig. 2. Solution $\mathbf{v}_{h_f}$ (left) and data z (right)

Table 3. Example 3

l	iterations	h_0	h_f	h^*
1/128	29	(0.3593, 0.3593)	(0.6405, 0.6405)	(0.6,0.6)
1/256	23	(0.3593 , 0.3593)	(0.6214, 0.6214)	(0.6,0.6)
1/512	21	(0.3613,0.3613)	(0.6055,0.6055)	(0.6,0.6)

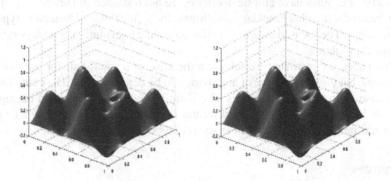

Fig. 3. Solution $\mathbf{v}_{h_f}$ (left) and data z (right)

Table 4. Example 4

l	iterations	h_0	h_f	h^*
1/128	63	(0.4296,0.4296)	(0.6477,0.6477)	(0.6,0.6)
1/256	53	(0.4296,0.4257)	(0.6195 ,0.6202)	(0.6,0.6)
1/512	44	(0.4277,0.4277)	(0.6042,0.6042)	(0.6,0.6)

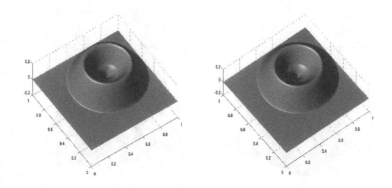

Fig. 4. Solution $\mathbf{v}_{h_f}$ (left) and data z (right)

with $r = |(x_1, x_2) - h^*|$ and $h^* = (0.6, 0.6)$. The corresponding figure is Fig. 4. Results are presented in table 4. Similar conclusions as for the previous examples are drawn.

8   Conclusion

In the numerical examples above, the experimental data z is given on approximately 50% of the domain. Except for the simple first example, the topological derivative is not able to find the correct position of the hole, while our algorithm does.

Self-adjoint extensions of elliptic operators are not restricted to Dirichlet conditions, and can be used with other boundary conditions, including those of Neumann type. Further research will be focused on applying this modeling to evolution boundary problems and spectral problems.

In this paper, we restricted ourselves to the case of a small ball for the numerical applications, but the general case of any shape can be derived easily. The case of several holes is also interesting, and requires additional work for the theoretical background. Indeed, in the expansion of the solution, the holes interact with each other and the expansion is more involved. Such an analysis can be found, e.g. in [14].

References

1. Allaire, G.: Shape Optimization by the Homogenization Method. Springer, New York (2002)
2. Bendsoe, M.P., Sigmund, O.: Topology Optimization. Theory, Methods and Applications. Springer, New York (2003)
3. Demkov, I.N., Ostrovsky, V.N.: A Method of Zero Radius Potential in Atomic Physics. Leningrad University, Leningrad (1975) (Russian)
4. Garreau, S., Guillaume, Ph., Masmoudi, M.: The topological asymptotic for PDE systems: the elasticity case. SIAM J. Control Optim. 39(6), 1756–1778 (2001)
5. Hackbusch, W.: Elliptic Differential Equations. Springer Series in Computationnal Mathematics, vol. 18. Springer, Berlin (1992)
6. Henrot, A., Pierre, M.: Variation et optimisation de formes: une analyse géométrique. No. 48, de Mathématiques et Applications. Springer, Heidelberg (2005)

7. Kamotskii, I.V., Nazarov, S.A.: Spectral problems in singularly perturbed domains and self-adjoint extensions of differential operators. Amer. Math. Soc. Transl. 199(2) (2000)
8. Kanaun, S.K., Levin, V.M.: The Effective Field Method in the Mechanics of Composite Materials, Izdatel'stvo Petrozavodskogo Universiteta, Petrozavodsk (1993) (Russian)
9. Laurain, A.: Singularly Perturbed Domains in Shape Optimization, PhD Thesis, Université Henri Poincaré, Nancy 1 (2006)
10. Nazarov, S.A.: Asymptotic conditions at a point, self adjoint extensions of operators, and the method of matched asymptotic expansions. American Mathematical Society Translations 198(2), 77–125 (1999)
11. Nazarov, S.A., Sokołowski, J.: Asymptotic analysis of shape functionals. Journal de Mathématiques pures et appliquées 82, 125–196 (2003)
12. Nazarov, S.A., Sokołowski, J.: Self adjoint extensions of differential operators in application to shape optimization. Comptes Rendus Mecanique 331(10), 667–672 (2003)
13. Nazarov, S.A., Sokołowski, J.: Selfadjoint extensions for elasticity system in application to shape optimization. To appear in Bulletin of the Polish Academy of Sciences – Mathematics
14. Nazarov, S.A., Sokołowski, J.: Self adjoint extensions for the Neumann Laplacian in application to shape optimization, Les prépublications de l'institut Elie Cartan 9 (2003)
15. Pavlov, B.S.: The theory of extension and explicitly soluble models. Uspehi Mat. Nauk 42(6), 99–131 (1987); Engl. Transl. in Soviet Math. Surveys 42(6), 127–168 (1987)
16. Sokołowski, J., Żochowski, A.: On the topological derivative in shape optimization. SIAM Journal on Control and Optimization 37(4), 1251–1272 (1999)
17. Sokołowski, J., Żochowski, A.: Optimality conditions for simultaneous topology and shape optimization. SIAM Journal on Control and Optimization 42(4), 1198–1221 (2003)

Fundamental Solutions and Optimal Control
of Neutral Systems

Kai Liu

Division of Statistics and Probability, Department of Mathematical Sciences,
The University of Liverpool, Peach Street, Liverpool, L69 7ZL, U.K.
k.liu@liv.ac.uk

Abstract. In this work, we shall consider standard optimal control problems for a class of neutral functional differential equations in Banach spaces. As the basis of a systematic theory of neutral models, the fundamental solution is constructed and a variation of constants formula of mild solutions is established. Necessary conditions in terms of the solutions of neutral adjoint systems are established to deal with the fixed time integral convex cost problem of optimality. Based on optimality conditions, the maximum principle for time varying control domain is presented.

1   Introduction

Let X be a separable Banach space with norm $\|\cdot\|_X$. For any fixed constant $r > 0$, we denote by $L_r^2 = L^2([-r,0];X)$ the space of all X-valued equivalence classes of measurable functions which are square integrable on $[-r,0]$. Let $\mathscr{X}$ denote the product Banach space $X \times L_r^2$ with the norm

$$\|\phi\|_{\mathscr{X}} = \sqrt{\|\phi_0\|_X^2 + \|\phi_1\|_{L_r^2}^2} \quad \text{for all} \quad \phi = (\phi_0, \phi_1) \in \mathscr{X}. \tag{1}$$

Consider the following neutral functional differential equation on X,

$$\frac{d}{dt}\left[y(t) - \int_{-r}^0 D(\theta)y(t+\theta)d\theta\right] = Ay(t) + \int_{-r}^0 d\eta(\theta)y(t+\theta) + f(t)$$
$$\text{for any } t > 0, \tag{2}$$
$$y(0) = \phi_0, \; y_0(\cdot) = \phi_1(\cdot), \;\; \phi = (\phi_0, \phi_1) \in \mathscr{X},$$

where $A : X \to X$ with domain $\mathscr{D}(A) \subset X$ is the infinitesimal generator of a C_0-semigroup $\{S(t); t \geq 0\}$ on X and $y_t(\theta) := y(t+\theta)$ for any $\theta \in [-r,0]$ and $t \geq 0$. Here $f(\cdot)$ is some properly given function in X and η is the Stieltjes measure given by

$$\eta(\tau) = -\sum_{i=1}^m \chi_{(-\infty,-r_i]}(\tau)A_i - \int_\tau^0 B(\theta)d\theta, \quad \tau \in [-r,0]. \tag{3}$$

It is assumed that $0 < r_1 < r_2 < \cdots < r_m \leq r$, $A_i \in \mathscr{L}(X)$, $i = 1, \cdots, m$, the family of all bounded, linear operators on X and $B(\cdot), D(\cdot) \in L^2([-r,0];\mathscr{L}(X))$, the Banach space of all $\mathscr{L}(X)$-valued equivalence classes of square integrable functions on $[-r,0]$.

A. Korytowski et al. (Eds.): System Modeling and Optimization, IFIP AICT 312, pp. 350–369, 2009.

The abstract formulation (2) in an infinite dimensional space has been well motivated both theoretically and practically by such systems as neutral partial functional differential equations. There exists an extensive literature which deals with various problems of the so-called distributed parameter systems with time delays in a Banach space (see the monograph [19], for instance, and references cited therein for a systematic statement). The system (2) in finite or infinite dimensions was considered in a systematic way by [8] and [19] in spaces of continuous functions, e.g., in $C([-r,0];X)$, the space of all continuous functions from $[-r,0]$ into X. The same phase spaces were also used by many works such as [7,9] and references cited therein. Although this choice certainly has its advantages, it is often useful to work in L^p space instead. There are at least two good reasons for this setting. First, it allows ones to work in a Hilbert space. This is particularly important in control, stability and optimality theory in which an adjoint theory for the system (2) is essentially needed, e.g., [1,5,6,11,17]. Unfortunately, as indicated in Hale and Lunel [9], it is generally difficult to have an adjoint theory in a continuous functions space setting. Second, the choice of L^p-phase space allows us to consider equations having discontinuous solutions. This is the case when ones consider discontinuous initial functions for (2) or the system is perturbed by some random resources with jumps (cf. [14]).

Although there is some work, e.g., [13], on optimal control of functional differential equations in infinite dimensions available in the existing literature, there exist however few results on the same topic for functional differential equations of neutral type. The current work will devote itself to exploring some basic material. We shall consider the fixed time integral convex cost problem for mild solutions of a class of functional differential equations of neutral type in Banach spaces. To this end, we first formulate and study a class of linear neutral systems to derive fundamental results. Precisely, as the basis of whole theory we shall construct fundamental solutions or Green operators and establish a variation of constants formula. Adjoint neutral systems are considered and the associated representation formulae of adjoint states in terms of fundamental solutions are established. All of these will allow us to present results on the existence of optimal controls, necessary conditions of optimality and maximum principle.

The following are some notations and terminologies to be used in the work. The symbol $\mathbb{R}_+$ denotes the set of all nonnegative numbers and $\mathbb{R}^n$ denotes the n-dimensional real vector space with the usual Euclidean norm $\|\cdot\|_{\mathbb{R}^n}$. For any $\lambda \in \mathbb{C}^1$, the symbols $\Re(\lambda)$ and $\Im(\lambda)$ denote the real and imaginary parts of complex number λ, respectively. For any separable Banach space U, we use U^* to denote its adjoint space and $\langle \cdot, \cdot \rangle_{U,U^*}$ the dual pairing, respectively. We use $\mathscr{L}(U,X)$ to denote the space consisting of all bounded linear operators T from U into X with domain U. When $X = U$, $\mathscr{L}(X,X)$ is denoted by $\mathscr{L}(X)$. Every operator norm is simply denoted by $\|\cdot\|$ when there is no danger of confusion. The symbols $\mathscr{D}(T)$ and $\mathscr{R}(T)$ will be used to denote the domain and range of operator T, respectively. For a closed linear operator A on a dense domain $\mathscr{D}(A) \subset X$ into X, its adjoint operator is denoted by A^*. Given an interval $E \subset \mathbb{R}^1$, the function χ_E denotes the characteristic function on the interval E. For a measurable function $f : \mathbb{R}^1 \to X$, its Laplace transform $\hat{f}$ is defined by

$$\hat{f}(\lambda) = \int_0^\infty e^{-\lambda t} f(t) dt$$

whenever the Bochner integral exists.

2 Fundamental Solutions

In this section, we shall consider a class of linear autonomous neutral functional differential equations on X which are defined formally by

$$\frac{d}{dt}\left[y(t) - \int_{-r}^0 D(\theta)y(t+\theta)d\theta\right] = Ay(t) + \int_{-r}^0 d\eta(\theta)y(t+\theta) \text{ for any } t > 0, \tag{4}$$
$$y(0) = \phi_0, \ y_0(\cdot) = \phi_1(\cdot), \quad \phi = (\phi_0, \phi_1) \in \mathscr{X},$$

where $A : X \to X$ and η are defined as in (2) and (3). Generally, it is quite restrictive to find a solution in the usual sense for the equation (4). Instead, it is hoped to consider an "integrated" form of the system (4). To this end, we further assume that for each i, $i = 1, \cdots, m$, and $\theta \in [-r, 0]$, $\mathscr{R}(D(\theta)) \subset \mathscr{D}(A)$ such that $AD(\cdot) \in L^2([-r,0]; \mathscr{L}(X))$.

Consider the following integral equation on X

$$y(t,\phi) = \int_{-r}^0 D(\theta)y(t+\theta,\phi)d\theta + S(t)\left[\phi_0 - \int_{-r}^0 D(\theta)\phi_1(\theta)d\theta\right]$$
$$+ \int_0^t S(t-s)\left[\int_{-r}^0 d\eta(\theta)y(s+\theta,\phi) + \int_{-r}^0 AD(\theta)y(s+\theta,\phi)d\theta\right]ds, \ \forall t > 0,$$
$$y(0,\phi) = \phi_0, \ y_0(\cdot,\phi) = \phi_1(\cdot), \quad \phi = (\phi_0,\phi_1) \in \mathscr{X}. \tag{5}$$

For simplicity, we sometimes denote $y(t,\phi)$ and $y_t(\cdot,\phi)$ by $y(t)$ and $y_t(\cdot)$, respectively, in the remainder of this work. The following existence and uniqueness of solutions of Equation (5) can be established in the spirit of Datko [5] and Wu [18].

Theorem 1. *For arbitrary $T \geq 0$, $\phi = (\phi_0, \phi_1) \in \mathscr{X}$, (i) there exists a unique solution $y(t,\phi) \in L^2([-r,T];X)$ of (5); (ii) for arbitrary $t \in [0,T]$, $\|y(t,\phi)\|_X \leq Ce^{\gamma t}\|\phi\|_{\mathscr{X}}$ almost everywhere for some constants $\gamma \in \mathbb{R}^1$ and $C > 0$.*

The solution $y(t,\phi)$ of the equation (5) is called a *mild solution* of (4). For any $x \in X$, let $\phi_0 = x$, $\phi_1(\theta) = 0$ for $\theta \in [-r,0)$ and $\phi = (x,0)$, we define the *fundamental solution* $G(t) \in \mathscr{L}(X), t \in \mathbb{R}^1$, of (5) with such an initial datum by

$$G(t)x = \begin{cases} y(t,\phi), & t \geq 0, \\ 0, & t < 0. \end{cases} \tag{6}$$

The term (6) implies that $G(t)$ is a unique solution of the equation

$$G(t) = S(t) + \int_{-r}^0 D(\theta)G(t+\theta)d\theta + \int_0^t S(t-s)\left[\int_{-r}^0 d\eta(\theta)G(s+\theta)\right.$$
$$\left. + \int_{-r}^0 AD(\theta)G(s+\theta)d\theta\right]ds, \quad \text{if } t \geq 0, \tag{7}$$
$$G(t) = O, \quad \text{if } t < 0,$$

where O is the null operator on X. It is immediate to see that $G(t)$ is strongly continuous on $\mathbb{R}^1$ and satisfies

$$\|G(t)\| \leq C \cdot e^{\gamma t}, \qquad t \geq 0, \tag{8}$$

for some $C > 0$ and $\gamma \in \mathbb{R}^1$.

2.1   Variation of Constants Formula

Consider a class of non-autonomous neutral functional differential equations on X

$$\frac{d}{dt}\left[y(t) - \int_{-r}^{0} D(\theta)y(t+\theta)d\theta\right] = Ay(t) + \int_{-r}^{0} d\eta(\theta)y(t+\theta) + f(t)$$

$$\text{for any } t > 0, \tag{9}$$

$$y(0) = \phi_0, \ y_0(\cdot) = \phi_1(\cdot), \quad \phi = (\phi_0, \phi_1) \in \mathscr{X},$$

where $A : X \to X$, η are defined as in the last subsection and $f(\cdot) \in L^2_{loc}(\mathbb{R}_+; X)$, the Fréchet space of functions which belong to $L^2([0,T];X)$ for any $T \geq 0$. Once again, we intend to consider the following integral equation of (9) on X,

$$y(t, f, \phi) = \int_{-r}^{0} D(\theta)y(t+\theta, f, \phi)d\theta + S(t)\left[\phi_0 - \int_{-r}^{0} D(\theta)\phi_1(\theta)d\theta\right]$$

$$+ \int_{0}^{t} S(t-s)\left[\int_{-r}^{0} d\eta(\theta)y(s+\theta, f, \phi)\right.$$

$$\left. + \int_{-r}^{0} AD(\theta)y(s+\theta, f, \phi)d\theta + f(s)\right]ds, \ \forall t > 0, \tag{10}$$

$$y(0, \phi) = \phi_0, \ y_0(\cdot, \phi) = \phi_1(\cdot), \quad \phi = (\phi_0, \phi_1) \in \mathscr{X}.$$

It is extremely useful to find an explicit representation for the solution $y(t, f, \phi)$ of (10) in applications, e.g., in the optimal control theory. This is possible if we restrict the initial data of (10) to some proper subset of $\mathscr{X}$. Indeed, let $W^{1,2}([-r,0];X)$ denote the Sobolev space of X-valued functions $x(\cdot)$ on $[-r, 0]$ such that $x(\cdot)$ and its distributional derivative belong to $L^2([-r,0];X)$, and define $\mathscr{W}^{1,2} = X \times W^{1,2}([-r,0];X)$.

The following *variation of constants formula* (11) provides a representation for solutions of (10) in terms of the fundamental solution $G(t) \in \mathscr{L}(X)$.

Theorem 2. *For arbitrary $\phi = (\phi_0, \phi_1) \in \mathscr{W}^{1,2}$, the solution $y(t, f, \phi)$ of (10) can be represented almost everywhere by*

$$y(t, f, \phi) = G(t)\phi_0 - V(t, 0)\phi_1(0) + \int_{-r}^{0} U(t, \theta)\phi_1(\theta)d\theta$$

$$+ \int_{-r}^{0} V(t, \theta)\phi_1'(\theta)d\theta + \int_{0}^{t} G(t-s)f(s)ds, \quad t \geq 0, \tag{11}$$

where for any $t \geq 0$, the kernels

$$U(t, \theta) = \int_{-r}^{\theta} G(t - \theta + \tau)d\eta(\tau) \in L^2([-r,0]; \mathscr{L}(X)), \qquad \theta \in [-r, 0], \tag{12}$$

and, similarly,

$$V(t,\theta) = \int_{-r}^{\theta} G(t-\theta+\tau)D(\tau)d\tau \in L^2([-r,0];\mathscr{L}(X)), \qquad \theta \in [-r,0]. \tag{13}$$

Proof. We first prove (11) by assuming $f \in L^2(\mathbb{R}_+;X) \cap L^1(\mathbb{R}_+;X)$. To this end, define

$$x(t) = G(t)\phi_0 - V(t,0)\phi_1(0) + \int_{-r}^{0} U(t,\theta)\phi_1(\theta)d\theta + \int_{-r}^{0} V(t,\theta)\phi_1'(\theta)d\theta$$
$$+ \int_{0}^{t} G(t-s)f(s)ds, \quad t \geq 0, \tag{14}$$

and $x(t) = \phi_1(t)$ for $t \in [-r,0)$. It is easy to see that $x(t) \in L^2([0,T];X)$ and $x(t)$ is almost everywhere continuous on $[0,T]$. For $\lambda \in \mathbb{C}^1$ with $\Re(\lambda)$ large enough, define

$$M(\lambda) = I - \int_{-r}^{0} e^{\lambda\theta}D(\theta)d\theta,$$
$$N(\lambda) = R(\lambda,A)\left[\int_{-r}^{0} e^{\lambda\theta}d\eta(\theta) + \int_{-r}^{0} e^{\lambda\theta}D(\theta)d\theta\right], \tag{15}$$

where I denotes the identity operator on X and $R(\lambda,A) = (\lambda I - A)^{-1}$ is the resolvent operator of A. From the structure of both $M(\lambda)$ and $N(\lambda)$, we see that $M(\lambda) \to I$ and $\|N(\lambda)\| \to 0$ as $\Re(\lambda) \to +\infty$. This implies that there exists a real number λ_0 such that for $\Re(\lambda) \geq \lambda_0$, both $M^{-1}(\lambda)$ and $[I - N(\lambda)M^{-1}(\lambda)]^{-1}$ exist. Therefore, we can apply the convolution theorem on Laplace transforms to (7) to obtain that

$$\hat{G}(\lambda) = R(\lambda,A) + \int_{-r}^{0} e^{\lambda\theta}D(\theta)d\theta \cdot \hat{G}(\lambda) + R(\lambda,A)\int_{-r}^{0} e^{\lambda\theta}d\eta(\theta) \cdot \hat{G}(\lambda)$$
$$+ R(\lambda,A)\int_{-r}^{0} e^{\lambda\theta}AD(\theta)d\theta \cdot \hat{G}(\lambda), \tag{16}$$

where $\hat{G}(\lambda)$ denotes the Laplace transform of $G(\cdot)$. This yields that for $\Re(\lambda) \geq \lambda_0$,

$$\hat{G}(\lambda) = M^{-1}(\lambda)[I - N(\lambda)M^{-1}(\lambda)]^{-1}R(\lambda,A). \tag{17}$$

Note that the Laplace transform $\hat{x}(\lambda)$ of $x(\cdot)$ makes sense for sufficiently large $\Re(\lambda)$. Therefore, we apply the convolution theorem on Laplace transforms to (14) and use Fubini's theorem to obtain that

$$\hat{x}(\lambda) = \hat{G}(\lambda)\left[\phi_0 - \int_{-r}^{0} D(\theta)\phi_1(\theta)d\theta + \left(\sum_{i=1}^{m} A_i e^{-\lambda r_i} \int_{-r_i}^{0} e^{-\lambda\tau}\phi_1(\tau)d\tau\right)\right.$$
$$+ \left(\int_{-r}^{0} e^{\lambda\theta}B(\theta)\int_{\theta}^{0} e^{-\lambda\tau}\phi_1(\tau)d\tau d\theta\right)$$
$$+ \int_{-r}^{0} \lambda e^{\lambda\theta}D(\theta)\int_{\theta}^{0} e^{-\lambda\tau}\phi_1(\tau)d\tau d\theta + \hat{f}(\lambda)\Bigg] \tag{18}$$
$$= \hat{G}(\lambda)\left[\phi_0 - \int_{-r}^{0} e^{\lambda\theta}D(\theta)\phi_1(0)d\theta + \int_{-r}^{0} e^{\lambda\theta}D(\theta)\int_{\theta}^{0} e^{-\lambda\tau}d\phi_1(\tau)d\theta\right.$$
$$+ \int_{-r}^{0} e^{\lambda\theta}d\eta(\theta)\int_{\theta}^{0} e^{-\lambda\tau}\phi_1(\tau)d\tau + \hat{f}(\lambda)\Bigg].$$

On the other hand, since $y(\cdot)$ satisfies the equation (10) and note that $y(t) = \phi_1(t)$ for $t \in [-r, 0)$, we can use Fubini's theorem again to calculate the Laplace transform $\hat{y}(\lambda)$ of $y(\cdot)$ which is given by

$$
\hat{y}(\lambda) = \int_{-r}^{0} e^{\lambda\theta}D(\theta)d\theta\hat{y}(\lambda) + \int_{-r}^{0} e^{\lambda\theta}D(\theta) \int_{\theta}^{0} e^{-\lambda\tau}\phi_1(\tau)d\tau d\theta
$$

$$
+ R(\lambda, A)\left\{\left[\phi_0 - \int_{-r}^{0} D(\theta)\phi_1(\theta)d\theta\right] + \left[\int_{-r}^{0} e^{\lambda\theta}d\eta(\theta) + \int_{-r}^{0} e^{\lambda\theta}AD(\theta)d\theta\right]\hat{y}(\lambda)\right.
$$

$$
+ \left[\int_{-r}^{0} e^{\lambda\theta}d\eta(\theta)\int_{\theta}^{0} e^{-\lambda\tau}\phi_1(\tau)d\tau + \right.
$$

$$
\left.\left.\int_{-r}^{0} e^{\lambda\theta}AD(\theta)\int_{\theta}^{0} e^{-\lambda\tau}\phi_1(\tau)d\tau d\theta\right] + \hat{f}(\lambda)\right\},
$$

$$\tag{19}$$

for sufficiently large $\Re(\lambda)$. In terms of (15), we can rewrite (19) as

$$
(M(\lambda) - N(\lambda))\hat{y}(\lambda)
$$

$$
= \left\{\left(I - \int_{-r}^{0} e^{\lambda\theta}D(\theta)d\theta\right) - R(\lambda, A)\left[\int_{-r}^{0} e^{\lambda\theta}d\eta(\theta) + \int_{-r}^{0} e^{\lambda\theta}AD(\theta)d\theta\right]\right\}\hat{y}(\lambda)
$$

$$
= \int_{-r}^{0} e^{\lambda\theta}D(\theta)\int_{\theta}^{0} e^{-\lambda\tau}\phi_1(\tau)d\tau d\theta + R(\lambda, A)\left[\phi_0 - \int_{-r}^{0} D(\theta)\phi_1(\theta)d\theta\right]
$$

$$
+ R(\lambda, A)\left[\int_{-r}^{0} d\eta(\theta)\int_{\theta}^{0} e^{\lambda(\theta-\tau)}\phi_1(\tau)d\tau + \right.
$$

$$
\left.\int_{-r}^{0} AD(\theta)\int_{\theta}^{0} e^{\lambda(\theta-\tau)}\phi_1(\tau)d\tau d\theta + \hat{f}(\lambda)\right],
$$

which immediately yields that

$$
\hat{y}(\lambda) = M^{-1}(\lambda)[I - N(\lambda)M^{-1}(\lambda)]^{-1}\left\{\int_{-r}^{0} e^{\lambda\theta}D(\theta)\int_{\theta}^{0} e^{-\lambda\tau}\phi_1(\tau)d\tau d\theta\right.
$$

$$
+ R(\lambda, A)\left[\phi_0 - \int_{-r}^{0} D(\theta)\phi_1(\theta)d\theta\right]
$$

$$
+ R(\lambda, A)\left[\int_{-r}^{0} e^{\lambda\theta}d\eta(\theta)\int_{\theta}^{0} e^{-\lambda\tau}\phi_1(\tau)d\tau + \right. \tag{20}
$$

$$
\left.\int_{-r}^{0} e^{\lambda\theta}AD(\theta)\int_{\theta}^{0} e^{-\lambda\tau}\phi_1(\tau)d\tau d\theta\right]
$$

$$
\left. + R(\lambda, A)\hat{f}(\lambda)\right\}.
$$

Then, by virtue of (17), it is immediate to see that for large $\Re(\lambda)$,

$$
\begin{aligned}
\hat{y}(\lambda) &= \hat{G}(\lambda)\left[(\lambda I - A)\left(\int_{-r}^{0} e^{\lambda\theta}D(\theta)\int_{\theta}^{0}e^{-\lambda\tau}\phi_1(\tau)d\tau d\theta\right)+\right.\\
&\quad \phi_0 - \int_{-r}^{0}D(\theta)\phi_1(\theta)d\theta\\
&\quad + \int_{-r}^{0}e^{\lambda\theta}d\eta(\theta)\int_{\theta}^{0}e^{-\lambda\tau}\phi_1(\tau)d\tau+\\
&\quad \left.\int_{-r}^{0}e^{\lambda\theta}AD(\theta)\int_{\theta}^{0}e^{-\lambda\tau}\phi_1(\tau)d\tau d\theta + \hat{f}(\lambda)\right]\\
&= \hat{G}(\lambda)\left[\phi_0 - \int_{-r}^{0}e^{\lambda\theta}D(\theta)\phi_1(0)d\theta + \int_{-r}^{0}e^{\lambda\theta}D(\theta)\int_{\theta}^{0}e^{-\lambda\tau}d\phi_1(\tau)d\theta\right.\\
&\quad \left.+ \int_{-r}^{0}e^{\lambda\theta}d\eta(\theta)\int_{\theta}^{0}e^{-\lambda\tau}\phi_1(\tau)d\tau + \hat{f}(\lambda)\right].
\end{aligned}
\tag{21}
$$

Therefore, from (18) it follows that

$$
\hat{y}(\lambda) = \hat{x}(\lambda)
$$

for sufficiently large $\Re(\lambda)$. By the uniqueness of Laplace transforms and the almost everywhere strong continuity of $y(t)$ and $x(t)$ on $\mathbb{R}_+$, we obtain that

$$
y(t) = x(t) \quad \text{for almost all } t \in \mathbb{R}_+,
$$

which proves the desired result. Lastly, we shall prove (11) for $f \in L^2_{loc}(\mathbb{R}^+;X)$. To this end, it suffices to prove (11) for $t \in [0,T]$ with any fixed $T \geq 0$. For a given $f \in L^2_{loc}(\mathbb{R}^+;X)$ and $T \geq 0$, we define the truncated function $f_T(t) = \chi_{[0,T]}(t)f(t)$. Then $f_T(\cdot) \in L^2(\mathbb{R}^+;X)\cap L^1(\mathbb{R}^1;X)$ and the corresponding solution $y_T(t)$ of (5) satisfies (11) for all $t \geq 0$. Since $y_T(t) = y(t)$ for $t \in [0,T]$, then (11) is true for all $f(\cdot) \in L^2_{loc}(\mathbb{R}^+;X)$. The proof is now complete. $\qquad\square$

3   Neutral Resolvent Operators

For each $\lambda \in \mathbb{C}^1$, we define the densely defined, closed linear operator $\Delta(\lambda,A,\eta,D)$ by

$$
\Delta(\lambda,A,\eta,D) = \lambda I - A - \int_{-r}^{0}e^{\lambda\theta}d\eta(\theta) - \int_{-r}^{0}\lambda e^{\lambda\theta}D(\theta)d\theta.
$$

The neutral resolvent set $\rho(A,\eta,D)$ is defined as the set of all values λ in $\mathbb{C}^1$ for which the operator $\Delta(\lambda,A,\eta,D)$ has a bounded inverse in X.

Proposition 1. *(i) Let $x \in X$, then $\int_0^t G(s)xds \in \mathscr{D}(A)$ for almost all $t \in \mathbb{R}_+$, and the relation*

$$
\begin{aligned}
A\int_0^t G(s)xds &= G(t)x - x - \int_0^t\int_{-r}^{0}d\eta(\theta)G(s+\theta)xds\\
&\quad - \int_{-r}^{0}D(\theta)G(t+\theta)xd\theta, \quad t \in \mathbb{R}_+, \quad x \in X,
\end{aligned}
\tag{22}
$$

holds almost everywhere.

(ii) Let $x \in \mathscr{D}(A)$, then $\int_0^t G(s)Axds \in X$ for almost all $t \in \mathbb{R}_+$, and the commutative relation

$$\int_0^t G(s)Axds = G(t)x - x - \int_0^t \int_{-r}^0 G(s+\theta)d\eta(\theta)xds$$

$$- \int_{-r}^0 G(t+\theta)D(\theta)xd\theta, \quad t \in \mathbb{R}_+, \quad x \in \mathscr{D}(A), \tag{23}$$

holds almost everywhere.

Proof. We first prove the claim (i). For any $x \in X$ and $\varepsilon > 0$, let $A(\varepsilon) = \varepsilon^{-1}(S(\varepsilon) - I)$. We calculate $A(\varepsilon)\int_0^t G(s)xds$ as follows:

$$A(\varepsilon)\int_0^t G(s)xds = \varepsilon^{-1}\int_0^t \left\{ S(\varepsilon)\left[S(s) + \int_{-r}^0 D(\theta)G(s+\theta)d\theta \right. \right.$$

$$+ \int_0^s S(s-u)\left(\int_{-r}^0 d\eta(\theta)G(u+\theta) + \right.$$

$$\left. \left. \int_{-r}^0 AD(\theta)G(u+\theta)d\theta \right)du \right] - G(s) \right\} xds$$

$$= \varepsilon^{-1}\int_0^t \left\{ G(s+\varepsilon) - G(s) + (S(\varepsilon) - I)\int_{-r}^0 D(\theta)G(s+\theta)d\theta - \right. \tag{24}$$

$$\int_{-r}^0 D(\theta)G(s+\theta+\varepsilon)d\theta$$

$$+ \int_{-r}^0 D(\theta)G(s+\theta)d\theta - \int_0^\varepsilon S(\varepsilon-v)\left[\int_{-r}^0 d\eta(\theta)G(s+v+\theta) \right.$$

$$\left. \left. + \int_{-r}^0 AD(\theta)G(s+v+\theta)d\theta \right]dv \right\} xds, \quad t \geq 0.$$

Note the strong continuity of $G(t)$ and it is not difficult for us to deduce that

$$\lim_{\varepsilon \to 0+} A(\varepsilon)\int_0^t G(s)xds = \lim_{\varepsilon \to 0+} \varepsilon^{-1}\int_0^\varepsilon \left(G(t+s) - G(s) \right)xds$$

$$+ \int_0^t \int_{-r}^0 AD(\theta)G(s+\theta)xd\theta ds - \lim_{\varepsilon \to 0+} \varepsilon^{-1}\left(\int_0^t \int_{-r}^0 D(\theta)G(s+\theta+\varepsilon)d\theta ds \right.$$

$$- \int_0^t \int_{-r}^0 D(\theta)G(s+\theta)d\theta \right)xds - \int_0^t \int_{-r}^0 d\eta(\theta)G(s+\theta)xds -$$

$$\int_0^t \int_{-r}^0 AD(\theta)G(s+\theta)xd\theta ds \tag{25}$$

$$= G(t)x - x - \lim_{\varepsilon \to 0+} \varepsilon^{-1}\left(\int_0^t \int_{-r}^0 D(\theta)G(s+\theta+\varepsilon)d\theta ds - \right.$$

$$\int_0^t \int_{-r}^0 D(\theta)G(s+\theta)d\theta ds \right)x$$

$$- \int_0^t \int_{-r}^0 d\eta(\theta)G(s+\theta)xds.$$

However, it is easy to see that

$$
\lim_{\varepsilon \to 0+} \varepsilon^{-1} \left(\int_0^t \int_{-r}^0 D(\theta) G(s+\theta+\varepsilon) d\theta ds - \int_0^t \int_{-r}^0 D(\theta) G(s+\theta) d\theta ds \right) x
$$
$$
= \int_{-r}^0 D(\theta) \lim_{\varepsilon \to 0+} \left\{ \int_0^\varepsilon \varepsilon^{-1} G(t+s+\theta) ds - \int_0^\varepsilon \varepsilon^{-1} G(s+\theta) ds \right\} x d\theta = \qquad (26)
$$
$$
\int_{-r}^0 D(\theta) G(t+\theta) x d\theta.
$$

Therefore, $\int_0^t G(s) x ds \in \mathscr{D}(A)$ for almost all $t \in \mathbb{R}_+$, and

$$
A \int_0^t G(s) x ds = G(t) x - x - \int_0^t \int_{-r}^0 d\eta(\theta) G(s+\theta) x ds -
$$
$$
\int_{-r}^0 D(\theta) G(t+\theta) x d\theta, \quad t \in \mathbb{R}_+, \qquad (27)
$$

almost everywhere. The proof of (i) is complete.

Next, we intend to prove the relation (23). Firstly, by definition note that for sufficiently large $\mathfrak{R}(\lambda)$,

$$
\Delta(\lambda, A, \eta, D) = (\lambda I - A) \left[I - R(\lambda, A) \int_{-r}^0 e^{\lambda\theta} d\eta(\theta) - R(\lambda, A) \int_{-r}^0 \lambda e^{\lambda\theta} D(\theta) d\theta \right],
$$

and by using (15), we have that

$$
\left\| R(\lambda, A) \int_{-r}^0 e^{\lambda\theta} d\eta(\theta) + R(\lambda, A) \int_{-r}^0 \lambda e^{\lambda\theta} D(\theta) d\theta \right\| \qquad (28)
$$
$$
= \| N(\lambda) - M(\lambda) + I \| \le \| N(\lambda) \| + \| I - M(\lambda) \| \to 0 \text{ as } \mathfrak{R}(\lambda) \to +\infty.
$$

Hence, the bounded inverse

$$
\left[I - R(\lambda, A) \int_{-r}^0 e^{\lambda\theta} d\eta(\theta) - R(\lambda, A) \int_{-r}^0 \lambda e^{\lambda\theta} D(\theta) d\theta \right]^{-1}
$$

exists for sufficiently large $\mathfrak{R}(\lambda)$. For such a $\mathfrak{R}(\lambda)$ it is easy to deduce that

$$
\hat{G}(\lambda) x = \left[I - R(\lambda, A) \int_{-r}^0 e^{\lambda\theta} d\eta(\theta) - R(\lambda, A) \int_{-r}^0 \lambda e^{\lambda\theta} D(\theta) d\theta \right]^{-1} R(\lambda, A) x, \quad \forall x \in X.
$$
$$
\tag{29}
$$

For any $x \in \mathscr{D}(A)$, it then follows from (29) that

$$\lambda \hat{G}(\lambda)x - x =$$

$$\lambda \left[I - R(\lambda,A) \int_{-r}^{0} e^{\lambda\theta} d\eta(\theta) - R(\lambda,A) \int_{-r}^{0} \lambda e^{\lambda\theta} D(\theta) d\theta \right]^{-1} R(\lambda,A)x$$

$$- \left[I - R(\lambda,A) \int_{-r}^{0} e^{\lambda\theta} d\eta(\theta) - R(\lambda,A) \int_{-r}^{0} \lambda e^{\lambda\theta} D(\theta) d\theta \right]^{-1} R(\lambda,A)(\lambda I - A)$$

$$\cdot \left[I - R(\lambda,A) \int_{-r}^{0} e^{\lambda\theta} d\eta(\theta) - R(\lambda,A) \int_{-r}^{0} \lambda e^{\lambda\theta} D(\theta) d\theta \right] x \tag{30}$$

$$= \left[I - R(\lambda,A) \int_{-r}^{0} e^{\lambda\theta} d\eta(\theta) - R(\lambda,A) \int_{-r}^{0} \lambda e^{\lambda\theta} D(\theta) d\theta \right]^{-1} R(\lambda,A)$$

$$\cdot \left\{ \lambda I - \left(\lambda I - A - \int_{-r}^{0} e^{\lambda\theta} d\eta(\theta) - \int_{-r}^{0} \lambda e^{\lambda\theta} D(\theta) d\theta \right) \right\} x$$

$$= \hat{G}(\lambda) \left(A + \int_{-r}^{0} e^{\lambda\theta} d\eta(\theta) + \int_{-r}^{0} \lambda e^{\lambda\theta} D(\theta) d\theta \right) x.$$

On the other hand, note that the following Laplace transform holds

$$\int_{0}^{\infty} e^{-\lambda t} \left(\int_{0}^{t} \left\{ G(s)Ax + \int_{-r}^{0} G(s+\theta) d\eta(\theta)x \right\} ds + \int_{-r}^{0} G(t+\theta) D(\theta) x d\theta \right) dt$$

$$= \lambda^{-1} \hat{G}(\lambda) \left(A + \int_{-r}^{0} e^{\lambda\theta} d\eta(\theta) + \int_{-r}^{0} \lambda e^{\lambda\theta} D(\theta) d\theta \right) x, \tag{31}$$

for sufficiently large $\Re(\lambda)$. Therefore, by (30), (31) and the uniqueness of Laplace transforms, we have that for all $x \in \mathscr{D}(A)$,

$$G(t)x - x = \int_{0}^{t} \left\{ G(s)Ax + \int_{-r}^{0} G(s+\theta) d\eta(\theta)x \right\} ds +$$

$$\int_{-r}^{0} G(t+\theta) D(\theta) x d\theta, \quad t \in \mathbb{R}_{+}, \tag{32}$$

almost everywhere. This completes the proof of (ii). $\square$

4 Adjoint Theory

In the sequel we shall assume that X is reflexive. We intend to establish an adjoint theory of neutral functional differential equations. Let $\psi^* = (\psi_0^*, \psi_1^*) \in \mathscr{X}^*$. The "formal" transposed neutral system of (4) on X^* is defined by

$$\frac{d}{dt} \left[y^*(t) - \int_{-r}^{0} D^*(\theta) y^*(t+\theta) d\theta \right] = A^* y^*(t) + \int_{-r}^{0} d\eta^*(\theta) y^*(t+\theta) + f^*(t),$$

$$t > 0, \tag{33}$$

$$y^*(0) = \psi_0^*, \quad y_0^*(\cdot) = \psi_1^*(\cdot), \psi^* = (\psi_0^*, \psi_1^*) \in \mathscr{X}^*,$$

where $\eta^*(\theta)$, $D^*(\theta)$ and A^* denote the adjoint operators of $\eta(\theta)$, $D(\theta)$ and A, respectively, and $f^* \in L^1([0,T];X^*)$. It is well known that A^* generates a C_0-semigroup

$S^*(t)$ on X^* which is the adjoint of $S(t)$, $t \geq 0$. Hence, we can construct a fundamental solution $G_*(t)$ which is characterized as the unique solution of

$$G_*(t) = S^*(t) + \int_{-r}^0 D^*(\theta)G_*(t+\theta)d\theta + \int_0^t S^*(t-s)\Big[\int_{-r}^0 d\eta^*(\theta)G_*(\theta+s)$$
$$+ \int_{-r}^0 A^*D^*(\theta)G_*(s+\theta)d\theta\Big]ds, \quad t \geq 0, \tag{34}$$
$$G_*(t) = O, \ t < 0.$$

We denote by $G^*(t)$ the adjoint of $G(t)$, $t \in \mathbb{R}^1$. The following theorem shows that $G^*(t) = G_*(t)$ for all $t \in \mathbb{R}^1$.

Theorem 3. *Let $G_*(t)$ be the solution of (34). Then*

$$G^*(t) = G_*(t) \quad \text{for almost all } t \in \mathbb{R}^1.$$

Proof. Since $G(t)$ satisfies (7), then

$$G^*(t) = S^*(t) + \int_{-r}^0 G^*(t+\theta)D^*(\theta)d\theta + \int_0^t \Big(\int_{-r}^0 G^*(s+\theta)d\eta^*(\theta)$$
$$+ \int_{-r}^0 G^*(s+\theta)D^*(\theta)A^*d\theta\Big)S^*(t-s)ds, \quad t \geq 0. \tag{35}$$

Note that $S^*(t)$ is strongly continuous on $\mathbb{R}_+$. Then by using (34), (35) and the Lebesgue dominated convergence theorem, $G^*(t)x^*$ and $G_*(t)x^*$ are of exponential order for each $x^* \in X^*$. Hence, both $G^*(t)$ and $G_*(t)$ are Laplace transformable. Taking Laplace transform on both sides of (34), we have for sufficiently large $\Re(\lambda)$ that

$$\hat{G}_*(\lambda) = R(\lambda, A^*) + \int_{-r}^0 e^{\lambda\theta}D^*(\theta)d\theta \cdot \hat{G}_*(\lambda) + R(\lambda, A^*)\int_{-r}^0 e^{\lambda\theta}d\eta^*(\theta) \cdot \hat{G}_*(\lambda)$$
$$+ R(\lambda, A^*)\int_{-r}^0 e^{\lambda\theta}A^*D^*(\theta)d\theta \cdot \hat{G}_*(\lambda), \tag{36}$$

where $R(\lambda, A^*)$ denotes the resolvent of A^*. Similarly to (15), ones can define

$$M_*(\lambda) = I - \int_{-r}^0 e^{\lambda\theta}D^*(\theta)d\theta,$$
$$N_*(\lambda) = R(\lambda, A^*)\Big[\int_{-r}^0 e^{\lambda\theta}d\eta^*(\theta) + \int_{-r}^0 e^{\lambda\theta}A^*D^*(\theta)d\theta\Big], \tag{37}$$

for arbitrary $\lambda \in \mathbb{C}^1$ with $\Re(\lambda)$ large enough. It is easy to see that $M_*(\lambda) \to I$ and $\|N_*(\lambda)\| \to 0$ as $\Re(\lambda) \to +\infty$. This implies that for sufficiently large $\Re(\lambda)$, both $M_*(\lambda)^{-1}$ and $[I - M_*(\lambda)^{-1}N_*(\lambda)]^{-1}$ exist. Therefore, by virtue of (37) and (36), we can rewrite $\hat{G}_*(\lambda)$ as

$$\hat{G}_*(\lambda) = [I - M_*(\lambda)^{-1}N_*(\lambda)]^{-1}M_*(\lambda)^{-1}R(\lambda, A^*). \tag{38}$$

On the other hand, the following equality holds

$$\hat{G}(\lambda)\left(\lambda I - A - \int_{-r}^{0} e^{\lambda\theta} d\eta(\theta) - \int_{-r}^{0} \lambda e^{\lambda\theta} D(\theta) d\theta\right) = I \tag{39}$$

for sufficiently large $\Re(\lambda)$. Substituting $\lambda = \bar{\lambda}$ (complex conjugate) into (39) and taking its adjoint, we obtain

$$\begin{aligned} I &= \left(\bar{\lambda}I - A - \int_{-r}^{0} e^{\bar{\lambda}\theta} d\eta(\theta) - \int_{-r}^{0} \bar{\lambda} e^{\bar{\lambda}\theta} D(\theta) d\theta\right)^{*} (\hat{G}(\bar{\lambda}))^{*} \\ &= \left(\lambda I - A^{*} - \int_{-r}^{0} e^{\lambda\theta} d\eta^{*}(\theta) - \int_{-r}^{0} \lambda e^{\lambda\theta} D^{*}(\theta) d\theta\right) \hat{G}^{*}(\lambda), \end{aligned} \tag{40}$$

so that

$$\hat{G}^{*}(\lambda) = \left(I - A^{*} - \int_{-r}^{0} e^{\lambda\theta} d\eta^{*}(\theta) - \int_{-r}^{0} \lambda e^{\lambda\theta} D^{*}(\theta) d\theta\right)^{-1}. \tag{41}$$

Note that we have

$$\begin{aligned} R(\lambda, A^{*})^{-1} M_{*}(\lambda) &\left[I - M_{*}(\lambda)^{-1} N_{*}(\lambda)\right] = R(\lambda, A^{*})^{-1} [M_{*}(\lambda) - N_{*}(\lambda)] \\ &= R(\lambda, A^{*})^{-1}\left\{I - \int_{-r}^{0} e^{\lambda\theta} D^{*}(\theta) d\theta - \right. \\ &\quad \left. R(\lambda, A^{*})\left[\int_{-r}^{0} e^{\lambda\theta} d\eta^{*}(\theta) + \int_{-r}^{0} e^{\lambda\theta} A^{*} D^{*}(\theta) d\theta\right]\right\} \\ &= \lambda I - A^{*} - (\lambda I - A^{*}) \int_{-r}^{0} e^{\lambda\theta} D^{*}(\theta) d\theta - \int_{-r}^{0} e^{\lambda\theta} d\eta^{*}(\theta) - \int_{-r}^{0} e^{\lambda\theta} A^{*} D^{*}(\theta) d\theta \\ &= \lambda I - A^{*} - \int_{-r}^{0} \lambda e^{\lambda\theta} D^{*}(\theta) d\theta - \int_{-r}^{0} e^{\lambda\theta} d\eta^{*}(\theta). \end{aligned} \tag{42}$$

Thus, ones have by virtue of (41) and (42) that

$$\hat{G}^{*}(\lambda) = \left[I - M_{*}(\lambda)^{-1} N_{*}(\lambda)\right]^{-1} M_{*}(\lambda)^{-1} R(\lambda, A^{*}),$$

which, together with (38), immediately implies that for sufficiently large $\Re(\lambda)$,

$$\hat{G}^{*}(\lambda) = \hat{G}_{*}(\lambda)$$

and then by the uniqueness of Laplace transforms,

$$G^{*}(t) = G_{*}(t), \quad t \in \mathbb{R}^{+},$$

almost everywhere. Since $G^{*}(t) = G_{*}(t) = 0$ if $t < 0$, the desired result follows now.

Note that the adjoint $B^{*}(\theta)$, $D^{*}(\theta)$ of $B(\theta)$, $D(\theta)$ defined in (33) satisfy

$$B^{*}(\cdot),\ D^{*}(\cdot) \in L^{2}([-r,0]; \mathscr{L}(X^{*})),$$

respectively. For arbitrary $\psi^* = (\psi_0^*, \psi_1^*) \in X^* \times W^{1,2}([-r,0];X^*)$, the mild solution $y^*(t,f^*,\psi^*)$ of the adjoint equation (33) exists and may be represented by

$$
y^*(t,f^*,\psi^*) = G^*(t)\psi_0^* + V_*(t,0)\psi_1^*(0) + \int_{-r}^{0} U_*(t,\theta)\psi_1^*(\theta)d\theta
$$
$$
+ \int_{-r}^{0} V_*(t,\theta)\psi_1^*(\theta)'d\theta + \int_0^t G^*(t-s)f^*(s)ds,
\tag{43}
$$

where

$$
U_*(t,\theta) = \int_{-r}^{\theta} G^*(t-\theta+\tau)d\eta^*(\tau) \in L^2([-r,0];\mathscr{L}(X^*)), \quad \theta \in [-r,0],
\tag{44}
$$

and

$$
V_*(t,\theta) = \int_{-r}^{\theta} G^*(t-\theta+\tau)D^*(\tau)d\theta \in L^2([-r,0];\mathscr{L}(X^*)), \quad \theta \in [-r,0].
\tag{45}
$$

Note that the operators $U_*(t,\theta)$ and $V_*(t,\theta)$ in (44) and (45) are not necessarily identical with the adjoints $U^*(t,\theta)$ and $V^*(t,\theta)$ of $U(t,\theta)$ and $V(t,\theta)$, respectively. In particular, by an argument similar as for Proposition 1 ones can easily obtain the following theorem.

Theorem 4. *(i) Let $x^* \in X^*$, then $\int_0^t G^*(s)x^*ds \in \mathscr{D}(A^*)$ for almost all $t \in \mathbb{R}_+$, and the relation*

$$
A^* \int_0^t G^*(s)x^*ds = G^*(t)x^* - x^* - \int_0^t \int_{-r}^{0} d\eta^*(\theta)G^*(s+\theta)x^*ds
$$
$$
- \int_{-r}^{0} D^*(\theta)G^*(t+\theta)x^*d\theta, \quad t \in \mathbb{R}_+, \quad x^* \in X^*,
\tag{46}
$$

holds almost everywhere.
(ii) Let $x^* \in \mathscr{D}(A^*)$, then $\int_0^t G^*(s)A^*x^*ds \in X^*$ for almost all $t \in \mathbb{R}_+$, and the commutative relation

$$
\int_0^t G^*(s)A^*x^*ds = G^*(t)x^* - x^* - \int_0^t \int_{-r}^{0} G^*(s+\theta)d\eta^*(\theta)x^*ds
$$
$$
- \int_{-r}^{0} G^*(t+\theta)D^*(\theta)x^*d\theta, \quad t \in \mathbb{R}_+, \quad x^* \in \mathscr{D}(A^*),
\tag{47}
$$

holds almost everywhere.

5   Optimal Control

Let $T > 0$ and U be a separable Banach space. Consider the following neutral hereditary controlled system on X:

$$
\frac{d}{dt}\left[y(t) - \int_{-r}^{0} D(\theta)y(t+\theta)d\theta\right] = Ay(t) + \int_{-r}^{0} d\eta(\theta)y(\theta+t) + f(t) + Q(t)u(t),
$$
$$
t \in [0,T],
\tag{48}
$$
$$
y(0) = \phi_0, \quad y_0 = \phi_1, \quad \phi = (\phi_0,\phi_1) \in \mathscr{W}^{1,2}, \quad u \in U_{ad},
$$

where A, η, D are given as in (4), $f \in L^2([0,T];X)$, $U_{ad} \subset L^2([0,T];U)$ and $Q \in L^\infty([0,T];\mathscr{L}(U,X))$.

The quantities $y(\cdot)$, $u(\cdot)$, Q and U_{ad} in (48) denote a system state, a control, a controller and a class of admissible controls, respectively. It is known by virtue of Theorem 2 that the following form

$$y(t) = G(t)\left[\phi_0 - \int_{-r}^0 D(\theta)\phi_1(\theta)d\theta\right] + \int_{-r}^0 U(t,\theta)\phi_1(\theta)d\theta + \int_{-r}^0 V(t,\theta)\phi_1'(\theta)d\theta$$

$$+ \int_0^t G(t-s)f(s)ds + \int_0^t G(t-s)Q(s)u(s)ds$$

$$= y(t,f,\phi) + \int_0^t G(t-s)Q(s)u(s)ds, \quad t \geq 0,$$

$$(49)$$

is the mild solution of (48) where $y(t,f,\phi)$ is given by (11).

5.1 Existence of Optimal Control

In what follows, the admissible set U_{ad} is assumed to be closed and convex in $L^2([0,T];U)$. Let $J = J(u)$ be the integral convex cost given by

$$J = R(y(T)) + \int_0^T \Big(P(y(t),t) + L(u(t),t)\Big)dt, \quad (50)$$

where $R: X \to \mathbb{R}^1$, $P: X \times [0,T] \to \mathbb{R}^1$ and $L: U \times [0,T] \to \mathbb{R}^1$. We are interested in the following control problem on the finite interval $I = [0,T]$: find a control $u \in U_{ad}$ which minimizes the cost J subject to (48).

Assumption A_1:

(1) $R: X \to \mathbb{R}^1$ is continuous and convex, and there exists a constant $c_0 > 0$ such that $R(x) \geq -c_0$ on X;
(2) $P: X \times [0,T] \to \mathbb{R}^1$ is measurable in $t \in [0,T]$ for each $x \in X$ and continuous, convex in $x \in X$ for $t \in [0,T]$, and there exists a constant $c_1 > 0$ such that $P(x,t) \geq -c_1$ on $X \times [0,T]$;
(3) $L: U \times [0,T] \to \mathbb{R}^1$ satisfies that for any $u \in U_{ad}$, $L(u(t),t)$ is integrable on $[0,T]$ and the functional $\Gamma: U_{ad} \to \mathbb{R}^1$ given by

$$\Gamma(u) = \int_0^T L(u(t),t)dt$$

is continuous and convex. Moreover, there exists a monotone increasing function $\theta_0 \in C(\mathbb{R}^+;\mathbb{R}^1)$ such that $\lim_{r\to\infty}\theta_0(r) = \infty$ and

$$\Gamma(u) = \int_0^T L(u(t),t)dt \geq \theta_0(\|u\|_{L^2([0,T];U)}) \quad \text{for } u \in U_{ad}.$$

Theorem 5. *Assume that the assumption A_1 is satisfied. Then there exists a control $u_0 \in U_{ad}$ that minimizes the cost J in (50).*

Proof. Let $\{u_n\}$ be a minimizing sequence of J such that

$$\inf_{u \in U_{ad}} J(u) = \lim_{n \to \infty} J(u_n) = m_0.$$

By virtue of A_1, it follows that

$$J(u) \geq \theta_0(\|u\|_{L^2([0,T];U)}) - c_0 - c_1 T \quad \text{for } u \in U_{ad}.$$

Hence, a standard argument with $\lim_{r \to \infty} \theta_0(r) = \infty$ yields that the minimizing sequence $\{u_n\}$ is bounded in $L^2([0,T];U)$, which, together with the closedness of U_{ad}, implies that there exists a subsequence (which we denote it again by $\{u_n\}$) of $\{u_n\}$ and a $u_0 \in U_{ad}$ such that

$$u_n \to u_0 \quad \text{weakly in} \quad L^2([0,T];U). \tag{51}$$

We denote $y^{u_n}(t)$ and $y^{u_0}(t)$ the mild solutions of (48) corresponding to u_n and u_0, respectively. For any fixed $x^* \in X^*$ and $t \in [0,T]$, since $G(t) = 0$ if $t < 0$, then we have that for any $t \geq 0$,

$$\langle y^{u_n}(t), x^* \rangle_{X,X^*} = \langle y(t,f,\phi), x^* \rangle_{X,X^*} + \int_0^t \langle u_n(s), Q^*(s)G^*(t-s)x^* \rangle_{U,U^*} ds, \tag{52}$$

where $y(t,f,\phi)$ is the mild solution of (48) corresponding to $Q(\cdot) = 0$. Since $Q(t) \in \mathscr{L}(U,X)$ and $G(t)$ is piecewise strongly continuous on $[0,T]$, it is easy to see that the function $Q^*(\cdot)G^*(t-\cdot)x^*$ belongs to $L^2([0,T];U^*)$. Hence, by virtue of (51) and (52), it follows that

$$\begin{aligned}
\langle y^{u_n}(t), x^* \rangle_{X,X^*} &\to \langle y(t,f,\phi), x^* \rangle_{X,X^*} + \int_0^t \langle u_0(s), Q^*(s)G^*(t-s)x^* \rangle_{U,U^*} ds \\
&= \langle y(t,f,\phi), x^* \rangle_{X,X^*} + \left\langle \int_0^t G(t-s)Q(s)u_0(s)ds, x^* \right\rangle_{X,X^*} \\
&= \langle y^{u_0}(t), x^* \rangle_{X,X^*} \quad \text{as } n \to \infty,
\end{aligned} \tag{53}$$

i.e.,

$$y^{u_n}(t) \to y^{u_0}(t) \quad \text{weakly in } X \text{ as } n \to \infty. \tag{54}$$

It is well known that continuity plus convexity imply weak lower semi-continuity. Then the condition (1) in Assumption A_1 and (54) with $t = T$ imply

$$\varliminf_{n \to \infty} R(y^{u_n}(T)) \geq R(y^{u_0}(T)). \tag{55}$$

In a similar way, we have

$$\varliminf_{n \to \infty} P(y^{u_n}(t),t) \geq P(y^{u_0}(t),t), \qquad t \in [0,T]. \tag{56}$$

It follows via Fatou's lemma that

$$\varliminf_{n \to \infty} \int_0^T P(y^{u_n}(t),t)dt \geq \int_0^T \varliminf_{n \to \infty} P(y^{u_n}(t),t)dt \geq \int_0^T P(y^{u_0}(t),t), \qquad t \in [0,T]. \tag{57}$$

As for the term $\int_0^T L(u_n(t),t)dt$, it is clear from (3) in Assumption $\mathbf{A}_1$ that

$$\lim_{n\to\infty} \Gamma(u_n) \geq \Gamma(u_0) = \int_0^T L(u_0(t),t)dt. \tag{58}$$

Therefore, by (55), (57) and (58) we have

$$\begin{aligned}
m_0 = \inf_{u\in U_{ad}} J(u) &\geq \varliminf_{n\to\infty} R(y^{u_n}(T)) + \varliminf_{n\to\infty} \int_0^T P(y^{u_n}(t),t)dt + \varliminf_{n\to\infty} \Gamma(u_n) \\
&\geq R(y^{u_0}(T)) + \int_0^T \left[P(y^{u_0}(t),t) + L(u_0(t),t) \right] dt \\
&= J(u_0) > -\infty,
\end{aligned} \tag{59}$$

so that $m_0 = J(u_0)$. This proves that u_0 is the optimal solution for J.

5.2 Optimality Condition

In this subsection, we shall seek necessary optimality conditions of the optimal solution u for J in (50). The existence of optimal solutions is assumed but not the closedness of U_{ad}.

Assumption $\mathbf{A}_2$:

(1) $R : X \to \mathbb{R}^1$ is continuous and Gâteau differentiable, and the Gâteau derivative $R'(x) \in X^*$ for each $x \in X$;
(2) $P : X \times [0,T] \to \mathbb{R}^1$ is measurable in $t \in [0,T]$ for each $x \in X$ and continuous, convex on X for $t \in [0,T]$, and furthermore there exist functions $\partial_x P : X \times [0,T] \to X^*$, $\theta_1 \in L^1([0,T];\mathbb{R}^1)$, $\theta_2 \in C(\mathbb{R}^+;\mathbb{R}^1)$ such that:
　(a) $\partial_x P(x,t)$ is measurable in $t \in [0,T]$ for each $x \in X$ and continuous in $x \in X$ for $t \in [0,T]$ and the value $\partial_x P(x,t)$ is the Gâteau derivative of $P(x,t)$ in the first argument for $(x,t) \in X \times [0,T]$, and
　(b) $\|\partial_x P(x,t)\|_{X^*} \leq \theta_1(t) + \theta_2(\|x\|_X)$ for $(x,t) \in X \times [0,T]$;
(3) $L : U \times [0,T] \to \mathbb{R}^1$ is measurable in $t \in [0,T]$ for each $z \in U$ and continuous, convex on U for each $t \in [0,T]$. Moreover, there exist function $\theta_3(\cdot) \in L^1([0,T];\mathbb{R}^1)$ and constant $M > 0$ such that

$$|L(z,t)| \leq \theta_3(t) + M\|z\|_U^2 \quad \text{for } (z,t) \in U \times [0,T].$$

Lemma 1. [12] *Consider the function*

$$J(v) = J_1(v) + J_2(v)$$

for any $v \in L^2([0,T];U)$ where we assume that the functions $J_i(v)$, $i = 1,2$, are continuous and convex. Further assume that the function $v \to J_1(v)$ is differentiable. Then the unique element u in U_{ad} satisfying $J(u) = \inf_{v\in U_{ad}} J(v)$ is characterized by

$$J_1'(u)(v-u) + J_2(v) - J_2(u) \geq 0 \quad \text{for all } v \in U_{ad}. \tag{60}$$

Now we are in a position to state one of the main theorems in this section.

Theorem 6. *Suppose that the assumption* $\mathbf{A}_2$ *holds and* $u \in U_{ad}$ *is an optimal solution for* J *in (50). Then the integral inequality*

$$\int_0^T \langle v(t) - u(t), -Q^*(t)p(t) \rangle_{U,U^*} dt + \int_0^T \Big(L(v(t),t) - L(u(t),t)\Big) dt \geq 0$$

$$\text{for all } v \in U_{ad}$$
(61)

holds, where

$$p(t) = -G^*(T-t)R'(y^u(T)) - \int_t^T G^*(s-t)\partial_x P(y^u(s),s)ds \qquad (62)$$

satisfies that for $t \in [0,T]$,

$$\frac{d}{dt}\Big[p(t) - \int_{-r}^0 D^*(\theta)p(t+\theta)d\theta\Big] + A^*p(t) +$$

$$\int_{-r}^0 d\eta^*(\theta)p(t-\theta) - \partial_x P(y^u(t),t) = 0, \qquad (63)$$

$$p(T) = -R'(y^u(T)), \qquad p(t) = 0, \ t \in (T, T+r],$$

in the weak sense.

Proof. By virtue of the assumption $\mathbf{A}_2$, we have by Lebesgue's dominated convergence theorem that

$$(J - \Gamma)'(u)(v - u) = \Big\langle \int_0^T G(T-s)Q(s)(v(s) - u(s))ds, R'(y^u(T)) \Big\rangle_{X,X^*}$$

$$+ \int_0^T \Big\langle \int_0^s G(s-\tau)Q(\tau)(v(\tau) - u(\tau))d\tau, \partial_x P(y^u(s),s) \Big\rangle_{X,X^*} ds.$$
(64)

Note that all integrals in (64) are well defined by making use of the assumption $\mathbf{A}_2$. The first term of (64) can be written as

$$\int_0^T \langle v(s) - u(s), Q^*(s)G^*(T-s)R'(y^u(T)) \rangle_{U,U^*} ds. \qquad (65)$$

Using the standard Fubini lemma, the second term of (64) is transformed as

$$\int_0^T \int_0^s \langle G(s-\tau)Q(\tau)(v(\tau) - u(\tau)), \partial_x P(y^u(s),s) \rangle_{X,X^*} d\tau ds$$

$$= \int_0^T \langle v(\tau) - u(\tau), Q^*(\tau) \int_\tau^T G^*(s-\tau)\partial_x P(y^u(s),s)ds \rangle_{U,U^*} d\tau.$$
(66)

If we define $p(t)$ by (62) and apply Lemma 1 to the mapping $J = (J - \Gamma) + \Gamma$, then the relations (64), (65) and (66) yield the inequality (61). For the last statement, note that by virtue of (43), the function

$$-G^*(t)R'(y^u(0)) - \int_0^t G^*(t-s)\partial_x P(y^u(T-s),T-s)ds, \ t \in [0,T],$$

satisfies

$$\frac{d}{dt}\left[y^*(t) - \int_{-r}^0 D^*(\theta)y^*(t+\theta)d\theta\right] = A^*y^*(t) + \int_{-r}^0 d\eta^*(\theta)y^*(t+\theta)$$
$$- \partial_x P(y^u(T-t), T-t), \quad t \in [0,T], \tag{67}$$
$$y^*(0) = -R'(y^u(0)), \quad y^*(\theta) = 0, \quad \theta \in [-r,0).$$

A change of variable $t \to T - t$ yields the desired result. The proof is now complete. □

5.3 Maximum Principle

In view of Theorem 6, we can obtain from (61) the following "integral" maximum principle:

$$\max_{v \in U_{ad}} \int_0^T \left(\langle v(t), Q^*(t)p(t)\rangle_{U,U^*} - L(v(t),t)\right)dt =$$
$$\int_0^T \left(\langle u(t), Q^*(t)p(t)\rangle_{U,U^*} - L(u(t),t)\right)dt. \tag{68}$$

It is possible to improve this result to establish the so-called "pointwise" maximum principle for the convex cost (50). To this end, the assumption $\mathbf{A}_2$ is assumed at the moment. Let the admissible set U_{ad} be

$$U_{ad} = \left\{u \in L^2([0,T];U) : u(t) \in U(t), t \in [0,T]\right\}, \tag{69}$$

where the (time varying) control domain $U(t) \subset U, t \in [0,T]$, satisfies

Assumption $\mathbf{A}_3$:

(1) $U(t)$ is closed and convex in U for each $t \in [0,T]$;
(2) For any $t \in [0,T], z \in \text{Int}\,U(t)$, the interior of $U(t)$, there exists an $\varepsilon_0 > 0$ such that

$$z \in \left(\bigcap_{s \in (t,t+\varepsilon_0)} U(s)\right) \cup \left(\bigcap_{s \in (t-\varepsilon_0,t)} U(s)\right). \tag{70}$$

Theorem 7. *Let $u \in U_{ad}$ be an optimal solution for J in (50). Then*

$$\max_{z \in U(t)} \left\{\langle Q(t)z, p(t)\rangle_{X,X^*} - L(z,t)\right\} = \langle Q(t)u(t), p(t)\rangle_{X,X^*} - L(u(t),t), \quad t \in [0,T],$$

where $p(t)$ is given by (62).

Proof. Let $t \in [0,T]$ and $z \in \text{Int}\,U(t)$. Since z satisfies (70), we suppose, for instance, $z \in \bigcap_{s \in (t,t+\varepsilon_0)} U(s)$. Then it is easy to see that for any $\varepsilon > 0$ the function

$$v_\varepsilon(s) = \begin{cases} u(s), & s \in [0,t) \text{ or } (t+\varepsilon,T], \\ z, & s \in [t,t+\varepsilon], \end{cases} \tag{71}$$

belongs to U_{ad} for the u in (61). Substituting v_ε for v in (61) and dividing the resulting inequality by ε, we obtain

$$\frac{1}{\varepsilon}\int_t^{t+\varepsilon}\left\{\langle z-u(s),-Q^*(s)p(s)\rangle_{U,U^*}+\Big(L(z,s)-L(u(s),s)\Big)\right\}ds\geq 0. \qquad (72)$$

Since all the integrands in (72) are integrable on $[0,T]$ by virtue of the assumption A_2, the Lebesgue density theorem can apply. Then by letting $\varepsilon\to 0$ in (72), we have

$$\langle z,Q^*(t)p(t)\rangle_{U,U^*}-L(z,t)\leq\langle u(t),Q^*(t)p(t)\rangle_{U,U^*}-L(u(t),t), \qquad t\in[0,T]. \qquad (73)$$

Since the duality pairing $\langle z,Q^*(t)p(t)\rangle_{U,U^*}$ is continuous in z, we see from (73) that the maximum principle is true for such $t\in[0,T]$. $\qquad\qquad\square$

References

1. Ahmed, N.U., Teo, K.L.: Optimal Control of Distributed Parameter Systems. North-Holland, New York (1981)
2. Bernier, C., Manitius, A.: On semigroups in $R^n\times L^p$ corresponding to differential equations with delays. Canadian J. Math. 30, 897–914 (1978)
3. Bhat, K., Koivo, H.: Modal characterization of controllability and observability for time delay systems. IEEE Trans. Auto. Control AC-21, 292–293 (1976)
4. Datko, R.: Linear autonomous neutral differential equations in a Banach space. J. Differential Eqns. 25, 258–274 (1977)
5. Datko, R.: Representation of solutions and stability of linear differential-difference equations in a Banach space. J. Differential Eqns. 29, 105–166 (1978)
6. Delfour, M., Mitter, S.: Controllability, observability and optimal feedback control of affine hereditary differential systems. SIAM J. Control & Optim. 10, 298–327 (1972)
7. Diekmann, O., Gils, S., Lunel, S., Walther, H.: Delay Equations: Functional-, Complex- and Nonlinear Analysis. In: Applied Mathematical Sciences, vol. 110. Springer, New York (1995)
8. Hale, J.: Theory of Functional Differential Equations. Springer, Heidelberg (1977)
9. Hale, J., Lunel, S.: Introduction to Functional Differential Equations. Applied Mathematical Sciences, vol. 99. Springer, Heidelberg (1993)
10. Ito, K., Tarn, T.: A linear quadratic optimal control for neutral systems. Nonlinear Analysis 9, 699–727 (1985)
11. Jeong, J.: Stabilizability of retarded functional differential equation in Hilbert spaces. Osaka J. Math. 28, 347–365 (1991)
12. Lions, J.L.: Optimal Control of Systems Governed by Partial Differential Equations. Springer, Heidelberg (1971)
13. Nakagiri, S.: Optimal control of linear retarded systems in Banach spaces. J. Math. Anal. Appl. 120, 169–210 (1986)
14. Peszat, S., Zabczyk, J.: Stochastic Partial Differential Equations with Lévy Noise: An Evolution Equation Approach. In: Encyclopedia of Mathematics and its Applications. Cambridge Univ. Press, Cambridge (2007)
15. Salamon, D.: Control and Observation of Neutral Systems. Research Notes in Mathematics, vol. 91. Pitman Advanced Publishing Program, Boston (1984)

16. Tanabe, H.: Equations of Evolution. Pitman, New York (1979)
17. Travis, C., Webb, G.: Existence and stability for partial functional differential equations. Trans. Amer. Math. Soc. 200, 395–418 (1974)
18. Wu, J.: Semigroup and integral form of a class of partial differential equations with infinite delay. Differential and Integral Equations 6, 1325–1351 (1991)
19. Wu, J.: Theory and Applications of Partial Functional Differential Equations. In: Applied Mathematical Sciences, vol. 119. Springer, Berlin (1996)

Approximate Subgradient Methods for Lagrangian Relaxations on Networks

Eugenio Mijangos

University of the Basque Country, Department of Applied Mathematics,
Statistics and Operations Research, Spain
eugenio.mijangos@ehu.es

Abstract. Nonlinear network flow problems with linear/nonlinear side constraints can be solved by means of Lagrangian relaxations. The dual problem is the maximization of a dual function whose value is estimated by minimizing approximately a Lagrangian function on the set defined by the network constraints. We study alternative stepsizes in the approximate subgradient methods to solve the dual problem. Some basic convergence results are put forward. Moreover, we compare the quality of the computed solutions and the efficiency of these methods.

1 Introduction

Consider the nonlinearly constrained network flow problem (**NCNFP**)

$$\text{minimize} \quad f(x) \tag{1}$$
$$\text{subject to} \quad x \in \mathscr{F} \tag{2}$$
$$c(x) \leq 0, \tag{3}$$

where:

- $\mathscr{F} = \{x \in \mathbf{R}^n \mid Ax = b,\ 0 \leq x \leq \bar{x}\}$, where A is a node-arc incidence $m \times n$-matrix, b is the production/demand m-vector, x are the flows on the arcs of the network represented by A, and $\bar{x}$ is the vector of capacity bounds imposed on the flow of each arc.
- The side constraints (3) are defined by $c : \mathbf{R}^n \to \mathbf{R}^r$, such that $c = [c_1, \cdots, c_r]^T$, where $c_i(x)$ is linear or nonlinear and twice continuously differentiable on $\mathscr{F}$ for all $i = 1, \cdots, r$.
- $f : \mathbf{R}^n \to \mathbf{R}$ is nonlinear and twice continuously differentiable on $\mathscr{F}$.

We focus on the primal problem **NCNFP** and its dual problem

$$\text{maximize} \quad q(\mu) = \min_{x \in \mathscr{F}} l(x, \mu) \tag{4}$$
$$\text{subject to} \quad \mu \in \mathscr{M}, \tag{5}$$

where the Lagrangian function is

$$l(x, \mu) = f(x) + \mu^T c(x) \tag{6}$$

A. Korytowski et al. (Eds.): System Modeling and Optimization, IFIP AICT 312, pp. 370–381, 2009.

and $\mathcal{M} = \{\mu \mid \mu \geq 0, \ q(\mu) > -\infty\}$. We assume throughout this paper that the constraint set $\mathcal{M}$ is closed and convex. Since q is concave on $\mathcal{M}$, it is continuous on $\mathcal{M}$. When exact values of q are used, we assume that for every $\mu \in \mathcal{M}$ some vector $x(\mu)$ that minimizes $l(x,\mu)$ over $x \in \mathcal{F}$ can be calculated, yielding a subgradient $c(x(\mu))$ of q at μ, which allows to solve **NCNFP** by using primal-dual methods, see [2]. Nevertheless, a substantial drawback of this kind of methods is the need to obtain at each iteration an exact solution to the subproblem included in (4). In this paper in order to allow for inexact solution of this minimization, we consider *approximate subgradient methods* [6,8,7] in the solution of this problem. The basic difference between these methods and the classical subgradient methods is that they replace the subgradients with inexact subgradients.

Given a scalar $\varepsilon \geq 0$ and a vector $\overline{\mu} \in \mathcal{M}$, we say that c is an ε-*subgradient* (approximate subgradient) at $\overline{\mu}$ if

$$q(\mu) \leq q(\overline{\mu}) + \varepsilon + c^T(\mu - \overline{\mu}), \qquad \forall \mu \in \mathbf{R}^r. \tag{7}$$

The set of all ε-subgradients at $\overline{\mu}$ is the ε-*subdifferential* at $\overline{\mu}$ (i.e. $\partial_\varepsilon q(\overline{\mu})$).

In our context, we minimize approximately $l(x,\mu^k)$ over $x \in \mathcal{F}$ by efficient techniques specialized for networks [15], obtaining a vector $x^k \in \mathcal{F}$ with

$$l(x^k, \mu^k) \leq \inf_{x \in \mathcal{F}} l(x, \mu^k) + \varepsilon_k. \tag{8}$$

As is shown in [2,8], the corresponding constraint vector, $c(x^k)$, is an ε_k-subgradient at μ^k. If we denote $q_{\varepsilon_k}(\mu^k) = l(x^k, \mu^k)$, by definition of $q(\mu^k)$ and using (8) we have

$$q(\mu^k) \leq q_{\varepsilon_k}(\mu^k) \leq q(\mu^k) + \varepsilon_k \qquad \forall k. \tag{9}$$

An approximate subgradient method is defined by

$$\mu^{k+1} = [\mu^k + \alpha_k c^k]^+, \tag{10}$$

where c^k is an approximate subgradient at μ^k, $[\cdot]^+$ denotes the projection on the closed convex set $\mathcal{M}$, and α_k is a positive scalar stepsize.

Different ways of computing the stepsize have been considered:

(a) Constant step rule (CSR) with Shor-type scaling [14].
(b) A variant of the constant step rule (VCSR) of Shor.
(c) Diminishing stepsize rule with scaling (DSRS) [13,5,14].
(d) The diminishing stepsize rule without scaling (DSR) suggested by Correa and Lemaréchal in [3].
(e) A dynamically chosen stepsize rule based on an estimation of the optimal value of the dual function by means of an adjustment procedure (DSAP) similar to that suggested by Nedić and Bertsekas in [12] for incremental subgradient methods.

The convergence of these methods was studied in the cited papers for the case of exact subgradients. The convergence of the approximate subgradient methods was analyzed by Kiwiel [6].

An alternative study of the convergence of some of these methods and their application in the solution of nonlinear networks was carried out in [8,7].

In this work some basic convergence results obtained by Shor [14] are extended to approximate subgradient methods. Moreover, we compare the quality of the computed solution and the efficiency of the approximate subgradient methods when using CSR, VCSR, DSRS, DSR, and DSAP over **NCNFP** problems.

This paper is organized as follows: Sect. 2 presents the stepsize rules with the corresponding convergence results; Sect. 3, the solution to the nonlinearly constrained network flow problem; Sect. 4 puts forward the numerical tests; and Sect. 5 displays the conclusions.

2   Stepsize Rules and Convergence Results

Throughout this section, we use the notation

$$q^* = \sup_{\mu \in \mathcal{M}} q(\mu), \quad \mathcal{M}^* = \{\mu \in \mathcal{M} \mid q(\mu) = q^*\}, \tag{11}$$

and $\|\cdot\|$ denotes the standard Euclidean norm.

Assumption 1 (subgradient boundedness). There exists a scalar $C > 0$ such that for $\mu^k \in \mathcal{M}$, $\varepsilon_k \geq 0$ and $c^k \in \partial_{\varepsilon_k} q(\mu^k)$, we have $\|c^k\| \leq C$, for $k = 0, 1, \dots$.

We say that $\overline{\mu}$ is an ε-*optimal solution* of the dual problem when $0 \in \partial_\varepsilon q(\overline{\mu})$, i.e. when $q(\overline{\mu}) \geq q^* - \varepsilon$.

In this paper various kinds of stepsize rules have been considered.

2.1   Constant Step Rule (CSR)

As is well known the classical scaling of Shor (see [14])

$$\alpha_k = \frac{s_k}{\|c^k\|} \tag{12}$$

with $s_k = s$ gives rise to an s-constant-step algorithm.

Note that constant stepsizes (i.e. $\alpha_k = s$ for all k) are unsuitable because the function q may be nondifferentiable at the optimal point and then $\{c^k\}$ does not necessarily tend to zero, even if $\{\mu^k\}$ converges to the optimal point, see [14].

Next, we show some basic convergence results when c^k is an approximate subgradient, which are similar to the results obtained by Shor [14] in the case of exact subgradients.

Proposition 1. *Consider the ε-subgradient iteration*

$$\mu^{k+1} = \left[\mu^k + \alpha_k c^k\right]^+, \tag{13}$$

where $\alpha_k = s_k / \|c^k\|$ and $s_k = s > 0$ for any k, and $c^k \in \partial q_{\varepsilon_k}(\mu^k)$, with $\lim_{k \to \infty} \varepsilon_k = \varepsilon \geq 0$. Then, for any $\delta > 0$ and any dual optimal solution $\mu^* \in \mathcal{M}^*$, either one can find $k = \overline{k}$, where $\mu^{\overline{k}}$ is an ε_k-optimal solution, or there exist an index $\overline{k}$ and a point $\overline{\mu} \in \mathcal{M}$ such that $q(\overline{\mu}) = q(\mu^{\overline{k}}) + \varepsilon_{\overline{k}}$ and $\|\overline{\mu} - \mu^*\| < \frac{s}{2}(1 + \delta)$.

Proof. Let $\mu^* \in \mathcal{M}^*$ and let $\delta > 0$ be given. If $c^{\bar{k}} = 0$ for some $\bar{k}$ then $\mu^{\bar{k}}$ is an ε_k-optimal solution.

When $c^k \neq 0$ for all $k = 0, 1, 2, \ldots$, by the nonexpansiveness of the projection operation we have

$$\|\mu^{k+1} - \mu^*\|^2 \leq \|\mu^k + s\frac{c^k}{\|c^k\|} - \mu^*\|^2$$

$$= \|\mu^k - \mu^*\|^2 + s^2 - 2s(\mu^* - \mu^k)^T \frac{c^k}{\|c^k\|}. \tag{14}$$

Let $a_k(\mu^*) = (\mu^* - \mu^k)^T \frac{c^k}{\|c^k\|}$, which is the distance from μ^* to the supporting hyperplane $H_k = \{\mu \in \mathcal{M} \mid (\mu - \mu^k)^T c^k = 0\}$, that is, $a_k(\mu^*) = \text{dist}(\mu^*, H_k)$.

On the other hand, we suppose that $q(\mu^k) + \varepsilon_k < q^*$, as otherwise $0 \in \partial_{\varepsilon_k} q(\mu^k)$ and μ^k is an ε_k-optimal solution. Therefore, as $q(\cdot)$ is continuous on $\mathcal{M}$ we can define the level set $L_k^{\varepsilon} = \{\mu \in \mathcal{M} \mid q(\mu) = q(\mu^k) + \varepsilon_k\}$, which is closed. Hence, the distance $b_k(\mu^*) = \text{dist}(\mu^*, L_k^{\varepsilon})$ is well defined.

Since the set L_k^{ε} and the point μ^* lie on the same side of H_k and any segment joining μ^* with a point of H_k passes through L_k^{ε}, we have $a_k(\mu^*) \geq b_k(\mu^*)$. Then, from (14) we obtain

$$\|\mu^{k+1} - \mu^*\|^2 \leq \|\mu^k - \mu^*\|^2 + s^2 - 2sb_k(\mu^*). \tag{15}$$

If $b_k(\mu^*) \geq \frac{s}{2}(1 + \delta)$ for all $k = 0, 1, 2, \ldots$, then

$$\|\mu^{k+1} - \mu^*\|^2 \leq \|\mu^k - \mu^*\|^2 - \delta s^2 \leq \|\mu^0 - \mu^*\|^2 - \delta(k+1)s^2, \tag{16}$$

for all k.

But $\|\mu^{k+1} - \mu^*\|^2 \geq 0$. Therefore, $\bar{k}$ exists such that

$$b_{\bar{k}}(\mu^*) = \text{dist}(\mu^*, L_{\bar{k}}^{\varepsilon}) < \frac{s}{2}(1 + \delta), \tag{17}$$

and, hence, there exists $\bar{\mu} \in L_{\bar{k}}^{\varepsilon}$ with $q(\bar{\mu}) = q(\mu^{\bar{k}}) + \varepsilon_{\bar{k}}$ that verifies $\|\bar{\mu} - \mu^*\| < \frac{s}{2}(1 + \delta)$. □

Corollary 1. *If the set $\mathcal{M}^*$ contains a sphere with radius $r > s/2$ and the ε-subgradient method is applied with $\alpha_k = s/\|c^k\|$, then there exists k^* such that μ^{k^*} is an ε_{k^*}-optimal solution.*

Proof. By Proposition 1, for any $\delta > 0$ there exists $\bar{k}$ such that $\bar{\mu} \in L_{\bar{k}}^{\varepsilon}$ where $q(\bar{\mu}) = q(\mu^{\bar{k}}) + \varepsilon_{\bar{k}}$ with $\|\mu^* - \bar{\mu}\| < \frac{s}{2}(1 + \delta)$ for any $\mu^* \in \mathcal{M}^*$.

Let $r > s/2$, then we take $\bar{\delta}$ such that $0 < \bar{\delta} < \frac{r - s/2}{s/2}$, for which some k^* must exist such that

$$\text{dist}(\mu^*, L_{k^*}^{\varepsilon}) < \frac{s}{2}(1 + \bar{\delta}) < r,$$

for which there exists $\widehat{\mu} \in L_{k^*}^{\varepsilon}$ such that

$$\|\mu^* - \widehat{\mu}\| < \frac{s}{2}(1 + \overline{\delta}) < r,$$

then $\widehat{\mu} \in S(\mu^*; r) \subset \mathcal{M}^*$, that is, $q(\widehat{\mu}) = q^*$.

Since $q(\widehat{\mu}) = q(\mu^{k^*}) + \varepsilon_{k^*}$ (by definition of $L_{k^*}^{\varepsilon}$), then $q(\mu^{k^*}) + \varepsilon_{k^*} = q^*$ and μ^{k^*} is an ε_{k^*}-optimal solution. $\qquad\square$

Note that if $\varepsilon_k = 0$ for all k, we have Corollary 2 of Theorem 2.1 in [14]. In this work by default $s = 100$.

2.2 Variant of the Constant Step Rule (VCSR)

Since c^k is an approximate subgradient, there can exist a k such that $c^k \in \partial_{\varepsilon_k} q(\mu^k)$ with $\|c^k\| = 0$, but ε_k not being sufficiently small. In order to overcome this trouble we have considered the following variant

$$\alpha_k = \frac{s}{\delta + \|c^k\|}, \tag{18}$$

where s and δ are positive constants. The following proposition shows its kind of convergence (see [7]).

Proposition 2. *Let Assumption 1 hold. Let the optimal set $\mathcal{M}^*$ be nonempty. Suppose that a sequence $\{\mu^k\}$ is calculated by the ε-subgradient method given by (10), with the stepsize (18), where $\sum_{k=1}^{\infty} \varepsilon_k < \infty$. Then*

$$q^* - \limsup_{k \to \infty} q_{\varepsilon_k}(\mu^k) < \frac{s}{2}(\delta + C). \tag{19}$$

Note that for very small values of δ the stepsize (18) is similar to Shor's classical scaling; in contrast, for big values of δ (with regard to $\sup\{\|c^k\|\}$) the stepsize (18) looks like a constant stepsize. As a result we have chosen by default $\delta = 10^{-12}$ with $s = 100$.

2.3 Diminishing Stepsize Rule with Scaling (DSRS)

It can be seen from the proof of Proposition 1 that at each iteration of (13) the reduction in the distance to the optimal set $\mathcal{M}^*$ is guaranteed only outside a certain neighborhood of that set, with the size of that neighborhood depending on the value of the steplength s. Therefore, to obtain standard convergence results it is necessary to require that s_k tends to zero. The reduction of steplengths, however, should not be too rapid. In particular, if the series $\sum_{k=1}^{\infty} s_k$ is convergent then the sequence $\{\mu^k\}$ has a limit, but this limit may lie outside $\mathcal{M}^*$. So for (13), with $\alpha_k = s_k / \|c^k\|$, we have arrived at the classical conditions:

$$s_k > 0, \quad \{s_k\} \to 0 \quad \text{as} \quad k \to \infty, \quad \text{and} \quad \sum_{k=1}^{\infty} s_k = \infty. \tag{20}$$

There are several alternative proofs of convergence of this method for exact subgradients [13,5]. Below we present our version of the proof of the convergence of the approximate subgradient method, which is based on Theorem 2.2 given by Shor in [14] for exact subgradients (see also [6]).

Proposition 3. *Let* $\{\varepsilon_k\} \to 0$. *Consider the* ε-*subgradient iteration (13) for* $\alpha_k = s_k/\|c^k\|$, *where* $s_k > 0$ *is such that* $\lim_{k\to\infty} s_k = 0$ *and* $\sum_{k=1}^{\infty} s_k = \infty$. *Assume that* $\mathcal{M}^*$ *is closed, bounded, and non-empty. Then either an index* $\bar{k}$ *exists such that* $\mu^{\bar{k}} \in \mathcal{M}^*$ *or else*

$$\lim_{k\to\infty} \text{dist}(\mu^k, \mathcal{M}^*) = 0, \qquad \lim_{k\to\infty} q(\mu^k) = q^*, \qquad (21)$$

and $\lim_{k\to\infty} q_{\varepsilon_k}(\mu^k) = q^*$.

Proof. Let $\mu^* \in \mathcal{M}^*$. If there exists a $\bar{k}$ such that $\mu^{\bar{k}} \in \mathcal{M}^*$, the proposition holds. Assume that this $\bar{k}$ does not exist, i.e $\mu^k \notin \mathcal{M}^*$ for all k. Then μ^k can be an ε_k-optimal solution or not. If μ^k is not an ε_k-optimal solution, like in the proof of Proposition 1 (see (15)), we obtain

$$\|\mu^{k+1} - \mu^*\|^2 \leq \|\mu^k - \mu^*\|^2 + s_k^2 - 2s_k b_k(\mu^*). \qquad (22)$$

For a fixed $a > 0$, consider the set $\{\mu \in \mathcal{M} \mid q(\mu) \geq q^* - a\}$ and its boundary Γ_{q^*-a}. By assumption, the set $\mathcal{M}^*$ is closed and bounded. Thus Γ_{q^*-a} is compact, as q is concave over $\mathcal{M}$ convex, and, hence, it is continuous (see Proposition B.9 in [2]).

Since $\{\varepsilon_k\} \to 0$, there exists N_ε, such that for all $k \geq N_\varepsilon$, $a > \varepsilon_k$. Furthermore, $\mathcal{M}^* \cap \Gamma_{q^*-a} = \emptyset$ and there exists a number

$$\rho(a) = \text{dist}(\Gamma_{q^*-a}, \mathcal{M}^*) = \min_{\mu^* \in \mathcal{M}^*, \lambda \in \Gamma_{q^*-a}} \|\lambda - \mu^*\|. \qquad (23)$$

Since $\{s_k\} \to 0$, one can find $N_{\rho(a)} \geq N_\varepsilon$ such that for all $k > N_{\rho(a)}$, $s_k < \rho(a)$ and $a > \varepsilon_k$.

If $q(\mu^k) < q^* - a$, then $b_k(\mu^*) > \rho(a)$ and from (22) we have

$$\|\mu^{k+1} - \mu^*\|^2 \leq \|\mu^k - \mu^*\|^2 - \rho(a)s_k, \quad \forall k > N_{\rho(a)}, \qquad (24)$$

as by adding the inequalities $s_k^2 < \rho(a)s_k$ and $-2s_k b_k(\mu^*) < -2s_k\rho(a)$ we obtain $s_k^2 - 2s_k b_k(\mu^*) < -\rho(a)s_k$ for all $k > N_{\rho(a)}$.

By adding (24), we have

$$\|\mu^{k+1} - \mu^*\|^2 \leq \|\mu^0 - \mu^*\|^2 - \rho(a)\sum_{i=1}^{k} s_i, \qquad (25)$$

and as $\sum_{k=1}^{\infty} s_k = \infty$, there must exist $N_a > N_{\rho(a)}$ such that for all $\bar{k} \geq N_a$ it holds $q(\mu^{\bar{k}}) \geq q^* - a$.

From here on both cases (when μ^k is an ε_k-optimal solution and when it is not) are unified. Note that if μ^k is an ε_k-optimal solution, we have $q(\mu^k) \geq q^* - \varepsilon_k > q^* - a$.

Define $d(a) = \max_{\lambda \in \Gamma_{q^*-a}} \{\min_{\mu^* \in \mathcal{M}^*} \|\lambda - \mu^*\|\}$. If $q(\mu^{\bar{k}}) \geq q^* - a$, then $\min_{\mu^* \in \mathcal{M}^*} \|\mu^{\bar{k}} - \mu^*\| \leq d(a)$.

By the nonexpansiveness of the projection operator for all k it holds

$$\|\mu^{k+1} - \mu^*\| \leq \left\|\left(\mu^k + s_k \frac{c^k}{\|c^k\|}\right) - \mu^*\right\| \leq \|\mu^k - \mu^*\| + s_k. \tag{26}$$

Therefore, for $k = \bar{k}$ we have

$$\min_{\mu^* \in \mathcal{M}^*} \|\mu^{\bar{k}+1} - \mu^*\| \leq \min_{\mu^* \in \mathcal{M}^*} \|\mu^{\bar{k}} - \mu^*\| + s_k \leq d(a) + s_k, \quad \forall \bar{k} \geq N_a. \tag{27}$$

On the other hand, for all $k > N_{\rho(a)}$ with $q(\mu^k) < q^* - a$, by (24), we have

$$\|\mu^{k+1} - \mu^*\| \leq \|\mu^k - \mu^*\|, \tag{28}$$

and hence,

$$\min_{\mu^* \in \mathcal{M}^*} \|\mu^{k+1} - \mu^*\| \leq \min_{\mu^* \in \mathcal{M}^*} \|\mu^k - \mu^*\|. \tag{29}$$

By combining (27) and (29) we obtain

$$\|\mu^{\bar{k}+1} - \mu^*\| \leq d(a) + \max_{\bar{k} > N_a}\{s_{\bar{k}}\} \tag{30}$$

for all $\bar{k} > N_a > N_{\rho(a)}$.

Since $d(a) \to 0$ as $a \to 0$, for all $\delta > 0$ there exists a_δ such that $d(a_\delta) \leq \delta/2$.

Next, one can find an index N_δ such that $q(\mu^k) \geq q^* - a_\delta$ and $s_k \leq \delta/2$ for all $k > N_\delta$. Therefore, for all $k > N_\delta$, by (30) we have

$$\min_{\mu^* \in \mathcal{M}^*} \|\mu^k - \mu^*\| \leq \delta. \tag{31}$$

This proves that $\lim_{k \to \infty} \left(\min_{\mu^* \in \mathcal{M}^*} \|\mu^k - \mu^*\|\right) = 0$.

By continuity of q, we have $\lim_{k \to \infty} q(\mu^k) = q^*$. Moreover, as $\{\varepsilon_k\} \to 0$, by the inequalities (9) we obtain $\lim_{k \to \infty} q_{\varepsilon_k}(\mu^k) = q^*$, which completes the proof.   □

An example of such a stepsize is

$$\alpha_k = \frac{s_k}{\|c^k\|}, \quad \text{with } s_k = s/\hat{k}, \tag{32}$$

for $\hat{k} = \lfloor k/m \rfloor + 1$. We use by default $s = 100$ and $m = 5$.

2.4   Diminishing Stepsize Rule (DSR)

The convergence of the subgradient method using a diminishing stepsize was shown by Correa and Lemaréchal, see [3]. Next, we consider the special case where c^k is an ε_k-subgradient and $\alpha_k = s_k$ in (13).

The following proposition is proved in [8].

Proposition 4. *Let the optimal set $\mathcal{M}^*$ be nonempty. Also, assume that the sequences $\{s_k\}$ and $\{\varepsilon_k\}$ are such that*

$$s_k > 0, \quad \sum_{k=0}^{\infty} s_k = \infty, \quad \sum_{k=0}^{\infty} s_k^2 < \infty, \quad \sum_{k=0}^{\infty} s_k \varepsilon_k < \infty. \tag{33}$$

Then, the sequence $\{\mu^k\}$, generated by the ε-subgradient method, where $c^k \in \partial_{\varepsilon_k} q(\mu^k)$ (with $\{\|c^k\|\}$ bounded), converges to some optimal solution.

An example of such a stepsize is

$$\alpha_k = s_k = s/\widehat{k}, \tag{34}$$

for $\widehat{k} = \lfloor k/m \rfloor + 1$. In this work we use by default $s = 100$ and $m = 5$.

An interesting alternative for the ordinary subgradient method is the *dynamic stepsize rule*

$$\alpha_k = \gamma_k \frac{q^* - q(\mu^k)}{\|c^k\|^2}, \tag{35}$$

with $c^k \in \partial q(\mu^k)$ and $0 < \underline{\gamma} \le \gamma_k \le \overline{\gamma} < 2$, [13,14].

Unfortunately, in most practical problems q^* and $q(\mu^k)$ are unknown. Then, the latter can be approximated by $q_{\varepsilon_k}(\mu^k) = l(x^k, \mu^k)$ and q^* replaced with an estimate q_{lev}^k. This leads to the stepsize rule

$$\alpha_k = \gamma_k \frac{q_{lev}^k - q_{\varepsilon_k}(\mu^k)}{\|c^k\|^2}, \tag{36}$$

where $c^k \in \partial_{\varepsilon_k} q(\mu^k)$ is bounded for $k = 0, 1, \dots$.

2.5   Dynamic Stepsize with Adjustment Procedure (DSAP)

An option to estimate q^* is to use the *adjustment procedure* suggested by Nedić and Bertsekas [12], but fitted for the ε-subgradient method

In this procedure q_{lev}^k is the best function value achieved up to the kth iteration, in our case $\max_{0 \le j \le k} q_{\varepsilon_j}(\mu^j)$, plus a positive amount δ_k, which is adjusted according to algorithm's progress.

The adjustment procedure obtains q_{lev}^k as follows:

$$q_{lev}^k = \max_{0 \le j \le k} q_{\varepsilon_j}(\mu^j) + \delta_k, \tag{37}$$

and δ_k is updated according to

$$\delta_{k+1} = \begin{cases} \rho \delta_k, & \text{if } q_{\varepsilon_{k+1}}(\mu^{k+1}) \ge q_{lev}^k, \\ \max\{\beta \delta_k, \delta\}, & \text{if } q_{\varepsilon_{k+1}}(\mu^{k+1}) < q_{lev}^k, \end{cases} \tag{38}$$

where δ_0, δ, β, and ρ are fixed positive constants with $\beta < 1$ and $\rho \ge 1$.

The convergence of the approximate subgradient method for this stepsize was analyzed in [8]; see also [6].

3   Solution to NCNFP

An algorithm is given below for solving **NCNFP**. This algorithm uses the approximate subgradient method described in Sect. 1.

The value of the dual function $q(\mu^k)$ is estimated by minimizing approximately $l(x,\mu^k)$ over $x \in \mathscr{F}$ (the set defined by the network constraints) so that the optimality tolerance, τ_x^k, becomes more rigorous as k increases, i.e. the minimization will be *asymptotically exact* [1]. In other words, we set $q_{\varepsilon_k}(\mu^k) = l(x^k,\mu^k)$, where x^k minimizes approximately the nonlinear network subproblem **NNS$_k$**

$$\underset{x\in\mathscr{F}}{\text{minimize }} l(x,\mu^k) \tag{39}$$

in the sense that this minimization stops when we obtain an x^k that verifies the KKT conditions with τ_x^k accuracy, which implies that the norm of the reduced gradient holds

$$\|Z^T\nabla_x l(x^k,\mu^k)\| \le \tau_x^k, \tag{40}$$

where $\lim_{k\to\infty}\tau_x^k = 0$ and Z represents the reduction matrix whose columns form a base of the null subspace generated by the rows of the matrix of active network constraints of this subproblem (including the active capacity constraints on the flows of each arc), see [11]. Let $\bar{x}^k$ be the minimizer of this subproblem approximated by x^k. Then, it can be proved (see [8]) that there exists a positive w, such that $l(x^k,\mu^k) \le l(\bar{x}^k,\mu^k) + w\tau_x^k$ for $k = 1,2,\dots$. If we set $\varepsilon_k = \omega\tau_x^k$, this inequality becomes (8). Moreover, as

$$\tau_x^{k+1} = \sigma\tau_x^k, \qquad \text{for a fixed } \sigma \in (0,1), \tag{41}$$

then $\sum_{k=1}^{\infty}\varepsilon_k < \infty$, and so $\lim_{k\to\infty}\varepsilon_k = 0$. Consequently, to solve this problem we can use the approximate subgradient methods with the stepsizes described in Sect. 2. We denote $q_{\varepsilon_k}(\mu^k) = l(x^k,\mu^k)$, which satisfies the inequality (9). In this work, $\sigma = 10^{-1}$ by default. Note that in this case, $\varepsilon_k = \tau_x^k\omega = 10^{-(k-1)}\tau_x^1\omega$.

Algorithm 1 (Approximate subgradient method for NCNFP)

Step 0: *Initialize.* Set $k = 1$, N_{max}, τ_x^1, ε_q, ε_μ and τ_μ. Set $\mu^1 = 0$.

Step 1: *Compute* the dual function estimate, $q_{\varepsilon_k}(\mu^k)$, by solving **NNS$_k$** with accuracy τ_x^k, then $x^k \in \mathscr{F}$ is an approximate solution, $q_{\varepsilon_k}(\mu^k) = l(x^k,\mu^k)$, and $c^k = c(x^k)$ is an ε_k-subgradient of q in μ^k.

Step 2: *Check the stopping rules* for μ^k.

T_1: Stop if $\max\limits_{i=1,\dots,r}\left\{(c_i^k)^+\right\} < \tau_\mu$, where $(c_i^k)^+ = \max\{0,c_i(x^k)\}$.

T_2: Stop if $\dfrac{|q^k - (q^{k-1} + q^{k-2} + q^{k-3})/3|}{1 + |q^k|} < \varepsilon_q$, where $q^l = q_{\varepsilon_l}(\mu^l)$.

T_3: Stop if $\dfrac{1}{5}\sum\limits_{i=0}^{4}\|\mu^{k-i} - \mu^{k-i-1}\|_\infty < \varepsilon_\mu$.

T_4: Stop if k reaches a prefixed value N_{max}.

If μ^k fulfils one of these tests, then it is deemed approximately optimal, and (x^k, μ^k) is an approximate primal-dual solution.

Step 3: *Update* the estimate μ^k by means of the iteration

$$\mu_i^{k+1} = \begin{cases} \mu_i^k + \alpha_k c_i^k, & \text{if } \mu_i^k + \alpha_k c_i^k > 0 \\ \\ 0, & \text{otherwise} \end{cases} \tag{42}$$

where α_k is computed using some stepsize rule. Go to Step 1.

In Step 0, for the stopping rules, $\tau_\mu = 10^{-5}$, $\varepsilon_q = 10^{-7}$, $\varepsilon_\mu = 10^{-3}$ and $N_{max} = 200$ have been taken. In addition, $\tau_x^1 = 10^{-2}$ by default.

Step 1 is carried out by the code PFNL (described in [9]), which is based on the specific procedures for nonlinear network flows [15] and the active set procedure [11], using a spanning tree as the basis matrix of the network constraints.

In Step 2, alternative heuristic tests have been used for practical purposes. T_1 checks the feasibility of x^k, as if the violation of the side constraints has been sufficiently reduced, then (x^k, μ^k) is an acceptable primal-dual solution for **NCNFP**. T_2 and T_3 mean that μ does not improve for the last iterations. T_4 is used to stop the algorithm when this is not able to find a good enough solution.

To obtain α_k in Step 3, we have used the iteration (10) (see Sect. 1) with the five stepsize rules considered in Sect. 2: CSR, VCSR, DSRS, DSR, and DSAP. In the implementation of DSAP we use $\rho = 2$, $\beta = 1/\rho$, $\delta_0 = 0.5\|(c^1)^+\|$, and $\delta = 10^{-7}|l(x^0, \mu^1)|$, where x^0 is the initial feasible point for Step 1 and $k = 1$.

The values given above have been heuristically chosen. The implementation in Fortran-77 of the previous algorithm, termed PFNRN05, was designed to solve large-scale nonlinear network flow problems with nonlinear side constraints.

4 Numerical Tests

In order to obtain a computational comparison of the performance of the stepsizes CSR, VCSR, DSRS, DSR, and DSAP, some computational tests are carried out, which consist in solving nonlinear network flow problems with nonlinear side constraints using PFNRN05 code with the alternative stepsizes, where the objective functions are strictly convex and the side-constraint functions are convex. Therefore, these problems have a unique primal solution x^* and the duality gap is zero. The numerical tests have been carried out on a Sun Enterprise 250 under UNIX.

The problems used in these tests were created by means of the DIMACS-random-network generators Rmfgen and Gridgen (see [4]). These generators provide linear flow problems in networks without side constraints. The side constraints are defined by convex quadratic functions and were generated through the *Dirnl* random generator described in [9,10].

These test problems have up to 4008 variables, 1200 nodes, and 1253 side constraints, see [7]. The objective functions are nonlinear and strictly convex, and are either Namur functions (**n1**) or polynomial functions (**e2**). The polynomial functions give rise to problems with a moderate number of superbasic variables (degrees of freedom) at the optimizer, whereas the Namur functions [15] generate a high number of superbasic variables. More details in [7].

In Table 1 we compare the quality of the solution by means of the value of maximum violation of the side constraints at the optimal solution, $c^* = \|[c(x^*)]^+\|_\infty$, and the efficiency by the CPU times (in seconds) used to compute the solution. Note that c^* offers information about the feasibility of this solution and, hence, about its duality gap.

Table 1. Comparison of the quality/efficiency for the stepsizes

Prob.	CSR	VCSR	DSRS	DSR	DSAP
c15e2	$10^{-5}/4.5$	$10^{-5}/4.4$	$10^{-4}/8.5$	$10^{-4}/2.9$	$10^{-8}/2.2$
c17e2	$10^{-5}/6.0$	$10^{-5}/7.2$	$10^{-4}/13.7$	$10^{-4}/6.3$	$10^{-6}/3.1$
c18e2	$10^{-5}/21.3$	$10^{-5}/30.7$	$10^{-4}/67.0$	$10^{-2}/24.3$	$10^{-5}/38.0$
c13n1	$10^{-6}/64.3$	$10^{-6}/54.5$	$10^{-6}/64.3$	$10^{-5}/84.6$	$10^{-6}/44.2$
c15n1	$10^{-6}/82.3$	$10^{-6}/61.4$	$10^{-6}/83.2$	$10^{-4}/426.9$	$10^{-8}/99.8$
c17n1	$10^{-5}/78.9$	$10^{-5}/75.2$	$10^{-5}/79.1$	$10^{-4}/361.1$	$10^{-6}/92.3$
c22e2	$10^{-5}/3.8$	$10^{-5}/3.9$	$10^{-4}/8.5$	$10^{-4}/2.5$	$10^{-6}/2.0$
c23e2	$10^{-5}/6.3$	$10^{-5}/6.2$	$10^{-4}/6.0$	$10^{-3}/5.0$	$10^{-7}/5.9$
c24e2	$10^{-5}/15.9$	$10^{-5}/15.9$	$10^{-4}/94.8$	$10^{-2}/40.4$	$10^{-9}/5.2$
c34e2	$10^{-6}/4.8$	$10^{-6}/5.3$	$10^{-6}/4.9$	–	$10^{-8}/5.1$
c35e2	$10^{-6}/2.0$	$10^{-6}/2.3$	$10^{-6}/2.0$	$10^{-7}/2.9$	$10^{-9}/2.4$
c38e2	$10^{-6}/8.2$	$10^{-5}/7.4$	$10^{-6}/8.4$	$10^{-5}/12.4$	$10^{-8}/8.0$
c42e2	$10^{-7}/19.0$	$10^{-7}/14.1$	$10^{-7}/18.9$	$10^{-7}/14.1$	$10^{-7}/14.5$
c47e2	$10^{-5}/404.9$	$10^{-5}/401.0$	$10^{-5}/448.0$	–	$10^{-7}/657.1$
c48e2	$10^{-5}/114.4$	$10^{-5}/132.2$	$10^{-5}/116.7$	–	$10^{-7}/257.6$

Table 1 points out that the quality of the solution computed by PFNRN05 for the stepsize DSR is lower than that of DSRS, whereas that of this stepsize is slightly lower than that of CSR and VCSR. Also, the quality of the solution obtained with the stepsize DSAP is clearly higher than that computed for the rest of stepsizes.

Regarding the efficiency of PFNRN05 for these stepsizes, we observe that the efficiency when we use DSAP is similar to that obtained for CSR, VCSR, and DSRS, whereas our code for DSR is less efficient and robust than for the other stepsizes.

5 Conclusions

In this work some basic convergence results of subgradient methods for the stepsizes CSR and DSRS have been extended to approximate subgradient methods. Moreover, in the numerical tests carried out over convex nonlinear problems of nonlinearly constrained networks we have observed that the quality of the solution obtained by PFNRN05 for the dynamic stepsize rule DSAP is higher than that obtained for the other stepsizes, while the efficiency is similar.

The results of the numerical tests encourage to carry out further experimentation with other kind of problems and to compare the efficiency with that of well-known codes.

References

1. Bertsekas, D.P.: Constrained Optimization and Lagrange Multiplier Methods. Academic Press, New York (1982)
2. Bertsekas, D.P.: Nonlinear Programming, 2nd edn. Athena Scientific, Belmont (1999)
3. Correa, R., Lemarechal, C.: Convergence of some algorithms for convex minimization. Mathematical Programming 62, 261–275 (1993)
4. DIMACS. The first DIMACS international algorithm implementation challenge: The benchmark experiments. Technical Report, DIMACS, New Brunswick, NJ, USA (1991)
5. Ermoliev, Y.M.: Methods for solving nonlinear extremal problems. Cybernetics 2, 1–17 (1966)
6. Kiwiel, K.: Convergence of approximate and incremental subgradient methods for convex optimization. SIAM Journal on Optimization 14(3), 807–840 (2004)
7. Mijangos, E.: A variant of the constant step rule for approximate subgradient methods over nonlinear networks. In: Gavrilova, M.L., Gervasi, O., Kumar, V., Tan, C.J.K., Taniar, D., Laganá, A., Mun, Y., Choo, H. (eds.) ICCSA 2006. LNCS, vol. 3982, pp. 757–766. Springer, Heidelberg (2006)
8. Mijangos, E.: Approximate subgradient methods for nonlinearly constrained network flow problems. Journal of Optimization Theory and Applications 128(1), 167–190 (2006)
9. Mijangos, E., Nabona, N.: The application of the multipliers method in nonlinear network flows with side constraints. Technical Report 96/10, Dept. of Statistics and Operations Research, Universitat Politècnica de Catalunya, Barcelona, Spain (1996), http://www.ehu.es/~mepmifee
10. Mijangos, E., Nabona, N.: On the first-order estimation of multipliers from Kuhn-Tucker systems. Computers and Operations Research 28, 243–270 (2001)
11. Murtagh, B.A., Saunders, M.A.: Large-scale linearly constrained optimization. Mathematical Programming 14, 41–72 (1978)
12. Nedić, A., Bertsekas, D.P.: Incremental subgradient methods for nondifferentiable optimization. SIAM Journal on Optimization 12, 109–138 (2001)
13. Poljak, B.T.: Introduction to Optimization. Optimization Software Inc., New York (1987)
14. Shor, N.Z.: Minimization Methods for Nondifferentiable Functions. Springer, Berlin (1985)
15. Toint, Ph.L., Tuyttens, D.: On large scale nonlinear network optimization. Mathematical Programming 48, 125–159 (1990)

A Sample Time Optimization Problem in a Digital Control System

Wojciech Mitkowski and Krzysztof Oprzędkiewicz

Institute of Automatics, AGH University of Science and Technology
wmi@ia.agh.edu.pl, kop@agh.edu.pl

Abstract. In the paper a phenomenon of the existence of a sample time minimizing the settling time in a digital control system is described. As a control plant an experimental heat object was used. The control system was built with the use of a soft PLC system SIEMENS SIMATIC. As the control algorithm a finite dimensional dynamic compensator was applied. During tests of the control system it was observed that there exists a value of the sample time which minimizes the settling time in the system. This phenomenon is tried to explain.

1 Introduction

In the paper a digital control system for an experimental heat control plant is considered. The control plant is shown in Fig. 1. It has the form of a thin copper rod 30 cm long with an electric heater of length Δx_u localized at one end and resistive temperature sensor of length Δx at the other end. The input signal of the system is the standard current signal $0 - 5$ [mA]. It is amplified to the range $0 - 1.5$ [A] and it is the input signal for the heater. The temperature of the rod is measured with the use of a resistance sensor. The signal from the sensor is transformed to the standard current signal $0 - 5$ [mA] with the use of a transducer.

The structure of the digital control system is shown in Fig. 2. In this figure $y^+(k) = y(kh), h > 0, k = 0, 1, 2, \ldots$ and $u(t) = u^+(k)$ for $t \in [kh, (k+1)h), h > 0$ denotes the sample time of D/A and A/D converters working synchronically. During tests of this control system it was observed that the settling time (after this time the difference between the seat point r and the process value $y(t)$ is stably smaller than 5%) is a function of the sample time $h > 0$, $t_k = kh, k = 0, 1, 2, \ldots$, and this function has a minimum. This means that there exists a value of the sample time minimizing the settling time, which is one of fundamental direct control cost functions, applied in the industrial practice.

This effect was observed during tests of a discrete, finite-dimensional dynamic compensator. The control algorithm is based on the mathematical model of the control plant shown in Fig. 1, the control algorithm was implemented at the SIEMENS multipanel based "soft PLC" system. The hardware and software scheme of this system is shown in Fig. 6. In the control system shown in Fig. 2, D/A and A/D converters work synchronically.

A. Korytowski et al. (Eds.): System Modeling and Optimization, IFIP AICT 312, pp. 382–396, 2009.

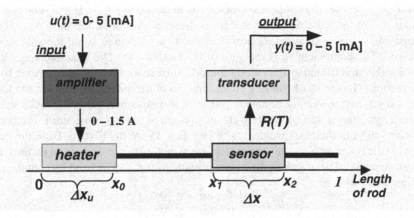

Fig. 1. The experimental heat control plant

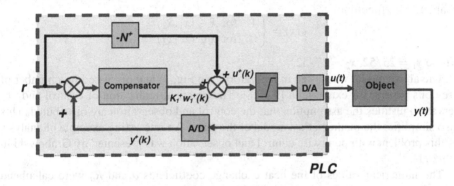

Fig. 2. The structure of the control system

In this paper the existence of the sample time minimizing the settling time is tried to explain. A method of optimal sample time determination for the considered control system is also proposed.

2 The Mathematical Model of the Control Plant

The mathematical model of the control plant we deal with is the following heat transfer equation:

$$\frac{\partial T(x,t)}{\partial t} = a\frac{\partial^2 T(x,t)}{\partial x^2} - R_a T(x,t) + b(x)u(t), \quad 0 \le x \le 1, \quad t \ge 0$$
$$\frac{\partial T(0,t)}{\partial x} = 0, \quad \frac{\partial T(1,t)}{\partial x} = 0, \quad t \ge 0$$
$$T(x,0) = 0, \quad 0 \le x \le 1 \tag{1}$$
$$y(t) = y_0 \int_0^1 T(x,t)c(x)\,dx$$

where $T(x,t)$ denotes the temperature of the rod at the point $x \in [0,1]$ and time moment t, a and R_a are the suitable heat transfer coefficients. All variables and constants are dimensionless. The coefficient R_a describes the heat exchange along the side surface of the rod. The coefficient y_0 is the gain of the slotted line. The heat exchange at the ends of rod is much smaller than along the side surface and it can be described by the homogeneous Neumann's boundary conditions. To make the model simpler, the length of the rod was assumed equal to one: $x \in [0,1]$. The diameter of the rod is much smaller than its length, that is why in equation (1) only length of rod x is considered. The control and observation in the plant we deal with (see Fig. 1) are distributed. Both the heater characteristic function $b(x)$ and sensor characteristic function $c(x)$ depend on the length of these elements. The length of the heater is equal to $\Delta x_u = 1/13$ and its characteristic function is as follows:

$$b(x) = \begin{cases} 1 & \text{for } x \in (0,x_0) \\ 0 & \text{for } x \notin (0,x_0) \end{cases} \tag{2}$$

where $x_0 = 1/13$. The temperature of the rod is measured in the segment $\Delta x = x_2 - x_1 = 1/13$ (distributed observation). The temperature sensor is described by the following characteristic function:

$$c(x) = \begin{cases} 1 & \text{for } x \in (x_1,x_2) \\ 0 & \text{for } x \notin (x_1,x_2) \end{cases} \tag{3}$$

where $x_1 = 25/52$, $x_2 = 27/52$.

The above model was tested in laboratory (see Fig. 3) and discussed in a number of previous papers, for example in [9,10,8]. The practical realization of control and observation justifies the assumption that the control and observation are distributed. This also simplifies the mathematical model of the plant. An interesting example of analysis of this problem with pointwise control and observation was presented by Grabowski in 1997 [1].

The numerical values of the heat exchange coefficients a and R_a were calculated with the use of the least squares method and the experimental step response of the plant (see [9]). They are equal to $a = 0.000945$, $R_a = 0.0271$. The idea of identification of the coefficients was to minimize (w.r.t. the parameters a and R_a) the following cost function:

$$J_0(a,R_a) = \sum_{k=1}^{S} [y(kh) - \tilde{y}(kh)]^2 \tag{4}$$

where $kh, k = 1,2,3,\ldots,S$ are discrete time moments, $y(kh)$ is the output of (1) for the control signal $u(t) = 1(t)$, $\tilde{y}(kh)$ is the step response of the real plant (see Fig. 1) measured at discrete time moments $kh, k = 1,2,3,\ldots,S$. During the identification experiment the sample time h was equal to 0.1 [s] and the number of samples S was equal to 3000.

The value of the steady-state gain y_0 was calculated via comparison of $y(kh)$ with $\tilde{y}(kh)$ after a suitably long time $t_s = Sh$. The value of this gain is equal to 25.7922.

Let $L(U,X)$ denote the space of linear continuous operators $S : U \to X$ with the following natural norm: $\|S\| = \sup\{\|Sv\|_X : \|v\|_U \leq 1\}$. Let $X = L^2(0,1;\mathbb{R})$ be a Hilbert space with scalar product $(p \mid d) = \int_0^1 p(x)d(x)dx$. The boundary problem (1) can be interpreted ([12], [11], p. 106; [2], p. 488) as the following differential equation in the Hilbert space $L^2(0,1;\mathbb{R})$:

$$\dot{T}(t) = AT(t) + Bu(t), \quad T(0) = 0, \quad 0 \leq t$$
$$y(t) = CT(t) \tag{5}$$

where $T(t) \in X = L^2(0,1;\mathbb{R})$, $u(t) \in \mathbb{R} = U$, $y(t) \in \mathbb{R} = Y$, $B \in L(U,X)$, $(Bu(t))(x) = b(x)u(t)$, $C \in L(X,Y)$, $CT(t) = y_0 (T(t)|c)$, A is a linear operator with the domain $D(A)$:

$$Aw = aw'' - R_a w, \quad w''(x) = d^2 w(x)/dx^2,$$
$$D(A) = \{w \in X : w'' \in X, w'(0) = 0, w'(1) = 0\} \tag{6}$$

The operator defined by (6) is self-adjoint and has a compact resolvent. This implies that A is a discrete operator. The spectrum of a discrete operator consists only of isolated eigenvalues and all eigenvalues have finite multiplicities. The operator given by (6) has a discrete spectrum with the simple eigenvalues:

$$\lambda_i = -i^2 a \pi^2 - R_a, \quad i = 0,1,2,3,\ldots \tag{7}$$

Eigenvectors associated with the eigenvalues (7) are defined as follows:

$$w_i(x) = \begin{cases} 1, & i = 0 \\ \sqrt{2}\cos(i\pi x), & i = 1,2,\ldots \end{cases} \tag{8}$$

The eigenvectors (8) build an orthonormal basis of the space $X = L^2(0,1;\mathbb{R})$.

If the basis of $X = L^2(0,1;\mathbb{R})$ is built by the set of eigenvectors (8), then the operators A, B and C in (5) can be expressed as the following infinite-dimensional matrices (see [13]):

$$A = diag\{\lambda_0, \lambda_1, \lambda_2, \ldots\}, \quad B = [b_0\ b_1\ b_2 \ldots]^T, \quad C = y_0[c_0\ c_1\ c_2 \ldots] \tag{9}$$

where $b_i = \int_0^1 b(x)w_i(x)dx = (b|w_i)$, $c_i = \int_0^1 c(x)w_i(x)dx = (c|w_i)$, $i = 0,1,2,\ldots$.

The operator (6) is the generator of an analytical, exponentially stable C_0- semigroup e^{At}, $t \geq 0$ in the space $X = L^2(0,1;\mathbb{R})$, where:

$$e^{At}w = \sum_{i=0}^{\infty} e^{\lambda_i t}(w|w_i)w_i \tag{10}$$

The analysis of (7) and (10) justifies the use of finite-dimensional approximation of (5). Instead of $i = 0, 1, 2, 3, \ldots$ we will use: $i = 0, 1, 2, 3, \ldots, N$. A suitable value of N for finite-dimensional approximation is $N = 25$ (see [9]). The respective finite-dimensional matrix representations of the operators A, B and C for the presented model (5) are as follows:

A = diag $(-0.0271, -0.0364, -0.0644, -0.1110, -0.1763, -0.2603, -0.3629,$
$-0.4841, -0.6240, -0.7826, -0.9598, -1.1556, -1.3702, -1.6033, -1.8551,$
$-2.1256, -2.4148, -2.7225, -3.0490, -3.3941, -3.7578, -4.1402, -4.5413,$
$-4.9610, -5.3993),$

$B = \text{col}(\quad 0.0769, \quad 0.1077, \quad 0.1046, \quad 0.0995, \quad 0.0926, \quad 0.0842, \quad 0.0745,$
$\quad 0.0638, \quad 0.0526, \quad 0.0412, \quad 0.0299, \quad 0.0190, \quad 0.0090, \quad 0.0000, -0.0077,$
$\quad -0.0139, -0.0187, -0.0218, -0.0234, -0.0235, -0.0223, -0.0200, -0.0168,$
$\quad -0.0130, -0.0087),$

$C^T = \text{col}(\quad 0.9920, \quad 0.0000, -1.3995, \quad 0.0000, \quad 1.3893, \quad 0.0000, -1.3724,$
$\quad 0.0000, \quad 1.3489, \quad 0.0000, -1.3191, \quad 0.0000, \quad 1.2832, \quad 0.0000, -1.2415,$
$\quad 0.0000, \quad 1.1944, \quad 0.0000, -1.1423, \quad 0.0000, \quad 1.0856, \quad 0.0000, -1.0248,$
$\quad 0.0000, \quad 0.9605).$

The form of operators (9) enables the decomposition of the system (5) into an infinite number of one-dimensional subsystems. Notice that the matrices B and C contain zero elements. This implies that the finite-dimensional approximation is both uncontrollable and unobservable, but the subsystems containing suitable nonzero elements of B and C are controllable and observable.

3 Digital Control System

The scheme of the digital control system is shown in Fig. 2. Both D/A and A/D converters work synchronically with the sample time $h > 0$ and their work can be described as follows: $y^+(k) = y(kh)$ and $u(t) = u^+(k)$ for $t \in [kh, (k+1)h)$, where $k = 0,1,2,\ldots$. The serial connection of a D/A converter, a continuous control plant and an A/D converter builds the discrete system described as follows (see, for example, [7], p. 140, 236):

$$T^+(k+1) = A^+ T^+(k) + B^+ u^+(k), \quad y^+(k) = C^+ T^+(k) \tag{11}$$

where $k = 0,1,2,\ldots$

$$A^+ = e^{Ah}, \quad B^+ = \int_0^h e^{At} B \, dt, \quad C^+ = C \tag{12}$$

and the matrices A, B and C are described by (9).

For the discrete system (11), the finite-dimensional discrete compensator proposed in [3] (p. 60), [4,5,6], or [7] (p. 236) can be built (see Fig. 2). The nonlinear element (saturation) is necessary to describe the technical realization of control, because real control signals are always bounded. The control system with the dynamic compensator and bounded control signal was discussed by Mitkowski and Oprzędkiewicz in 1999 [8].

Notice that the matrices B^+ and $C^+ = C$ contain zero elements. This implies that the system is neither controllable nor observable. The damping coefficients can be improved only in controllable and observable subsystems.

Let $T^+(k) = T(kh)$, where $k = 0,1,2,\ldots$. Equations (9) and (12) allow us to make the following decomposition of the system (11):

$$\begin{bmatrix} T_1^+(k+1) \\ T_2^+(k+1) \\ T_3^+(k+1) \end{bmatrix} = \begin{bmatrix} A_1^+ & 0 & 0 \\ 0 & A_2^+ & 0 \\ 0 & 0 & A_3^+ \end{bmatrix} \begin{bmatrix} T_1^+(k) \\ T_2^+(k) \\ T_3^+(k) \end{bmatrix} + \begin{bmatrix} B_1^+ \\ B_2^+ \\ B_3^+ \end{bmatrix} u^+(k),$$

$$y^+(k) = C_1^+ T_1^+(k) + C_2^+ T_2^+(k) + C_3^+ T_3^+(k) \tag{13}$$

where $T_i^+(k) \in X_i$, $i = 1,2,3$, $\dim X_1 < +\infty, \dim X_2 = p < +\infty, \dim X_3 = +\infty$.

For further considerations a finite-dimensional approximation with $N = 25$ and example value of sample time $h = 5$ [s] will be applied. Using the equations (12) we obtain the following values of matrices for the system after discretization:

$A^+ = $ diag (0.8733, 0.8335, 0.7247, 0.5740, 0.4141, 0.2722, 0.1629, 0.0889,
   0.0442, 0.0200, 0.0082, 0.0031, 0.0011, 0.0003, 0.0001, 0.0000, 0.0000,
   0.0000, 0.0000, 0.0000, 0.0000, 0.0000, 0.0000, 0.0000, 0.0000),

$B^+ = $ col (0.3597, 0.4924, 0.4471, 0.3818, 0.3077, 0.2354, 0.1718, 0.1201,
   0.0806, 0.0515, 0.0308, 0.0164, 0.0065, 0.0000,$-$0.0041,$-$0.0066,$-$0.0077,
   $-$0.0080,$-$0.0077,$-$0.0069,$-$0.0059,$-$0.0048,$-$0.0037,$-$0.0026,$-$0.0016).

As both D/A and A/D converters work synchronically, then $C^+ = C$.

In the scheme of Fig. 2, r denotes the seat point in the system, $y(t)$ denotes the system's output (the process value), $u(t)$ denotes the control signal, $y^+(k)$ and $u^+(k)$ denote discrete versions of control and output. The difference $y^+(k) - r$ describes the error in the control system. The compensator is described as follows:

$$\begin{bmatrix} w_1^+(k+1) \\ w_2^+(k+2) \end{bmatrix} = A_S^+ \cdot \begin{bmatrix} w_1^+(k) \\ w_2^+(k) \end{bmatrix} + B_S^+ \left[y^+(k) - r \right]$$

$$u^+(k) = K_1^+ w_1^+(k) + N^+ r$$

(14)

where

$$A_S^+ = \begin{bmatrix} A_1^+ - G_1^+ C_1^+ + B_1^+ K_1^+ & -G_1^+ C_2^+ \\ B_2^+ K_1^+ & A_2^+ \end{bmatrix}, \qquad B_S^+ = \begin{bmatrix} G_1^+ \\ 0 \end{bmatrix}$$

(15)

In (14) and (15) $w_1^+(k)$, $w_2^+(k)$ denote the state variables of the compensator, the constant N^+ should assure the steady-state error equal to zero (after a suitably long time $y(t) = r$ for the seat point $r = const$). The dynamics of the whole closed-loop control system is described by transient states, the same for increasing and decreasing values of r.

The whole closed-loop system of Fig. 2 for $r = 0$ and without the nonlinear element can be described as underneath:

$$\begin{bmatrix} e_1^+(k+1) \\ e_2^+(k+1) \\ T_1^+(k+1) \\ T_2^+(k+1) \\ T_3^+(k+1) \end{bmatrix} = \begin{bmatrix} A_1^+ - G_1^+ C_1^+ & -G_1^+ C_2^+ & 0 & 0 & -G_1^+ C_3^+ \\ 0 & A_2^+ & 0 & 0 & 0 \\ B_1^+ K_1^+ & 0 & A_1^+ + B_1^+ K_1^+ & 0 & 0 \\ B_2^+ K_1^+ & 0 & B_2^+ K_1^+ & A_2^+ & 0 \\ B_3^+ K_1^+ & 0 & B_3^+ K_1^+ & 0 & A_3^+ \end{bmatrix} \begin{bmatrix} e_1^+(k) \\ e_2^+(k) \\ T_1^+(k) \\ T_2^+(k) \\ T_3^+(k) \end{bmatrix}$$

$$e_i^+(k) = w_i^+(k) - T_i^+(k), \quad i = 1, 2$$

(16)

where $k = 0, 1, 2, \ldots$. Notice that the state operator:

$$A^+ = blocdiag(A_1^+, A_2^+, A_3^+)$$

(17)

of the system (13) is diagonal and it has a discrete spectrum containing the following eigenvalues:

$$\lambda_i^+ = e^{\lambda_i h}, \quad i = 0, 1, 2, \ldots,$$

(18)

where λ_i is expressed by (7), $h > 0$. This implies that $|\lambda_i^+| < 1$ and the discrete system (13) is asymptotically stable. The state operator of the discrete system (16) has the following form:

$$\tilde{A}^+ = blocdiag(A_1^+, A_2^+, A_1^+, A_2^+, A_3^+) + D \tag{19}$$

where D is a bounded operator (it describes a bounded perturbation of A^+). This implies that the operator $\tilde{A}^+$ has a discrete spectrum only.

The discrete system (16) with discrete operator $\tilde{A}^+$ is asymptotically stable, if its discrete spectrum is localized inside the unit circle: $|\lambda(\tilde{A}^+)| < 1$. The matrices A_1^+ and A_2^+ are finite-dimensional. The matrices G_1^+ and K_1^+ should assure the meeting of the following condition:

$$|\lambda(A_1^+ - G_1^+ C_1^+)| < 1 \quad \text{and} \quad |\lambda(A_1^+ + B_1^+ K_1^+)| < 1 \tag{20}$$

This condition can be fulfilled only if the suitable subsystems are controllable and observable.

In particular, the matrices G_1^+ and K_1^+ can be assigned to assure the eigenvalues of $A_1^+ - G_1^+ C_1^+$ and $A_1^+ + B_1^+ K_1^+$ equal to zero. Furthermore, when the dimension of A_1^+ is equal to 1 ($\dim X_1 = 1$ in equation (13)), the matrices G_1^+ and K_1^+, assuring eigenvalues equal to zero turn to real numbers equal to

$$G_1^+ = \frac{A_1^+}{C_1^+} = \frac{\lambda_0^+}{y_0 c_0}, \quad K_1^+ = -\frac{A_1^+}{B_1^+} = -\frac{\lambda_0^+}{b_0^+} \tag{21}$$

In (21), $\lambda_0^+ = e^{\lambda_0 h}$ denotes the first, most poorly damped element of the discrete system's spectrum, y_0, c_0 and b_0^+ denote suitable elements of output and control matrices for discrete system.

Denote the maximal eigenvalue of A_2^+ by $\lambda_{max}^+ = \max \lambda(A_2^+)$. Notice that if $\|C_3^+\| \to 0$ for $\dim X_2 \to +\infty$, then the bounded perturbation theorem implies the existence of such a $p < +\infty$ that the discrete system (16) is exponentially stable with a damping coefficient $\gamma \in (\lambda_{max}^+, \lambda_0^+)$.

Generally, the construction of feedback (14), (15) consists in calculating the matrices K_1^+ and G_1^+. It can be done with the use of finite-dimensional approach, for example LQ (see [7], pp. 71, 78). The assumed damping coefficient can be obtained by increasing the dimension of the subsystem X_2: $p = \dim X_2$. The constant N^+ is calculated to assure the steady-state error in the control system equal to zero.

4 The Sample Time Optimization

Assume for further considerations that $\dim X_1 = 1$ and $p = \dim X_2 = 2$. The subsystem X_1 is controllable and observable, because $b_0 \neq 0$ and $y_0 c_0 \neq 0$ (see (12)). Then from (21) we obtain:

$$G_1^+ = \frac{e^{-R_a h}}{y_0 c_0}, \quad K_1^+ = \frac{R_a e^{-R_a h}}{b_0 (e^{-R_a h} - 1)} \tag{22}$$

The coefficients G_1^+ and K_1^+ of the compensator (14) are functions both of plant's parameters and sample time $h > 0$.

5   Experiments

The general hardware and software scheme of the laboratory control system is shown in Fig. 3, the scheme of the control plant is shown in Fig. 1.

The control system with the compensator for the plant shown in Fig. 1 was implemented on the "soft PLC" platform shown in Fig. 6. During experiments a servo problem was tested. Example experimental plots of the seat point r, the control signal $u(t)$ and the output $y(t)$ are shown in Fig. 4. They were obtained after a step change of the seat point from the value $r = 0.25$ [mA] to $r = 1.75$ [mA]. The sample time h was equal to 723 [ms].

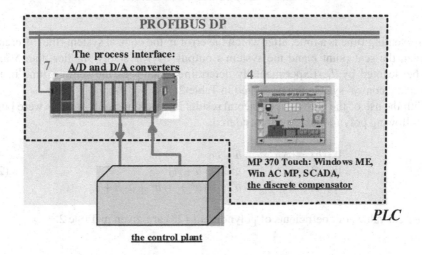

Fig. 3. The hardware and software scheme of the laboratory control system

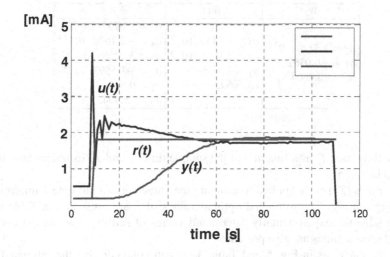

Fig. 4. Plots of $y(t)$, $u(t)$ and r in the considered control system

Table 1. Experimentally determined values of settling time T_c

Sample time h [ms]	settling time T_c[s]
300	52.37
400	50.52
500	49.17
600	48.35
700	48.01
800	48.17
900	48.73
1000	50.05

The settling time is a time, after which the error in the control system (the difference between the seat point r and the system's output $y(t)$) is stably smaller than 5%. It will be denoted by T_c. Experimentally determined values of the settling time in the considered control system are presented in Table 2.

With the use of the above experimental results interpolation polynomials were built. The following polynomials were considered:

$$W_2(h) = a_2 h^2 + a_1 h + a_0$$
$$W_3(h) = a_3 h^3 + a_2 h^2 + a_1 h + a_0 \qquad (23)$$
$$W_5(h) = a_5 h^5 + a_4 h^4 + a_3 h^3 + a_2 h^2 + a_1 h + a_0$$

Numerical values of coefficients of polynomials (23) are given in Table 2.

Table 2. Coefficients of interpolation polynomials (23)

W_2	W_3	W_5
$a_2 = 2.5220 \cdot 10^{-5}$ $a_1 = -0.0362$ $a_0 = 60.9604$	$a_3 = 1.8434 \cdot 10^{-9}$ $a_2 = 2.1626 \cdot 10^{-5}$ $a_1 = -0.0340$ $a_0 = 60.5650$	$a_5 = 9.0705 \cdot 10^{-14}$ $a_4 = -2.7670 \cdot 10^{-14}$ $a_3 = 3.2368 \cdot 10^{-7}$ $a_2 = -1.5593 \cdot 10^{-4}$ $a_1 = 0.0123$ $a_0 = 55.9942$

The settling time T_c as a function of the sample time h, and all its interpolations are presented in Fig. 5.

Polynomials (23) were applied to estimate the values of sample time h minimizing the settling time T_c. The estimated optimal values of h are presented in Table 4. In the same table the experimentally determined values of settling time for the optimal estimated sample times are also presented.

From the diagrams in Fig. 5 and Table 3 we can conclude that the estimated and experimental results are very close.

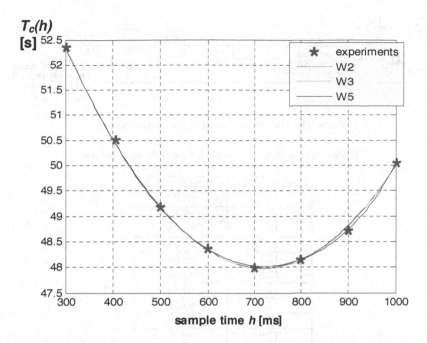

Fig. 5. The settling time T_c as a function of sample time h – experiments and all interpolation polynomials

Table 3. Values of sample time h and corresponding values of settling time T_c, determined experimentally and estimated with the use of polynomials (23)

Polynomial	W_2	W_3	W_5
Estimated optimal sample time [ms]	715	721	723
Estimated optimal settling time [s]	47.98	47.97	48.02
Experimentally determined settling time for estimated optimal sample time [s]	48.17	48.13	48.12

6 Simulations

The control system described in the previous section was also tested with the use of MATLAB/SIMULINK. During simulations the dependence of the settling time T_c on the sample time h was tested. The SIMULINK model of the control system is shown in Fig. 6. The simulation parameters were as follows: the stop time 300[s], numerical method: Ode 45 (Dormand Prince), variable step, minimal step size: 0.005[s], maximal step size: 0.01[s], initial step size: 0.01[s].

The settling time T_c as a function of the sample time h is shown in Table 4 and in Fig. 7.

The minimal value of the function shown in Fig. 7 is localized between $h = 12.6$ [s] and $h = 12.8$ [s].

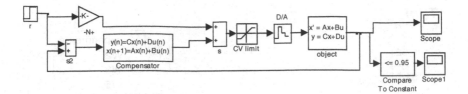

Fig. 6. The SIMULINK model of the control system

Table 4. The settling time as a function of the sample time

sample time h [s]	settling time T_c [s]
12.0	69.92
12.1	69.83
12.2	69.75
12.3	69.69
12.4	69.64
12.5	69.61
12.6	69.59
12.7	69.59
12.8	69.59
12.9	69.6
13.0	69.63
13.1	69.66
13.2	69.70

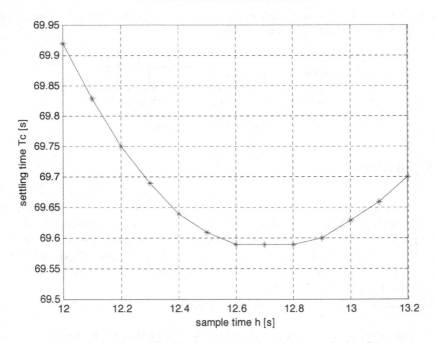

Fig. 7. The settling time T_c as a function of the sample time h

7 A Comparison with a Weighted Cost Function

Consider the cost function $J(h)$ proposed in [3], p. 61, [4] and [7], p. 241:

$$J(h) = \left\|G_1^+(h)\right\| + \left\|K_1^+(h)\right\| + \alpha h \tag{24}$$

This cost function is a weighted sum of the settling time, approximately expressed by the term αh and the norms $\|G_1^+\|$ and $\|K_1^+\|$ describing gain coefficients of the discrete compensator (14). For the particular case, if G_1^+ and K_1^+ are described by (22), the cost function (24) has the following form:

$$J(h) = \frac{e^{-R_a h}}{y_0 c_0} + \frac{R_a e^{-R_a h}}{b_0 \left(1 - e^{-R_a h}\right)} + \alpha h \tag{25}$$

where c_0 and b_0 are appropriate elements of the output and control matrices.

The cost function (25) depends on model's parameters and sample time $h > 0$. An example diagram of function (25) for the considered object parameters, weight coefficient $\alpha = 25$ and range of sample time h from 0.5 to 1.0 [s] is shown in Fig. 8.

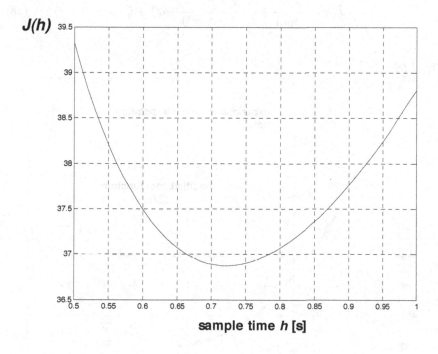

Fig. 8. The cost function (25) as a function of sample time h

The value h_{opt} of the sample time h minimizing the function shown in Fig. 8 was calculated numerically and it is equal to

$$h_{opt} = 0.7216 \tag{26}$$

The value of function (25) for the sample time h_{opt} is equal to

$$J(h_{opt}) = 36.8738 \qquad (27)$$

The comparison of the cost function (25) proposed theoretically and shown in Fig. 8 with the experimental results shown in Fig. 5 allows us to conclude that the value of the sample time h_{opt} minimizing the cost function (25) is equal to the value of h, which minimizes the interpolating polynomial W_3. The value of $J(h_{opt})$ is smaller than $T_c(h_{opt})$, because the function J does not express the settling time.

Furthermore, the general form of the dependence of the settling time T_c on the sample time h obtained with the use of simulation, described by Table 1 and shown in Fig. 7 is the same as the functions shown in Fig. 8 and 5. It also has a minimum, but the sample time minimizing the settling time is much longer.

Notice also that it is possible to modify the cost function (25) so that it directly expresses the settling time T_c. The proposed modification of $J(h)$ is described underneath. It changes the value of function without changing the localization of the minimum. Let us consider the cost function $\tilde{J}(h) = J(h) + \zeta$, $\zeta \in \mathbb{R}$. From (25) we obtain:

$$\tilde{J}(h) = \frac{e^{-R_a h}}{y_0 c_0} + \frac{R_a e^{-R_a h}}{b_0 \left(1 - e^{-R_a h}\right)} + \alpha h + \zeta \qquad (28)$$

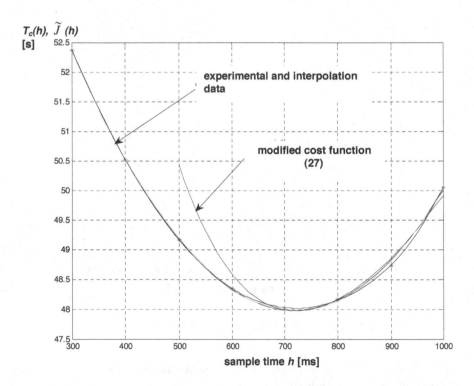

Fig. 9. Comparison of modified cost function $\tilde{J}(h)$ with experimental data and interpolation polynomial $W_3(h)$

Fig. 9 shows the diagrams of $\tilde{J}(h)$ for $\zeta = 11.1$ and the interpolation polynomial $W_3(h)$, which was obtained with the use of experimental data.

From the diagrams presented in Fig. 9 we can conclude that the modified cost function (28) for $h \in (650, 1000)$, obtained theoretically can be used as a good approximation of the settling time T_c as a function of sample time h in the considered laboratory control system.

8   Conclusions and Open Problems

The final conclusions of the paper can be formulated as follows:

1. During tests of the control system with the discrete dynamic compensator a new, interesting phenomenon was observed. The dependence between the sample time and the settling time was observed both in the real control system and during simulations. Additionally, the diagram of the cost function proposed in (25) is very similar to diagrams describing the dependence of the settling time on the sample time for real case and simulations.
2. An important open problem is to explain the significant difference between the value of sample time minimizing the settling time in simulation and experiments. This difference can be caused by the fact, that the equation (5) and its finite-dimensional approximation used to simulate the control plant in MATLAB are dimensionless in contrast to the real plant. Additionally, the input-output model of the considered system operates only with the use of the standard current signal, although the model describes the heat transfer process, whose measure is temperature. The dependence of the temperature of rod on input and output current signals is very complicated and a knowledge about it is not necessary for the synthesis of the control system for the plant. A correct scaling of the model used to simulations may be the solution of this problem.
3. In the construction of the finite-dimensional compensator the bounded perturbation theorem is applied (see [2], p. 497; [11], p. 79, 81). Relations (12) allow us to use this theorem to the discrete time case.
4. The modified cost function (28) can be applied as a measure of the settling time T_c in real control system. It sufficiently well describes the dependence of the settling time T_c on the model's parameters and the sample time $h > 0$.
5. The relation (28) was proposed after the analysis of the simple model of the considered control plant. It can be helpful to explain the phenomenon of the existence of optimal sample time h_{opt} observed during experiments with the use of real soft PLC control system, presented in Fig. 3. The determining of the parameters α and ξ in modified cost function (28) will be considered by authors in the future.
6. The simulation tests of the function we deal with also confirmed the existence of the sample time minimizing the settling time in the system, but the localization of minimum is significantly different from theoretical and experimental case.
7. As another model of the considered control plant the transfer function with delay can be also considered. It has the following, known form: $G(s) = Ke^{-\tau s}/(Ts+1)$. Its parameters: delay time τ and time constant T are relatively simple to identify. During an analysis of this model it was observed that $\alpha \approx 1.2\,\tau$ and $\zeta \approx 0.5\tau$.

Acknowledgements. This work was supported by the Ministry of Science and Higher Education of Poland under the Research Project NN 514414034.

References

1. Grabowski, P.: An example of identification of a parabolic system. Models with pointwise observation/control. Prace z Automatyki, pp. 181–186. Wydawnictwa AGH, Kraków (1997) (in Polish)
2. Kato, T.: Perturbation Theory for Linear Operators. Springer, Berlin (1980)
3. Mitkowski, W.: Stabilization of linear autonomous systems. Automatyka 35 (1984) (in Polish)
4. Mitkowski, W.: Stabilization of the linear parabolic system with the use of discrete dynamic compensator. Elektrotechnika 4(2), 189–197 (1985) (in Polish)
5. Mitkowski, W.: Feedback stabilization of second order evolution equations with damping by discrete-time input-output data. In: IMACS-IFAC Symp. on Modelling and Simulation for Control of Lumped and Distributed Parameter Systems, Lille, France, pp. 355–358 (1986)
6. Mitkowski, W.: Stabilization of infinite-dimensional linear systems with the use of dynamic feedback. Arch. Automatyki i Telemechaniki 33(4), 515–528 (1988) (in Polish)
7. Mitkowski, W.: Stabilization of Dynamical Systems: WNT (1991) (in Polish)
8. Mitkowski, W., Oprzędkiewicz, K.: Finite dimensional dynamic feedback with bounded control. Automatyka 3(1), 235–242 (1999)
9. Oprzędkiewicz, K.: Modeling of parabolic problems in MATLAB/SIMULINK. In: Proc. of III Conf. Comp. Simulation, Mexico City, pp. 45–49 (1995)
10. Oprzędkiewicz, K.: An example of parabolic system identification. Elektrotechnika 16(2), 99–106 (1997) (in Polish)
11. Pazy, A.: Semigroups of Linear Operators and Applications to Partial Differential Equations. Springer, New York (1983)
12. Sakawa, Y.: Feedback stabilization of linear diffusion systems. SIAM J. Control and Optimization 31(5), 667–676 (1983)
13. Triggiani, R.: On the stabilization problem in Banach space. J. Math. Anal. Appl. 52(3), 383–403 (1975)

Level Set Method for Shape and Topology Optimization of Contact Problems

Andrzej Myśliński

Systems Research Institute, ul. Newelska 6, 01-447 Warsaw, Poland
myslinsk@ibspan.waw.pl

Abstract. This paper deals with simultaneous topology and shape optimization of elastic contact problems. The structural optimization problem for an elastic contact problem is formulated. Shape as well as topological derivatives formulae of the cost functional are provided using material derivative and asymptotic expansion methods, respectively. These derivatives are employed to formulate necessary optimality condition for simultaneous shape and topology optimization and to calculate a descent direction in numerical algorithm. Level set based numerical algorithm for the solution of this optimization problem is proposed. Numerical examples are provided and discussed.

1 Introduction

The paper is concerned with the formulation of a necessary optimality condition and the numerical solution of a structural optimization problem for an elastic body in unilateral contact with a rigid foundation. The contact with a given friction, described by Coulomb law, is assumed to occur at a portion of the boundary of the body. The displacement field of the body in unilateral contact is governed by an elliptic variational inequality of the second order. The results concerning the existence, regularity and finite-dimensional approximation of solutions to contact problems are given in [10]. The structural optimization problem for the elastic body in contact consists in finding such topology as well as such shape of the boundary of the domain occupied by the body that the normal contact stress along a contact boundary is minimized. It is assumed that the volume of the body is bounded.

Shape optimization of static elastic contact problems is considered, among others, in [10,20], where necessary optimality conditions, results concerning convergence of finite-dimensional approximation and numerical results are provided. Material derivative method is employed in monograph [20] to calculate the sensitivity of solutions to contact problems as well as the derivatives of domain depending functionals with respect to variations of the boundary of the domain occupied by the body. Necessary optimality conditions for shape optimization of elastic contact problems are formulated also in monograph [15]. In this monograph contact problems are considered in the mixed variational formulation and the results of numerical experiments are reported. Shape optimization of a dynamic contact problem with Coulomb friction and heat flow is considered in [13]. In this paper the material derivative method is employed to formulate

A. Korytowski et al. (Eds.): System Modeling and Optimization, IFIP AICT 312, pp. 397–410, 2009.

a necessary optimality condition. The finite element method for the spatial derivatives and the finite difference method for the time derivatives are employed to discretize the optimization problem. The level set based method is applied to find numerically the optimal solution.

Topology optimization deals with the optimal material distribution within the body resulting in its optimal shape [22]. The topological derivative is employed to account variations of the solutions to state equations or shape functionals with respect to emerging of small holes in the interior of the domain occupied by the body. The notion of the topological derivative and results concerning its application in optimization of elastic structures are reported in the series of papers [2,6,7,9,14,21,22,23]. Among others, paper [23] deals with the calculation of topological derivatives of solutions to Signorini and elastic contact problems. Asymptotic expansion method combined with transformation of energy functional are employed to calculate these derivatives. Simultaneous shape and topology optimization of Signorini and elastic contact problems with or without friction are analyzed in papers [6] and [14], [16] respectively. In these papers the level set method is incorporated in the numerical algorithm.

In structural optimization the level set method [4,12,19,25] is employed in numerical algorithms for tracking the evolution of the domain boundary on a fixed mesh and finding an optimal domain. This method is based on an implicit representation of the boundaries of the optimized structure. A level set model describes the boundary of the body as an isocontour of a scalar function of a higher dimensionality. The evolution of the boundary of the domain is governed by Hamilton-Jacobi equation. While the shape of the structure may undergo major changes, the level set function remains simple in its topology. Level set methods are numerically efficient and robust procedures for the tracking of interfaces, which allows domain boundary shape changes in the course of iteration. Applications of the level set methods in structural optimization can be found, among others, in [1,2,14,16,25]. The speed vector field in Hamilton-Jacobi equation driving the propagation of the level set function is given by the Eulerian derivative of an appropriately defined cost functional with respect to the variations of the free boundary. Recently, in the series of papers [8,9,11,18,26,27] different numerical improvements of the level set method employed for the numerical solution of the structural optimization problems are proposed and numerically tested.

This paper deals with topology and shape optimization of elastic contact problems. The optimization problem for elastic contact problem is formulated. Shape as well as topological derivatives formulae of the cost functional are provided using material derivative [20] and asymptotic expansion [22] methods, respectively. These derivatives are employed to formulate necessary optimality condition for simultaneous shape and topology optimization. Level set based numerical algorithm for the solution of the shape optimization problem is proposed. The finite element and finite difference methods [10] are used as the discretization methods. Numerical examples are provided and discussed.

The paper extends the author's previous results in the field of structural optimization of contact systems contained in [15,16] by considering besides shape also topology optimization of these systems or developing optimality conditions, respectively.

2   Problem Formulation

Consider deformations of an elastic body occupying two-dimensional domain Ω with the smooth boundary Γ. Assume $\Omega \subset D$ where D is a bounded smooth hold-all subset of R^2. The body is subject to body forces $f(x) = (f_1(x), f_2(x)), x \in \Omega$. Moreover, surface tractions $p(x) = (p_1(x), p_2(x)), x \in \Gamma$, are applied to a portion Γ_1 of the boundary Γ. We assume that the body is clamped along the portion Γ_0 of the boundary Γ, and that the contact conditions are prescribed on the portion Γ_2, where $\Gamma_i \cap \Gamma_j = \emptyset, i \neq j, i, j = 0, 1, 2,$ $\Gamma = \bar{\Gamma}_0 \cup \bar{\Gamma}_1 \cup \bar{\Gamma}_2$.

We denote by $u = (u_1, u_2), u = u(x), x \in \Omega$, the displacement of the body and by $\sigma(x) = \{\sigma_{ij}(u(x))\}, i, j = 1, 2$, the stress field in the body. Consider elastic bodies obeying Hooke's law, i.e., for $x \in \Omega$ and $i, j, k, l = 1, 2$

$$\sigma_{ij}(u(x)) = a_{ijkl}(x)e_{kl}(u(x)). \tag{1}$$

We use here and throughout the paper the summation convention over repeated indices [10]. The strain $e_{kl}(u(x)), k, l = 1, 2$, is defined by:

$$e_{kl}(u(x)) = \frac{1}{2}(u_{k,l}(x) + u_{l,k}(x)), \tag{2}$$

where $u_{k,l}(x) = \frac{\partial u_k(x)}{\partial x_l}$. The stress field σ satisfies the system of equations [10]

$$-\sigma_{ij}(x)_{,j} = f_i(x) \quad x \in \Omega, i, j = 1, 2, \tag{3}$$

where $\sigma_{ij}(x)_{,j} = \frac{\partial \sigma_{ij}(x)}{\partial x_j}, i, j = 1, 2$. The following boundary conditions are imposed

$$u_i(x) = 0 \text{ on } \Gamma_0, \, i = 1, 2, \tag{4}$$
$$\sigma_{ij}(x)n_j = p_i \text{ on } \Gamma_1, \, i, j = 1, 2, \tag{5}$$
$$u_N \leq 0, \, \sigma_N \leq 0, \, u_N\sigma_N = 0 \text{ on } \Gamma_2, \tag{6}$$
$$|\sigma_T| \leq 1, \, u_T\sigma_T + |u_T| = 0 \text{ on } \Gamma_2, \tag{7}$$

where $n = (n_1, n_2)$ is the unit outward versor to the boundary Γ. Here $u_N = u_i n_i$ and $\sigma_N = \sigma_{ij}n_in_j, i, j = 1, 2$, represent the normal components of displacement u and stress σ, respectively. The tangential components of displacement u and stress σ are given by $(u_T)_i = u_i - u_N n_i$ and $(\sigma_T)_i = \sigma_{ij}n_j - \sigma_N n_i, i, j = 1, 2$, respectively. $|u_T|$ denotes the Euclidean norm in R^2 of the tangent vector u_T. Recall [10], (6) - (7) describe Signorini non-penetration condition and Coulomb friction law, respectively. For the sake of simplicity it is assumed that the tangential contact stress is bounded by 1, i.e., the product of the static friction coefficient and given normal contact stress is equal to 1. The equality in (7) can be written in the equivalent form as $u_T\sigma_T \leq 0$ and $(1 - |\sigma_T|)u_T = 0$. Therefore (7) describes friction phenomenon including sliding. The results concerning the existence of solutions to (1) - (7) can be found in [10,20].

2.1 Variational Formulation of Contact Problem

Let us formulate contact problem (3) - (7) in variational form. Denote by V_{sp} and K the space and set of kinematically admissible displacements:

$$V_{sp} = \{z \in [H^1(\Omega)]^2 : z_i = 0 \text{ on } \Gamma_0, i = 1,2\}, \tag{8}$$

$$K = \{z \in V_{sp} : z_N \le 0 \text{ on } \Gamma_2\}. \tag{9}$$

$H^1(\Omega)$ denotes Sobolev space of square integrable functions and their first derivatives; $[H^1(\Omega)]^2 = H^1(\Omega) \times H^1(\Omega)$. Denote also by Λ the set

$$\Lambda = \{\zeta \in L^2(\Gamma_2) : |\zeta| \le 1\}.$$

Variational formulation of problem (3) - (7) has the form: *find a pair* $(u, \lambda) \in K \times \Lambda$ *satisfying*

$$\int_\Omega a_{ijkl} e_{ij}(u) e_{kl}(\varphi - u) dx - \int_\Omega f_i(\varphi_i - u_i) dx - \int_{\Gamma_1} p_i(\varphi_i - u_i) ds$$
$$+ \int_{\Gamma_2} \lambda(\varphi_T - u_T) ds \ge 0 \quad \forall \varphi \in K, \tag{10}$$

$$\int_{\Gamma_2} (\zeta - \lambda) u_T ds \le 0 \quad \forall \zeta \in \Lambda, \tag{11}$$

$i, j, k, l = 1, 2$. Function λ is interpreted as a Lagrange multiplier corresponding to term $|u_T|$ in equality constraint in (7) [10,20]. This function is equal to tangent stress along the boundary Γ_2, i.e., $\lambda = \sigma_{T|\Gamma_2}$. Function λ belongs to the space $H^{-1/2}(\Gamma_2)$, i.e., the space of traces on the boundary Γ_2 of functions from the space $H^1(\Omega)$. Here, following [10] function λ is assumed to be more regular, i.e., $\lambda \in L^2(\Gamma_2)$. The results concerning the existence of solutions to system (10) - (11) can be found, among others, in [10].

2.2 Optimization Problem

Before formulating a structural optimization problem for (10) - (11) let us introduce the set U_{ad} of admissible domains. Denote by $Vol(\Omega)$ the volume of the domain Ω equal to

$$Vol(\Omega) = \int_\Omega dx. \tag{12}$$

Domain Ω is assumed to satisfy the volume constraint of the form

$$Vol(\Omega) - Vol^{giv} \le 0, \tag{13}$$

where the constant $Vol^{giv} = const_0 > 0$ is given. In the case of shape optimization of problem (10) - (11) the optimized domain Ω is assumed to satisfy equality volume condition, i.e., (13) is assumed to be satisfied as equality. In the case of topology optimization Vol^{giv} is assumed to be the initial domain volume and (13) is satisfied in the form $Vol(\Omega) = r_{fr} Vol^{giv}$ with $r_{fr} \in (0,1)$ [22]. The set U_{ad} has the following form

$$U_{ad} = \{\Omega : E \subset \Omega \subset D \subset R^2 : \Omega \text{ is Lipschitz continuous,}$$
$$\Omega \text{ satisfies condition (13)}, P_D(\Omega) \le const_1\}, \tag{14}$$

where $E \subset R^2$ is a given domain such that Ω as well as all perturbations of it satisfy $E \subset \Omega$. $P_D(\Omega) = \int_\Gamma dx$ is the perimeter of a domain Ω in D [5], [20, p. 126]. The perimeter constraint is added in (14) to ensure the compactness of the set U_{ad} in the square integrable topology of characteristic functions as well as the existence of optimal domains. The constant $const_1 > 0$ is assumed to exist. The set U_{ad} is assumed to be nonempty. In order to define a cost functional we shall also need the following set M^{st} of auxiliary functions

$$M^{st} = \{\phi = (\phi_1, \phi_2) \in [H^1(D)]^2 : \phi_i \leq 0 \text{ on } D, i = 1, 2, \| \phi \|_{[H^1(D)]^2} \leq 1\}, \qquad (15)$$

where the norm $\| \phi \|_{[H^1(D)]^2} = (\sum_{i=1}^2 \| \phi_i \|_{H^1(D)}^2)^{1/2}$.

In order to formulate an optimization problem we have to define the cost functional. Measurements and engineering practice indicate that when two surfaces are in contact a large stress on the contact boundary occurs. Usually, the normal contact stress σ_N attains maximal values in the middle of the contact area. The goal of structural engineers is to reduce this maximal value of the stress as much as possible. Thus, the cost functional $S(\Gamma_2) = \max_{x \in \Gamma_2} | \sigma_N(x) |$ is natural criterion of optimization directly reflecting the design objectives. Unfortunately, the optimization problem with the cost functional $S(\Gamma_2)$ is nonsmooth and difficult for analysis and numerical solution [15]. This is the reason why the criterion of maximal contact stress is approximated by integral, differentiable functionals. Recall from [15] the cost functional approximating the normal contact stress on the contact boundary

$$J_\phi(u(\Omega)) = \int_{\Gamma_2} \sigma_N(u)\phi_N(x)ds, \qquad (16)$$

depending on the auxiliary given bounded function $\phi(x) \in M^{st}$. Function ϕ in most cases, including finite-dimensional spaces, is chosen piecewise constant or piecewise linear in a hold-all domain D. The integral (16) is nonnegative for all $\phi \in K$. For given ϕ, the bigger is the normal contact stress on the boundary, the bigger is the value of the cost functional. This integral is also related to the strain energy of the body (for details see [10,15]). σ_N and ϕ_N are the normal components of the stress field σ corresponding to a solution u satisfying system (10) - (11) and the function ϕ, respectively.

Consider the following structural optimization problem: *for a given function $\phi \in M^{st}$, find a domain $\Omega^\star \in U_{ad}$ such that*

$$J_\phi(u(\Omega^\star)) = \min_{\Omega \in U_{ad}} J_\phi(u(\Omega)). \qquad (17)$$

The existence of an optimal domain $\Omega^\star \in U_{ad}$ follows by standard arguments (see [5,20]).

3   Optimality Conditions

3.1   Shape Derivative

Consider variations of domain $\Omega \subset D$ with respect to the boundary Γ only. Assume that in (14) volume condition is satisfied as equality, i.e., constant volume condition holds.

Let τ be a given parameter such that $0 \le \tau < \tau_0$, τ_0 is prescribed, and $V = V(x, \tau)$, $x \in \Omega$, be a given admissible velocity field. The set of admissible velocity fields V consists of vector fields regular enough (C^k class, $k \ge 1$, for details see [20]) with respect to x and τ and such that on the boundary ∂D of D either $V = 0$ at singular points of this boundary or normal component $V \cdot n$ of V equals $V \cdot n = 0$ at points of this boundary where the outward unit normal field n exists. Therefore the perturbations of domain Ω are governed by the transformation

$$T(\tau, V) : \bar{D} \to \bar{D},$$

i.e., $\Omega_\tau = T(\tau, V)(\Omega)$ [20]. Since only small perturbations of Ω are considered, this transformation can have the form of perturbation of the identity operator I in R^2. An example of such transformation is $T(\tau, \tilde{V}) = I + \tau \tilde{V}(x)$, where $\tilde{V}$ denotes a smooth vector field defined on R^2 [20].

The Euler derivative of the domain functional $J_\phi(\Omega)$ is defined as

$$dJ_\phi(\Omega, V) = \lim_{\tau \to 0^+} \frac{J_\phi(\Omega_\tau) - J_\phi(\Omega)}{t}. \tag{18}$$

In [15], using the material derivative approach [20], the Euler derivative of the cost functional (16) has been calculated and a necessary optimality condition for the shape optimization problem (17) has been formulated. This Euler derivative has the form

$$dJ_\phi(u(\Omega); V) = \int_\Gamma (\sigma_{ij} e_{kl}(\phi + p^{adt}) - f \cdot \phi) V(0) \cdot n \, ds$$
$$- \int_{\Gamma_1} [\frac{\partial(p \cdot (p^{adt} + \phi))}{\partial n} + \kappa p \cdot (p^{adt} + \phi)] V(0) \cdot n \, ds \tag{19}$$
$$+ \int_{\Gamma_2} [\lambda(p_T^{adt} + \phi_T) + q^{adt} u_T] \kappa V(0) \cdot n \, ds,$$

where $i, j, k, l = 1, 2$, $V(0) = V(x, 0)$, the displacement $u \in V_{sp}$ and the stress $\lambda \in \Lambda$ satisfy state system (10) - (11). κ denotes the mean curvature of the boundary Γ. The adjoint functions $p^{adt} \in K_1$ and $q^{adt} \in \Lambda_1$ satisfy for $i, j, k, l = 1, 2$, the following system

$$\int_\Omega a_{ijkl} e_{ij}(\phi + p^{adt}) e_{kl}(\varphi) dx + \int_{\Gamma_2} q^{adt} \varphi_T ds = 0, \quad \forall \varphi \in K_1, \tag{20}$$

and

$$\int_{\Gamma_2} \zeta(p_T^{adt} + \phi_T) ds = 0, \quad \forall \zeta \in \Lambda_1, \tag{21}$$

where the cones K_1 and Λ_1 are given by [15,20]

$$K_1 = \{\xi \in V_{sp} : \xi_N = 0 \text{ on } A^{st}\}, \tag{22}$$
$$\Lambda_1 = \{\zeta \in L^2(\Gamma_2) : \zeta(x) = 0 \text{ on } B_1 \cup B_2 \cup B_1^+ \cup B_2^+\}, \tag{23}$$

while the coincidence set $A^{st} = \{x \in \Gamma_2 : u_N = 0\}$. Moreover $B_1 = \{x \in \Gamma_2 : \lambda(x) = -1\}$, $B_2 = \{x \in \Gamma_2 : \lambda(x) = +1\}$, $\tilde{B}_i = \{x \in B_i : u_N(x) = 0\}$, $i = 1, 2$, $B_i^+ = B_i \setminus \tilde{B}_i$, $i = 1, 2$. The necessary optimality condition is formulated in [15].

Lemma 1. *Let $\Omega^\star \in U_{ad}$ be an optimal solution to the problem (17). Then there exist Lagrange multipliers $\mu_1 \in R$, associated with the constant volume constraint and $\mu_2 \in R$, $\mu_2 \ge 0$, associated with the finite perimeter constraint such that for all admissible* *vector fields V and such that all perturbations $\delta\Omega \in U_{ad}$ of domain $\Omega \in U_{ad}$ satisfy $E \subset \Omega \cup \delta\Omega \subset D$, at any optimal solution $\Omega^\star \in U_{ad}$ to the shape optimization problem (17) the following conditions are satisfied:*

$$dJ_\phi(u(\Omega^\star);V) + \mu_1 \int_{\Gamma^\star} V(0) \cdot nds + \mu_2 dP_D(\Omega^\star;V) \geq 0, \tag{24}$$

$$\mu_1 \left(\int_{\Omega^\star} dx - const_0 \right) = 0, (\mu_2^\sim - \mu_2)(P_D(\Omega^\star) - const_1) \leq 0, \tag{25}$$

$$\forall \mu_2^\sim \in R, \ \mu_2^\sim \geq 0, \tag{26}$$

where $u(\Omega^\star)$ denotes the solution to (10) - (11) in the domain $\Omega^\star$, $\Gamma^\star = \partial\Omega^\star$, the Euler derivative $dJ_\phi(u(\Omega^\star);V)$ is given by (19) and $dP_D(\Omega;V)$ denotes the Euler derivative of the finite perimeter functional $P_D(\Omega)$ (see [20, p. 126]). The given constant $const_0 > 0$ and constant $const_1 > 0$ are the same as in (14).

3.2 Topological Derivative

Classical shape optimization is based on the perturbation of the boundary of the initial shape domain. The initial and final shape domains have the same topology. The aim of the topological optimization is to find an optimal shape without any a priori assumption about the structure's topology.

The value of the goal functional (16) can be minimized by the topology variation of the domain Ω. The topology variations of geometrical domains are defined as functions of a small parameter ρ such that $0 < \rho < R, R > 0$ given. They are based on the creation of a small hole

$$B(x,\rho) = \{z \in R^2 : |x - z| < \rho\} \tag{27}$$

of radius ρ at a point $x \in \Omega$ in the interior of the domain Ω. The Neumann boundary conditions are prescribed on the boundary ∂B of the hole. Denote the perturbed domain by $\Omega_\rho = \Omega \setminus \overline{B(x,\rho)}$.

The topological derivative $TJ_\phi(\Omega,x)$ of the domain functional $J_\phi(\Omega)$ at $\Omega \subset R^2$ is a function depending on a center x of the small hole and is defined by [1,17,22]

$$TJ_\phi(\Omega,x) = \lim_{\rho \to 0^+} [J_\phi(\Omega \setminus \overline{B(x,\rho)}) - J_\phi(\Omega)]/\pi\rho^2. \tag{28}$$

This derivative is calculated by the asymptotic expansion method [22]. To minimize the cost functional $J_\phi(\Omega)$ the holes have to be created at the points of domain Ω where TJ_ϕ is negative.

The formulae for topological derivatives of cost functionals for plane elasticity systems or contact problems are provided, among others, in papers [6,7,21,23]. Using the methodology from [22] as well as the results of differentiability of solutions to variational inequalities [20], we can calculate the formulae of the topological derivative $TJ_\phi(\Omega;x_0)$ of the cost functional (16) at a point $x_0 \in \Omega$. This derivative is equal to

$$TJ_\phi(u(\Omega),x_0) = -[f(\phi + w^{adt}) + \frac{1}{E}(a_u a_{w^{adt}+\phi} + 2b_u b_{w^{adt}+\phi} \cos 2\delta)]|_{x=x_0}$$
$$- \int_{\Gamma_2} (s^{adt} u_T + \lambda(w_T^{adt} + \phi_T))\kappa ds, \tag{29}$$

where $a_{\tilde{\beta}} = \sigma_I(\tilde{\beta}) + \sigma_{II}(\tilde{\beta})$, $b_{\tilde{\beta}} = \sigma_I(\tilde{\beta}) - \sigma_{II}(\tilde{\beta})$, and either $\tilde{\beta} = "u"$ or $\tilde{\beta} = "w^{adt} + \phi"$, $\sigma_I(u)$ and $\sigma_{II}(u)$ denote principal stresses for displacement u, δ is the angle between principal stresses directions [22]. E denotes Young modulus.

In order to obtain formula (29), the plane elasticity system (3) - (5) is written in the polar coordinate system aligned with the principal stress directions. The asymptotic expansions of displacement, strain and stress with respect to parameter ρ in the ring adjacent to the hole $B(x, \rho)$ hold [21]. Taking into account these asymptotic expansions, the regularity of solutions to the state system (3) - (7), calculating the derivatives of the cost functional (16) and solutions to the state system (3) - (7) with respect to parameter ρ as well as using the results of differentiability of solutions to variational inequalities we obtain (29). The integral term in (29) follows from the assumption that tangent displacement and stress functions along Γ_2 are dependent on the parameter ρ.

The adjoint state $(w_\rho^{adt}, s_\rho^{adt}) \in K_1 \times \Lambda_1$ satisfies system (20) - (21) in domain Ω_ρ rather than Ω, i.e.,

$$\int_{\Omega_\rho} a_{ijkl} e_{ij}(\phi + w_\rho^{adt}) e_{kl}(\varphi) dx + \int_{\Gamma_2} s_\rho^{adt} \varphi_T ds = 0, \ \forall \varphi \in K_1, \tag{30}$$

and

$$\int_{\Gamma_2} \zeta(w_{\rho T}^{adt} + \phi_T) ds = 0, \quad \forall \zeta \in \Lambda_1, \tag{31}$$

where $w_\rho^{adt}|_{\rho=0} = w^{adt}(x_0)$. By standard arguments [5,20,21] it can be shown that if $\Omega^\star \in U_{ad}$ is an optimal domain to the problem (17) it satisfies for all $x_0 \in \Omega^\star$ the necessary optimality condition of the form (24) - (26) with topological derivative (29) rather than Euler derivative (19) in (24), and inequality in (25) rather than equality as well as with Lagrange multiplier $\mu_1 \geq 0$.

3.3 Domain Differential

Finally consider the variation of the functional (16) resulting both from the nucleation of the internal small hole as well as from the boundary variations. In order to take into account these perturbations, in [21] the notion of the domain differential of the domain functional has been introduced. The domain differential $DJ_\phi(\Omega; V, x_0)$ of the shape functional (16) at $\Omega \subset R^2$ in direction V and at point $x_0 \in \Omega$ is defined as

$$DJ_\phi(\Omega; V, x_0)(\tau, \rho) = \tau dJ_\phi(\Omega, V) + \pi \rho^2 T J_\phi(\Omega, x_0). \tag{32}$$

This differential completely characterizes the variation of the cost functional $J_\phi(\Omega)$ with respect to the simultaneous shape and topology perturbations (for details see [21]). The shape derivative $dJ_\phi(u(\Omega), V)$ and the topological derivative $TJ_\phi(u(\Omega), x_0)$ are provided by (19) and (29) respectively. They depend on the solution u to the state system (10) - (11). Using standard arguments [20] we can show

Lemma 2. *Let $\Omega^\star \in U_{ad}$ be an optimal solution to the problem (17). Then there exist Lagrange multipliers $\mu_1 \in R$, $\mu_1 \geq 0$, associated with the volume constraint and $\mu_2 \in R$, $\mu_2 \geq 0$, associated with the finite perimeter constraint such that for all admissible vector* fields V, for all admissible pairs (ρ, τ) of parameters and for all $x_0 \in \Omega^\star$ and such that all perturbations $\delta\Omega \in U_{ad}$ of domain $\Omega \in U_{ad}$ satisfy $E \subset \Omega \cup \delta\Omega \subset D$, at any optimal solution $\Omega^\star \in U_{ad}$ to the shape and topology optimization problem (17) the following conditions are satisfied:

$$DJ_\phi(u(\Omega^\star);V,x_0)(\tau,\rho) + \mu_1 \int_{\Gamma^\star} V(0) \cdot n ds + \mu_2 dP_D(\Omega^\star;V) \geq 0, \qquad (33)$$

$$(\mu_1^\sim - \mu_1)(\int_{\Omega^\star} dx - const_0) \leq 0, \forall \mu_1^\sim \in R, \ \mu_1^\sim \geq 0, \qquad (34)$$

$$(\mu_2^\sim - \mu_2)(P_D(\Omega^\star) - const_1) \leq 0, \forall \mu_2^\sim \in R, \ \mu_2^\sim \geq 0, \qquad (35)$$

where $u(\Omega^\star)$ denotes the solution to (10) - (11) in the domain $\Omega^\star$, $\Gamma^\star = \partial\Omega^\star$, the domain differential $DJ_\phi(u(\Omega^\star);V,x_0)(\tau,\rho)$ is given by (32) and $dP_D(\Omega;V)$ denotes the derivative of the finite perimeter functional $P_D(\Omega)$ (see [1,6],[20, p. 126]). The given constant $const_0 > 0$ and constant $const_1 > 0$ are the same as in (14).

4   Shape Representation by Level Set Method

In the paper the level set method [19] is employed to solve numerically problem (17). Consider the evolution of a domain Ω under a velocity field V. Let $t > 0$ denote the time variable. Under the mapping $T(t,V)$ we have

$$\Omega_t = T(t,V)(\Omega) = (I + tV)(\Omega), \ t > 0. \qquad (36)$$

By Ω_t^- we denote the interior of the domain Ω_t and by Ω_t^+ we denote the outside of the domain Ω_t. The domain Ω_t and its boundary $\partial\Omega_t$ are defined by a function $\Phi = \Phi(x,t) : R^2 \times [0,t_0) \to R$ satisfying

$$\begin{cases} \Phi(x,t) = 0, & \text{if } x \in \partial\Omega_t, \\ \Phi(x,t) < 0, & \text{if } x \in \Omega_t^-, \\ \Phi(x,t) > 0, & \text{if } x \in \Omega_t^+, \end{cases} \qquad (37)$$

i.e., the boundary $\partial\Omega_t$ is the level curve of the function Φ. Recall [19], the gradient of the implicit function is defined as $\nabla\Phi = (\frac{\partial\Phi}{\partial x_1}, \frac{\partial\Phi}{\partial x_2})$, the local unit outward normal n to the boundary is equal to $n = \frac{\nabla\Phi}{|\nabla\Phi|}$, the mean curvature $\kappa = \nabla \cdot n$. In the level set approach, Heaviside function $H(\Phi)$ and Dirac function $\delta(\Phi)$ are used to transform integrals from domain Ω into domain D. These functions are defined as

$$H(\Phi) = 1 \text{ if } \Phi \geq 0, \quad H(\Phi) = 0 \text{ if } \Phi < 0, \qquad (38)$$

$$\delta(\Phi) = H'(\Phi), \delta(x) = \delta(\Phi(x)) \mid \nabla\Phi(x) \mid, \ x \in D. \qquad (39)$$

Assume that velocity field V is known for every point x lying on the boundary $\partial\Omega_t$, i.e., with $\Phi(x,t) = 0$. Therefore the equation governing the evolution of the interface in $D \times [0,t_0]$ has the form [19]

$$\Phi_t(x,t) + V(x,t) \cdot \nabla_x \Phi(x,t) = 0, \qquad (40)$$

where Φ_t denotes a partial derivative of Φ with respect to the time variable t. Equation (40) is known as Hamilton-Jacobi equation.

4.1   Structural Optimization Problem in Domain D

Using the notion of the level set function (37) as well as functions (38) and (39) structural optimization problem (17) may be reformulated in the following way: *for a given function $\phi \in M^{st}$, find function Φ such that*

$$J_\phi(u(\Phi^\star)) = \min_{\Phi \in U_{ad}^\Phi} J_\phi(u(\Phi)) \tag{41}$$

where

$$J_\phi(u(\Phi)) = \int_D \sigma_N(u)\phi_N(x)\delta(\Phi) \mid \nabla\Phi \mid ds, \tag{42}$$

$$U_{ad}^\Phi = \{\Phi: \Phi \text{ satisfies } (37), Vol(\Phi) \leq Vol^{giv}, P_D(\Phi) \leq const_1\}, \tag{43}$$

$$Vol(\Phi) = \int_D H(\Phi)dx, \tag{44}$$

$$P_D(\Phi) = \int_D \delta(\Phi) \mid \nabla\Phi \mid dx. \tag{45}$$

Moreover, a pair $(u,\lambda) \in K \times \Lambda$ satisfies system

$$\int_D a_{ijkl}e_{ij}(u)e_{kl}(\varphi - u)H(\Phi)dx - \int_D f_i(\varphi_i - u_i)H(\Phi)dx$$

$$- \int_D p_i(\varphi_i - u_i)\delta(\Phi) \mid \nabla\Phi \mid dx \tag{46}$$

$$+ \int_D \lambda(\varphi_T - u_T)\delta(\Phi) \mid \nabla\Phi \mid dx \geq 0 \forall \varphi \in K,$$

$$\int_D (\zeta - \lambda)u_T\delta(\Phi) \mid \nabla\Phi \mid dx \leq 0 \,\forall \zeta \in \Lambda, \tag{47}$$

while V_{sp} and K are defined by (8) and (9), respectively, on domain D rather than Ω and $i,j,k,l = 1,2$.

5   Level Set Based Numerical Algorithm

The topological derivative can provide better prediction of the structure topology with different levels of material volume than the method based on updating the shape of initial structure containing many regularly distributed holes [1,25]. Our approach is based on the application of the topological derivative to predict the structure topology and substitute material according to the material volume constraint and next to optimize the structure topology to merge the unreasonable material interfaces and to change the shape of material boundary. For the sake of simplicity in the description of the algorithm we omit the bounded perimeter constraint in (14). Therefore the level set method combined with the shape or topological derivatives results in the following conceptual algorithm (A1) to solve structural optimization problem (17):

Step 1: Choose: a computational domain D such that $\Omega \subset D$, an initial level set function $\Phi^0 = \Phi_0$ representing $\Omega^0 = \Omega$, function $\phi \in M^{st}$, parameters r^0, $\varepsilon_1, \varepsilon_2, \tilde{\mu}_1^0 = \mu_1^0 = 0$, $k = n = 0$.
Step 2: Calculate the solution (u^n, λ^n) to the state system (46) - (47).

Step 3: Calculate the solution $((w^{adt})^n, (s^{adt})^n)$ to the adjoint system (30) - (31) as well as the topological derivative $TJ_\phi(\Omega^n, x)$ of the cost functional (16) given by (29).

Step 4: For given $\tilde{\mu}_1^n$ set $\Omega^{n+1} = \{x \in \Omega^n : TJ_\phi(\Omega, x) \geq \chi_{n+1}\}$ where χ_{n+1} is chosen in such a way that $Vol(\Omega^{n+1}) = m_{n+1}$, $m_{n+1} = qm_n$. Fill the void part $D \setminus \Omega^{n+1}$ with a very weak material with Young modulus $E^w = 10^{-5}E$. Update $\tilde{\mu}_1^{n+1} = \tilde{\mu}_1^n + r^n(Vol_1^{giv})$, $r^n > 0$, $Vol_1^{giv} = Vol(\Omega^{n+1}) - r_{fr}Vol^{giv}$. If $| \tilde{\mu}_1^{n+1} - \tilde{\mu}_1^n | \leq \varepsilon_1$ then set $\Omega^k = \Omega^{n+1}$ and go to Step 5. Otherwise set $n = n+1$, go to Step 2.

Step 5: Calculate the solution $((p^{adt})^k, (q^{adt})^k)$ to the adjoint system (20) - (21). Calculate the shape derivative $dJ_\phi(u(\Omega^k))$ of the cost functional (16) given by (19).

Step 6: For given μ_1^k solve the level set equation (40) to calculate the level set function Φ^{k+1}.

Step 7: Set Ω^{k+1} equal to the zero level set of function Φ^{k+1}. Calculate $\mu_1^{k+1} = \mu_1^k + r^k(Vol(\Omega^{k+1}) - Vol_1^{giv})$, $r^k > 0$. If $| \mu_1^{k+1} - \mu_1^k | \leq \varepsilon_2$ then Stop. Otherwise set $k = k+1$, $\Omega^n = \Omega^{k+1}$, and go to Step 2.

Let us remark that having localized a small hole one can consider to perform further the size optimization of the radius of the existing hole rather than shape optimization. However such approach, confining to hole change only inside the optimized domain and not allowing for the change of the shape of the external boundary of the body seems to be not versatile. Level set approach allowing for tracking changes of internal and external interfaces of the optimized domain on a fixed mesh, including the merging of holes, seems to be more suitable in simultaneous shape and topology optimization of domains.

State (10) - (11) and adjoint (20) - (21) systems are discretized using finite element method [10]. Displacement and stress functions in state system (10) - (11) are approximated by piecewise bilinear functions in domain D and piecewise constant functions on the boundary Γ_2 respectively. Similar approximation is used to discretize the adjoint system (20) - (21) or (30) - (31). These systems are solved using the primal-dual algorithm with active set strategy [3,24]. In level set approach these state and adjoint systems are transferred from domain Ω into fixed hold-all domain D using the regularized Heaviside and Dirac functions. Finite difference method is employed to discretize Hamilton-Jacobi equation and the explicit up-wind scheme is used to solve it. For more details concerning implementation of this algorithm see [16].

6   Numerical Methods and Example

The discretized structural optimization problem (17) is solved numerically. The numerical algorithms described in the previous sections have been used. The algorithm is programmed in Matlab environment. As an example a body occupying 2D domain

$$\Omega = \{(x_1, x_2) \in R^2 : 0 \leq x_1 \leq 8 \wedge 0 < v(x_1) \leq x_2 \leq 4\}, \tag{48}$$

is considered. The boundary Γ of the domain Ω is divided into three pieces

$$\Gamma_0 = \{(x_1, x_2) \in R^2 : x_1 = 0, 8 \wedge 0 < v(x_1) \leq x_2 \leq 4\}, \tag{49}$$

$$\Gamma_1 = \{(x_1, x_2) \in R^2 : 0 \leq x_1 \leq 8 \wedge x_2 = 4\}, \tag{50}$$

$$\Gamma_2 = \{(x_1, x_2) \in R^2 : 0 \leq x_1 \leq 8 \wedge v(x_1) = x_2\}. \tag{51}$$

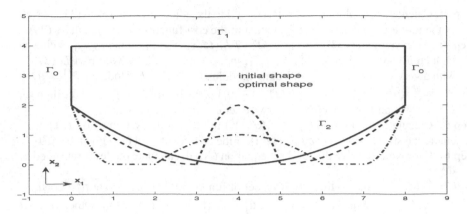

Fig. 1. Shape optimization – optimal domain

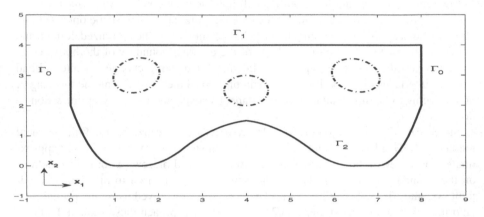

Fig. 2. Simultaneous topology and shape optimization – optimal domain

The domain Ω and the boundary Γ_2 depend on the function v. This function is the variable subject to shape optimization [13,15,16]. The initial position of the boundary Γ_2 is given as in Fig. 1. The computations are carried out for the elastic body characterized by the Poisson's ratio $v = 0.29$, the Young modulus $E = 2.1 \cdot 10^{11} \text{N/m}^2$. The body is loaded by boundary traction $p_1 = 0$, $p_2 = -5.6 \cdot 10^6$ N along Γ_1, body forces $f_i = 0$, $i = 1,2$. Auxiliary function ϕ is selected as piecewise constant (or linear) on D and is approximated by a piecewise constant (or bilinear) functions. The computational domain $D = [0,8] \times [0,4]$ is selected. Domain D is discretized with a fixed rectangular mesh of 24×12.

Fig. 1 displays the optimal solution of shape optimization problem (17). Fig. 2 presents the optimal domain obtained by solving topological and shape optimization problem (17) in the computational domain D using algorithm (A1) and employing the optimality condition (33) - (35). The areas of holes denoted by dotted lines appear in the central part of the body and near the fixed edges. These areas result from the merging of small holes from which the material has been removed into the bigger ones. Recall that the application of formula (29) allows to insert infinitesimal holes only. Although the shape of the optimal contact boundary Γ_2 is similar to the optimal shape obtained in the case of shape optimization only (see Fig. 2) but the obtained shape of the boundary Γ_2 is not so strongly changed comparing to the initial one as the optimal shape obtained in the case of shape optimization only. The obtained normal contact stress is almost constant along the optimal shape boundary and has been significantly reduced comparing to the initial one.

7   Conclusions

The structural optimization problem for elastic contact problem with the prescribed friction is solved numerically in the paper. The topological derivative method as well as the level set approach combined with the shape gradient method are used. The friction term complicates both the form of the gradients of the cost or penalty functionals as well as numerical process.

Obtained preliminary numerical results seem to be in accordance with physical reasoning. They indicate that the proposed numerical algorithm allows for significant improvements of the structure from one iteration to the next. They also indicate the future research direction aiming at better reconciliation in one algorithm holes nucleation flexibility and geometric update fidelity.

References

1. Allaire, G., Jouve, F., Toader, A.: Structural optimization using sensitivity analysis and a level set method. Journal of Computational Physics 194(1), 363–393 (2004)
2. Burger, M., Hackl, B., Ring, W.: Incorporating topological derivatives into level set methods. Journal of Computational Physics 194(1), 344–362 (2004)
3. Bergonioux, M., Kunisch, K.: Augmented Lagrangian techniques for elliptic state constrained optimal control problems. SIAM Journal on Control and Optimization 35, 1524–1543 (1997)
4. Chopp, H., Dolbow, J.: A hybrid extended finite element / level set method for modelling phase transformations. International Journal for Numerical Methods in Engineering 54, 1209–1232 (2002)
5. Delfour, M., Zolesio, J.-P.: Shapes and Geometries: Analysis, Differential Calculus and Optimization. SIAM Publications, Philadelphia (2001)
6. Fulmański, P., Laurain, A., Scheid, J.F., Sokołowski, J.: A Level Set Method in Shape and Topology Optimization for Variational Inequalities. Int. J. Appl. Math. Comput. Sci. 17, 413–430 (2007)
7. Garreau, S., Guillaume, Ph., Masmoudi, M.: The topological asymptotic for PDE systems: the elasticity case. SIAM Journal on Control Optimization 39, 1756–1778 (2001)
8. Gomes, A., Suleman, A.: Application of spectral level set methodology in topology optimization. Structural Multidisciplinary Optimization 31, 430–443 (2006)
9. de Gournay, F.: Velocity extension for the level set method and multiple eigenvalue in shape optimization. SIAM Journal on Control and Optimization 45(1), 343–367 (2006)
10. Haslinger, J., Mäkinen, R.: Introduction to Shape Optimization. Theory, Approximation, and Computation. SIAM Publications, Philadelphia (2003)

11. He, L., Kao, Ch.Y., Osher, S.: Incorporating topological derivatives into shape derivatives based level set methods. Journal of Computational Physics 225, 891–909 (2007)
12. Hintermüller, M., Ring, W.: A level set approach for the solution of a state-constrained optimal control problem. Numerische Mathematik 98, 135–166 (2004)
13. Myśliński, A.: Level set method for shape optimization of contact problems. In: Neittaanmäki, P., Jyväskylä. (eds.) Proceedings of European Congress on Computational Methods in Applied Sciences and Engineering, Finland (2004)
14. Myśliński, A.: Topology and shape optimization of contact problems using a level set method. In: Herskovits, J., Mazorche, S., Canelas, A. (eds.) Proceedings of VI World Congresses of Structural and Multidisciplinary Optimization, Rio de Janeiro, Brazil (2005)
15. Myśliński, A.: Shape Optimization of Nonlinear Distributed Parameter Systems. Academic Printing House EXIT, Warsaw (2006)
16. Myśliński, A.: Level Set Method for Optimization of Contact Problems. Engineering Analysis with Boundary Elements 32, 986–994 (2008)
17. Novotny, A.A., Feijóo, R.A., Padra, C., Tarocco, E.: Topological derivative for linear elastic plate bending problems. Control and Cybernetics 34(1), 339–361 (2005)
18. Norato, J.A., Bendsoe, M.P., Haber, R., Tortorelli, D.A.: A topological derivative method for topology optimization. Structural Multidisciplinary Optimization 33, 375–386 (2007)
19. Osher, S., Fedkiw, R.: Level Set Methods and Dynamic Implicit Surfaces. Springer, New York (2003)
20. Sokołowski, J., Zolesio, J.-P.: Introduction to Shape Optimization. Shape Sensitivity Analysis. Springer, Berlin (1992)
21. Sokołowski, J., Żochowski, A.: Optimality conditions for simultaneous topology and shape optimization. SIAM Journal on Control 42(4), 1198–1221 (2003)
22. Sokołowski, J., Żochowski, A.: On topological derivative in shape optimization. In: Lewiński, T., Sigmund, O., Sokołowski, J., Żochowski, A. (eds.) Optimal Shape Design and Modelling, pp. 55–143. Academic Printing House EXIT, Warsaw (2004)
23. Sokołowski, J., Żochowski, A.: Modelling of topological derivatives for contact problems. Numerische Mathematik 102(1), 145–179 (2005)
24. Stadler, G.: Semismooth Newton and augmented Lagrangian methods for a simplified friction problem. SIAM Journal on Optimization 15(1), 39–62 (2004)
25. Wang, M.Y., Wang, X., Guo, D.: A level set method for structural topology optimization. Computer Methods in Applied Mechanics and Engineering 192, 227–246 (2003)
26. Xia, Q., Wang, M.Y., Wang, S., Chen, S.: Semi-Lagrange method for level set based structural topology and shape optimization. Multidisciplinary Structural Optimization 31, 419–429 (2006)
27. Wang, S.Y., Lim, K.M., Khao, B.C., Wang, M.Y.: An extended level set method for shape and topology optimization. Journal of Computational Physics 221, 395–421 (2007)

Second Order Sufficient Optimality Conditions for a Control Problem with Continuous and Bang-Bang Control Components: Riccati Approach

Nikolai P. Osmolovskii[1,2,3,*] and Helmut Maurer[4,**]

[1] Systems Research Institute, Polish Academy of Sciences,
ul. Newelska 6, 01-447 Warszawa, Poland
[2] Politechnika Radomska, 26-600 Radom, ul. Malczewskiego 20 A, Poland
[3] Akademia Podlaska, ul. 3 Maja 54, 08-110 Siedlce, Poland
osmolovski@ap.siedlce.pl
[4] Westfälische Wilhelms-Universität Münster,
Institut für Numerische und Angewandte Mathematik, Einsteinstr. 62,
D-48149 Münster, Germany
maurer@math.uni-muenster.de

Abstract. Second order sufficient optimality conditions for bang-bang control problems in a very general form have been obtained in [15,21,13,12,1]. These conditions require the positive definiteness (coercivity) of an associated quadratic form on the finite-dimensional critical cone. In the present paper, we investigate similar conditions for optimal control problems with a control variable having two components: a continuous unconstrained control appearing nonlinearly and a bang-bang control appearing linearly and belonging to a convex polyhedron. The coercivity of the quadratic form can be verified by checking solvability of an associated matrix Riccati equation. The results are applied to an economic control problem in optimal production and maintenance, where existing sufficient conditions fail to hold.

1 Introduction

The classical sufficient second order optimality conditions for an optimization problem with constraints require that the second variation of the Lagrangian be positive definite on the cone of critical directions. In this paper, we investigate sufficient quadratic optimality conditions of such type for optimal control problems with a vector control variable having two components: a *continuous* unconstrained control *appearing nonlinearly* in the control system and a *bang-bang* control *appearing linearly* and belonging to a convex polyhedron.

In the pure bang-bang case, where all control components appear linearly, second order necessary and sufficient optimality conditions in a very general form have been obtained in Milyutin, Osmolovskii[15], Osmolovskii [21], Agrachev, Stefani, Zezza [1], and Maurer, Osmolovskii [13,12,24]. Two alternative approaches were developed to establish sufficiency: (1) check the positive definiteness of an associated quadratic form

---

* Supported by the grants RFBR 08-01-00685 and NSh-3233.2008.1.
** Supported by the Deutsche Forschungsgemeinschaaft under grant MA 691/18-1.

A. Korytowski et al. (Eds.): System Modeling and Optimization, IFIP AICT 312, pp. 411–429, 2009.

on the *finite-dimensional* critical cone; (2) verify second-order sufficient conditions for an induced *finite-dimensional optimization problem* with respect to switching times and free final time. Second order sufficient optimality conditions for control problems where the control variable appears nonlinearly, more precisely, where the strict Legendre condition holds, have been given, e.g., in Malanowski [8], Maurer [9], Maurer, Pickenhain [14], Milyutin, Osmolovskii [15], and Zeidan [28]. Here, one way to check the positive definiteness of the quadratic form is by means of the solvability of an associated Riccati equation.

In this paper, we investigate a class of control problems having two control components, a continuous and a bang-bang component. Such control problems are frequently encountered in practice. Our aim is to obtain second-order sufficient optimality conditions for this class of control problems. It does not come as a surprise that the proof techniques for obtaining sufficient conditions combine the above mentioned methods in the pure bang-bang case and the case where the strict Legendre condition holds. For simplicity, we shall consider control problems on a fixed time interval.

In Sect. 2, we give a statement of the control problem with continuous and bang-bang control components (main problem), formulate the minimum principle (first order necessary optimality condition) and introduce the notion of bounded-strong local minimum. In Sect. 3, we present second-order sufficient optimality conditions (SSC) for a bounded-strong minimum in the problem. The main result in Theorem 1 is stated without proof, which is very similar to the proof in the bang-bang case; cf. Milyutin, Osmolovskii [15] and Osmolovskii, Maurer [24], Part 1. Details of the proof will be published elsewhere. In Sect. 4, we give criteria for the positive definiteness of the quadratic form on the critical cone in terms of solutions to a matrix Riccati equation which may be discontinuous at the switching times. In Sect. 5, the main result in Theorem 2 is applied to an economic control problem for optimal production and maintenance which was introduced by Cho, Abad and Parlar [4]. We will show that the numerical solution obtained in Maurer, Kim, and Vossen [10] satisfies the second order test derived in Sect. 4 while existing sufficiency results fail to hold.

2   Control Problem on a Fixed Time Interval

2.1   The Main Problem

Let $x(t) \in \mathbb{R}^{d(x)}$ denote the state variable and $u(t) \in \mathbb{R}^{d(u)}$, $v(t) \in \mathbb{R}^{d(v)}$ the control variables in the time interval $t \in [t_0, t_f]$ with fixed initial time t_0 and final time t_f. We shall refer to the following optimal control problem (1)-(4) as the *main problem*:

$$\text{Minimize} \quad \mathscr{J}(x(\cdot), u(\cdot), v(\cdot)) = J(x(t_0), x(t_f)) \tag{1}$$

subject to the constraints

$$\dot{x}(t) = f(t, x(t), u(t), v(t)), \quad u(t) \in U, \quad (t, x(t), v(t)) \in \mathscr{Q}, \tag{2}$$

$$F(x(t_0), x(t_f)) \leq 0, \quad K(x(t_0), x(t_f)) = 0, \quad (x(t_0), x(t_f)) \in \mathscr{P}, \tag{3}$$

where the control variable u appears *linearly* in the system dynamics,

$$f(t, x, u, v) = a(t, x, v) + B(t, x, v)u. \tag{4}$$

Here, F, K, a are column-vector functions, B is a $d(x) \times d(u)$ matrix function, $\mathscr{P} \subset \mathbb{R}^{2d(x)}$, $\mathscr{Q} \subset \mathbb{R}^{1+d(x)+d(v)}$ are open sets and $U \subset \mathbb{R}^{d(u)}$ is a convex polyhedron. The functions J, F, K are assumed to be twice continuously differentiable on $\mathscr{P}$ and the functions a, B are twice continuously differentiable on $\mathscr{Q}$. The dimensions of F, K are denoted by $d(F), d(K)$. By $\Delta = [t_0, t_f]$ we shall denote the interval of control and use the abbreviations

$$x_0 = x(t_0), \quad x_f = x(t_f), \quad p = (x_0, x_f). \tag{5}$$

A process

$$\Pi = \{(x(t), u(t), v(t)) \mid t \in [t_0, t_f]\} \tag{6}$$

is said to be *admissible*, if $x(\cdot)$ is absolutely continuous, $u(\cdot), v(\cdot)$ are measurable and bounded on Δ and the triple of functions $(x(t), u(t), v(t))$ together with the endpoints $p = (x(t_0), x(t_f))$ satisfies the constraints (2) and (3).

Definition 1. An admissible process Π affords a *Pontryagin local minimum*, if for each compact set $\mathscr{C} \subset \mathscr{Q}$ there exists $\varepsilon > 0$ such that $\mathscr{J}(\tilde{\Pi}) \geq \mathscr{J}(\Pi)$ for all admissible processes $\tilde{\Pi} = \{(\tilde{x}(t), \tilde{u}(t), \tilde{v}(t)) \mid t \in [t_0, t_1]\}$ such that:

(a) $\max_{\Delta} |\tilde{x}(t) - x(t)| < \varepsilon$;

(b) $\int_{\Delta} |\tilde{u}(t) - u(t)| dt < \varepsilon$; $\int_{\Delta} |\tilde{v}(t) - v(t)| dt < \varepsilon$;

(c) $(t, \tilde{x}(t), \tilde{v}(t)) \in \mathscr{C}$ a.e. on Δ.

2.2 First Order Necessary Optimality Conditions

Let

$$\Pi = \{(x(t), u(t), v(t)) \mid t \in [t_0, t_f]\} \tag{7}$$

be a fixed admissible process such that the control $u(t)$ is a *piecewise constant* function and the control $v(t)$ is a *continuous* function on the interval $\Delta = [t_0, t_f]$. In order to make the notations simpler, we do not use such symbols and indices as zero, hat or asterisk to distinguish this trajectory from others. Denote by

$$\theta = \{t_1, \ldots, t_s\}, \quad t_0 < t_1 < \cdots < t_s < t_f, \tag{8}$$

the finite set of all *discontinuity* points (jump points) of the control $u(t)$. Then $\dot{x}(t)$ is a piecewise continuous function whose discontinuity points belong to θ, and hence $x(t)$ is a piecewise smooth function on Δ. Henceforth, we shall use the notation

$$[u]^k = u^{k+} - u^{k-} \tag{9}$$

to denote the jump of the function $u(t)$ at the point $t_k \in \theta$, where

$$u^{k-} = u(t_k - 0), \quad u^{k+} = u(t_k + 0) \tag{10}$$

are the left hand and the right hand values of the control $u(t)$ at t_k, respectively. Similarly, we denote by $[\dot{x}]^k$ the jump of the function $\dot{x}(t)$ at the point t_k.

Let us formulate first-order necessary conditions of a Pontryagin minimum for the process Π in the form of the Pontryagin minimum principle. To this end we introduce the Pontryagin or Hamiltonian function

$$H(t,x,\psi,u,v) = \psi f(t,x,u,v) = \psi a(t,x,v) + \psi B(t,x,v)u, \qquad (11)$$

where ψ is a row vector of dimension $d(\psi) = d(x)$, while x, u, f, F and K are column vectors. The row vector of dimension $d(u)$,

$$\sigma(t,x,\psi,v) = \psi B(t,x,v), \qquad (12)$$

will be called the *switching function for the u-component* of the control. Denote by l the endpoint Lagrange function

$$l(\alpha_0,\alpha,\beta,p) = \alpha_0 J(p) + \alpha F(p) + \beta K(p), \quad p = (x_0,x_f), \qquad (13)$$

where α and β are row-vectors with $d(\alpha) = d(F), d(\beta) = d(K)$, and α_0 is a number. We introduce a tuple of Lagrange multipliers

$$\lambda = (\alpha_0,\alpha,\beta,\psi(\cdot)) \qquad (14)$$

such that

$$\psi(\cdot) : \Delta \to \mathbb{R}^{d(x)} \qquad (15)$$

is continuous on Δ and continuously differentiable on each interval of the set $\Delta \setminus \theta$. In the sequel, we shall denote first or second order partial derivatives by subscripts referring to the variables.

Denote by M_0 the set of the normalized tuples λ satisfying the minimum principle conditions for the process Π:

$$\alpha_0 \geq 0, \ \ \alpha \geq 0, \ \ \alpha F(p) = 0, \ \ \alpha_0 + \sum \alpha_i + \sum |\beta_j| = 1, \qquad (16)$$

$$\dot{\psi} = -H_x, \ \ \forall t \in \Delta \setminus \theta, \qquad (17)$$

$$\psi(t_0) = -l_{x_0}, \ \ \psi(t_f) = l_{x_f}, \qquad (18)$$

$$H(t,x(t),\psi(t),u,v) \geq H(t,x(t),\psi(t),u(t),v(t))$$
$$\text{for all } t \in \Delta \setminus \theta, \, u \in U, \, v \in \mathbb{R}^{d(v)} \text{ such that } (t,x(t),v) \in \mathscr{Q}. \qquad (19)$$

The derivatives l_{x_0} and l_{x_f} are taken at the point $(\alpha_0,\alpha,\beta,p)$, where $p = (x(t_0),x(t_f))$, and the derivative H_x is evaluated at the point

$$(t,x(t),\psi(t),u(t),v(t)), \quad t \in \Delta \setminus \theta. \qquad (20)$$

The condition $M_0 \neq \emptyset$ constitutes a first-order necessary condition of a Pontryagin minimum for the process Π which is called the *Pontryagin minimum principle*, cf., Pontryagin et al. [25], Hestenes [7], Milyutin, Osmolovskii [15]. The set M_0 is a finite-dimensional compact set and the projector $\lambda \mapsto (\alpha_0,\alpha,\beta)$ is injective on M_0.

In the sequel, it will be convenient to use the abbreviation (t) for indicating all arguments $(t,x(t),\psi(t),u(t),v(t))$, e.g.,

$$H(t) = H(t,x(t),\psi(t),u(t),v(t)), \quad \sigma(t) = \sigma(t,x(t),\psi(t),v(t)). \qquad (21)$$

Let $\lambda = (\alpha_0, \alpha, \beta, \psi(\cdot)) \in M_0$. It is well-known that $H(t)$ is a continuous function. In particular, $[H]^k = H^{k+} - H^{k-} = 0$ holds for each $t_k \in \theta$, where $H^{k-} := H(t_k - 0)$ and $H^{k+} := H(t_k + 0)$. We shall denote by H^k the common value of H^{k-} and H^{k+}:

$$H^k := H^{k-} = H^{k+}. \tag{22}$$

The relations

$$[H]^k = 0, \quad [\psi]^k = 0, \quad k = 1,\ldots,s, \tag{23}$$

constitute the Weierstrass-Erdmann conditions for a broken extremal. We formulate one more condition of this type which will be important for the statement of second order conditions for extremal with jumps in the control. Namely, for $\lambda \in M_0$ and $t_k \in \theta$ consider the function

$$(\Delta_k H)(t) = H(t,x(t),\psi(t),u^{k+},v(t_k)) - H(t,x(t),\psi(t),u^{k-},v(t_k))$$
$$= \sigma(t,x(t),\psi(t),v(t_k))[u]^k. \tag{24}$$

Proposition 1. *For each $\lambda \in M_0$ the following equalities hold*

$$\frac{d}{dt}(\Delta_k H)\big|_{t=t_k-0} = \frac{d}{dt}(\Delta_k H)\big|_{t=t_k+0}, \quad k=1,\ldots,s. \tag{25}$$

Consequently, for each $\lambda \in M_0$ the function $(\Delta_k H)(t)$ has a derivative at the point $t_k \in \theta$. In the sequel, we will consider the quantities

$$D^k(H) = -\frac{d}{dt}(\Delta_k H)(t_k), \quad k=1,\ldots,s. \tag{26}$$

Then the minimum condition (19) implies the following property:

Proposition 2. *For each $\lambda \in M_0$ the following conditions hold:*

$$D^k(H) \geq 0, \quad k=1,\ldots,s. \tag{27}$$

Note that the value $D^k(H)$ also can be written in the form

$$D^k(H) = -H_x^{k+}H_\psi^{k-} + H_x^{k-}H_\psi^{k+} - [H_t]^k = \psi^{k+}\dot{x}^{k-} - \psi^{k-}\dot{x}^{k+} + [\psi_0]^k \tag{28}$$

where H_x^{k-} and H_x^{k+} are the left- and the right-hand values of the function $H_x(t)$ at t_k, respectively, $[H_t]^k$ is the jump of the function $H_t(t)$ at t_k, etc., and $\psi_0(t) = -H(t)$.

2.3 Integral Cost Function, Unessential Variables, Bounded-Strong Minimum

It is well-known that any control problem with a cost functional in integral form

$$\mathscr{J} = \int_{t_0}^{t_f} f_0(t,x(t),u(t),v(t))\,dt \tag{29}$$

can be reduced to the form (1) by introducing a new state variable y defined by the state equation

$$\dot{y} = f_0(t,x,u,v), \quad y(t_0) = 0. \tag{30}$$

This yields the cost function $\mathscr{J} = y(t_f)$. The control variable u is assumed to appear linearly in the function f_0,

$$f_0(t,x,u,v) = a_0(t,x,v) + B_0(t,x,v)u. \tag{31}$$

The component y is called an *unessential* component in the augmented problem. The general definition of an unessential component is as follows.

Definition 2. The i-th component x_i of the state vector x is called unessential if the function f does not depend on x_i and if the functions F, J, K are affine in $x_{i0} = x_i(t_0)$ and $x_{if} = x_i(t_f)$.

In the following, let $\underline{x}$ denote the vector of all essential components of state vector x.

Definition 3. The process Π affords a *bounded-strong minimum*, if for each compact set $\mathscr{C} \subset \mathscr{Q}$ there exists $\varepsilon > 0$ such that $\mathscr{J}(\widetilde{\Pi}) \geq \mathscr{J}(\Pi)$ for all admissible processes $\widetilde{\Pi} = \{(\tilde{x}(t), \tilde{u}(t), \tilde{v}(t)) \mid t \in [t_0, t_f]\}$ such that

(a) $|\tilde{x}(t_0) - x(t_0)| < \varepsilon$,

(b) $\max_{\Delta} |\underline{\tilde{x}}(t) - \underline{x}(t)| < \varepsilon$,

(c) $(t, \tilde{x}(t), \tilde{v}(t)) \in \mathscr{C}$ a.e. on Δ.

The *strict* bounded-strong minimum is defined in a similar way, with the non-strict inequality $\mathscr{J}(\widetilde{\Pi}) \geq \mathscr{J}(\Pi)$ replaced by the strict one and the process $\widetilde{\Pi}$ required to be different from Π.

3 Quadratic Sufficient Optimality Conditions

In this section, we shall formulate a quadratic sufficient optimality condition for a bounded-strong minimum (Definition 3) for given control process. This quadratic condition is based on the properties of a quadratic form on the so-called critical cone, whose elements are first order variations along a given process Π.

3.1 Critical Cone

For a given process Π we introduce the space $\mathscr{Z}^2(\theta)$ and the *critical cone* $\mathscr{K} \subset \mathscr{Z}^2(\theta)$. Denote by $P_\theta W^{1,2}(\Delta, \mathbb{R}^{d(x)})$ the space of piecewise continuous functions $\bar{x}(\cdot) : \Delta \to \mathbb{R}^{d(x)}$, which are absolutely continuous on each interval of the set $\Delta \setminus \theta$ and have a square integrable first derivative. By $L^2(\Delta, \mathbb{R}^{d(v)})$ we denote the space of square integrable functions $\bar{v}(\cdot) : \Delta \to \mathbb{R}^{d(v)}$. For each $\bar{x} \in P_\theta W^{1,2}(\Delta, \mathbb{R}^{d(x)})$ and for $t_k \in \theta$ we set

$$\bar{x}^{k-} = \bar{x}(t_k - 0), \quad \bar{x}^{k+} = \bar{x}(t_k + 0), \quad [\bar{x}]^k = \bar{x}^{k+} - \bar{x}^{k-}. \tag{32}$$

Let $\bar{z} = (\xi, \bar{x}, \bar{v})$, where $\xi \in \mathbb{R}^s$, $\bar{x} \in P_\theta W^{1,2}(\Delta, \mathbb{R}^{d(x)})$, $\bar{v} \in L^2(\Delta, \mathbb{R}^{d(v)})$. Thus,

$$\bar{z} \in \mathscr{Z}^2(\theta) := \mathbb{R}^s \times P_\theta W^{1,2}(\Delta, \mathbb{R}^{d(x)}) \times L^2(\Delta, \mathbb{R}^{d(v)}). \tag{33}$$

For each $\bar{z}$ we set

$$\bar{x}_0 = \bar{x}(t_0), \quad \bar{x}_f = \bar{x}(t_f), \quad \bar{p} = (\bar{x}_0, \bar{x}_f). \tag{34}$$

The vector $\bar{p}$ is considered as a column vector. Denote by

$$I_F(p) = \{ i \in \{1, \ldots, d(F)\} \mid F_i(p) = 0 \} \tag{35}$$

the set of indices of all active endpoint inequalities $F_i(p) \leq 0$ at the point $p = (x(t_0), x(t_f))$. Denote by $\mathscr{K}$ the set of all $\bar{z} \in \mathscr{Z}^2(\theta)$ satisfying the following conditions:

$$J'(p)\bar{p} \leq 0, \quad F_i'(p)\bar{p} \leq 0 \; \forall i \in I_F(p), \quad K'(p)\bar{p} = 0, \tag{36}$$
$$\dot{\bar{x}}(t) = f_x(t)\bar{x}(t) + f_v(t)\bar{v}(t), \tag{37}$$
$$[\bar{x}]^k = [\dot{x}]^k \xi_k, \quad k = 1, \ldots, s, \tag{38}$$

where $p = (x(t_0), x(t_f))$ and $[\dot{x}]^k = \dot{x}(t_k + 0) - \dot{x}(t_k - 0)$. It is obvious that $\mathscr{K}$ is a convex cone in the Hilbert space $\mathscr{Z}^2(\theta)$ with finitely many faces. We call $\mathscr{K}$ the *critical cone*.

3.2 Quadratic Form

Let us introduce a quadratic form on the critical cone $\mathscr{K}$ defined by the conditions (36)-(38). For each $\lambda \in M_0$ and $\bar{z} \in \mathscr{K}$ we set

$$\Omega(\lambda, \bar{z}) = \langle l_{pp}(p)\bar{p}, \bar{p} \rangle + \sum_{k=1}^{s} (D^k(H)\xi_k^2 - [\psi]^k \bar{x}_{av}^k \xi_k)$$
$$+ \int_{t_0}^{t_f} \left(\langle H_{xx}(t)\bar{x}(t), \bar{x}(t) \rangle + 2\langle H_{xv}(t)\bar{v}(t), \bar{x}(t) \rangle + \langle H_{vv}(t)\bar{v}(t), \bar{v}(t) \rangle \right) dt, \tag{39}$$

where

$$l_{pp}(p) = l_{pp}(\alpha_0, \alpha, \beta, p), \quad p = (x(t_0), x(t_f)), \quad \bar{x}_{av}^k = \frac{1}{2}(\bar{x}^{k-} + \bar{x}^{k+}), \tag{40}$$

$$H_{xx}(t) = H_{xx}(t, x(t), \psi(t), u(t), v(t)), \quad \text{etc.} \tag{41}$$

Note that the functional $\Omega(\lambda, \bar{z})$ is linear in λ and quadratic in $\bar{z}$.

3.3 Quadratic Sufficient Optimality Conditions

Denote by M_0^+ the set of all $\lambda \in M_0$ satisfying the conditions:

(a) $H(t, x(t), \psi(t), u, v) > H(t, x(t), \psi(t), u(t), v(t))$ for all $t \in \Delta \setminus \theta$, $u \in U$, $v \in \mathbb{R}^{d(v)}$, such that $(t, x(t), v) \in \mathscr{Q}$ and $(u, v) \neq (u(t), v(t))$;

(b) $H(t_k, x(t_k), \psi(t_k), u, v) > H^k$ for all $t_k \in \theta$, $u \in U$, $v \in \mathbb{R}^{d(v)}$ such that $(t_k, x(t_k), v) \in \mathscr{Q}$, $(u, v) \neq (u(t_k - 0), v(t_k))$, $(u, v) \neq (u(t_k + 0), v(t_k))$, where $H^k := H^{k-} = H^{k+}$.

Let Arg $\min\limits_{\tilde{u} \in U} \sigma(t)\tilde{u}$ be the set of points $v \in U$ where the minimum of the linear function $\sigma(t)\tilde{u}$ is attained.

Definition 4. For a given admissible process Π with a piecewise constant control $u(t)$ and continuous control $v(t)$ we say that $u(t)$ is a strict bang-bang control, if the set M_0 is nonempty and there exists $\lambda \in M_0$ such that

$$\text{Arg} \min_{\tilde{u} \in U} \sigma(t)\tilde{u} = [u(t-0), u(t+0)], \tag{42}$$

where $[u(t-0), u(t+0)]$ denotes the line segment spanned by the vectors $u(t-0)$ and $u(t+0)$.

If $\dim(u) = 1$, then the strict bang-bang property is equivalent to $\sigma(t) \neq 0$ for all $t \in \Delta \setminus \theta$. For $\dim(u) > 1$ the strict bang-bang property requires that two or more control components do not switch simultaneously and the components of the switching vector function vanish only at the switching points. If the set M_0^+ is nonempty, then, obviously, $u(t)$ is a strict bang-bang control.

Definition 5. An element $\lambda \in M_0$ is said to be *strictly Legendrian* if the following conditions are satisfied:

(a) for each $t \in \Delta \setminus \theta$ the quadratic form $\langle H_{vv}(t, x(t), \psi(t), u(t), v(t))\bar{v}, \bar{v}\rangle$ is positive definite on $\mathbb{R}^{d(v)}$;
(b) for each $t_k \in \theta$ the quadratic form $\langle H_{vv}(t_k, x(t_k), \psi(t_k), u(t_k-0), v(t_k))\bar{v}, \bar{v}\rangle$ is positive definite on $\mathbb{R}^{d(v)}$;
(c) for each $t_k \in \theta$ the quadratic form $\langle H_{vv}(t_k, x(t_k), \psi(t_k), u(t_k+0), v(t_k))\bar{v}, \bar{v}\rangle$ is positive definite on $\mathbb{R}^{d(v)}$;
(d) $D^k(\Pi) > 0$ for all $t_k \in \theta$.

Let $\text{Leg}_+(M_0^+)$ be the set of all strictly Legendrian elements $\lambda \in M_0^+$ and put

$$\bar{\gamma}(\bar{z}) = \langle \xi, \xi\rangle + \langle \bar{x}(t_0), \bar{x}(t_0)\rangle + \int_{t_0}^{t_f} \langle \bar{v}(t), \bar{v}(t)\rangle \, dt. \tag{43}$$

Theorem 1. *Let the following Condition $\mathscr{B}$ be fulfilled for the process Π:*

(a) *the set $\text{Leg}_+(M_0^+)$ is nonempty, hence, in particular $u(t)$ is a strict bang-bang control;*
(b) *there exists a nonempty compact set $M \subset \text{Leg}_+(M_0^+)$ and a number $C > 0$ such that $\max\limits_{\lambda \in M} \Omega(\lambda, \bar{z}) \geq C\bar{\gamma}(\bar{z})$ for all $\bar{z} \in \mathscr{K}$.*

Then Π is a strict bounded-strong minimum.

Remark. If the set $\text{Leg}_+(M_0^+)$ is nonempty and $\mathscr{K} = \{0\}$, then (b) is fulfilled automatically. This case can be considered as a *first order* sufficient optimality condition for a strict bounded-strong minimum.

The proof of Theorem 1 is very similar to the proof of the sufficient quadratic optimality condition for the pure bang-bang case given in Milyutin, Osmolovskii [15] Theorem 12.2, p. 302 and Osmolovskii, Maurer [24], Part 1. The proof is based on the SSC for broken extremals in the general problem of calculus of variations; see Osmolovskii [21].

4   Riccati Approach

The following question suggests itself from a numerical point of view: how does the numerical check of the quadratic sufficient optimality conditions in Theorem 1 proceed? For simplicity, we shall assume that (a) the initial value $x(t_0)$ is fixed and (b) there are no endpoint constraints of inequality type. Assumptions (a) and (b) will simplify the boundary conditions for the solution of the associated Riccati equation. Thus we consider a problem:

$$\text{Minimize } J(x(t_f)) \tag{44}$$

under the constraints

$$x(t_0) = x_0, \quad K(x(t_f)) = 0, \quad \dot{x} = f(t,x,u,v), \quad u \in U, \tag{45}$$

where

$$f(t,x,u,v) = a(t,x,v) + B(t,x,v)u, \tag{46}$$

$U \subset \mathbb{R}^{d(u)}$ is a convex polyhedron and J, K, B are C^2-functions. In the sequel, we shall assume that there exists $\lambda \in M_0$ such that $\alpha_0 > 0$.

4.1   Critical Cone $\mathscr{K}$ and Quadratic Form Ω

For this problem, the critical cone is a subspace which is defined by the relations

$$\bar{x}(t_0) = 0, \quad K_{x_f}(p)\bar{x}(t_f) = 0, \tag{47}$$

$$\dot{\bar{x}}(t) = f_x(t)\bar{x}(t) + f_v(t)\bar{v}(t), \quad [\bar{x}]^k = [\dot{x}]^k \xi_k, \quad k = 1,\ldots,s. \tag{48}$$

These relations imply that $J'(p)\bar{p} = 0$ since $\alpha_0 > 0$. Hence, the quadratic form is given by

$$\Omega(\lambda,\bar{z}) = \langle l_{x_f x_f}(p)\bar{x}_f, \bar{x}_f \rangle + \sum_{k=1}^{s} (D^k(H)\xi_k^2 - 2[\dot{\psi}]^k \bar{x}_{av}^k \xi_k)$$

$$+ \int_{t_0}^{t_f} \left(\langle H_{xx}(t)\bar{x}(t), \bar{x}(t) \rangle + 2\langle H_{xv}(t)\bar{v}(t), \bar{x}(t) \rangle + \langle H_{vv}(t)\bar{v}(t), \bar{v}(t) \rangle \right) dt, \tag{49}$$

where, by definition, $\bar{x}_f = \bar{x}(t_f)$. We assume that there exists $\lambda \in M_0^+$ such that

$$D^k(H) > 0, \quad k = 1,\ldots,s, \tag{50}$$

and the strengthened Legendre condition is satisfied with respect to v:

$$\langle H_{vv}(t)\bar{v}, \bar{v} \rangle \geq c\langle \bar{v}, \bar{v} \rangle \quad \forall \bar{v} \in \mathbb{R}^{d(v)}, \quad \forall t \in [t_0, t_f] \setminus \theta \quad (c > 0). \tag{51}$$

From now on we shall fix $\lambda \in M_0^+$ with these properties.

4.2 Q-Transformation of Ω on $\mathscr{K}$

Set $n = d(x)$ and let $Q(t)$ be a symmetric $n \times n$ matrix on $[t_0, t_f]$ with piecewise continuous entries that are absolutely continuous on each interval of the set $[t_0, t_f] \setminus \theta$. For each $\bar{z} \in \mathscr{K}$ we obviously have

$$\int_{t_0}^{t_f} \frac{d}{dt} \langle Q\bar{x}, \bar{x} \rangle \, dt = \langle Q\bar{x}, \bar{x} \rangle \Big|_{t_0}^{t_f} - \sum_{k=1}^{s} [\langle Q\bar{x}, \bar{x} \rangle]^k, \tag{52}$$

where $[\langle Q\bar{x}, \bar{x} \rangle]^k$ is the jump of the function $\langle Q\bar{x}, \bar{x} \rangle$ at the point $t_k \in \theta$. Using the equation $\dot{\bar{x}} = f_x \bar{x} + f_v \bar{v}$ and the initial condition $\bar{x}(t_0) = 0$, we obtain

$$-\langle Q(t_f)\bar{x}_f, \bar{x}_f \rangle + \sum_{k=1}^{s} [\langle Q\bar{x}, \bar{x} \rangle]^k$$

$$+ \int_{t_0}^{t_f} \left(\langle \dot{Q}\bar{x}, \bar{x} \rangle + \langle Q(f_x\bar{x} + f_v\bar{v}), \bar{x} \rangle + \langle Q\bar{x}, f_x\bar{x} + f_v\bar{v} \rangle \right) dt = 0. \tag{53}$$

Adding this zero term to the form $\Omega(\lambda, \bar{z})$ in (49) we get

$$\Omega(\lambda, \bar{z}) = \langle (l_{x_f x_f} - Q(t_f))\bar{x}_f, \bar{x}_f \rangle + \sum_{k=1}^{s} \left(D^k(H)\xi_k^2 - 2[\psi]^k \bar{x}_{av}^k \xi_k + [\langle Q\bar{x}, \bar{x} \rangle]^k \right)$$

$$+ \int_{t_0}^{t_f} \left(\langle (H_{xx} + \dot{Q} + Qf_x + f_x^T Q)\bar{x}, \bar{x} \rangle + \langle (H_{xv} + Qf_v)\bar{v}, \bar{x} \rangle \right. \tag{54}$$

$$\left. + \langle (H_{vx} + f_v^T Q)\bar{x}, \bar{v} \rangle + \langle H_{vv}(t)\bar{v}(t), \bar{v}(t) \rangle \right) dt.$$

We call this formula the *Q-transformation of Ω on $\mathscr{K}$*.

4.3   Transformation of Ω on $\mathscr{K}$ to Perfect Squares

In order to transform the integral term in $\Omega(\lambda, \bar{z})$ to a *perfect square* we assume that $Q(t)$ satisfies the following matrix Riccati equation; cf. also [9,14,28]:

$$\dot{Q} + Qf_x + f_x^T Q + H_{xx} - (H_{xv} + Qf_v)H_{vv}^{-1}(H_{vx} + f_v^T Q) = 0. \tag{55}$$

Then the integral term in Ω can be written as

$$\int_{t_0}^{t_f} \langle H_{vv}^{-1}\bar{h}, \bar{h} \rangle \, dt, \quad \text{where } \bar{h} = (H_{vx} + f_v^T Q)\bar{x} + H_{vv}\bar{v}. \tag{56}$$

A remarkable fact is that the terms

$$\omega_k := D^k(H)\xi_k^2 - 2[\psi]^k \bar{x}_{av}^k \xi_k + [\langle Q\bar{x}, \bar{x} \rangle]^k \tag{57}$$

can also be transformed to perfect squares if the matrix $Q(t)$ satisfies a special *jump condition* at each point $t_k \in \theta$. This jump condition was obtained in Osmolovskii, Lempio [22]. Namely, for each $k = 1, \ldots, s$ put

$$Q^{k-} = Q(t_k - 0), \quad Q^{k+} = Q(t_k + 0), \quad [Q]^k = Q^{k+} - Q^{k-}, \tag{58}$$

$$q_{k-} = ([\dot{x}]^k)^T Q^{k-} - [\psi]^k, \quad b_{k-} = D^k(H) - (q_{k-})[\dot{x}]^k, \tag{59}$$

where $[\dot{x}]^k$ is a column vector, while q_{k-}, $([\dot{x}]^k)^T$ and $[\psi]^k$ are row vectors, and b_{k-} is a number. We shall assume that

$$b_{k-} > 0, \ k = 1, \ldots, s, \tag{60}$$

holds and that Q satisfies the following jump conditions

$$[Q]^k = (b_{k-})^{-1} (q_{k-})^T (q_{k-}), \tag{61}$$

where (q_{k-}) is a row vector, $(q_{k-})^T$ is a column vector and hence $(q_{k-})^T (q_{k-})$ is a symmetric $n \times n$ matrix. Then one can show (see [22]) that

$$\omega_k = (b_{k-})^{-1} ((b_{k-})\xi_k + (q_{k-})(\bar{x}^{k+}))^2 = (b_{k-})^{-1} (D^k(H)\xi_k + (q_{k-})(\bar{x}^{k-}))^2. \tag{62}$$

Therefore, we obtain the following transformation of the quadratic form $\Omega = \Omega(\lambda, \bar{z})$ to perfect squares on the critical cone $\mathcal{K}$:

$$\Omega = \langle (l_{x_f x_f} - Q(t_f))\bar{x}_f, \bar{x}_f \rangle + \sum_{k=1}^{s} (b_{k-})^{-1} (D^k(H)\xi_k + (q_{k-})(\bar{x}^{k-}))^2$$

$$+ \int_{t_0}^{t_f} \langle H_{vv}^{-1}\bar{h}, \bar{h} \rangle \, dt, \tag{63}$$

where

$$\bar{h} = (H_{vx} + f_v^T Q)\bar{x} + H_{vv}\bar{v}. \tag{64}$$

In addition, let us assume that

$$\langle (l_{x_f x_f} - Q(t_f))\bar{x}_f, \bar{x}_f \rangle \geq 0 \tag{65}$$

for all $\bar{x}_f \in \mathbb{R}^{d(x)}$ such that

$$K_{x_f}(x(t_f))\bar{x}_f = 0. \tag{66}$$

Then, obviously, $\Omega(\lambda, \bar{z}) \geq 0$ on $\mathcal{K}$. Let us show now that $\Omega(\lambda, \bar{z}) > 0$ for each nonzero element $\bar{z} \in \mathcal{K}$. This means that $\Omega(\lambda, \bar{z})$ is *positive definite* on the critical cone $\mathcal{K}$ since $\Omega(\lambda, \bar{z})$ is a Legendrian quadratic form.

Assume that $\Omega(\lambda, \bar{z}) = 0$ for some element $\bar{z} \in \mathcal{K}$. Then, for this element, the following equations hold

$$\bar{x}(t_0) = 0, \tag{67}$$

$$D^k(H)\xi_k + (q_{k-})(\bar{x}^{k-}) = 0, \quad k = 1, \ldots, s, \tag{68}$$

$$\bar{h}(t) = 0 \text{ a.e. in } \Delta. \tag{69}$$

From the last equation we get

$$\bar{v} = -H_{vv}^{-1}(H_{vx} + f_v^T Q)\bar{x}. \tag{70}$$

Using this formula in the equation

$$\dot{\bar{x}} = f_x \bar{x} + f_v \bar{v},$$

we see that $\bar{x}$ is a solution of the linear equation

$$\dot{\bar{x}} = (f_x - f_v H_{vv}^{-1}(H_{vx} + f_v^T Q))\bar{x}. \tag{71}$$

This equation together with initial condition $\bar{x}(t_0) = 0$ implies that

$$\bar{x}(t) = 0 \text{ for all } t \in [t_0, t_1). \tag{72}$$

Consequently, $\bar{x}^{1-} = 0$, and then $\xi_1 = 0$ by virtue of (68) with $k = 1$. The equality $\xi_1 = 0$ together with jump condition $[\bar{x}]^1 = [\dot{x}]^1 \xi_1$ imply that $[\bar{x}]^1 = 0$, i.e., $\bar{x}$ is continuous at t_1. Consequently, $\bar{x}^{1+} = 0$. From the last condition and equation (71) it follows that

$$\bar{x}(t) = 0 \quad \text{for all } t \in (t_1, t_2). \tag{73}$$

Repeating this argument we obtain

$$\xi_1 = \xi_2 = \ldots = \xi_s = 0, \quad \bar{x}(t) = 0 \ \forall t \in [t_0, t_f]. \tag{74}$$

Then from (70) it follows that $\bar{v} = 0$. Consequently, we have $\bar{z} = 0$ and thus have proved the following theorem.

Theorem 2. *Assume that there exists a symmetric matrix $Q(t)$, defined on $[t_0, t_f]$, such that*

(a) *$Q(t)$ is piecewise continuous on $[t_0, t_f]$ and continuously differentiable on each interval of the set $[t_0, t_f] \setminus \theta$;*
(b) *$Q(t)$ satisfies the Riccati equation*

$$\dot{Q} + Q f_x + f_x^T Q + H_{xx} - (H_{xv} + Q f_v) H_{vv}^{-1}(H_{vx} + f_v^T Q) = 0 \tag{75}$$

on each interval of the set $[t_0, t_f] \setminus \theta$;
(c) *at each point $t_k \in \theta$ matrix $Q(t)$ satisfies the jump condition*

$$[Q]^k = (b_{k-})^{-1}(q_{k-})^T(q_{k-}), \tag{76}$$

where
$$q_{k-} = ([\dot{x}]^k)^T Q^{k-} - [\dot{\psi}]^k, \quad b_{k-} = D^k(H) - (q_{k-})[\dot{x}]^k > 0; \tag{77}$$

(d) *$\langle (l_{x_f x_f} - Q(t_f))\bar{x}_f, \bar{x}_f \rangle \geq 0$ for all $\bar{x}_f \in \mathbb{R}^{d(x)}$ such that*

$$K_{x_f}(x(t_f))\bar{x}_f = 0. \tag{78}$$

Then $\Omega(\lambda, \bar{z})$ is positive definite on the subspace $\mathcal{H}$.

Remark. No strict inequality is imposed in the boundary condition (d), since the initial state is fixed. This property easily follows from a perturbation argument; cf., e.g., [11]. When endpoint conditions include inequality constraints of the form $F(x(t_f)) \leq 0$, then the inequality (d) has to be checked on the cone of elements $\bar{x}_f \in \mathbb{R}^{d(x)}$ satisfying $K_{x_f}(x(t_f))\bar{x}_f = 0$ and $F_{i,x_f}(x(t_f))\bar{x}_f \leq 0$ if $F_i(x(t_f)) = 0$.

In some problems, it is more convenient to integrate the Riccati equation (75) backwards from $t = t_f$. A similar proof shows that we can replace condition (c) in Theorem 2 by the following condition:

(c+)   at each point $t_k \in \theta$ the matrix $Q(t)$ satisfies the jump condition

$$[Q]^k = (b_{k+})^{-1}(q_{k+})^T(q_{k+}),$$
$$\text{where } q_{k+} = ([\dot{x}]^k)^T Q^{k+} - [\dot{\psi}]^k, \quad b_{k+} = D^k(H) + (q_{k+})[\dot{x}]^k > 0. \tag{79}$$

In the next section, we shall discuss an optimal control problem in economics, where all conditions in Theorem 2 can be verified numerically. Let us mention, however, that Theorem 2 is not applicable to the minimum-fuel orbit transfer problem in Oberle, Taubert [16], since the strict Legendre condition (51) does not hold along the zero thrust arc. Nevertheless, Rosendahl [26] has succeeded in deriving second-order sufficient conditions for those controls that belong to a given control structure. For that purpose, the Riccati approach in [11] is extended to multiprocess control problems that are induced by the given control structures.

5   Numerical Example: Optimal Control of Production and Maintenance

Cho, Abad and Parlar [4] have introduced an optimal control model where a dynamic maintenance problem is incorporated into a production control problem to simultaneously compute optimal production and maintenance policies. In this model, the dynamics is linear with respect to both production and maintenance control, whereas the cost functional is quadratic with respect to production control and linear with respect to maintenance control. Hence, the model fits into the more general type of control problems considered in (1)-(4). Recently, a detailed numerical analysis for different final times and two types of cost functionals has been given in Maurer, Kim, Vossen [10]. For a certain range of final times, the production control is continuous while the maintenance control is bang-bang. The aim in this section is to show that the sufficient conditions in Theorem 2 are satisfied for the computed solutions.

The notations for state and control variables are different from [4,10] and are chosen in conformity with those in the preceding sections: $x_1(t)$: inventory level at time $t \in [0, t_f]$ with fixed final time $t_f > 0$; $x_2(t)$: proportion of good units of end items produced at time t; $v(t)$: scheduled production rate (control); $m(t)$: preventive maintenance rate to reduce the proportion of defective units produced (control); $\alpha(t)$: obsolescence rate of the process performance in the absence of maintenance; $s(t)$: demand rate; $\rho > 0$: discount rate. The dynamics of the process is given by

$$\dot{x}_1(t) = x_2(t)v(t) - s(t), \qquad\qquad x_1(0) = x_{10} > 0,$$
$$\dot{x}_2(t) = -\alpha(t)x_2(t) + (1 - x_2(t))m(t), \quad x_2(0) = x_{20} > 0, \tag{80}$$

with the following bounds on the control variables,

$$0 \le v(t) \le V, \quad 0 \le m(t) \le M \quad \text{for } 0 \le t \le t_f. \tag{81}$$

Since all demands must be satisfied, the following state constraint is imposed:

$$0 \le x_1(t) \quad \text{for } 0 \le t \le t_f. \tag{82}$$

Computations show that this state constraint is automatically satisfied if we impose the boundary condition

$$x_1(t_f) = 0. \tag{83}$$

The optimal control problem then is to *maximize* the total discounted profit plus the salvage value of $x_2(t_f)$,

$$J(x_1, x_2, m, v) = \int_0^{t_f} [ws - hx_1(t) - rv(t)^2 - cm(t)]e^{-\rho t}\, dt + bx_2(t_f)e^{-\rho t_f}, \tag{84}$$

under the constraints (80)-(83). For later computations, the values of constants are chosen as in Cho et al. [4]:

$$s(t) \equiv 4, \ \alpha(t) \equiv 2, \ x_{10} = 3, \ x_{20} = 1, \ V = 3, \ M = 4,$$
$$\rho = 0.1, \ w = 8, \quad h = 1, \quad c = 2.5, \ b = 10, \ r = 2. \tag{85}$$

In the discussion of the *minimum principle* (16)-(19), we will not use the *current value* Hamiltonian function as in [4,10] but will work with the Hamiltonian (Pontryagin) function (11):

$$H(t, x_1, x_2, \psi_1, \psi_2, m, v) = e^{-\rho t}(-ws + hx_1 + rv^2 + cm)$$
$$+ \psi_1(x_2 v - s) + \psi_2(-\alpha x_2 + (1 - x_2)m), \tag{86}$$

where ψ_1, ψ_2 denote the adjoint variables. The adjoint equations (17) and transversality conditions (18) yield in view of $x_1(t_f) = 0$ and the salvage term in the cost functional:

$$\dot{\psi}_1 = -he^{-\rho t}, \quad \psi_1(t_f) = \mu,$$
$$\dot{\psi}_2 = -\psi_1 v + \psi_2(\alpha + m), \quad \psi_2(t_f) = -be^{-\rho t_f}. \tag{87}$$

The multiplier μ is not known a priori and will be computed later. We choose a time horizon for which the control constraint $0 \le v(t) \le 3$ does not become active. Hence, the minimization in (19) leads to the equation $0 = H_v = 2re^{-\rho t}v + \psi_1 x_2$, which yields the control

$$v = -\psi_1 x_2 e^{\rho t}/2r. \tag{88}$$

Since the maintenance control enters the Hamiltonian linearly, the control m is determined by the sign of the switching function

$$\sigma^m(t) = H_m = e^{-\rho t}c + \psi_2(t)(1 - x_2(t)) \tag{89}$$

as the policy

$$m(t) = \begin{cases} M, & \text{if} \quad \sigma^m(t) < 0 \\ 0, & \text{if} \quad \sigma^m(t) > 0 \\ \text{singular, if} & \sigma^m(t) \equiv 0 \quad \text{for } t \in I_{\text{sing}} \subset [0, t_f] \end{cases} . \tag{90}$$

For the final time $t_f = 1$ which was considered in [4] and [10], the maintenance control contains a singular arc. But the computations in [10] show that for final times $t_f \in [0.15, 0.98]$ the maintenance control has two bang-bang arcs:

$$m(t) = \begin{cases} 0, & \text{for} \quad 0 \le t \le t_1 \\ M = 4, & \text{for} \quad t_1 < t \le t_f \end{cases} . \tag{91}$$

Let us study the control problem with final time $t_f = 0.9$ in more detail. To compute a solution candidate, we apply nonlinear programming methods to the discretized control problem with a large number N of grid points $\tau_i = i \cdot t_f / N$, $i = 0, 1, ..., N$; cf. [2,3]. We use the modeling language AMPL of Fourer et al. [6], the interior point optimization code IPOPT of Wächter et al. [27] and the integration method of Heun.

For $N = 5000$ grid points, the computed state, control and adjoint functions are displayed in Fig. 1. We find the following values for the switching time, functional value and some selected state and adjoint variables:

$$\begin{aligned} t_1 &= 0.65691, & J &= 26.705, \\ x_1(t_1) &= 0.84924, & x_2(t_1) &= 0.226879, \\ x_1(t_f) &= 0., & x_2(t_f) &= 0.574104, \\ \psi_1(0) &= -7.8617, & \psi_2(0) &= -4.70437, \\ \psi_1(t_1) &= -8.4975, & \psi_2(t_1) &= -3.2016, \\ \psi_1(t_f) &= -8.72313, & \psi_2(t_f) &= -9.13931. \end{aligned} \tag{92}$$

Let us apply now the second-order sufficient conditions in Theorem 2. Observe first that the sufficiency theorem in Feichtinger and Hartl [5], p. 36, Satz 2.2, is not applicable here. The assumptions in this theorem require that the minimized Hamiltonian $H^{\min}(t, x, \psi(t))$ be *convex* in the state variable $x = (x_1, x_2)$. However, using the minimizing control $v = -\psi_1 x_2 e^{\rho t} / 2r$ from (88), we obtain

$$H^{\min}(t, x, \psi(t)) = -\frac{e^{\rho t}}{4r} \psi_1(t)^2 x_2^2 + L(x), \tag{93}$$

where $L(x)$ denotes a linear function in the variable x. Since $\psi_1(t) \ne 0$ for $t \in [0, t_f]$, the minimized Hamiltonian is *strictly concave* in the variable x_2. Hence, the sufficiency theorem in [5], Satz 2.2, cannot be used here.

Now we compute the quantities needed in Theorem 2 and (79). The derivative of the switching function $\sigma^m(t) = e^{-\rho t} c + \psi_2(t)(1 - x_2(t))$ in (89) is given by

$$\dot{\sigma}^m = -\rho e^{-\rho t} c - \psi_1 v(1 - x_2) + \psi_2 \alpha, \quad v = -\psi_1 x_2 e^{\rho t} / 2r. \tag{94}$$

Inserting the values given in (92) we get

$$D^1(H) = -4\dot{\sigma}^m(t_1) = 27.028 > 0, \quad \sigma^m(t) \ne 0 \quad \forall t \ne t_1. \tag{95}$$

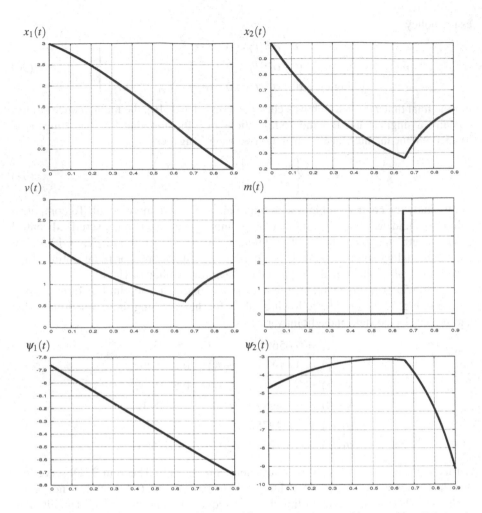

Fig. 1. From above: state variables $x_1(t)$ and $x_2(t)$; control variables $v(t)$ and $m(t)$; adjoint variables $\psi_1(t)$ and $\psi_2(t)$

Hence, the maintenance control $m(\cdot)$ is a *strict bang-bang control*; see Fig 2.
    Now we evaluate the Riccati equation

$$\dot{Q} = -Qf_x - f_x^T Q - H_{xx} + (H_{xv} + Qf_v)(H_{vv})^{-1}(H_{vx} + f_v^T Q) \tag{96}$$

for the symmetric 2×2-matrix $Q = \begin{pmatrix} q_{11} & q_{12} \\ q_{12} & q_{22} \end{pmatrix}$. Computing the expressions

$$f_x = \begin{pmatrix} 0 & v \\ 0 & -(\alpha+m) \end{pmatrix}, \ f_v = \begin{pmatrix} x_2 \\ 0 \end{pmatrix}, \ H_{xx} = \mathbf{0}, \ H_{xv} = (0, \psi_1)^T, \ H_{vv} = 2re^{-\rho t}, \tag{97}$$

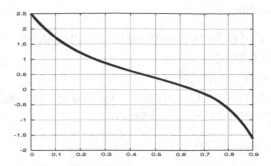

Fig. 2. Switching function $\sigma^m(t)$

the matrix Riccati equation (96) yields the following ODE system:

$$\dot{q}_{11} = q_{11}^2 x_2^2 e^{\rho t}/2r, \tag{98}$$

$$\dot{q}_{12} = -q_{11}v + q_{12}(\alpha + m) + e^{\rho t}q_{11}x_2(\psi_1 + q_{12}x_2)/2r, \tag{99}$$

$$\dot{q}_{22} = -2q_{12}v + 2q_{22}(\alpha + m) + e^{\rho t}(\psi_1 + q_{12}x_2)^2/2r. \tag{100}$$

Since equations (98) and (99) are satisfied by zero functions q_{11} and q_{12}, we can try to find a solution to the Riccati system with $q_{11}(t) = q_{12}(t) \equiv 0$ on $[0,t_f]$. Then (100) reduces to the *linear equation*

$$\dot{q}_{22} = 2q_{22}(\alpha + m) + e^{\rho t}\psi_1^2/2r. \tag{101}$$

Obviously, this linear equation has always a solution. The remaining difficulty is to satisfy the jump and boundary conditions in Theorem 2 (c) and (d). Instead of condition (c) we will verify conditions (c+) and (79) which are more convenient for the backward integration of (100). The boundary conditions in Theorem 2 (d) show that the initial value $Q(0)$ can be chosen arbitrarily while the terminal condition imposes the sign condition $q_{22}(t_f) \leq 0$, since $x_2(t_f)$ is free. We shall take the boundary condition

$$q_{22}(t_f) = 0. \tag{102}$$

Using the computed values in (92), we solve the linear equation (101) with terminal condition (102). At the switching time t_1 we obtain the value

$$q_{22}(t_1) = -1.5599. \tag{103}$$

Next, we evaluate the jump in the state and adjoint variables and will check conditions (79). We get

$$([\dot{x}]^1)^T = (0, M(1 - x_2(t_1))), \quad [\psi]^1 = (0, M\psi_2(t_1)), \tag{104}$$

which yield the quantities

$$
\begin{aligned}
q_{1+} &= ([\dot{x}]^1)^T Q^{1+} - [\dot{\psi}]^1 = (0, \, M(1 - x_2(t_1))q_{22}(t_1+) - M\psi_2(t_1)) \\
&= (0, 8.2439), \tag{105} \\
b_{1+} &= D^1(H) + (q_{1+})[\dot{x}]^1 \\
&= D^1(H) + M^2(1 - x_2(t_1))((1 - x_2(t_1))q_{22}(t_1+) - \psi_2(t_1)) \\
&= 27.028 + 133.55 = 160.58 > 0. \tag{106}
\end{aligned}
$$

Then the jump condition in (79),

$$
[Q]^1 = (b_{1+})^{-1}(q_{1+})^T(q_{1+}) = \begin{pmatrix} 0 & 0 \\ 0 & [q_{22}]^1 \end{pmatrix}, \tag{107}
$$

reduces to a jump condition for $q_{22}(t)$ at t_1. However, we do not need to evaluate this jump condition explicitly because the linear equation (101) has a solution regardless of the value $q_{22}(t_1-)$. Hence, we conclude from Theorem 2 that the numerical solution characterized by (92) and displayed in Fig. 1 provides a strict bounded-strong minimum.

Acknowledgements. We are indebted to Kazimierz Malanowski for helpful comments.

References

1. Agrachev, A.A., Stefani, G., Zezza, P.L.: Strong optimality for a bang-bang trajectory. SIAM J. Control and Optimization 41, 991–1014 (2002)
2. Betts, J.T.: Practical Methods for Optimal Control Using Nonlinear Programming. Advances in Design and Control. SIAM, Philadelphia (2001)
3. Büskens, C., Maurer, H.: SQP-methods for solving optimal control problems with control and state constraints: adjoint variables, sensitivity analysis and real-time control. J. of Computational and Applied Mathematics 120, 85–108 (2000)
4. Cho, D.I., Abad, P.L., Parlar, M.: Optimal production and maintenance decisions when a system experiences age-dependent deterioration. Optimal Control Applic. and Methods 14, 153–167 (1993)
5. Feichtinger, G., Hartl, R.F.: Optimale Kontrolle ökonomischer Prozesse. de Gruyter Verlag, Berlin (1986)
6. Fourer, R., Gay, D.M., Kernighan, B.W.: AMPL: A Modeling Language for Mathematical Programming. Duxbury Press, Brooks-Cole Publishing Company (1993)
7. Hestens, M.: Calculus of Variations and Optimal Control Theory. John Wiley, New York (1966)
8. Malanowski, K.: Stability and sensitivity analysis for optimal control problems with control-state constraints. Dissertationes Mathematicae, Institute of Mathematics, Polish Academy of Sciences (2001)
9. Maurer, H.: First and second order sufficient optimality conditions in mathematical programming and optimal control. Mathematical Programming Study 14, 163–177 (1981)
10. Maurer, H., Kim, J.-H.R., Vossen, G.: On a state-constrained control problem in optimal production and maintenance. In: Deissenberg, C., Hartl, R.F. (eds.) Optimal Control and Dynamic Games, Applications in Finance, Management Science and Economics, pp. 289–308. Springer, Heidelberg (2005)

11. Maurer, H., Oberle, H.J.: Second order sufficient conditions for optimal control problems with free final time: The Riccati approach. SIAM J. Control and Optimization 41, 380–403 (2002)
12. Maurer, H., Osmolovskii, N.P.: Second order sufficient conditions for time optimal bang-bang control problems. SIAM J. Control and Optimization 42, 2239–2263 (2004)
13. Maurer, H., Osmolovskii, N.P.: Second order optimality conditions for bang-bang control problems. Control and Cybernetics 32(3), 555–584 (2003)
14. Maurer, H., Pickenhain, S.: Second-order sufficient conditions for control problems with mixed control-state constraints. Journal of Optimization Theory and Applications 86, 649–667 (1995)
15. Milyutin, A.A., Osmolovskii, N.P.: Calculus of Variations and Optimal Control. Translations of Mathematical Monographs, vol. 180. American Mathematical Society, Providence (1998)
16. Oberle, H.J., Taubert, K.: Existence and multiple solutions of the minimum-fuel orbit transfer problem. J. Optimization Theory and Applications 95, 243–262 (1997)
17. Osmolovskii, N.P.: High-order necessary and sufficient conditions for Pontryagin and bounded-strong minima in the optimal control problems. Dokl. Akad. Nauk SSSR, Ser. Cybernetics and Control Theory 303, 1052–1056 (1988); English transl. Sov. Phys. Dokl. 33(12), 883–885 (1988)
18. Osmolovskii, N.P.: Quadratic conditions for nonsingular extremals in optimal control (A theoretical treatment). Russian J. of Mathematical Physics 2, 487–516 (1995)
19. Osmolovskii, N.P.: Second order conditions for broken extremal. In: Ioffe, A., Reich, S., Shafir, I. (eds.) Calculus of Variations and Optimal Control (Technion 1998), pp. 198–216. Chapman and Hall/CRC, Boca Raton (2000)
20. Osmolovskii, N.P.: Second-order sufficient conditions for an extremum in optimal control. Control and Cybernetics 31(3), 803–831 (2002)
21. Osmolovskii, N.P.: Quadratic optimality conditions for broken extremals in the general problem of calculus of variations. Journal of Math. Science 123(3), 3987–4122 (2004)
22. Osmolovskii, N.P., Lempio, F.: Jacobi-type conditions and Riccati equation for broken extremal. Journal of Math. Science 100(5), 2572–2592 (2000)
23. Osmolovskii, N.P., Lempio, F.: Transformation of quadratic forms to perfect squares for broken extremals. Journal of Set Valued Analysis 10, 209–232 (2002)
24. Osmolovskii, N.P., Maurer, H.: Equivalence of second order optimality conditions for bang-bang control problems. Part 1: Main results. Control and Cybernetics 34, 927–950 (2005); Part 2: Proofs, variational derivatives and representations. Control and Cybernetics 36, 5–45 (2007)
25. Pontryagin, L.S., Boltyanskii, V.G., Gamkrelidze, R.V., Mishchenko, E.F.: The Mathematical Theory of Optimal Processes. Fizmatgiz, Moscow; English translation: Pergamon Press, New York (1964)
26. Rosendahl, R.: Second order sufficient conditions for space-travel optimal control problems. Talk presented at the 23rd IFIP TC7 Conference on Systems Modelling and Optimization, Cracow, Poland (2007)
27. Wächter, A., et al.: http://projects.coin-or.org/Ipopt
28. Zeidan, V.: The Riccati equation for optimal control problems with mixed state control problems: necessity and sufficiency. SIAM Journal on Control and Optimization 32, 1297–1321 (1994)

Shape Sensitivity Analysis for Compressible Navier-Stokes Equations

Pavel I. Plotnikov[1], Evgenya V. Ruban[1], and Jan Sokołowski[2,*]

[1] Lavrentyev Institute of Hydrodynamics, Siberian Division of Russian Academy of Sciences,
Lavrentyev pr. 15, Novosibirsk 630090, Russia
plotnikov@hydro.nsc.ru
[2] Institut Elie Cartan, Laboratoire de Mathématiques, Université Henri Poincaré Nancy 1,
B.P. 239, 54506 Vandoeuvre lés Nancy Cedex, France
Jan.Sokolowski@iecn.u-nancy.fr

Abstract. In the paper compressible, stationary Navier-Stokes (N-S) equations are considered. The model is well-posed, there exist weak solutions in bounded domains, subject to inhomogeneous boundary conditions. The shape sensitivity analysis is performed for N-S boundary value problems, in the framework of small perturbations of the so-called *approximate solutions*. The approximate solutions are determined from Stokes problem and the small perturbations are given by solutions to the full nonlinear model. Such solutions are unique. The differentiability of the specific solutions with respect to the coefficients of differential operators implies the shape differentiability of the drag functional. The shape gradient of the drag functional is derived in the classical and useful for computations form, an appropriate adjoint state is introduced to this end. The proposed method of shape sensitivity analysis is general, and can be used to establish the well-posedness for distributed and boundary control problems as well as for inverse problems in the case of the state equations in the form of compressible Navier-Stokes equations.

1 Introduction

Shape optimization for compressible Navier-Stokes equations (N-S) is important for applications [9] and it is investigated from numerical point of view in the field of scientific computations, however the mathematical analysis of such problems is not available in the existing literature. One of the reasons is the lack of the existence results for inhomogeneous boundary value problems for such equations. We refer the reader to the chapter [21] for the state of art and some new results in this domain.

The results established in the paper lead in particular to the first order optimality conditions for a class of shape optimization problems for compressible Navier-Stokes equations.

---

* The research was partially supported by the grant N514 021 32/3135 of the Polish Ministry of Education.

A. Korytowski et al. (Eds.): System Modeling and Optimization, IFIP AICT 312, pp. 430–447, 2009.

Our results for Fourier-Navier-Stokes (F-N-S) and N-S boundary value problems can be presented according to the following plan.

- *Mathematical modeling, well posedness of solutions to the boundary value problems.* The most general setting for such analysis is introduced in [20] and covers the F-N-S boundary value problems in bounded domains with inhomogeneous boundary conditions. We point out that in [18] the diatomic gases are considered and the existence of solutions for the mathematical models is shown. The shape differentiability of solutions is proved in [19] for the Navier-Stokes boundary value problems in bounded domains with inhomogeneous boundary conditions.
- *The drag functional* is minimized, however the same approach can be used for more general problems of shape optimization including the lift maximization and the density distribution optimization at the outlet of the flow domain.
- *Framework for the shape sensitivity analysis.* The new results are derived for small perturbations of the approximate solutions to compressible N-S equations. In [19] the shape sensitivity analysis is performed with respect to the adjugate matrix defined for the Jacobi matrix of a given domain transformation mapping. Our approach allows for substantial simplification of the sensitivity analysis compared to the existing results obtained in the case of incompressible fluids by using the velocity or perturbation of identity methods of shape sensitivity analysis.
- *Material derivatives* of solutions to compressible N-S equations in the fixed domain setting are obtained in [19]. The shape differentiability of solutions for compressible N-S boundary value problems is shown with respect to weak norms, i.e., in the negative Sobolev spaces for the hyperbolic component, that is, the transport equation, however the obtained material derivatives are sufficiently regular in order to obtain the shape gradients given by some functions, and such a result is actually very useful for possible application of numerical methods of shape optimization of the level set type – since the shape gradients are the coefficients of the non linear hyperbolic equation.
- *Shape gradient* of the drag functional is determined by means of the complicated adjoint state, and we observe that the expression obtained is sufficiently smooth and given by a function, it implies that e.g., the level set method can be employed for numerical solution of the shape optimization for the drag minimization.

The shape optimization for compressible Navier-Stokes equations is an important branch of the research, e.g. in aerodynamics. The main difficulty in analysis of such optimization problems is the mathematical modeling, i.e., the lack of the existence results for inhomogeneous boundary value problems in bounded domains [18]. The authors proved the existence of an optimal shape for drag minimization in three spatial dimensions under the Mosco convergence of admissible domains and assuming that the family of admissible domains is nonempty [17]. This is a result on the compactness of the set of solutions to Navier-Stokes equations for the admissible family of obstacles, we refer the reader to [14]-[17] for further details. The shape differentiability of solutions to N-S equations with respect to boundary perturbations is shown in [19], and leads to the optimality system for the shape optimization problem under considerations.

2 Compressible, Stationary Fourier-Navier-Stokes Equations

The most general results on the existence of solutions to N-S equations from the point of view of drag minimization are established in [20] for the complete model including the heat conduction. The model is formulated in the following way.

Modeling. Let us consider the following set of equations with the state variables: $\mathbf{u}$ is the velocity field in the bounded domain Ω in three spatial dimensions, $\rho > 0$ stands for the mass density, and $\vartheta \geq 0$ is the temperature

$$\Delta \mathbf{u} + \lambda \nabla \mathrm{div} \mathbf{u} = k \, \mathrm{div}(\rho \mathbf{u} \otimes \mathbf{u}) + \omega \nabla(\rho(1+\vartheta)) + \rho \mathbf{g} \quad \text{in } \Omega, \tag{1}$$

$$\mathrm{div}(\rho \mathbf{u}) = 0 \quad \text{in } \Omega, \tag{2}$$

$$\Delta \vartheta = k\gamma^{-1}\left(\rho \mathbf{u} \nabla \vartheta + (\gamma-1)(1+\vartheta)\rho \mathrm{div} \mathbf{u}\right) - k\omega^{-1}(1-\gamma^{-1})D \quad \text{in } \Omega, \tag{3}$$

where $k = R$, $\omega = R/(\gamma \varepsilon^2)$, R is the Reynolds number, ε is the Mach number, λ is the viscosity ratio, γ is the adiabatic constant, $\mathbf{g}$ denotes the dimensionless mass force, and the dissipative function D is defined by the equality

$$D = \frac{1}{2}(\nabla \mathbf{u} + \nabla \mathbf{u}^*)^2 + (\lambda - 1)\mathrm{div} \mathbf{u}^2. \tag{4}$$

The governing equations should be supplemented with the boundary conditions. The velocity of the gas coincides with a given vector field $\mathbf{U} \in C^\infty(\mathbb{R}^3)^3$ on the surface $\partial \Omega$. In this framework, the boundary of the flow domain is divided into three subsets: the inlet Σ_{in}, outgoing set Σ_{out}, and characteristic set Σ_0 defined by the equalities

$$\Sigma_{\mathrm{in}} = \left\{x \in \overline{\Sigma} : \mathbf{U} \cdot \mathbf{n} < 0\right\}, \quad \Sigma_{\mathrm{out}} = \left\{x \in \overline{\Sigma} : \mathbf{U} \cdot \mathbf{n} > 0\right\}, \tag{5}$$

$\Sigma_0 = \{x \in \partial \Omega : \mathbf{U} \cdot \mathbf{n} = 0\}$, where $\mathbf{n}$ stands for the unit outward normal to $\partial \Omega$.

The state variables satisfy the boundary conditions

$$\mathbf{u} = \mathbf{U}, \quad \vartheta = 0 \text{ on } \partial \Omega, \quad \rho = g \text{ on } \Sigma_{\mathrm{in}}, \tag{6}$$

in which g is a given positive function.

Emergent vector field conditions. The existence of solutions to the mathematical model can be established under geometrical conditions related to the characteristic set $\Gamma \subset \Sigma_0$ and the boundary data $\mathbf{U}$ for the velocity field, see Fig. 1 for a specific geometry of the bounded flow domain Ω with an obstacle S. Note that the boundary of the obstacle is not important for such a condition. The emergent vector field condition is known in the theory of PDE's for the oblique derivative problems, and it is introduced and exploited in our papers for the compressible N-S equations with the hyperbolic component for the mass transport. In our case the condition allows us to construct in an appropriate way the solutions to the mass transport equation.

Assume that a characteristic set $\Gamma \subset \partial \Omega$ and a given vector field $\mathbf{U}$ satisfy the following conditions, referred to as the *emergent vector field conditions*.

Emergent vector field conditions: The set Γ is a a closed C^∞ one-dimensional manifold. Moreover, there is a positive constant c such that

$$\mathbf{U} \cdot \nabla(\mathbf{U} \cdot \mathbf{n}) > c > 0 \text{ on } \Gamma. \tag{7}$$

Since the vector field $\mathbf{U}$ is tangent to $\partial\Omega$ on Γ, the quantity in the left-hand side of (7) is well defined.

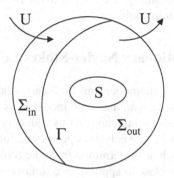

Fig. 1. Characteristic set $\Gamma \subset \partial B$ on the exterior boundary of the flow domain Ω

This condition is obviously fulfilled for all strictly convex domains and constant vector fields. It has a simple geometric interpretation, that $\mathbf{U} \cdot \mathbf{n}$ only vanishes up to the first order at Γ, and for each point $P \in \Gamma$, the vector $\mathbf{U}(P)$ points to the part of $\partial\Omega$ where $\mathbf{U}$ is an exterior vector field.

Boundary value problem. We use the approximate solutions of the F-N-S boundary value problems in order to show the existence, uniqueness and the stability of solutions to F-N-S boundary value problems. The method is general and well suited in the framework of mathematical modeling in the shape optimization, in the optimal control and in solution of inverse problems.

Let us consider the following boundary value problem for the incompressible Navier-Stokes equations

$$\Delta\mathbf{u}_0 - \nabla p_0 = k\operatorname{div}(\mathbf{u}_0 \otimes \mathbf{u}_0), \quad \operatorname{div}\mathbf{u}_0 = 0 \text{ in } \Omega, \quad \mathbf{u}_0 = \mathbf{U} \text{ on } \partial\Omega, \quad \Pi p_0 = p_0. \tag{8}$$

In our notations Π is the projection,

$$\Pi u = u - \frac{1}{\operatorname{meas}\Omega} \int_\Omega u\, dx. \tag{9}$$

It is well known that for each $\mathbf{U} \in C^\infty(\Omega)$ satisfying the orthogonality conditions

$$\int_{\partial\Omega} \mathbf{U} \cdot \mathbf{n}\, ds = 0 \tag{10}$$

and all sufficiently small k, this problem has a unique C^∞-solution. The triple $(\rho_0, \mathbf{u}_0, \vartheta_0) =: (1, \mathbf{u}_0, 0)$ is an approximate solution for small Mach numbers.

Theorem 1. *There exist positive constants* k^*, ω^* *such that for each fixed* $k \in [0, k^*]$, *and all* $\omega > \omega^*$, *the stationary N-S-F problem has a solution* $(\mathbf{u}_\omega, \rho_\omega, \vartheta_\omega) \in (X^{1+s,r})^3 \times X^{s,r} \times X^{1+s,r}$ *such that*

$$\|\mathbf{u}_\omega - \mathbf{u}_0\|_{1+s,r} + \|\rho_\omega - 1\|_{s,r} + \|\vartheta_\omega\|_{1+s,r} \to 0 \ \text{as} \ \omega \to \infty. \tag{11}$$

Proof is given in [20]. Applications of Theorem 1 to the existence of optimal shapes, as well as to the shape differentiability of solutions to the F-N-S boundary value problem are presented in the forthcoming publications.

3   Compressible, Stationary Navier-Stokes Equations

We restrict ourselves to the inhomogeneous boundary value problems for compressible, stationary Navier-Stokes equations. Such modeling is considered in [14]-[19]. In particular, the well-posedness for inhomogeneous boundary value problems of elliptic-hyperbolic type is shown in [19]. Analysis is performed for small perturbations of the approximate solutions, which are determined from the Stokes problem. The existence and uniqueness of solutions close to approximate solution are proved, and in addition, the differentiability of solutions with respect to the coefficients of differential operators is shown in [19]. The results on the well-posedness of nonlinear problem are interesting on their own, and are used to obtain the shape differentiability of the drag functional for incompressible Navier-Stokes equations. The shape gradient of the drag functional is derived in the classical and useful for computations form, an appropriate adjoint state is introduced to this end. The shape derivatives of solutions to the Navier-Stokes equations are given by smooth functions, however the shape differentiability is shown in a weak norm. The method of analysis proposed in [19] is general, and can be used to establish the well-posedness for distributed and boundary control problems as well as for inverse problems in the case of the state equations in the form of compressible Navier-Stokes equations. The differentiability of solutions to the Navier-Stokes equations with respect to the data leads to the first order necessary conditions for a broad class of optimization problems.

4   Drag Minimization

We present an example of shape optimization in aerodynamics. Mathematical analysis of the drag minimization problem for compressible Navier-Stokes equations can be found, e.g., in [17] on the domain continuity of solutions, and in [19] on the shape differentiability of the drag functional.

Mathematical model in the form of N-S equations. We assume that the viscous gas occupies the double-connected domain $\Omega = B \backslash S$, where $B \subset \mathbb{R}^3$ is a hold-all domain with the smooth boundary $\Sigma = \partial B$, and $S \subset B$ is a compact obstacle. Furthermore, we assume that the velocity of the gas coincides with a given vector field $\mathbf{U} \in C^\infty(\mathbb{R}^3)^3$ on the surface Σ. In this framework, the boundary of the flow domain Ω is divided into the three subsets (see (5)), inlet Σ_{in}, outgoing set Σ_{out}, and Σ_0. In its turn, the compact

$\Gamma = \Sigma_0 \cap \Sigma$ splits the surface Σ into three disjoint parts $\Sigma = \Sigma_{in} \cup \Sigma_{out} \cup \Gamma$. The problem is to find the velocity field $\mathbf{u}$ and the gas density ρ satisfying the following equations along with the boundary conditions

$$\Delta \mathbf{u} + \lambda \nabla \mathrm{div} \mathbf{u} = R\rho \mathbf{u} \cdot \nabla \mathbf{u} + \frac{R}{\varepsilon^2} \nabla p(\rho) \text{ in } \Omega, \quad \mathrm{div}\,(\rho \mathbf{u}) = 0 \text{ in } \Omega, \tag{12}$$

$$\mathbf{u} = \mathbf{U} \text{ on } \Sigma, \quad \mathbf{u} = 0 \text{ on } \partial S, \quad \rho = \rho_0 \text{ on } \Sigma_{in}, \tag{13}$$

where the pressure $p = p(\rho)$ is a smooth, strictly monotone function of the density, ε is the Mach number, R is the Reynolds number, λ is the viscosity ratio, and ρ_0 is a positive constant.

Drag minimization. One of the main applications of the theory of compressible viscous flows is the optimal shape design in aerodynamics. The classical example is the problem of minimization of the drag of airfoil traveling in the atmosphere with uniform speed $\mathbf{U}_\infty$. Recall that in our framework the hydrodynamical force acting on the body S is defined by the formula,

$$\mathbf{J}(S) = -\int_{\partial S} (\nabla \mathbf{u} + (\nabla \mathbf{u})^* + (\lambda - 1)\mathrm{div}\mathbf{u}\mathbf{I} - \frac{R}{\varepsilon^2} p\mathbf{I}) \cdot \mathbf{n} dS. \tag{14}$$

In a frame attached to the moving body the drag is the component of $\mathbf{J}$ parallel to $\mathbf{U}_\infty$,

$$J_D(S) = \mathbf{U}_\infty \cdot \mathbf{J}(S), \tag{15}$$

and the lift is the component of $\mathbf{J}$ in the direction orthogonal to $\mathbf{U}_\infty$. For the fixed data, the drag can be regarded as a functional depending on the shape of the obstacle S. The minimization of the drag and the maximization of the lift are among shape optimization problems of some practical importance. We show the shape differentiability of the drag functional with respect to the boundary variations.

5   Shape Sensitivity Analysis

We start with description of our framework for shape sensitivity analysis, or more general, for well-posedness of compressible N-S equations. The detailed proofs of the results presented in the section are given in the forthcoming paper [19]. To this end we choose the vector field $\mathbf{T} \in C^2(\mathbb{R}^3)^3$ vanishing in the vicinity of Σ, and define the mapping

$$y = x + \varepsilon \mathbf{T}(x), \tag{16}$$

which describes the perturbation of the shape of the obstacle. We refer the reader to [25] for more general framework and results in shape optimization. For small ε, the mapping $x \mapsto y$ takes diffeomorphically the flow region Ω onto $\Omega_\varepsilon = B \setminus S_\varepsilon$, where the perturbed obstacle $S_\varepsilon = y(S)$. Let $(\bar{\mathbf{u}}_\varepsilon, \bar{\rho}_\varepsilon)$ be solutions to problem (12) in Ω_ε. After substituting $(\bar{\mathbf{u}}_\varepsilon, \bar{\rho}_\varepsilon)$ into the formulae for $\mathbf{J}$, the drag becomes the function of the parameter ε. Our aim is, in fact, to prove that this function is well-defined and differentiable at $\varepsilon = 0$. This leads to the first order shape sensitivity analysis for solutions to compressible Navier-Stokes equations. It is convenient to reduce such an analysis to the analysis of dependence of solutions with respect to the coefficients of the governing equations. To this end, we introduce the functions $\mathbf{u}_\varepsilon(x)$ and $\rho_\varepsilon(x)$ defined in the unperturbed domain Ω by the formulae

$$\mathbf{u}_\varepsilon(x) = \mathbf{N}\bar{\mathbf{u}}_\varepsilon(x + \varepsilon\mathbf{T}(x)), \quad \rho_\varepsilon(x) = \bar{\rho}_\varepsilon(x + \varepsilon\mathbf{T}(x)), \tag{17}$$

where

$$\mathbf{N}(x) = [\det(\mathbf{I} + \varepsilon\mathbf{T}'(x))(\mathbf{I} + \varepsilon\mathbf{T}'(x))]^{-1}. \tag{18}$$

is the adjugate matrix of the Jacobi matrix $\mathbf{I} + \varepsilon\mathbf{T}'$. Furthermore, we also use the notation $\mathfrak{g}(x) = \sqrt{\det\mathbf{N}}$. It is easy to see that the matrix $\mathbf{N}(x)$ depends analytically upon the small parameter ε and

$$\mathbf{N} = \mathbf{I} + \varepsilon\mathbf{D}(x) + \varepsilon^2\mathbf{D}_1(\varepsilon, x), \tag{19}$$

where $\mathbf{D} = \operatorname{div}\mathbf{T}\mathbf{I} - \mathbf{T}'$. Calculations show that for $\mathbf{u}_\varepsilon, \rho_\varepsilon$, the following boundary value problem is obtained

$$\Delta\mathbf{u}_\varepsilon + \nabla\left(\lambda\mathfrak{g}^{-1}\operatorname{div}\mathbf{u}_\varepsilon - \frac{R}{\varepsilon^2}p(\rho_\varepsilon)\right) = \mathscr{A}\mathbf{u}_\varepsilon + R\mathscr{B}(\rho_\varepsilon, \mathbf{u}_\varepsilon, \mathbf{u}_\varepsilon) \text{ in } \Omega, \tag{20}$$

$$\operatorname{div}\left(\rho_\varepsilon\mathbf{u}_\varepsilon\right) = 0 \text{ in } \Omega, \tag{21}$$

$$\mathbf{u}_\varepsilon = \mathbf{U} \text{ on } \Sigma, \quad \mathbf{u}_\varepsilon = 0 \text{ on } \partial S, \tag{22}$$

$$\rho_\varepsilon = \rho_0 \text{ on } \Sigma_{\text{in}}. \tag{23}$$

Here, the linear operator $\mathscr{A}$ and the nonlinear mapping $\mathscr{B}$ are defined in terms of $\mathbf{N}$,

$$\mathscr{A}(\mathbf{u}) = \Delta\mathbf{u} - \mathbf{N}^{-1}\operatorname{div}\left(\mathfrak{g}^{-1}\mathbf{N}\mathbf{N}^*\nabla(\mathbf{N}^{-1}\mathbf{u})\right), \tag{24}$$

$$\mathscr{B}(\rho, \mathbf{u}, \mathbf{w}) = \rho(\mathbf{N}^*)^{-1}\left(\mathbf{u}\nabla(\mathbf{N}^{-1}\mathbf{w})\right). \tag{25}$$

The specific structure of the matrix $\mathbf{N}$ does not play any particular role in the further analysis. Therefore, we consider a general problem of the existence, uniqueness and dependence on coefficients of the solutions to equations (20)-(23) under the assumption that $\mathbf{N}$ is a given matrix-valued function which is close, in an appropriate norm, to the identity mapping $\mathbf{I}$ and coincides with $\mathbf{I}$ in the vicinity of Σ. By abuse of notations, we write simply $\mathbf{u}$ and ρ instead of $\mathbf{u}_\varepsilon$ and ρ_ε, when studying the well-posedness and dependence on $\mathbf{N}$. Before formulation of main results we write the governing equation in more transparent form using the change of unknown functions proposed, e.g., in [13]. To do so we introduce *the effective viscous pressure*

$$q = \frac{R}{\varepsilon^2}p(\rho) - \lambda\mathfrak{g}^{-1}\operatorname{div}\mathbf{u}, \tag{26}$$

and rewrite equations (20)-(23) in the equivalent form

$$\Delta\mathbf{u} - \nabla q = \mathscr{A}(\mathbf{u}) + R\mathscr{B}(\rho, \mathbf{u}, \mathbf{u}) \text{ in } \Omega, \tag{27}$$

$$\operatorname{div}\mathbf{u} = a\sigma_0 p(\rho) - \frac{gq}{\lambda} \quad \text{in } \Omega, \tag{28}$$

$$\mathbf{u}\cdot\nabla\rho + g\sigma_0 p(\rho)\rho = \frac{gq}{\lambda}\rho \quad \text{in } \Omega, \tag{29}$$

$$\mathbf{u} = \mathbf{U} \quad \text{on } \Sigma, \quad \mathbf{u} = 0 \quad \text{on } \partial S, \tag{30}$$

$$\rho = \rho_0 \quad \text{on } \Sigma_{\text{in}}, \tag{31}$$

where $\sigma_0 = R/(\lambda\varepsilon^2)$. We point out that the solutions to the compressible N-S equations are determined in the form (34). In the new variables $(\mathbf{u}, q, \rho)$ the expression for the force $\mathbf{J}$ reads

$$\mathbf{J} = -\int_\Omega [g^{-1}(\mathbf{N}^*\nabla(\mathbf{Nu}) + \nabla(\mathbf{Nu})^*\mathbf{N} - \operatorname{div}\mathbf{u}) - q - R\rho\mathbf{u}\otimes\mathbf{u}]\mathbf{N}^*\nabla\eta\,dx, \tag{32}$$

where $\eta \in C^\infty(\Omega)$ is an arbitrary function, which is equal to 1 in an open neighborhood of the obstacle S and 0 in a vicinity of Σ. The value of $\mathbf{J}$ is independent of the choice of the function η.

6 Perturbations of the Approximate Solutions

We assume that $\lambda \gg 1$ and $R \ll 1$, which corresponds to almost incompressible flow with low Reynolds number. In such a case, the *approximate solutions* to problem (27)-(31) can be chosen in the form $(\rho_0, \mathbf{u}_0, q_0)$, where ρ_0 is a constant in boundary condition (31), and $(\mathbf{u}_0, q_0)$ is a solution to the boundary value problem for the Stokes equations,

$$\Delta\mathbf{u}_0 - \nabla q_0 = 0, \ \operatorname{div}\mathbf{u}_0 = 0 \ \text{in } \Omega, \ \mathbf{u}_0 = \mathbf{U} \ \text{on } \Sigma, \ \mathbf{u}_0 = 0 \ \text{on } \partial S, \ \Pi q_0 = q_0, \tag{33}$$

where Π is the projector (9). Equations (33) can be obtained as the limit of equations (27)-(31) for the passage $\lambda \to \infty$, $R \to 0$. It follows from the standard elliptic theory that for the boundary $\partial\Omega \in C^\infty$, we have $(\mathbf{u}_0, q_0) \in C^\infty(\Omega)$. We look for solutions to problem (27)-(31) in the form

$$\mathbf{u} = \mathbf{u}_0 + \mathbf{v}, \quad \rho = \rho_0 + \varphi, \quad q = q_0 + \lambda\sigma_0 p(\rho_0) + \pi + \lambda m, \tag{34}$$

with the unknown functions $\vartheta = (\mathbf{v}, \pi, \varphi)$ and the unknown constant m. Substituting (34) into (27)-(31) we obtain the following boundary problem for ϑ,

$$\Delta\mathbf{v} - \nabla\pi = \mathscr{A}(\mathbf{u}) + R\mathscr{B}(\rho, \mathbf{u}, \mathbf{u}) \quad \text{in } \Omega, \tag{35}$$

$$\operatorname{div}\mathbf{v} = g\left(\frac{\sigma}{\rho_0}\varphi - \Psi[\vartheta] - m\right) \quad \text{in } \Omega, \tag{36}$$

$$\mathbf{u}\cdot\nabla\varphi + \sigma\varphi = \Psi_1[\vartheta] + mg\rho \quad \text{in } \Omega, \tag{37}$$

$$\mathbf{v} = 0 \text{ on } \partial\Omega, \quad \varphi = 0 \text{ on } \Sigma_{\text{in}}, \quad \Pi\pi = \pi, \tag{38}$$

where

$$\Psi_1[\vartheta] = \mathfrak{g}\left(\rho\Psi[\vartheta] - \frac{\sigma}{\rho_0}\varphi^2\right) + \sigma\varphi(1 - \mathfrak{g}), \tag{39}$$

$$\Psi[\vartheta] = \frac{q_0 + \pi}{\lambda} - \frac{\sigma}{p'(\rho_0)\rho_0}H(\varphi), \tag{40}$$

$$\sigma = \sigma_0 p'(\rho_0)\rho_0, \tag{41}$$

$$H(\varphi) = p(\rho_0 + \varphi) - p(\rho_0) - p'(\rho_0)\varphi, \tag{42}$$

the vector field $\mathbf{u}$ and the function ρ are given by (34). Finally, we specify the constant m. In our framework, in contrast to the case of homogeneous boundary problem, the solution to such a problem is not trivial. Note that, since $\operatorname{div}\mathbf{v}$ is of the null mean value, the right-hand side of equation (37) must satisfy the compatibility condition

$$m\int_\Omega \mathfrak{g}\,dx = \int_\Omega \mathfrak{g}\left(\frac{\sigma}{\rho_0}\varphi - \Psi[\vartheta]\right)dx, \tag{43}$$

which formally determines m. This choice of m leads to essential mathematical difficulties. To make this issue clear note that in the simplest case $\mathfrak{g} = 1$ we have $m = \rho_0^{-1}\sigma(\mathbf{I} - \Pi)\varphi + O(|\vartheta|^2, \lambda^{-1})$, and the principal linear part of the governing equations (35)-(38) becomes

$$\begin{pmatrix} \Delta & -\nabla & 0 \\ \operatorname{div} & 0 & -\frac{\sigma}{\rho_0} \\ 0 & 0 & \mathbf{u}\nabla + \sigma \end{pmatrix}\begin{pmatrix} \mathbf{v} \\ \pi \\ \varphi \end{pmatrix} + \begin{pmatrix} 0 \\ m \\ -m\rho_0 \end{pmatrix} \sim \begin{pmatrix} \Delta\mathbf{v} - \nabla\pi \\ \operatorname{div}\mathbf{v} - \frac{\sigma}{\rho_0}\Pi\varphi \\ \mathbf{u}\nabla\varphi + \sigma\Pi\varphi \end{pmatrix} \tag{44}$$

Hence, the question of solvability of the linearized equations derived for (35)-(38), (46), (47) can be reduced to the question of solvability of the boundary value problem for nonlocal transport equation

$$\mathbf{u}\nabla\varphi + \sigma\Pi\varphi = f, \tag{45}$$

which is very difficult because of the loss of maximum principle. In fact, this question is concerned with the problem of the control of the total gas mass in compressible flows. Recall that the absence of the mass control is the main obstacle for proving the global solvability of inhomogeneous boundary problems for compressible Navier-Stokes equations, we refer to [8] for discussion. In order to cope with this difficulty we write the compatibility condition in a sophisticated form, which allows us to control the total mass of the gas. To this end we introduce the auxiliary function ζ satisfying the equations

$$-\operatorname{div}(\mathbf{u}\zeta) + \sigma\zeta = \sigma\mathfrak{g} \text{ in } \Omega, \quad \zeta = 0 \text{ on } \Sigma_{\text{out}}, \tag{46}$$

and fix the constant m as follows

$$m = \varkappa\int_\Omega (\rho_0^{-1}\Psi_1[\vartheta]\zeta - \mathfrak{g}\Psi[\vartheta])\,dx, \quad \varkappa = \left(\int_\Omega \mathfrak{g}(1 - \zeta - \rho_0^{-1}\zeta\varphi)\,dx\right)^{-1}. \tag{47}$$

In this way the auxiliary function ζ becomes an integral part of the solution to problem (35)-(38), (46), (47).

7 Function Spaces

In this section we assemble some technical results which are used throughout the paper. Function spaces play a central role, and we recall some notations, fundamental definitions and properties, which are classical. The proofs of some results given here can be found, e.g. in [19].

For our applications we need the results in three spatial dimensions, however the results are presented for the dimension $d \geq 2$.

Let Ω be the whole space $\mathbb{R}^d$ or a bounded domain in $\mathbb{R}^d$ with the boundary $\partial\Omega$ of class C^1. For an integer $l \geq 0$ and for an exponent $r \in [1,\infty)$, we denote by $H^{l,r}(\Omega)$ the Sobolev space endowed with the norm $\|u\|_{H^{l,r}(\Omega)} = \sup_{|\alpha|\leq l}\|\partial^\alpha u\|_{L^r(\Omega)}$. For real $0 < s < 1$, the fractional Sobolev space $H^{s,r}(\Omega)$ is obtained by the interpolation between $L^r(\Omega)$ and $H^{1,r}(\Omega)$, and consists of all measurable functions with the finite norm

$$\|u\|_{H^{s,r}(\Omega)} = \|u\|_{L^r(\Omega)} + |u|_{s,r,\Omega} \tag{48}$$

where

$$|u|_{s,r,\Omega}^r = \int_{\Omega\times\Omega} |x-y|^{-d-rs}|u(x)-u(y)|^r dxdy. \tag{49}$$

In the general case, the Sobolev space $H^{l+s,r}(\Omega)$ is defined as the space of measurable functions with the finite norm $\|u\|_{H^{l+s,r}(\Omega)} = \sup_{|\alpha|\leq l}\|\partial^\alpha u\|_{H^{s,r}(\Omega)}$. For $0 < s < 1$, the Sobolev space $H^{s,r}(\Omega)$ is, in fact the interpolation space $[L^r(\Omega), H^{1,r}(\Omega)]_{s,r}$.

Furthermore, the notation $H_0^{l,r}(\Omega)$, with an integer l, stands for the closed subspace of the space $H^{l,r}(\Omega)$ of all functions $u \in L^r(\Omega)$ which being extended by zero outside of Ω belong to $H^{l,r}(\mathbb{R}^d)$.

Denote by $\mathscr{H}_0^{0,r}(\Omega)$ and $\mathscr{H}_0^{1,r}(\Omega)$ the subspaces of $L^r(\mathbb{R}^d)$ and $H^{1,r}(\mathbb{R}^d)$, respectively, of all functions vanishing outside of Ω. Obviously $\mathscr{H}_0^{1,r}(\Omega)$ and $H_0^{1,r}(\Omega)$ are isomorphic topologically and algebraically and we can identify them. However, we need the interpolation spaces $\mathscr{H}_0^{s,r}(\Omega)$ for non-integers, in particular for $s = 1/r$.

Definition 1. For all $0 < s \leq 1$ and $1 < r < \infty$, we denote by $\mathscr{H}_0^{s,r}(\Omega)$ the interpolation space $[\mathscr{H}_0^{0,r}(\Omega), \mathscr{H}_0^{1,r}(\Omega)]_{s,r}$ endowed with one of two equivalent norms [19] defined by interpolation method.

It follows from the definition of interpolation spaces that $\mathscr{H}_0^{s,r}(\Omega) \subset H^{s,r}(\mathbb{R}^d)$ and for all $u \in \mathscr{H}_0^{s,r}(\Omega)$,

$$\|u\|_{H^{s,r}(\mathbb{R}^d)} \leq c(r,s)\|u\|_{\mathscr{H}_0^{s,r}(\Omega)}, \quad u = 0 \text{ outside } \Omega. \tag{50}$$

In other words, $\mathscr{H}_0^{s,r}(\Omega)$ consists of all elements $u \in H^{s,r}(\Omega)$ such that the extension $\bar{u}$ of u by 0 outside of Ω has the finite $[\mathscr{H}_0^{0,r}(\Omega), \mathscr{H}_0^{1,r}(\Omega)]_{s,r}$-norm. We identify u and $\bar{u}$ for the elements $u \in \mathscr{H}_0^{s,r}(\Omega)$. With this identification it follows that $H_0^{1,r}(\Omega) \subset \mathscr{H}_0^{s,r}(\Omega)$ and the space $C_0^\infty(\Omega)$ is dense in $\mathscr{H}_0^{s,r}(\Omega)$. It is worthy to note that for $0 < s < 1$ and for $1 < r < \infty$, the function $\bar{u}$ belongs to the space $H^{s,r}(\mathbb{R}^d)$ if and only if $u \in H^{s,r}(\Omega)$ and dist $(x,\partial\Omega)^{-s}u \in L^r(\Omega)$. We also point out that the interpolation space $\mathscr{H}_0^{s,r}(\Omega)$ coincides with the Sobolev space $H_0^{s,r}(\Omega)$ for $s \neq 1/r$. Recall that the standard space $H_0^{s,r}(\Omega)$ is the completion of $C_0^\infty(\Omega)$ in the $H^{s,r}(\Omega)$-norm.

Embedding theorems. For $sr > d$ and $0 \le \alpha < s - r/d$, the embedding $H^{s,r}(\Omega) \hookrightarrow C^\alpha(\Omega)$ is continuous and compact. In particular, for $sr > d$, the Sobolev space $H^{s,r}(\Omega)$ is a commutative Banach algebra, i.e. for all $u, v \in H^{s,r}(\Omega)$,

$$\|uv\|_{H^{s,r}(\Omega)} \le c(r,s)\|u\|_{H^{s,r}(\Omega)}\|v\|_{H^{s,r}(\Omega)}. \tag{51}$$

If $sr < d$ and $t^{-1} = r^{-1} - d^{-1}s$, then the embedding $H^{s,r}(\Omega) \hookrightarrow L^t(\Omega)$ is continuous. In particular, for $\alpha \le s$, $(s-\alpha)r < d$ and $\beta^{-1} = r^{-1} - d^{-1}(s-\alpha)$,

$$\|u\|_{H^{\alpha,\beta}(\Omega)} \le c(r,s,\alpha,\beta,\Omega)\|u\|_{H^{s,r}(\Omega)}. \tag{52}$$

It follows from (50) that all the embedding inequalities remain true for the elements of the interpolation space $\mathscr{H}_0^{s,r}(\Omega)$.

Duality. We define

$$\langle u,v \rangle = \int_\Omega uv\,dx \tag{53}$$

for any functions such that the right hand side makes sense. For $r \in (1,\infty)$, each element $v \in L^{r'}(\Omega)$, $r' = r/(r-1)$, determines the functional L_v of $(\mathscr{H}_0^{s,r}(\Omega))'$ by the identity $L_v(u) = \langle u,v \rangle$. We introduce the $(-s,r')$-norm of an element $v \in L^{r'}(\Omega)$ to be by definition the norm of the functional L_v, that is

$$\|v\|_{\mathscr{H}^{-s,r'}(\Omega)} = \sup_{\substack{u \in \mathscr{H}_0^{s,r}(\Omega) \\ \|u\|_{\mathscr{H}_0^{s,r}(\Omega)}=1}} |\langle u,v \rangle|. \tag{54}$$

Let $\mathscr{H}^{-s,r'}(\Omega)$ denote the completion of the space $L^{r'}(\Omega)$ with respect to $(-s,r')$-norm. For an integer s, $\mathscr{H}^{-s,r'}(\Omega)$ is topologically and algebraically isomorphic to $(H_0^{s,r}(\Omega))'$. The same conclusion holds true for all $s \in (0,1)$. Moreover, we can identify $\mathscr{H}^{-s,r'}(\Omega)$ with the interpolation space $[L^{r'}(\Omega), H_0^{-1,r'}(\Omega)]_{s,r}$, see e.g., [19]. With this denotations we have the duality principle

$$\|u\|_{\mathscr{H}_0^{s,r}(\Omega)} = \sup_{\substack{v \in C_0^\infty(\Omega) \\ \|v\|_{\mathscr{H}^{-s,r'}(\Omega)}=1}} |\langle u,v \rangle|. \tag{55}$$

With applications to the theory of Navier-Stokes equations in mind, we introduce the smaller dual space defined as follows. We identify the function $v \in L^{r'}(\Omega)$ with the functional $L_v \in (H^{s,r}(\Omega))'$ and denote by $\mathbb{H}^{-s,r'}(\Omega)$ the completion of $L^{r'}(\Omega)$ in the norm

$$\|v\|_{\mathbb{H}^{-s,r'}(\Omega)} := \sup_{\substack{u \in H^{s,r}(\Omega) \\ \|u\|_{H^{s,r}(\Omega)}=1}} |\langle u,v \rangle|. \tag{56}$$

In the sense of this identification the space $C_0^\infty(\Omega)$ is dense in the interpolation space $\mathbb{H}^{-s,r}(\Omega)$. It follows immediately from the definition that

$$\mathbb{H}^{-s,r}(\Omega) \subset (H^{s,r}(\Omega))' \subset \mathcal{H}^{-s,r}(\Omega). \tag{57}$$

For an arbitrary bounded domain $\Omega \subset \mathbb{R}^3$ with a Lipschitz boundary, we introduce the Banach spaces

$$X^{s,r} = H^{s,r}(\Omega) \cap H^{1,2}(\Omega), \tag{58}$$
$$Y^{s,r} = H^{s+1,r}(\Omega) \cap H^{2,2}(\Omega), \tag{59}$$
$$Z^{s,r} = \mathcal{H}^{s-1,r}(\Omega) \cap L^2(\Omega) \tag{60}$$

equipped with the norms

$$\|u\|_{X^{s,r}} = \|u\|_{H^{s,r}(\Omega)} + \|u\|_{H^{1,2}(\Omega)}, \tag{61}$$
$$\|u\|_{Y^{s,r}} = \|u\|_{H^{1+s,r}(\Omega)} + \|u\|_{H^{2,2}(\Omega)}, \tag{62}$$
$$\|u\|_{Z^{s,r}} = \|u\|_{\mathcal{H}^{s-1,r}(\Omega)} + \|u\|_{L^2(\Omega)}. \tag{63}$$

It can be easily seen that the embeddings $Y^{s,r} \hookrightarrow X^{s,r} \hookrightarrow Z^{s,r}$ are compact and for $sr > 3$, each of the spaces $X^{s,r}$ and $Y^{s,r}$ is a commutative Banach algebra.

8   Existence and Uniqueness Theory

Denote by E the closed subspace of the Banach space $Y^{s,r}(\Omega)^3 \times X^{s,r}(\Omega)^2$ in the following form

$$E = \{\vartheta = (\mathbf{v}, \pi, \varphi) : \mathbf{v} = 0 \text{ on } \partial\Omega, \quad \varphi = 0 \text{ on } \Sigma_{\text{in}}, \quad \Pi\pi = \pi\}, \tag{64}$$

and denote by $\mathscr{B}_\tau \subset E$ the closed ball of radius τ centered at 0. Next, note that for $sr > 3$, elements of the ball $\mathscr{B}_\tau$ satisfy the inequality

$$\|\mathbf{v}\|_{C^1(\Omega)} + \|\pi\|_{C(\Omega)} + \|\varphi\|_{C(\Omega)} \le c_e(r, s, \Omega)\|\vartheta\|_E \le c_e\tau, \tag{65}$$

where the norm in E is defined by

$$\|\vartheta\|_E = \|\mathbf{v}\|_{Y^{s,r}(\Omega)} + \|\pi\|_{X^{s,r}(\Omega)} + \|\varphi\|_{X^{s,r}(\Omega)}. \tag{66}$$

Theorem 2. *Assume that the surface Σ and given vector field $\mathbf{U}$ satisfy emergent field conditions. Furthermore, let σ^*, τ^* be given constants determined in [19], and let positive numbers r, s, σ satisfy the inequalities*

$$1/2 < s \le 1, \quad 1 < r < 3/(2s-1), \quad sr > 3, \quad \sigma > \sigma^*. \tag{67}$$

Then there exists $\tau_0 \in (0, \tau^*]$, depending only on $\mathbf{U}, \Omega, r, s, \sigma$, such that for all

$$\tau \in (0, \tau_0], \quad \lambda^{-1}, R \in (0, \tau^2], \quad \|\mathbf{N} - \mathbf{I}\|_{C^2(\Omega)} \le \tau^2, \tag{68}$$

problem (35)-(38), (46), (47), with $\mathbf{u}_0$ given by (33), has a unique solution $\vartheta \in B_\tau$. Moreover, the auxiliary function ζ and the constants $\varkappa, m$ admit the estimates

$$\|\zeta\|_{X^{s,r}} + |\varkappa| \le c, \quad |m| \le c\tau < 1, \tag{69}$$

where the constant c depends only on $\mathbf{U}, \Omega, r, s$ *and* σ.

9   Material Derivatives of Solutions

Theorem 2 guarantees the existence and uniqueness of solutions to problem (35)-(38), (46), (47) for all $\mathbf{N}$ close to the identity matrix $\mathbf{I}$. The totality of such solutions can be regarded as the mapping from $\mathbf{N}$ to the solution of the Navier-Stokes equations. The natural question is the smoothness properties of this mapping, in particular its differentiability. With application to shape optimization problems in mind, we consider the particular case where the matrices $\mathbf{N}$ depend on the small parameter ε and have representation (19). We assume that C^1 norms of the matrix-valued functions $\mathbf{D}$ and $\mathbf{D}_1(\varepsilon)$ in (19) have a majorant independent of ε. By virtue of Theorem 2, there are the positive constants ε_0 and τ such that for all sufficiently small R, λ^{-1} and $\varepsilon \in [0, \varepsilon_0]$, problem (35)-(38), (46), (47) with $\mathbf{N} = \mathbf{N}(\varepsilon)$ has a unique solution $\vartheta(\varepsilon) = (\mathbf{v}(\varepsilon), \pi(\varepsilon), \varphi(\varepsilon))$, $\zeta(\varepsilon), m(\varepsilon)$, which admits the estimate

$$\|\vartheta(\varepsilon)\|_E + |m(\varepsilon)| \le c\tau, \quad \|\zeta(\varepsilon)\|_{X^{s,r}} \le c, \tag{70}$$

where the constant c is independent of ε, and the Banach space E is defined by (64). Denote the solution for $\varepsilon = 0$, $(\vartheta(0), m(0), \zeta(0))$ by (ϑ, m, ζ), and define the finite differences with respect to ε

$$(\mathbf{w}_\varepsilon, \omega_\varepsilon, \psi_\varepsilon) = \varepsilon^{-1}(\vartheta - \vartheta(\varepsilon)), \ \xi_\varepsilon = \varepsilon^{-1}(\zeta - \zeta(\varepsilon)), \ n_\varepsilon = \varepsilon^{-1}(m - m(\varepsilon)). \tag{71}$$

Formal calculations show that the limit $(\mathbf{w}, \omega, \psi, \xi, n) = \lim_{\varepsilon \to 0}(\mathbf{w}_\varepsilon, \omega_\varepsilon, \psi_\varepsilon, \xi_\varepsilon, n_\varepsilon)$ is a solution to linearized equations

$$\Delta \mathbf{w} - \nabla \omega = R\mathscr{C}_0(\mathbf{w}, \psi) + \mathscr{D}_0(\mathbf{D}) \text{ in } \Omega,$$

$$\operatorname{div} \mathbf{w} = b_{21}^0 \psi - b_{22}^0 \omega + b_{23}^0 n + b_{30}^0 \eth \text{ in } \Omega,$$

$$\mathbf{u}\nabla\psi + \sigma\psi = -\mathbf{w} \cdot \nabla\varphi + b_{11}^0 \psi + b_{12}^0 \omega + b_{13}^0 n + b_{10}^0 \eth \text{ in } \Omega,$$

$$-\operatorname{div}(\mathbf{u}\xi) + \sigma\xi = \operatorname{div}(\zeta\mathbf{w}) + \sigma\eth \text{ in } \Omega, \tag{72}$$

$$\mathbf{w} = 0 \text{ on } \partial\Omega, \ \psi = 0 \text{ on } \Sigma_{\text{in}}, \ \xi = 0 \text{ on } \Sigma_{\text{out}},$$

$$\omega - \Pi\omega = 0, \ n = \varkappa \int_\Omega \left(b_{31}^0 \psi + b_{32}^0 \omega + b_{34}^0 \xi + b_{30}^0 \eth \right) dx,$$

where $\eth = 1/2 \operatorname{Tr} \mathbf{D}$, the variable coefficients b_{ij}^0 and the operators $\mathscr{C}_0, \mathscr{D}_0$, are defined by the formulae

$$b_{11}^0 = \Psi[\vartheta] - \rho H'(\varphi) + m - \frac{2\sigma}{\rho_0}\varphi,$$

$$b_{12}^0 = \lambda^{-1}\rho,$$

$$b_{13}^0 = \rho,$$

$$b_{10}^0 = \rho\Psi[\vartheta] - \frac{\sigma}{\rho_0}\varphi^2 - \sigma\varphi + m\rho,$$

$$b_{21}^0 = \frac{\sigma}{\rho_0}\psi_0 + H'(\varphi),$$

$$b_{22}^0 = -\lambda^{-1}, \qquad (73)$$

$$b_{23}^0 = -1,$$

$$b_{20}^0 = \sigma\varphi\rho_0^{-1} - \Psi[\vartheta] - m,$$

$$b_{31}^0 = \rho_0^{-1}\varsigma\left(\Psi[\vartheta] - \rho H'(\varphi) - \frac{2\sigma}{\rho_0}\varphi\right) - H'(\varphi) + m\rho_0^{-1}\varsigma,$$

$$b_{32}^0 = (\lambda\rho_0)^{-1}\rho\varsigma b_{12}^0 + \lambda^{-1},$$

$$b_{34}^0 = \rho_0^{-1}\Psi_1[\vartheta] + m(1 + \rho_0^{-1}\varphi)$$

$$b_{30}^0 = \rho_0^{-1}\varsigma(\vartheta_0 - m\rho) + \Psi[\vartheta] - m(1 - \varsigma - \rho_0^{-1}\varsigma\varphi),$$

$$\mathscr{C}_0(\psi, \mathbf{w}) = R\psi\mathbf{u}\nabla\mathbf{u} + R\rho\mathbf{w}\nabla\mathbf{u}, + R\rho\mathbf{u}\nabla\mathbf{w}, \qquad (74)$$

$$\mathscr{D}_0(\mathbf{D}) = R\mathbf{u}\nabla(\mathbf{D}\mathbf{u}) + RD^*(\mathbf{u}\nabla\mathbf{u}) \qquad (75)$$

$$+ \operatorname{div}\left((\mathbf{D} + \mathbf{D}^*)\nabla\mathbf{u} - \frac{1}{2}\mathrm{Tr}\mathbf{D}\nabla\mathbf{u}\right) - \mathbf{D}\Delta\mathbf{u} - \Delta(\mathbf{D}\mathbf{u}).$$

The justification of the formal procedure meets serious problems, since the smoothness of solutions to problem (35)-(38), (46), (47) is not sufficient for the well-posedness of problem (72) in the standard weak formulation. In order to cope with this difficulty we define *very weak solutions* to problem (72). The construction of such solutions is based on the following lemma [19]. The lemma is given in $\mathbb{R}^d$, for our application $d = 3$.

Lemma 1. *Let $\Omega \subset \mathbb{R}^d$ be a bounded domain with the Lipschitz boundary, let exponents s and r satisfy the inequalities $sr > d$, $1/2 \le s \le 1$ and $\varphi, \varsigma \in H^{s,r}(\Omega) \cap H^{1,2}(\Omega)$, $\mathbf{w} \in \mathscr{H}_0^{1-s,r}(\Omega) \cap H_0^{1,2}(\Omega)$. Then there is a constant c depending only on s, r and Ω, such that the trilinear form*

$$\mathfrak{B}(\mathbf{w}, \varphi, \varsigma) = -\int_\Omega \varsigma \mathbf{w} \cdot \nabla\varphi\, dx \qquad (76)$$

satisfies the inequality

$$|\mathfrak{B}(\mathbf{w}, \varphi, \varsigma)| \le c\|\mathbf{w}\|_{\mathscr{H}_0^{1-s,r}(\Omega)}\|\varphi\|_{H^{s,r}(\Omega)}\|\varsigma\|_{H^{s,r}(\Omega)}, \qquad (77)$$

and can be continuously extended to $\mathfrak{B} : \mathscr{H}_0^{1-s,r}(\Omega)^d \times H^{s,r}(\Omega)^2 \to \mathbb{R}$. In particular, we have $\varsigma\nabla\varphi \in \mathscr{H}^{s-1,r}(\Omega)$ and $\|\varsigma\nabla\varphi\|_{H^{1-s,r}(\Omega)} \le c\|\varphi\|_{H^{s,r}(\Omega)}\|\varsigma\|_{H^{s,r}(\Omega)}.$

Definition 2. The vector field $\mathbf{w} \in \mathscr{H}_0^{1-s,r'}(\Omega)^3$, functionals $(\omega, \psi, \xi) \in \mathbb{H}^{-s,r'}(\Omega)^3$ and constant n are said to be a weak solution to problem (72), if $\langle \omega, 1 \rangle = 0$ and the identities

$$
\begin{aligned}
&\int_\Omega \mathbf{w}\left(\mathbf{H} - R\rho \nabla \mathbf{u} \cdot \mathbf{h} + R\rho \nabla \mathbf{h}^* \mathbf{u}\right) dx - \mathfrak{B}(\mathbf{w}, \varphi, \varsigma) - \mathfrak{B}(\mathbf{w}, \upsilon, \zeta) \\
&+ \left\langle \omega, G - b_{12}^0 \varsigma - b_{22}^0 g - \varkappa b_{32}^0 \right\rangle + \left\langle \psi, F - b_{11}^0 \varsigma - b_{21}^0 g - \varkappa b_{31}^0 - R\mathbf{u} \cdot \nabla \mathbf{u} \cdot \mathbf{h} \right\rangle \\
&+ \left\langle \xi, M - \varkappa b_{34}^0 \right\rangle + n\left(1 - \left\langle 1, b_{13}^0 \varsigma \right\rangle\right) \\
&= \left\langle \eth, b_{10}^0 \varsigma + b_{20}^0 g + \varkappa b_{30}^0 + \sigma \upsilon \right\rangle + \left\langle \mathscr{D}_0, \mathbf{h} \right\rangle.
\end{aligned}
\tag{78}
$$

hold true for all $(\mathbb{H}, G, F, M) \in (C^\infty(\Omega))^6$ such that $G = \Pi G$. Here $\eth = 1/2\,\mathrm{Tr}\,\mathbf{D}$, the test functions $\mathbf{h}, g, \varsigma, \upsilon$ are defined by the solutions to adjoint problems

$$
\Delta \mathbf{h} - \nabla g = \mathbb{H}, \quad \mathrm{div}\,\mathbf{h} = G, \quad \mathscr{L}^* \varsigma = F, \quad \mathscr{L} \upsilon = M \text{ in } \Omega,
\tag{79}
$$

$$
\mathbf{h} = 0 \text{ on } \partial\Omega, \quad \Pi g = g, \varsigma = 0 \text{ on } \Sigma_{\mathrm{out}}, \quad \upsilon = 0 \text{ on } \Sigma_{\mathrm{in}}.
\tag{80}
$$

We are now in a position to formulate the third main result of this paper.

Theorem 3. *Under the above assumptions,*

$$
\mathbf{w}_\varepsilon \to \mathbf{w} \text{ weakly in } \mathscr{H}_0^{1-s,r'}(\Omega), \quad n_\varepsilon \to n \text{ in } \mathbb{R},
$$

$$
\psi_\varepsilon \to \psi, \quad \omega_\varepsilon \to \omega, \quad \xi_\varepsilon \to \xi \, (*)\text{-weakly in } \mathbb{H}^{-s,r'}(\Omega) \text{ as } \varepsilon \to 0,
\tag{81}
$$

where the limits, vector field $\mathbf{w}$, functionals ψ, ω, ξ, and the constant n are given by the weak solution to problem (72).

Note that the matrices $\mathbf{N}(\varepsilon)$ defined by equalities (18) meet all requirements of Theorem 3, and in the special case we have in representation (19)

$$
\mathbf{D}(x) = \mathrm{div}\,\mathbf{T}(x)\,\mathbf{I} - \mathbf{T}'(x).
\tag{82}
$$

Therefore, Theorem 3, together with the formulae (15) and (32), imply the existence of the shape derivative for the drag functional at $\varepsilon = 0$. Straightforward calculations lead to the following result.

Theorem 4. *Under the assumptions of Theorem 3, there exists the shape derivative*

$$
\frac{d}{d\varepsilon} J_D(S_\varepsilon)\Big|_{\varepsilon=0} = L_e(\mathbf{T}) + L_u(\mathbf{w}, \omega, \psi),
\tag{83}
$$

where the linear forms L_e and L_u are defined by the equalities

$$
\begin{aligned}
L_e(\mathbf{T}) = &\int_\Omega \mathrm{div}\,\mathbf{T}(\nabla \mathbf{u} + \nabla \mathbf{u}^* - \mathrm{div}\,\mathbf{u}\mathbf{I})\mathbf{U}_\infty\, dx \\
&- \int_\Omega \left[\nabla \mathbf{u} + \nabla \mathbf{u}^* - \mathrm{div}\,\mathbf{u} - q\mathbf{I} - R\rho \mathbf{u} \otimes \mathbf{u}\right] \mathbf{D}\nabla \eta \cdot \mathbf{U}_\infty\, dx \\
&- \int_\Omega \left[\mathbf{D}^*\nabla \mathbf{u} + \nabla \mathbf{u}^*\mathbf{D} + \nabla(\mathbf{D}\mathbf{u}) + \nabla(\mathbf{D}\mathbf{u})^*\right]\nabla \eta \cdot \mathbf{U}_\infty\, dx
\end{aligned}
\tag{84}
$$

and

$$L_u(\mathbf{w}, \omega, \psi) = \int_\Omega \mathbf{w} \left[\Delta\eta\mathbf{U}_\infty + R\rho(\mathbf{u}\cdot\nabla\eta)\mathbf{U}_\infty + R\rho(\mathbf{u}\cdot\mathbf{U}_\infty)\nabla\eta \right] dx$$
$$+ \langle \omega, \nabla\eta\cdot\mathbf{U}_\infty \rangle + R\langle \psi, (\mathbf{u}\cdot\nabla\eta)(\mathbf{u}\cdot\mathbf{U}_\infty) \rangle. \tag{85}$$

While L_e depends directly on the vector field $\mathbf{T}$, the linear form L_u depends on the weak solution $(\mathbf{w}, \psi, \omega)$ to problem (72), thus depends on the *direction* $\mathbf{T}$ in a very implicit manner, which is inconvenient for applications. In order to cope with this difficulty, we define the *adjoint state* $\mathbf{Y} = (\mathbf{h}, g, \varsigma, \upsilon, l)^\top$ given as a solution to the linear equation

$$\mathfrak{L}\mathbf{Y} - \mathfrak{U}\mathbf{Y} - \mathfrak{V}\mathbf{Y} = \Theta, \tag{86}$$

supplemented with boundary conditions (80). Here the operators $\mathfrak{L}, \mathfrak{U}, \mathfrak{V}$ and the vector field Θ are defined by

$$\mathfrak{L} = \begin{pmatrix} \Delta & -\nabla & 0 & 0 & 0 \\ \operatorname{div} & 0 & 0 & 0 & 0 \\ 0 & 0 & \mathscr{L}^* & 0 & 0 \\ 0 & 0 & 0 & \mathscr{L} & 0 \\ 0 & 0 & -\mathbb{B}_{13} & 0 & 1 \end{pmatrix}, \tag{87}$$

$$\mathfrak{U} = \begin{pmatrix} 0 & 0 & -\nabla\varphi & -\varsigma\nabla & 0 \\ 0 & 0 & \Pi_{21} & 0 & 0 \\ 0 & 0 & 0 & 0 & 0 \\ 0 & 0 & 0 & 0 & 0 \\ 0 & 0 & 0 & 0 & 0 \end{pmatrix}, \tag{88}$$

$$\mathfrak{V} = \begin{pmatrix} R\rho(\nabla\mathbf{u} - \mathbf{u}\nabla) & 0 & 0 & 0 & 0 \\ 0 & -\lambda^{-1}\Pi & 0 & 0 & \varkappa\Pi b_{32}^0 \\ R\mathbf{u}\cdot\nabla\mathbf{u} & b_{12}^0 & b_{11}^0 & 0 & \varkappa b_{31}^0 \\ 0 & 0 & 0 & 0 & \varkappa b_{34}^0 \\ 0 & 0 & 0 & 0 & 0 \end{pmatrix}, \tag{89}$$

$$\Theta = (\Delta\eta\mathbf{U}_\infty + R\rho(\nabla\eta\otimes\mathbf{U}_\infty + \mathbf{U}_\infty\otimes\nabla\eta)\mathbf{u}, \Pi(\nabla\eta\cdot\mathbf{U}_\infty), R(\mathbf{u}\nabla\eta)(\mathbf{u}\mathbf{U}_\infty), 0, 0), \tag{90}$$

$$\Pi_{2i}(\cdot) = \Pi(b_{2i}^0(\cdot)), \quad \mathbb{B}_{13}(\cdot) = \langle 1, b_{13}^0(\cdot) \rangle. \tag{91}$$

The following theorem guarantees the existence of the adjoint state and gives the expression of the shape derivative for the drag functional in terms of the vector field $\mathbf{T}$.

Theorem 5. *Assume that a given solution* $\vartheta \in \mathscr{B}_\tau$, $(\zeta, m) \in X^{s,r} \times \mathbb{R}$ *to problem (35)-(38), (46), (47) meets all requirements of Theorem 2. Then there exists a positive constant* τ_1 *(depending only on* $\mathbf{U}, \Omega$ *and* r, s*) such that, if* $\tau \in (0, \tau_1]$ *and* $R\lambda^{-1} \leq \tau_1^2$, *then there exists a unique solution* $\mathbf{Y} \in (Y^{s,r})^3 \times (X^{s,r})^3 \times \mathbb{R}$ *to problem (86), (80). The form* L_u *has the representation*

$$L_u(\mathbf{w}, \psi, \omega) = \int_\Omega \left[\operatorname{div}\mathbf{T}(b_{10}^0\varsigma + b_{20}^0 g + \sigma\upsilon + \varkappa b_{30}^0 l) + \mathscr{D}_0(\operatorname{div}\mathbf{T} - \mathbf{T}')\mathbf{h} \right] dx \tag{92}$$

where the coefficients b_{ij}^0 *and the operator* $\mathscr{D}_0$ *are defined by the formulae (73), (75).*

References

1. Bello, J.A., Fernandez-Cara, E., Lemoine, J., Simon, J.: The differentiability of the drag with respect to variations of a Lipschitz domain in a Navier-Stokes flow. SIAM J. Control. Optim. 35(2), 626–640 (1997)
2. Feireisl, E.: Dynamics of Viscous Compressible Fluids. Oxford University Press, Oxford (2004)
3. Feireisl, E., Novotný, A.H., Petzeltová, H.: On the domain dependence of solutions to the compressible Navier-Stokes equations of a barotropic fluid. Math. Methods Appl. Sci. 25(2), 1045–1073 (2002)
4. Feireisl, E.: Shape optimization in viscous compressible fluids. Appl. Math. Optim. 47, 59–78 (2003)
5. Heywood, J.G., Padula, M.: On the uniqueness and existence theory for steady compressible viscous flow. In: Fundamental Directions in Mathematical Fluids Mechanics. Adv. Math. Fluids Mech., pp. 171–189. Birkhauser, Basel (2000)
6. Kweon, J.R., Kellogg, R.B.: Compressible Navier-Stokes equations in a bounded domain with inflow boundary condition. SIAM J. Math. Anal. 28(1), 94–108 (1997)
7. Kweon, J.R., Kellogg, R.B.: Regularity of solutions to the Navier-Stokes equations for compressible barotropic flows on a polygon. Arch. Ration. Mech. Anal. 163(1), 36–64 (2000)
8. Lions, P.L.: Mathematical Topics in Fluid Dynamics. Compressible Models, vol. 2. Oxford Science Publication, Oxford (1998)
9. Mohammadi, B., Pironneau, O.: Shape optimization in fluid mechanics. Ann. Rev. Fluid Mech. 36, 255–279 (2004)
10. Novotný, A., Padula, M.: Existence and uniqueness of stationary solutions for viscous compressible heat conductive fluid with large potential and small non-potential external forces. Siberian Math. Journal 34, 120–146 (1993)
11. Novotný, A., Straškraba, I.: Introduction to the Mathematical Theory of Compressible Flow. Oxford Lecture Series in Mathematics and its Applications, vol. 27. Oxford University Press, Oxford (2004)
12. Oleinik, O.A., Radkevich, E.V.: Second Order Equation with Non-negative Characteristic Form. American Math. Soc., Providence. Plenum Press, New York (1973)
13. Padula, M.: Existence and uniqueness for viscous steady compressible motions. Arch. Rational Mech. Anal. 97(1), 1–20 (1986)
14. Plotnikov, P.I., Sokołowski, J.: On compactness, domain dependence and existence of steady state solutions to compressible isothermal Navier-Stokes equations. J. Math. Fluid Mech. 7(4), 529–573 (2005)
15. Plotnikov, P.I., Sokołowski, J.: Concentrations of solutions to time-discretizied compressible Navier -Stokes equations. Communications in Mathematical Phisics 258(3), 567–608 (2005)
16. Plotnikov, P.I., Sokołowski, J.: Stationary boundary value problems for Navier-Stokes equations with adiabatic index $\nu < 3/2$. Doklady Mathematics 70(1), 535–538 (2004); Translated from Doklady Akademii Nauk 397(1-6) (2004)
17. Plotnikov, P.I., Sokołowski, J.: Domain dependence of solutions to compressible Navier-Stokes equations. SIAM J. Control Optim. 45(4), 1147–1539 (2006)
18. Plotnikov, P.I., Sokołowski, J.: Stationary solutions for Navier-Stokes equations for diatomic gases. Russian Mathematical Surveys 62(3) (2007) (RAS, Uspekhi Mat. Nauk 62:3 117148)
19. Plotnikov, P.I., Ruban, E.V., Sokołowski, J.: Inhomogeneous boundary value problems for compressible Navier-Stokes equations: well-posedness and sensitivity analysis. SIAM J. Math. Analysis 40(3), 1152–1200 (2008)
20. Plotnikov, P.I., Ruban, E.V., Sokołowski, J.: Inhomogeneous boundary value problems for compressible Navier-Stokes and transport equations. Journal de Mathématiques Pures et Appliquées (to appear)

21. Plotnikov, P.I., Sokołowski, J.: Stationary boundary value problems for compressible Navier-Stokes equations. In: Chipot, M. (ed.) Handbook of Differential Equations, vol. 6, pp. 313–410. Elsevier, Amsterdam (2008)
22. Schlichting, H.: Boundary-layer Theory. McGraw-Hill Series in Mechanical Engineering. McGraw-Hill, New York (1955)
23. Simon, J.: Domain variation for drag in Stokes flow. In: Control Theory of Distributed Parameter Systems and Applications. LNCIS, vol. 159, pp. 28–42. Springer, Heidelberg (1991)
24. Slawig, T.: A formula for the derivative with respect to domain variations in Navier-Stokes flow based on an embedding domain method. SIAM J. Control Optim. 42(2), 495–512 (2003)
25. Sokołowski, J., Zolésio, J.-P.: Introduction to Shape Optimization. Shape Sensitivity Analysis. Springer Series in Computational Mathematics, vol. 16. Springer, Heidelberg (1992)

Conservative Control Systems Described
by the Schrödinger Equation

Salah E. Rebiai

Department of Mathematics, Faculty of Sciences,
University of Batna, 05000 Batna, Algeria
rebiai@hotmail.com

Abstract. An important subclass of well-posed linear systems is formed by the conservative systems. A conservative system is a system for which a certain energy balance equation is satisfied both by its trajectories and those of its dual system. In Malinen et al. [10], a number of algebraic characterizations of conservative linear systems are given in terms of the operators appearing in the state space description of the system. Weiss and Tucsnak [20] identified by a detailed argument a large class of conservative linear systems described by a second order differential equation in a Hilbert space and an output equation, and they may have unbounded control and observation operators. In this paper, we give two examples of conservative linear control systems described by the linear Schrödinger equation on an n-dimensional domain with boundary control and boundary observation. These examples do not fit into the framework of [20].

1   Introduction

Abstract representation of particular classes of infinite dimensional systems has received considerable attention in the literature. For partial differential equations subject to control either acting on the boundary of, or else as a point control within, a multidimensional bounded domain we refer to Balakrishnan [1] and Washburn [17] for parabolic problems and to Triggiani [15], Lasiecka and Triggiani [4], Flandoli et al. [3] for hyperbolic and Petrowsky-type problems. For functional differential equations with delays in control and observation, see Salamon [12] and Delfour [2]. In [13], Salamon introduced the class of well-posed infinite dimensional linear systems. The aim was to provide a unifying abstract framework to formulate and solve control problems for systems described by functional and partial differential equations. Roughly speaking, a well-posed linear system is a linear time invariant system such that on any finite time interval, the operator from the initial state and the input function to the final state and the output function is bounded. An important subclass of well-posed linear systems is formed by the conservative systems. A conservative system is a system for which a certain energy balance equation is satisfied both by its trajectories and those of its dual system. In Malinen et al. [10], a number of algebraic characterizations of conservative linear systems are given in terms of the operators appearing in the state space description of the system.

Weiss and Tucsnak [20] identified by a detailed argument a large class of conservative linear systems described by a second order differential equation in a Hilbert space

A. Korytowski et al. (Eds.): System Modeling and Optimization, IFIP AICT 312, pp. 448–458, 2009.
© IFIP International Federation for Information Processing 2009

and an output equation, and they may have unbounded control and observation operators.

In this paper, we give two examples of conservative control systems described by the linear Schrödinger equation with boundary control and boundary observation. These examples do not fit into the framework of [20].

The following notations are used. Let X be a Hilbert space, then:

- $C^n(0,+\infty;X)$: the space of n times continuously differentiable X-valued functions on $[0,+\infty)$, $n \in \mathbb{N}$.
- $BC^n(0,+\infty;X)$: the space of those $f \in C^n(0,+\infty;X)$ for which $f, f', ..., f^{(n)}$ are all bounded on $[0,+\infty)$.
- $H^n(0,+\infty;X)$: the Sobolev space of X-valued functions with $n \in \mathbb{N}$.

2   Some Concepts from Conservative Well-Posed Linear Systems

In this section, we first gather some basic facts about admissible control and observation operators and about well-posed linear systems. For more details we refer to Falandoli et al. [3] (though here the expressions "admissible" and "well-posed" are not stated explicitly), Salamon [13], Weiss [18] and Staffans [14]. Then, we recall the definition and some basic properties of conservative systems, see Weiss et al. [19], Tucsnak and Weiss [16], Weiss and Tucsnak [20] for further details.

Let X be a Hilbert space and $A: D(A) \to X$ be the generator of a strongly continuous semigroup $\mathbb{S}$ on X. We define the Hilbert space X_1 as $D(A)$ with the norm $\|y\|_1 = \|(\beta I - A)y\|$ where $\beta \in \rho(A)$ is fixed. The Hilbert space X_{-1} is the completion of X with respect to the norm $\|y\|_{-1} = \|(\beta I - A)^{-1}y\|$. It is known that $X_{-1} = D(A^*)'$, the dual space with respect to the pivot space X. We have

$$X_1 \subset X \subset X_{-1} \tag{1}$$

with continuous dense injections.

$\mathbb{S}$ extends to a semigroup on X_{-1} denoted by the same symbol. The generator of this semigroup is an extension of A whose domain is X so that $A: X \to X_{-1}$.

Definition 1. Let U be a Hilbert space. The operator B is said to be an admissible control operator of $\mathbb{S}$ if the input maps $\{\Phi_t\}_{t \geq 0}$ are bounded from $L^2(0,+\infty;U)$ to X for all finite $t \geq 0$ where

$$\Phi_t u := \int_0^t \mathbb{S}(t-\tau)Bu(\tau)d\tau. \tag{2}$$

If y is a solution of

$$y'(t) = Ay(t) + Bu(t), \, t \geq 0, \tag{3}$$

which is an equation in X_{-1} with $y(0) \in X$ and $u \in L^2(0,+\infty;U)$, then $y(t) \in X$ for all $t \geq 0$. In this case y is a continuous X-valued function of t and we have for all $t \geq 0$

$$y(t) = \mathbb{S}(t)y(0) + \int_0^t \mathbb{S}(t-\tau)Bu(\tau)d\tau. \tag{4}$$

The admissible control operator is called infinite-time admissible if for any $u \in L^2(0, +\infty; U)$ the map

$$t \rightarrow \Phi_t u \tag{5}$$

from $[0, +\infty)$ to X is bounded.

Definition 2. Let Y be another Hilbert space. The operator $C \in \mathscr{L}(X_1, Y)$ is called an admissible observation operator for $\mathbb{S}$ if for every $t > 0$ there exists a $k_t \geq 0$ such that

$$\int_0^t \|C\mathbb{S}(\tau)y_0\|^2 d\tau \leq k_t \|y_0\|^2, \ \forall y_0 \in D(A). \tag{6}$$

The admissibility of C means that there is a continuous operator

$$\Psi : X \rightarrow L_{loc}^2(0, +\infty; Y) \tag{7}$$

such that

$$(\Psi y_0)(t) = C\mathbb{S}(t)y_0, \ \forall y_0 \in D(A). \tag{8}$$

The operator Ψ is completely determined by (8) because $D(A)$ is dense in X.

C is said to be an infinite-time admissible observation operator for $\mathbb{S}$ if there exists $K > 0$ such that

$$\int_0^{+\infty} \|C\mathbb{S}(\tau)y_0\|^2 d\tau \leq K \|y_0\|^2, \ \forall y_0 \in D(A). \tag{9}$$

The following duality result holds.

Theorem 1. *(Salamon [13], Staffans [14]) C is an (infinite-time) admissible observation operator for $\mathbb{S}$ if and only if C^* is an (infinite-time) admissible control operator for the adjoint semigroup $\mathbb{S}^*$.*

Remark 1. In view of the above theorem, we see that Definition 1 is equivalent to the Hypothesis (H1) in Flandoli et al. [3].

Remark 2. For PDE systems with boundary control such as the multidimensional Schrödinger equation with Dirichlet control, the admissibility of B is a sharp trace regularity result not obtainable by the standard trace theory. It is established by PDE hard analysis energy methods (see Lasiecka and Triggiani [5] and [6] and the references therein).

Theorem 2. *If $B \in \mathscr{L}(U, X_{-1})$ is an admissible control operator for $\mathbb{S}$ and $C \in \mathscr{L}(X_1, Y)$ is an admissible operator for $\mathbb{S}$, then the transfer functions of the system Σ given by the triple (A, B, C) are solutions of*

$$G : \rho(A) \rightarrow \mathscr{L}(U, Y) \tag{10}$$

of

$$G(s) - G(\beta) = -(s - \beta)C(sI - A)^{-1}(\beta I - A)^{-1}B \tag{11}$$

for s and β in $\rho(A)$.

Definition 3. The system Σ given by the triple (A,B,C) is said to be a well-posed linear system if $B \in \mathcal{L}(U,X_{-1})$ is an admissible control operator for $\mathbb{S}$ and $C \in \mathcal{L}(X_1,Y)$ is an admissible operator for $\mathbb{S}$, and its transfer functions are bounded on some right-half plane.

For a well-posed linear system the operator Σ_t from the initial state and the input function to the final state and the output function is bounded on any finite time interval $[0,t]$. The input and output functions u and z are locally L^2 with values in U and Y respectively. The state trajectory y is an X-valued function.

The boundedness property mentioned above means that for every $t > 0$, there is a $c_t > 0$ such that

$$\|y(t)\|^2 + \int_0^t \|z(\tau)\|^2 d\tau \le c_t(\|y(0)\|^2 + \int_0^t \|u(\tau)\|^2 d\tau). \tag{12}$$

Remark 3. The class of well-posed linear systems includes many systems described by partial differential equations or delay differential equations. The formal resemblance to finite dimensional systems is one of its main advantages. Much work has been done on this class of systems, see Staffans [14] and the references therein. There are however important systems that do not belong to this class, see Lasiecka and Triggiani [6,7].

Definition 4. A well-posed linear system Σ given by the triple (A,B,C) is called conservative if for every $t \ge 0$, the operator

$$\Sigma_t : X \times L^2(0,t;U) \to X \times L^2(0,+\infty;Y) \tag{13}$$

is unitary. This means that for every $t \ge 0$, the following statements are true:

(i) Σ_t is an isometry, i.e.

$$\|y(t)\|^2 + \int_0^t \|z(\tau)\|^2 d\tau = \|y(0)\|^2 + \int_0^t \|u(\tau)\|^2 d\tau \tag{14}$$

(ii) Σ_t is onto.

Theorem 3. *(Weiss and Tucsnak [20]) The system Σ is conservative if and only if the balance equation (14) or its differential form*

$$\frac{d}{dt}\|y(t)\|^2 = \|u(t)\|^2 - \|z(t)\|^2 \tag{15}$$

holds for all state trajectories of Σ as well as for all state trajectories of the dual system Σ_d for suitable initial state and input function.

In Tucsnak and Weiss [16], the authors investigated conditions under which such systems are exponentially stable or strongly stable. It turns out that these properties are equivalent to certain controllability and observability properties.

3   The Schrödinger Equation with Dirichlet-Type Boundary Feedback

Let Ω be an open bounded domain in $\mathbb{R}^n$ with C^2-boundary $\Gamma = \overline{\Gamma_0 \cup \Gamma_1}$ where Γ_0 and Γ_1 are disjoint parts of Γ with $\Gamma_1 \neq \varnothing$.

Let $G: H^{-1}(\Omega) \to H_0^1(\Omega)$ be the operator defined by:

$$Gf = \varphi \text{ if and only if } \varphi \in H_0^1(\Omega) \text{ and } -\Delta\varphi = f. \tag{16}$$

We consider the system described by the equations

$$y_t(x,t) = i\Delta y(x,t) \text{ in } \Omega \times (0,+\infty) \tag{17}$$

$$y(x,0) = y_0(x) \text{ in } \Omega \tag{18}$$

$$y = 0 \text{ on } \Gamma_0 \times (0,+\infty) \tag{19}$$

$$y + \frac{i}{2}\frac{\partial(Gy)}{\partial v} = u \text{ on } \Gamma_1 \times (0,+\infty). \tag{20}$$

In (20), $v = (v_1,...,v_n)$ is the unit outward normal on Γ.

The input of this system is the function u in (20). The output associated with this system is

$$z = y - \frac{i}{2}\frac{\partial(Gy)}{\partial v} \text{ on } \Gamma_1 \times (0,+\infty). \tag{21}$$

Remark 4. The system (17)-(20) without the term $i\frac{\partial(Gy)}{\partial v}$ has been considered in Lasiecka and Triggiani [5] where $H^{-1}(\Omega)$ is identified as the space of optimal regularity and exact controllability and uniform stabilization results have been established on this space, the latter via the dissipative feedback $u = -i\frac{\partial(Gy)}{\partial v}$.

The precise statement of well-posedness and conservativity of the system described by (17)-(21) is given in the following theorem.

Theorem 4. *The equations (17)-(21) determine a conservative linear system Σ with input and output space $U = L^2(\Gamma_1)$ and state space $X = H^{-1}(\Omega)$.*
If

$$y_0 \in Z_D = \{f \in L^2(\Omega): \Delta f \in H^{-1}(\Omega), f|_\Gamma \in L^2(\Gamma) \text{ and } f = 0 \text{ on } \Gamma_0\} \tag{22}$$

and the compatibility condition

$$y_0(x) + \frac{i}{2}\frac{\partial(Gy_0)}{\partial v}(x) = u(x,0) \text{ for } x \in \Gamma_1 \tag{23}$$

holds, then (17)-(21) have a unique solution y,z satisfying

$$y \in BC(0,+\infty;Z_D) \cap BC^1(0,+\infty;H^{-1}(\Omega)), \ z \in H^1(0,+\infty;U). \tag{24}$$

Proof. We proceed as in Lasiecka and Triggiani [5] to rewrite the system (17)-(21) into an abstract form. Let $A_D: D(A_D) \to X$ be defined by

$$D(A_D) = H_0^1(\Omega) \text{ and } A_D\varphi = -\Delta\varphi, \ \forall\varphi \in D(A_D) \tag{25}$$

A_D is self-adjoint, positive and boundedly invertible.

Let $D \in \mathscr{L}(L^2(\Gamma_1), L^2(\Omega))$ be the Dirichlet map given by

$$g = Dv \Longleftrightarrow \{ \Delta g = 0 \text{ in } \Omega,\ g|_{\Gamma_0} = 0,\ g|_{\Gamma_1} = v \} \tag{26}$$

Define the operator $B \in \mathscr{L}(U, D(A_D)')$, by

$$Bu = iA_D Du \tag{27}$$

Then

$$B^* \varphi = iD^* \varphi = i\frac{\partial (G\varphi)}{\partial v} \tag{28}$$

Using these operators, we can formulate (17)-(21) as an abstract system of the form

$$y'(t) = (-iA_D - \frac{1}{2}BB^*)y(t) + Bu(t) \tag{29}$$

$$y(0) = y_0 \tag{30}$$

$$z(t) = -B^*y(t) + u(t) \tag{31}$$

To continue we need the following results.

Lemma 1. *(Lasiecka and Triggiani [5]) The operator $A_1 = -iA_D - \frac{1}{2}BB^*$ is the generator of a strongly continuous contraction semigroup $\mathbb{S}$ on X.*

Lemma 2. *(Salamon [13], Weiss & Tucsnak [20]) Let $u \in H^2(0, +\infty; U)$ and $y_0 \in X$ satisfy the compatibility condition*

$$A_1 y_0 + Bu(0) \in X. \tag{32}$$

Then the initial value problem (29), (30) has a unique solution y given by

$$y(t) = \mathbb{S}(t)y_0 + \int_0^t \mathbb{S}(t - \tau)Bu(\tau)d\tau, \tag{33}$$

which satisfies

$$y \in C^1(0, +\infty; X) \cap C(0, +\infty; Z), \tag{34}$$

where

$$Z = D(A_1) + (\beta I - A)^{-1}BU. \tag{35}$$

Lemma 3. *Under the assumptions of Lemma 2, we have the identity*

$$\frac{d}{dt}\|y(t)\|^2 = \|u(t)\|^2 - \|z(t)\|^2 \tag{36}$$

Proof. Taking the inner product of both sides of (29) with $y(t)$, we obtain

$$\frac{d}{dt}\|y(t)\|^2 = 2Re\langle A_1 y(t) + Bu(t), y(t)\rangle. \tag{37}$$

From the expression of A_1, we obtain

$$\frac{d}{dt}\|y(t)\|^2 = -\|B^*u(t)\|^2 + 2Re\langle Bu(t), y(t)\rangle \tag{38}$$

Using now the formula

$$\|z(t)\|^2 = \|B^*u(t)\|^2 - 2Re\langle Bu(t), y(t)\rangle + \|u(t)\|^2 \tag{39}$$

we get the desired identity.   □

Lemma 4. *B is an infinite-time admissible control operator for* $\mathbb{S}$.

Proof. Suppose first that

$$u \in H_L^2(0, +\infty; U) = \{v \in H^2(0, +\infty; U); v(0) = 0\} \tag{40}$$

and define $y(t) = \Phi_t u$ for all $t \geq 0$. Then $A_1 y(0) + Bu(0) \in X$.

It follows from Lemma 2, that $y(.) \in C^1(0, +\infty; X)$. Integrating the identity in Lemma 3 on $[0, t]$ we get that for all $u \in H_L^2(0, +\infty; U)$

$$\|\Phi_t u\|^2 \leq \int_0^t \|u(\tau)\|^2 d\tau \leq \|u\|_{L^2(0, +\infty; U)}^2. \tag{41}$$

Since $H_L^2(0, +\infty; U)$ is dense in $L^2(0, +\infty; U)$, we conclude that B is infnite-time admissible.   □

Lemma 5. *Let C be the restriction of* $-B^*$ *to* $D(A_1)$. *Then C is an infinite-time admissible operator for* $\mathbb{S}$.

Proof. Let $y_0 \in D(A_1)$ and take $y(t) = \mathbb{S}(t)y_0$. It follows from Lemma 2 that $y \in C^1(0, +\infty; X)$. Integrating (36) with $u = 0$ on $[0, t]$, we get

$$\int_0^t \|z(\tau)\|^2 d\tau \leq \|y_0\|^2. \tag{42}$$

But $z(t) = -B^*y(t) = Cy(t)$. Therefore the estimate

$$\int_0^t \|C\mathbb{S}(\tau)y_0\|^2 d\tau \leq \|y_0\|^2 \tag{43}$$

holds for every $t > 0$.   □

From the previous results, we deduce that the equations (17)-(21) define a well-posed linear system Σ with state space $X = H^{-1}(\Omega)$ and input and output space $U = L^2(\Gamma_1)$.

The dual system of Σ denoted by Σ_d is described by

$$y_d'(t) = (iA_D - \frac{1}{2}BB^*)y_d(t) - Bu_d(t) \tag{44}$$

$$z(t) = B^*y_d(t) + u_d(t) \tag{45}$$

where $u_d(t)$, $y_d(t)$ and $z_d(t)$ are the input, state and output of Σ_d at some $t \geq 0$.

Proceeding as in the proof of Lemma 3, one can show that the state trajectory and the output function of Σ_d satisfy the energy balance equation (36). Hence the conservativity of Σ.

Now, notice the following:

- The space $Z = D(A_1) + (\beta I - A)^{-1}D$ introduced in Lemma 1 is given in this case by $Z = D(A_D) + DU$ and coincides with Z_D.
- The condition (23) is equivalent to $A_1 y(0) + Bu(0) \in X$.
- The transfer function $G(s)$ of Σ satisfies, because of the conservativity of Σ,

$$\|G(s)\| \leq 1 \text{ for all } s \in \mathbb{C}_0. \tag{46}$$

Using these facts together with Lemma 1, Lemma 4 and Lemma 5, we conclude that for every $y_0 \in Z_D$ and every $u \in H^1(0, +\infty; U)$ such that

$$y_0(x) + \frac{i}{2}\frac{\partial(Gy_0)}{\partial v}(x) = u(x,0) \text{ for } x \in \Gamma_1 \tag{47}$$

the state trajectory y and the output function z satisfy the smoothness boundedness conditions

$$y \in BC(0, +\infty; Z_D) \cap BC^1(0, +\infty; H^{-1}(\Omega)), \ z \in H^1(0, +\infty; U). \tag{48}$$

(Proposition 4.6 in Weiss and Tucsnak [20]).   □

Remark 5. Assume the following additional condidtions on the triple $\{\Omega, \Gamma_0, \Gamma_1\}$: there exists a real vector field $h(x) \in [C^1(\overline{\Omega})]^n$ such that

$$Re(\int_{\Omega} H(x)v(x).\overline{v(x)}d\Omega) \geq \rho \int_{\Omega} |v(x)|^2 d\Omega, \forall v \in [L^2(\Omega)]^n \text{ for some } \rho > 0,$$
$$\text{where } H(x) = (\frac{\partial h_i(x)}{\partial x_j}), \ i,j = 1,...,n \tag{49}$$

and

$$h.v \leq 0 \text{ on } \Gamma_0. \tag{50}$$

Then the semigroup $\mathbb{S}$ is exponentially stable (Lasiecka and Triggiani [5]). Thus, from Russell [11], Tucsnak and Weiss [16], we conclude that the pair (A_1, B) is exactly controllable in finite time and the pair (A_1, C) is exactly observable in finite time.

4 The Schrödinger Equation with Neumann-Type Boundary Feedback

In this section, we suppose that the boundary Γ is of class C^2 and satisfies

$$\Gamma = \Gamma_0 \cup \Gamma_1 \text{ with } \overline{\Gamma_1} \cap \overline{\Gamma_0} = \emptyset \tag{51}$$

where both Γ_0 and Γ_1 are nonempty.

Let $a(.)$ be an $L^\infty(\Gamma_1)$ such that $a(x) \neq 0$ for all $x \in \Gamma_1$.
We are interested in the linear system described by

$$y_t(x,t) = i\Delta y(x,t) \text{ in } \Omega \times (0,+\infty) \tag{52}$$

$$y(x,0) = y_0(x) \text{ in } \Omega \tag{53}$$

$$y(x,t) = 0 \text{ on } \Gamma_0 \times (0,+\infty) \tag{54}$$

$$\frac{\partial}{\partial v}y(x,t) - i|a(x)|^2 y(x,t) = \sqrt{2}a(x)u(x,t) \text{ on } \Gamma_1 \times (0,+\infty) \tag{55}$$

$$\frac{\partial}{\partial v}y(x,t) + i|a(x)|^2 y(x,t) = \sqrt{2}a(x)z(x,t) \text{ on } \Gamma_1 \times (0,+\infty) \tag{56}$$

Remark 6. Lasiecka et al. [9] have considered the system (52)-(55) with $u = 0$. They have proved under an additional assumption on the triple $\{\Omega, \Gamma_0, \Gamma_1\}$ that the energy decays exponentially to zero in the uniform topology of $L^2(\Omega)$.

Theorem 5. *The equations (52)-(56) determine a conservative linear system Σ with input and output space $U = L^2(\Gamma_1)$ and state space $X = L^2(\Omega)$. If*

$$y_0 \in Z_N = \{f \in H^1_{\Gamma_0}(\Omega) : \Delta f \in L^2(\Omega), \frac{\partial f}{\partial v} \in aL^2(\Gamma_1)\}, u \in H^1(0,+\infty;U)\} \tag{57}$$

and the compatibility condition

$$\frac{\partial}{\partial v}y(x,0) - i|a(x)|^2 y(x,0) = \sqrt{2}a(x)u(x,0) \text{ for } x \in \Gamma_1 \tag{58}$$

holds. Then (52)-(56) have a unique solution y, z satisfying

$$y \in BC(0,+\infty;Z_N) \cap BC^1(0,+\infty;L^2(\Omega)), z \in H^1(0,+\infty;U). \tag{59}$$

Proof. The equations (52)-(56) can be written as an abstract system of the form (see Lasiecka et al. [9] and Lasiecka and Triggiani [8])

$$y'(t) = A_2 y(t) + Bu(t) \tag{60}$$

$$y(0) = y_0 \tag{61}$$

$$z(t) = -B^* y(t) + u(t) \tag{62}$$

where

- $A_2 = -iA_N - \frac{1}{2}BB^*$;
- The operator $A_N : D(A_N) \rightarrow L^2(\Omega)$ is defined by

$$A_N\varphi = -\Delta\varphi, \; \varphi \in D(A_N) = \{\varphi \in Z_N : \frac{\partial\varphi}{\partial v} = 0 \text{ on } \Gamma_1\}; \tag{63}$$

- $B \in \mathscr{L}(U;D(A_2)')$ is defined by

$$Bu = i\sqrt{2}A_N N(a(x)u); \tag{64}$$

- $N \in \mathscr{L}(L^2(\Gamma_1);L^2(\Omega))$ is the Neuman map given by

$$g = Nv \Longleftrightarrow \{\Delta g = 0 \text{ in } \Omega, g|_{\Gamma_0} = 0, \frac{\partial g}{\partial v}\Big|_{\Gamma_1} = v\}; \tag{65}$$

- $B^*\varphi = -i\sqrt{2}\bar{a}(x)\varphi.$

Now, the theorem can be established by following the steps of the proof of Theorem 4. □

Remark 7. Following Lasiecka et al. [9] and Lasiecka and Triggiani [8], the assertions of Remark 5 hold also for the system (52)-(56).

Acknowledgements. The author is indebted to the anonymous referee for the constructive criticisms and suggestions that led to a significant improvement of the paper.

References

1. Balakrishnan, A.V.: Applied Functional Analysis, 2nd edn. Springer, Heidelberg (1981)
2. Delfour, M.C.: The linear quadratic optimal control problem with delays in the state and control variables: a state space approach. SIAM J. Control Optim. 24, 835–883 (1986)
3. Flandoli, F., Lasiecka, I., Triggiani, R.: Algebraic Riccati equations with non-smoothing observation arising in hyperbolic and Euler-Bernoulli boundary control problems. Annali Matem. Pura Appl. CLii, 307–382 (1988)
4. Lasiecka, I., Triggiani, R.: A cosine operator approach to modelling $L_2(0, T; L_2(\Gamma))$ boundary input hyperbolic equations. Appl. Math. Optimiz. 7, 35–83 (1981)
5. Lasiecka, I., Triggiani, R.: Optimal regularity, exact controllability and uniform stabilization of Schrödinger equations with Dirichlet control. Different. and Integral Eqs. 5, 521–535 (1992)
6. Lasiecka, I., Triggiani, R.: $L_2(\Sigma)$-regularity of the boundary to boundary operator B^*L for hyperbolic and Petrowski PDEs. Abstract and Applied Analysis 19, 1061–1139 (2003)
7. Lasiecka, I., Triggiani, R.: The operator B^*L for the wave equation with Dirichlet control. Abstract and Applied Analysis 20, 625–634 (2004)
8. Lasiecka, I., Triggiani, R.: Well-posedness and sharp uniform decay rates at the $L_2(\Omega)$-Level of the Schrödinger equation with nonlinear boundary dissipation. J. Evol. Equ. 6, 485–537 (2006)
9. Lasiecka, I., Triggiani, R., Zhang, X.: Global uniqueness, observability and stabilization of non-conservative Schrödinger equation via pointwise Carleman estimates. Part II: $L^2(\Omega)$-estimates. J. Inverse Ill-Posed Problems 12, 1–49 (2004)
10. Malinen, J., Staffans, O.J., Weiss, G.: When is a linear system conservative? Quart. Appl. Math. 64, 61–91 (2006)
11. Russell, D.L.: Exact boundary value controllability theorems for wave and heat processes in star complemented regions. In: Roxin, E.O., Sternberg, L. (eds.) Differential Games and Control Theory. Marcel Dekker, New York (1974)
12. Salamon, D.: Control and Observation of Neutral Systems. Pitman, London (1984)
13. Salamon, D.: Infinite dimensional linear systems with unbounded control and observation: a functional analytic approach. Trans. Amer. Math. Soc. 300, 383–431 (1987)
14. Staffans, O.: Well-posed linear systems. Cambridge University Press, Cambridge (2005)
15. Triggiani, R.: A cosine operator approach to modeling $L_2(0, T; L_2(\Gamma))$-boundary input problems for hyperbolic systems. In: Proceedings of the 8th Conference on Differential Equations and Optimization Techniques. LNCIS, vol. 6, pp. 380–390. University of Würzburg, Germany (1977)
16. Tucsnak, M., Weiss, G.: How to get a conservative well-posed linear system out of thin air. Part II. Controllability and stability. SIAM J. Control Optim. 42, 907–935 (2003)

17. Washburn, D.: A bound on the boundary input map for parabolic equations with applications to time optimal control. SIAM J. Control Optim. 17, 652–671 (1979)
18. Weiss, G.: Regular linear systems with feedback. Mathematics Control, Signal & Systems 7, 25–57 (1994)
19. Weiss, G., Staffans, O., Tucsnak, M.: Well-posed linear systems – a survey with emphasis on conservative systems. Internat. J. Appl. Math. Comput. Sci. 11, 7–34 (2001)
20. Weiss, G., Tucsnak, M.: How to get a conservative well-posed linear system out of thin air. Part I. Well-posedness and energy balance. ESAIM. Control, Optim. Calc. Var. 9, 247–274 (2003)

Topological Derivatives in Plane Elasticity

Jan Sokołowski[1] and Antoni Żochowski[2,*]

[1] Institut Elie Cartan, Laboratoire de Mathématiques,
Université Henri Poincaré Nancy 1,
B.P. 239, 54506 Vandoeuvre lés Nancy Cedex, France
Jan.Sokolowski@iecn.u-nancy.fr
[2] Systems Research Institute of the Polish Academy of Sciences,
ul. Newelska 6, 01-447 Warszawa, Poland
zochowsk@ibspan.waw.pl

Abstract. We present a method for construction of the topological derivatives in plane elasticity. It is assumed that a hole is created in the subdomain of the elastic body which is filled out with isotropic material. The asymptotic analysis of elliptic boundary value problems in singularly perturbed geometrical domains is used in order to derive the asymptotics of the shape functionals depending on the solutions to the boundary value problems. Our method allows for the asymptotic expansions of arbitrary order, since the explicit solutions to the boundary value problems are obtained by the method of elastic potentials. Some numerical results are presented to show the applicability of the proposed method in numerical analysis of elliptic problems.

1 Introduction

One of the most important applications of the topological derivatives of shape functionals is elasticity, in particular in the fields of optimal design in structural mechanics and the numerical solution for inverse problems of detection of small imperfections. The mathematical theory of asymptotic analysis of elliptic boundary value problems in singularly perturbed domains, is considered in [6] and [10]. The method of compound asymptotic expansions in the framework of the asymptotic analysis leads to the asymptotic expansions of solutions and to the topological derivatives of the shape functionals as it is described in details, e.g., in the paper [12] for boundary value problems of linearized elasticity. The concept of topological derivatives of shape functionals [18] is derived in the framework of the method of compound asymptotic expansions [10], one of the techniques used in the asymptotic analysis of the boundary value problems in singularly perturbed geometrical domains. The so-called truncation method is described, e.g., in [9] (see [3] for further developments). The asymptotic analysis in impedance imaging and the theory of composite materials can be found, e.g., in [1].

We present here the results on asymptotics of the shape functionals for the specific class of the elliptic boundary value problems. Let there be given an elastic body which occupies the reference domain $\Omega \subset \mathbb{R}^d$, $d = 2,3$, with the material properties defined by the Hooke's tensor $\mathscr{C}_{ijkl}$, $i,j,k,l = 1,\ldots,d$. We assume that there is a ball

---

* The research is supported by the grant N51402132/3135 of the Polish Ministry of Education.

A. Korytowski et al. (Eds.): System Modeling and Optimization, IFIP AICT 312, pp. 459–475, 2009.
© IFIP International Federation for Information Processing 2009

$B_R(x) \subset \Omega, R > 0$, with the center $x \in \Omega$, filled with an isotropic material characterized by its Lame coefficients λ, μ. We investigate the asymptotics for $\rho \to 0$ of the displacement and the stress fields in the body $\Omega_\rho = \Omega \setminus \overline{B_\rho(x)}$ due to the creation of a small hole $B_\rho(x) \subset B_R(x)$ of the radius $R > \rho \to 0$. We also perform the asymptotic analysis of some shape functionals depending on the solution of the elasticity boundary value problems in $\Omega_\rho = \Omega \setminus \overline{B_\rho(x)}$ for $\rho \to 0$. It seems that the imposed condition on the isotropy of $B_R(x)$ cannot be avoided since for the specific application of the existing methods of asymptotic analysis we need the knowledge of fundamental solution of the elliptic operator in the region $B_R(x)$. In order to obtain the required asymptotics in the whole domain we employ [21] a domain decomposition technique combined with the fine analysis of the properties of the Steklov-Poincaré operator $\mathscr{A}_\rho, \rho \geq 0$, defined in the ball $B_R(x)$ as well as in the ring $C(R, \rho) = B_R(x) \setminus \overline{B_\rho(x)}$.

The paper contains the complete mathematical tools which are used to derive the form of topological derivatives for the specific class of composite elastic materials in two spatial dimensions.

2   Topological Derivatives of Shape Functionals in Isotropic Elasticity

We are going to present the results which can be obtained for 2D boundary value problems of linear elasticity. The results for 3D are not in the same explicit form. The same type of results on topological derivatives is derived for the contact problems by means of the asymptotic analysis combined with the domain decomposition technique [21].

We briefly introduce the concept of the topological derivative for an arbitrary shape functional. The topological derivative denoted by $\mathscr{T}_\Omega$ of a shape functional $\mathscr{J}(\Omega)$ is introduced in [18] in order to characterize the infinitesimal variation of $\mathscr{J}(\Omega)$ with respect to the infinitesimal variation of the topology of the domain Ω. The topological derivative allows us to derive the new optimality condition in the interior of an optimal domain, if such a domain exists and if the shape functional under studies admits the topological derivatives, for the shape optimization problem:

$$\mathscr{J}(\Omega^*) = \inf_\Omega \mathscr{J}(\Omega). \tag{1}$$

The optimal domain Ω^* is characterized by the first order condition [17] defined on the boundary of the optimal domain Ω^*, $d\mathscr{J}(\Omega^*; V) \geq 0$ for all admissible vector fields V, and by the following optimality condition defined in the interior of the domain Ω^*:

$$\mathscr{T}_{\Omega^*}(x) \geq 0 \text{ in } \Omega^*. \tag{2}$$

The other use of the topological derivative is connected with approximating the influence of the holes in the domain on the values of integral functionals of solutions, which allows us, e.g., to solve a class of shape inverse problems.

In general terms the notion of the *topological* derivative (TD) has the following meaning. Assume that $\Omega \subset \mathbb{R}^N$ is an open set and that there is given a shape functional

$$\mathscr{J} : \Omega \setminus K \to \mathbb{R} \tag{3}$$

for any compact subset $K \subset \overline{\Omega}$. We denote by $B_\rho(x), x \in \Omega$, the ball of radius $\rho > 0$, $B_\rho(x) = \{y \in \mathbb{R}^N \mid \|y - x\| < \rho\}$, $\overline{B_\rho(x)}$ is the closure of $B_\rho(x)$, and assume that there exists the following limit

$$\mathcal{T}(x) = \lim_{\rho \downarrow 0} \frac{\mathcal{J}(\Omega \setminus \overline{B_\rho(x)}) - \mathcal{J}(\Omega)}{|B_\rho(x)|}. \tag{4}$$

The function $\mathcal{T}(x), x \in \Omega$, is called the topological derivative of $\mathcal{J}(\Omega)$, and provides the information on the infinitesimal variation of the shape functional $\mathcal{J}$ if a small hole is created at $x \in \Omega$. This definition is suitable for Neumann–type boundary conditions on ∂B_ρ.

In many cases this characterization is constructive [5,2,3,8,12,14,15], i.e. TD can be evaluated for shape functionals depending on solutions of partial differential equations defined in the domain Ω.

2.1   Problem Setting for Elasticity Systems

We introduce the elasticity system in a form convenient for the evaluation of topological derivatives. Let us consider the elasticity equations in $\mathbb{R}^N$, where $N = 2$ for 2D and $N = 3$ for 3D,

$$\begin{cases} \operatorname{div} \sigma(u) = 0 \text{ in } \Omega \\ u = g \text{ on } \Gamma_D \\ \sigma(u)n = T \text{ on } \Gamma_N \end{cases} \tag{5}$$

and the same system in the domain with the spherical cavity $B_\rho(x_0) \subset \Omega$ centered at $x_0 \in \Omega$, $\Omega_\rho = \Omega \setminus \overline{B_\rho(x_0)}$,

$$\begin{cases} \operatorname{div} \sigma_\rho(u_\rho) = 0 \text{ in } \Omega_\rho \\ u_\rho = g \text{ on } \Gamma_D \\ \sigma_\rho(u_\rho)n = T \text{ on } \Gamma_N \\ \sigma_\rho(u_\rho)n = 0 \text{ on } \partial B_\rho(x_0) \end{cases} \tag{6}$$

where n is the unit outward normal vector on $\partial \Omega_\rho = \partial \Omega \cup \partial B_\rho(x_0)$. Assuming that $0 \in \Omega$, we can consider the case $x_0 = 0$.

Here u and u_ρ denote the displacement vectors fields, g is a given displacement on the fixed part Γ_D of the boundary, t is a traction prescribed on the loaded part Γ_N of the boundary. In addition, σ is the Cauchy stress tensor given, for $\xi = u$ (eq. 5) or $\xi = u_\rho$ (eq. 6), by

$$\sigma(\xi) = D\nabla^s \xi, \tag{7}$$

where $\nabla^s(\xi)$ is the symmetric part of the gradient of vector field ξ, that is

$$\nabla^s(\xi) = \frac{1}{2}\left(\nabla \xi + \nabla \xi^T\right), \tag{8}$$

and D is the elasticity tensor,

$$D = 2\mu \mathbb{I} + \lambda (I \otimes I), \tag{9}$$

with

$$\mu = \frac{E}{2(1+v)}, \quad \lambda = \frac{vE}{(1+v)(1-2v)} \quad \text{and} \quad \lambda = \lambda^* = \frac{vE}{1-v^2}, \tag{10}$$

E being the Young's modulus, v the Poisson's ratio and λ^* the particular case for plane stress. In addition, I and $I\!\!I$ respectively are the second and fourth order identity tensors. Thus, the inverse of D is

$$D^{-1} = \frac{1}{2\mu} \left[I\!\!I - \frac{\lambda}{2\mu + N\lambda} (I \otimes I) \right]. \tag{11}$$

The first shape functional under consideration depends on the displacement field,

$$J_u(\rho) = \int_{\Omega_\rho} F(u_\rho) d\Omega, \quad F(u_\rho) = (H u_\rho \cdot u_\rho)^p, \tag{12}$$

where F is a C^2 function. It is also useful for further applications in the framework of elasticity to introduce the yield functional of the form

$$J_\sigma(\rho) = \int_{\Omega_\rho} S\sigma(u_\rho) \cdot \sigma(u_\rho) d\Omega, \tag{13}$$

where S is an isotropic fourth-order tensor. Isotropicity means here that S may be expressed as follows

$$S = 2mI\!\!I + l(I \otimes I), \tag{14}$$

where l, m are real constants. Their values may vary for particular yield criteria. The following assumption assures that J_u, J_o are well defined for solutions of the elasticity system.

(CONDITION A) The domain Ω has piecewise smooth boundary, which may have reentrant corners with $\alpha < 2\pi$ created by the intersection of two planes. In addition, g, t must be compatible with $u \in H^1(\Omega; \mathbb{R}^N)$.

The interior regularity of u in Ω is determined by the regularity of the right hand side of the elasticity system. For simplicity the following notation is used for functional spaces,

$$H^1_g(\Omega_\rho) = \{ \psi \in [H^1(\Omega_\rho)]^N \mid \psi = g \text{ on } \Gamma_D \}, \tag{15}$$

$$H^1_{\Gamma_D}(\Omega_\rho) = \{ \psi \in [H^1(\Omega_\rho)]^N \mid \psi = 0 \text{ on } \Gamma_D \}, \tag{16}$$

$$H^1_{\Gamma_D}(\Omega) = \{ \psi \in [H^1(\Omega)]^N \mid \psi = 0 \text{ on } \Gamma_D \}, \tag{17}$$

here we use the convention that, e.g., $H^1_g(\Omega_\rho)$ stands for the Sobolev space of vector functions $[H^1_g(\Omega_\rho)]^N$.

The weak solutions to the elasticity systems are defined in the standard way.
Find $u_\rho \in H^1_g(\Omega_\rho)$ such that, for every $\phi \in H^1_{\Gamma_D}(\Omega)$,

$$\int_{\Omega_\rho} D\nabla^s u_\rho \cdot \nabla^s \phi \, d\Omega = \int_{\Gamma_N} T \cdot \phi \, dS. \tag{18}$$

We introduce the adjoint state equations in order to simplify the form of shape derivatives of functionals J_u, J_σ. For the functional J_u they take on the variational form: Find $w_\rho \in H^1_{\Gamma_D}(\Omega_\rho)$,

$$\int_{\Omega_\rho} D\nabla^s w_\rho \cdot \nabla^s \phi \, d\Omega = -\int_{\Omega_\rho} F'_u(u_\rho) \cdot \phi \, d\Omega, \tag{19}$$

for every $\phi \in H^1_{\Gamma_D}(\Omega)$, whose Euler-Lagrange equation reads

$$\begin{cases} \text{div } \sigma_\rho(w_\rho) = F'_u(u_\rho) & \text{in } \Omega_\rho \\ w_\rho = 0 & \text{on } \Gamma_D \\ \sigma_\rho(w_\rho)n = 0 & \text{on } \Gamma_N \\ \sigma_\rho(w_\rho)n = 0 & \text{on } \partial B_\rho(x_0) \end{cases} \tag{20}$$

while $v_\rho \in H^1_{\Gamma_D}(\Omega_\rho)$ is the adjoint state for J_σ and satisfies for all test functions $\phi \in H^1_{\Gamma_D}(\Omega)$ the following integral identity:

$$\int_{\Omega_\rho} D\nabla^s v_\rho \cdot \nabla^s \phi \, d\Omega = -2\int_{\Omega_\rho} DS\sigma(u_\rho) \cdot \nabla^s \phi \, d\Omega, \tag{21}$$

whose associated Euler-Lagrange equation becomes

$$\begin{cases} \text{div } \sigma_\rho(v_\rho) = -2\text{div } \left(DS\sigma_\rho(u_\rho)\right) & \text{in } \Omega_\rho \\ v_\rho = 0 & \text{on } \Gamma_D \\ \sigma_\rho(v_\rho)n = -2DS\sigma_\rho(u_\rho)n & \text{on } \Gamma_N \\ \sigma_\rho(v_\rho)n = -2DS\sigma_\rho(u_\rho)n & \text{on } S_\rho(x_0) = \partial B_\rho(x_0) \end{cases} \tag{22}$$

Remark 1. We observe that DS can be written as

$$DS = 4\mu m \mathbb{I} + \gamma(I \otimes I) \tag{23}$$

where

$$\gamma = \lambda l N + 2\left(\lambda m + \mu l\right). \tag{24}$$

Thus, when $\gamma = 0$, the boundary condition on $\partial B_\rho(x_0)$ in equation (22) becomes homogeneous and the yield criteria must satisfy the constraint

$$\frac{m}{l} = -\left(\frac{\mu}{\lambda} + \frac{N}{2}\right), \tag{25}$$

which is satisfied for the energy shape functional. In this particular case, tensor S is given by

$$S = \frac{1}{2}D^{-1} \implies \gamma = 0 \text{ and } 2m + l = \frac{1}{2E}, \tag{26}$$

which implies that the adjoint solution associated to J_σ can be explicitly obtained, such that $v_\rho = -(u_\rho - g)$.

2.2 Topological Derivatives in 2D Elasticity

We recall here the results derived in [18] for the 2D case. The principal stresses associated with the displacement field u are denoted by $\sigma_I(u)$, $\sigma_{II}(u)$, the trace of the stress tensor $\sigma(u)$ is denoted by $\mathrm{tr}\sigma(u) = \sigma_I(u) + \sigma_{II}(u)$. The shape functionals J_u, J_σ are defined in the same way as presented before, with the tensor S isotropic (that is similar to D). The weak solutions to the elasticity system as well as adjoint equations are defined in standard way. Then, from the expansions presented in the Appendix, we may formulate the following result [18]:

Theorem 1. *The expressions for the topological derivatives of the functionals J_u, J_σ have the form*

$$\mathscr{T}J_u(x_0) = -\left[F(u) + \frac{1}{E}\left(a_u a_w + 2b_u b_w \cos 2\delta\right)\right]_{x=x_0}$$
$$= -\left[F(u) + \frac{1}{E}\left(4\sigma(u)\cdot\sigma(w) - \mathrm{tr}\sigma(u)\mathrm{tr}\sigma(w)\right)\right]_{x=x_0} \tag{27}$$

$$\mathscr{T}J_\sigma(x_0) = -\left[\eta(a_u^2 + 2b_u^2) + \frac{1}{E}\left(a_u a_v + 2b_u b_v \cos 2\delta\right)\right]_{x=x_0}$$
$$= -\left[\eta(4\sigma(u)\cdot\sigma(u) - (\mathrm{tr}\sigma(u))^2) \right. \tag{28}$$
$$\left. + \frac{1}{E}\left(4\sigma(u)\cdot\sigma(v) - \mathrm{tr}\sigma(u)\mathrm{tr}\sigma(v)\right)\right]_{x=x_0}$$

Some of the terms in (27), (28) require explanation. According to equation (24) for $N = 2$, constant η is given by

$$\eta = l + 2\left(m + \gamma\frac{v}{E}\right). \tag{29}$$

Furthermore, we denote

$$\begin{aligned}
a_u &= \sigma_I(u) + \sigma_{II}(u), & b_u &= \sigma_I(u) - \sigma_{II}(u), \\
a_w &= \sigma_I(w) + \sigma_{II}(w), & b_w &= \sigma_I(w) - \sigma_{II}(w), \\
a_v &= \sigma_I(v) + \sigma_{II}(v), & b_v &= \sigma_I(v) - \sigma_{II}(v).
\end{aligned} \tag{30}$$

δ denotes the angle between principal stress directions for displacement fields u and w in (27), and for displacement fields u and v in (28).

Remark 2. For the energy stored in a 2D elastic body, tensor S is given by eq. (26), $\gamma = 0$ and $\eta = 1/(2E)$. Thus, since $v = -(u - g)$, we obtain the following well-known result

$$\mathscr{T}J_\sigma(x_0) = \frac{1}{2E}\left[4\sigma(u)\cdot\sigma(u) - (\mathrm{tr}\sigma(u))^2\right]_{x=x_0}. \tag{31}$$

3   Topological Derivatives for Contact Problems

In order to describe the domain decomposition method applied to the asymptotic analysis, and introduce the Steklov-Poincaré operators for the rings $C(R,\rho)$, $\rho \geq 0$, we present the related results for the two dimensional frictionless contact problems. Such problems are non smooth, therefore, in general, only the first term of the exterior asymptotic expansion of solutions can be derived. However, this leads to the topological derivatives of some shape functionals. We change the notation, compared to the previous sections, in particular $\mathbf{u}$ stands now for the displacement vector, and $\sigma(\mathbf{u})$ is the corresponding stress tensor.

We consider the isotropic two dimensional elasticity problem in plane stress formulation, the isotropy is in fact required only in the vicinity of a small hole. On a part Γ_u of $\partial\Omega$ we assume that the body is clamped $\mathbf{u} = 0$, the part Γ_g is loaded $\sigma(\mathbf{u}).\mathbf{n} = \mathbf{g}$ and on the part Γ_c there is the frictionless contact

$$u_n \geq 0, \qquad \sigma_n \leq 0,$$
$$\sigma_n u_n = 0, \qquad \sigma_\tau = \sigma.\mathbf{n} - \sigma_n \mathbf{n} = 0. \tag{32}$$

Here $u_n = u_i n_i$, $\sigma_n = n_i \sigma_{ij} n_j$, $\sigma.\mathbf{n} = \{\sigma_{ij} n_j\}_{i=1,2}$. We define also the ring $C(R,\rho) = B(R) \setminus \overline{B(\rho)}$ with $R > \rho$ and such that $B(R) \subset \Omega$, as well as $\Omega(r) = \Omega \setminus \overline{B(r)}$.

For such a problem it is impossible to evaluate topological derivatives of shape functionals by means of adjoint variables without additional assumptions on the strict complementarity type for the unknown solution. Therefore, we propose a method for computing the perturbation, caused by the hole $B(\rho)$, of the solution itself.

The bilinear form corresponding to the elastic energy may be written as

$$a(\rho; \mathbf{u}, \mathbf{v}) = \frac{1}{2} \int_{\Omega(\rho)} \sigma(\mathbf{u}) : \varepsilon(\mathbf{v}) \, dx \tag{33}$$

($\sigma : \varepsilon = \sigma_{ij}\varepsilon_{ij}$) for $\mathbf{u}, \mathbf{v} \in \mathbf{H}^1(\Omega)$ and the work of external forces is

$$L(\mathbf{u}) = \int_{\Gamma_g} \mathbf{u}^\top \mathbf{g} \, ds. \tag{34}$$

The method of the domain decomposition type is based on the analysis of the Steklov-Poincaré operator $\mathscr{A}_\rho$ defined in the following way. Consider the boundary value problem

$$\mathscr{L}\mathbf{w} = 0 \text{ in } C(R,\rho), \quad \sigma_n(\mathbf{w}) = 0 \text{ on } \partial B(\rho), \quad \mathbf{w} = \mathbf{v} \text{ on } \partial B(R). \tag{35}$$

Then we set

$$\mathscr{A}_\rho \mathbf{v} = \sigma_n(\mathbf{w}) \text{ on } \partial B(R). \tag{36}$$

Thus $\mathscr{A}_\rho$ is a mapping

$$\mathscr{A}_\rho : \mathbf{H}^{1/2}(\partial B(R)) \mapsto \mathbf{H}^{-1/2}(\partial B(R)). \tag{37}$$

It can be demonstrated constructively that

$$\mathscr{A}_\rho = \mathscr{A}_0 + \rho^2 \mathscr{A}_1 + \rho^4 \mathscr{A}_2 + \dots \tag{38}$$

in the linear operator norm corresponding to (37). Using this notation we have

$$a(\rho;u,u) = \frac{1}{2}\int_{\Omega(R)}\sigma(\mathbf{u}):\varepsilon(\mathbf{u})\,dx + \frac{1}{2}\int_{C(R,\rho)}\sigma(\mathbf{u}):\varepsilon(\mathbf{u})\,dx \qquad (39)$$

as well as

$$\frac{1}{2}\int_{C(R,\rho)}\sigma(\mathbf{u}):\varepsilon(\mathbf{u})\,dx = \frac{1}{2}\langle\mathscr{A}_\rho\mathbf{u},\mathbf{u}\rangle_{\partial B(R)}$$

$$= \frac{1}{2}\langle\mathscr{A}_0\mathbf{u},\mathbf{u}\rangle_{\partial B(R)} + \frac{1}{2}\rho^2\langle\mathscr{A}_1\mathbf{u},\mathbf{u}\rangle_{\partial B(R)} + \mathscr{R}(\mathbf{u},\mathbf{u}) \qquad (40)$$

where $\mathscr{R}(\mathbf{u},\mathbf{u})$ is of the order $O(\rho^4)$ on bounded sets in $\mathbf{H}^{1/2}(\partial B(R))$. With $\mathscr{A}_1$ we associate the bilinear form

$$b(\mathbf{u},\mathbf{v}) = \frac{1}{2}\langle\mathscr{A}_1\mathbf{u},\mathbf{u}\rangle_{\partial B(R)}. \qquad (41)$$

It is sufficient to consider the following approximation of the energy bilinear form in order to construct one term exterior approximation of the solution to the contact problem

$$a(\rho;\mathbf{u},\mathbf{u}) := a(0;\mathbf{u},\mathbf{u}) + \rho^2 b(\mathbf{u},\mathbf{u}). \qquad (42)$$

Denote by $\mathbf{H}^1_{\Gamma_u}(\Omega) = \{\mathbf{v} \in \mathbf{H}^1(\Omega) \mid \mathbf{v} = 0 \text{ on } \Gamma_u\}$ the Sobolev space, and let K be the convex cone

$$K = \{\mathbf{v} \in \mathbf{H}^1_{\Gamma_u}(\Omega) \mid v_n \geq 0 \text{ on } \Gamma_c\}. \qquad (43)$$

Recall that the following variational inequality furnishes the weak solutions to our contact problem in $\Omega(\rho)$

$$\mathbf{u} \in K: a(\rho;\mathbf{u},\mathbf{u}-\mathbf{v}) \geq L(\mathbf{v}-\mathbf{u}) \quad \forall \mathbf{v} \in K. \qquad (44)$$

Taking into account the approximation (42) and using abstract results on the differentiability of metric projection onto the polyhedric convex sets in Dirichlet space [16] we have the following result.

Theorem 2. *For ρ sufficiently small we have on $\Omega(R)$ the following expansion of the solution $\mathbf{u}$ with respect to the parameter ρ at $0+$,*

$$\mathbf{u} = \mathbf{u}_0 + \rho^2\mathbf{q} + o(\rho^2) \text{ in } \mathbf{H}^1(\Omega(R)), \qquad (45)$$

where the topological derivative $\mathbf{q}$ of the solution $\mathbf{u}$ to the contact problem is given by the unique solution of the following variational inequality

$$\mathbf{q} \in \mathscr{S}_K(\mathbf{u}): a(0;\mathbf{q},\mathbf{v}-\mathbf{q}) + b(\mathbf{u},\mathbf{v}-\mathbf{q}) \geq 0 \quad \forall \mathbf{v} \in \mathscr{S}_K(\mathbf{u}), \qquad (46)$$

where

$$\mathscr{S}_K(\mathbf{u}) = \{\mathbf{v} \in \mathbf{H}^1_{\Gamma_u}(\Omega) \mid v_n \leq 0 \text{ on } \Xi(\mathbf{u}), a(0;\mathbf{u},\mathbf{v}) = 0\}. \qquad (47)$$

The coincidence set $\Xi(\mathbf{u}) = \{\mathbf{x} \in \Gamma_c \mid u_n(\mathbf{x}) = 0\}$ is well defined [16] for any $\mathbf{u} \in \mathbf{H}^1(\Omega)$, and $\mathbf{u}_0 \in K$ is the solution of (44) for $\rho = 0$.

4   Complex Variable Method

In order to find an exact form of the Steklov-Poincaré operator in plane elasticity we need an analytic form of the solution for the elasticity system in the ring, with general displacement condition on the outer boundary and traction free inner boundary, parameterized by the (small) inner radius ρ. Let us assume for simplicity that the center of the ring lies at origin of the coordinate system, and take polar coordinates (r, θ) with $\mathbf{e}_r$ pointing outwards and $\mathbf{e}_\theta$ perpendicularly in the counter-clockwise direction. Then the displacement on the outer boundary $r = R$ may be given in the form of a Fourier series

$$2\mu(u_r + iu_\theta) = \sum_{k=-\infty}^{k=+\infty} U_k e^{ik\theta}. \tag{48}$$

The regularity condition for the boundary data translate into some inequalities for coefficients U_k, as will be made precise later.

The solution in the ring must be compared with the solution in the full circle, so we will have to construct it as well. Probably the best tool for obtaining both exact solutions is the complex variable method, described in [11]. It states that for plane domains with one hole these solutions have the form

$$\sigma_{rr} - i\sigma_{r\theta} = 2\Re\phi' - e^{2i\theta}(\bar{z}\phi'' + \psi'),$$
$$\sigma_{rr} + i\sigma_{\theta\theta} = 4\Re\phi', \tag{49}$$
$$2\mu(u_r + iu_\theta) = e^{-i\theta}(\kappa\phi - z\bar{\phi}' - \bar{\psi}),$$

where ϕ, ψ are given by complex series

$$\phi = A\log(z) + \sum_{k=-\infty}^{k=+\infty} a_k z^k,$$
$$\psi = -\kappa\bar{A}\log(z) + \sum_{k=-\infty}^{k=+\infty} b_k z^k. \tag{50}$$

Here μ is the Lame constant, v is the Poisson ratio, $\kappa = 3 - 4v$ in the plain strain case, and $\kappa = (3 - v)/(1 + v)$ for plane stress.

Now we can substitute displacement condition for $r = R$ into

$$2\mu(u_r + iu_\theta) = 2\kappa Ar\log(r)\frac{1}{z} - \bar{A}\frac{1}{r}z +$$
$$+ \sum_{p=-\infty}^{p=+\infty} [\kappa ra_{p+1} - (1-p)\bar{a}_{1-p}r^{-2p+1} - \bar{b}_{-(p+1)}r^{-2p-1}]z^p \tag{51}$$

and obtain the infinite system of linear equations

$$p = -1: 2\kappa Ar\log(r) + (\kappa a_0 - \bar{b}_0) - 2\bar{a}_2 r^2 = U_{-1}$$
$$p = 1: -\bar{A} + \kappa r^2 a_2 - \bar{b}_{-2}\frac{1}{r^2} = U_1 \tag{52}$$
$$p \notin \{-1, 1\}: \kappa r^{p+1}a_{p+1} - (1-p)\bar{a}_{1-p}r^{-p+1} - \bar{b}_{-(p+1)}r^{-(p+1)} = U_p.$$

The traction-free condition

$$\sigma . \mathbf{e}_r = [\sigma_{rr}, \sigma_{r\theta}]^\top \tag{53}$$

on some circle means $\sigma_{rr} = \sigma_{r\theta} = 0$. Hence, assuming $r := \rho$, we have another infinite system

$$p = -1: \ 2A + 2\bar{a}_2 r^2 + 2\frac{1}{r^2}b_{-2} = 0$$

$$p = 1: \ (\kappa+1)\frac{1}{r^2}\bar{A} = 0 \tag{54}$$

$$p \notin \{-1,1\}: \ (1+p)a_{p+1} + \bar{a}_{1-p}r^{-2p} + \frac{1}{r^2}b_{p-1} = 0.$$

Denote $d_0 = \kappa a_0 - \bar{b}_0$ since a_0, b_0 appear only in this combination. Using (52) we may recover the solution for the full circle. Because in this case the singularities must vanish, we have $b_{-k} = a_{-k} = A = 0$ for $k = 1,2,\ldots$ and comparing the same powers of r:

$$d_0^0 = U_{-1} + \frac{2}{\kappa}\bar{U}_1, \quad \Re a_1^0 = \frac{1}{(\kappa-1)R}\Re U_0, \quad \Im a_1^0 = \frac{1}{(\kappa+1)R}\Im U_0$$

$$a_k^0 = \frac{1}{\kappa R^k}U_{k-1}, \quad b_k^0 = -\frac{1}{R^k}[(k+2)\frac{1}{\kappa}U_{k+1} + \bar{U}_{-(k+1)}], \quad k > 1. \tag{55}$$

Now let us repeat the same procedure for the ring. Now the singularities may be present, because 0 does not belong to the domain. Hence, from (52) for $r = R$ and (54) for $r = \rho$ we obtain $A = 0$ and the formulas

$$d_0 = A_{-1} + \frac{2R^4}{\kappa R^4 + \rho^4}\bar{U}_1, \qquad a_2 = \frac{R^2}{\kappa R^4 + \rho^4}U_1$$

$$\Re a_1 = \frac{R}{(\kappa-1)R^2 + 2\rho^2}\Re U_0, \qquad \Im a_1 = \frac{1}{\kappa+1}\Im A_0 \tag{56}$$

$$b_{-1} = -\frac{2\rho^2 R}{(\kappa-1)R^2 + 2\rho^2}\Re U_0, \qquad b_{-2} = -\frac{\rho^4 R^2}{\kappa R^4 + \rho^4}\bar{U}_1$$

The rest of the coefficients will be computed later. However, we may at this stage compare the results with known solutions for the uniformly stretched circle or ring obtained in another way. In such a case $U_0 = 2\mu u_r(R)$ does not vanish and, for the full circle, $\psi = 0$, $\phi = a_1^0 z$ with

$$a_1^0 = \frac{2\mu}{(\kappa-1)R}u_r(R). \tag{57}$$

For the ring we have $\phi = a_1 z$, $\psi = b_{-1}\frac{1}{z}$ where

$$a_1 = \frac{1}{(\kappa-1) + 2\rho^2}2\mu u_R(1), \quad b_{-1} = -\frac{2\rho^2}{(\kappa-1) + 2\rho^2}2\mu u_R(1). \tag{58}$$

After substitutions we obtain, in both cases, the same results as given in [7]. Similarly the comparison with the solution for the ring with displacement conditions on both boundaries, obtained in [4] also using complex method, confirms the correctness of the formulas.

There remains to compute the rest of the coefficients a_k, b_k for the case of the ring. Taking $p = -k$, $k = 2, 3, \ldots$ in conditions on both boundaries gives the system

$$\kappa a_{-(k-1)} R^{-(k-1)} - (k+1)\bar{a}_{k+1} R^{k+1} - \bar{b}_{k-1} R^{k-1} = U_{-k}$$

$$-(k-1)a_{-(k-1)}\rho^2 + \bar{a}_{k+1}\rho^{2(k+1)} + b_{-(k+1)} = 0,$$

$$(59)$$

while $p = +k$, $k = 2, 3, \ldots$ results in

$$\kappa a_{k+1} R^{k+1} + (k-1)\bar{a}_{-(k-1)} R^{-(k-1)} - \bar{b}_{-(k+1)} R^{-(k+1)} = U_k$$

$$(k+1)a_{k+1}\rho^{2(k+1)} + \bar{a}_{-(k-1)}\rho^2 + b_{k-1}\rho^{2k} = 0.$$

$$(60)$$

These systems may be represented in a recursive form, convenient for numerical computations and further analysis. Namely,

$$S_k(\rho) \cdot \begin{bmatrix} a_{k+1} \\ b_{k-1} \end{bmatrix} = \begin{bmatrix} U_k \\ \bar{U}_{-k} \end{bmatrix}$$

$$(61)$$

where S_k has entries

$$S_k(\rho)_{11} = \kappa R^{k+1} - (k^2 - 1)R^{1-k}\rho^{2k} + k^2 R^{-(k+1)}\rho^{2(k+1)}$$

$$S_k(\rho)_{12} = -(k-1)(R^{1-k}\rho^{2(k-1)} - R^{-(k+1)}\rho^{2k})$$

$$S_k(\rho)_{21} = -(k+1)(R^{k+1} + \kappa R^{1-k}\rho^{2k})$$

$$S_k(\rho)_{22} = -R^{k-1} - \kappa R^{1-k}\rho^{2(k-1)}$$

$$(62)$$

as well as

$$\begin{bmatrix} a_{-(k-1)} \\ b_{-(k+1)} \end{bmatrix} = T_k(\rho) \cdot \begin{bmatrix} \bar{a}_{k+1} \\ \bar{b}_{k-1} \end{bmatrix},$$

$$(63)$$

where

$$T_k(\rho) = \begin{bmatrix} -(k+1)\rho^{2k} , & -\rho^{2(k-1)} \\ -k^2\rho^{2(k+1)} , & -(k-1)\rho^{2k} \end{bmatrix}.$$

$$(64)$$

In fact the formulas (63), (61) are correct also for $k = 0, 1$ and in the limit $\rho \longrightarrow 0+$, but the derivation must separate these cases.

Thus, for given $k > 1$ and using some initial a_k, b_k obtained earlier, we may first compute a_{k+1}, b_{k-1} using (61) and then $a_{-(k-1)}, b_{-(k+1)}$ from (63).

We may now use the above results for the asymptotic analysis of the solution. To simplify the formulas, we assume $R = 1$, which means only rescaling and does not diminish generality (in general case ρ would be replaced by ρ/R). Then by direct computation we get the following bounds for the differences between the coefficients on the full circle and the ring. For the initial values of k they read

$$d_0 - d_0^0 = -\rho^4 \frac{2}{\kappa(\kappa R^4 + \rho^4)}\bar{U}_1$$

$$a_1 - a_1^0 = -\rho^2 \frac{2}{(\kappa - 1)R((\kappa - 1)R^2 + 2\rho^2)}\Re U_0$$

$$a_2 - a_2^0 = -\rho^4 \frac{1}{\kappa R^2(\kappa R^4 + \rho^4)}U_1$$

$$(65)$$

and for higher values

$$|a_3 - a_3^0| \leq \Lambda \left(|U_2|\rho^4 + |U_{-2}|\rho^2 \right) \tag{66}$$

and for $k = 4, 5, \ldots$

$$|a_k - a_k^0| \leq \Lambda \left(|U_{k-1}|\rho^{3(k-1)/2} + |U_{1-k}|\rho^{3(k-2)/2} \right) \tag{67}$$

where the exponent $k/2$ has been used to counteract the growth of k^2 in terms like $k^2\rho^{k/2}$. Similarly

$$|b_1 - b_1^0| \leq \Lambda \left(|U_2|\rho^4 + |U_{-2}|\rho^2 \right) \tag{68}$$

and for $k = 2, 3, \ldots$

$$|b_k - b_k^0| \leq \Lambda \left(|U_{k+1}|\rho^{3(k+1)/2} + |U_{-(k+1)}|\rho^{3k/2} \right). \tag{69}$$

From relation (63) we get further estimates

$$\begin{aligned} |a_{-k}| &\leq \Lambda\rho^{2k} \left(|U_{k+1}| + |U_{-(k+1)}| \right), \quad k = 1, 2, \ldots \\ |b_{-k}| &\leq \Lambda\rho^{2(k-1)} \left(|U_{k-1}| + |U_{1-k}| \right), \quad k = 3, 4, \ldots \end{aligned} \tag{70}$$

Here Λ is a constant independent from ρ and U_i. Observe that the corrections proportional to ρ^2 are present only in a_1, b_1, a_3, b_{-1}, a_{-1}. The rest is of the order at least $O(\rho^3)$ (in fact $O(\rho^4)$).

These estimates may be translated into the following theorem concerning the solution of the elasticity system in the ring.

Theorem 3. *The condition*

$$\|\mathbf{u}\|_{\mathbf{H}^{1/2}(\partial B(R))} \leq \Lambda_0 \tag{71}$$

which in terms of U_i means

$$\sum_{k=-\infty}^{k=+\infty} \sqrt{1+k^2}\, |U_k|^2 \leq \Lambda_0 \tag{72}$$

ensures that the expression for elastic energy concentrated in the ring splits into the one corresponding to the full circle, correction proportional to ρ^2 and the rest, which is uniformly of the order $\Lambda_0\rho^3$.

4.1 Numerical Illustration

We shall show two solutions corresponding to different boundary conditions on the outer boundary, obtained using the representations derived above in terms of (in these particular cases finite) complex series.

Rugby-like deformation. Let us take $u_r = s_0 \cos^2\theta = \frac{1}{2}s_0 + \frac{1}{2}s_0 \cos 2\theta$. Hence

$$[U_k, k \in \mathbb{Z}] = [\ldots, \frac{1}{2}\mu s_0, 0, U_0 = \mu s_0, 0, \frac{1}{2}\mu s_0, \ldots]. \tag{73}$$

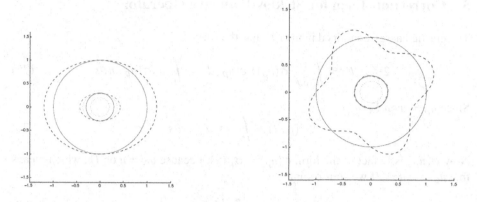

Fig. 1. Rugby-like and bubble-like distortions

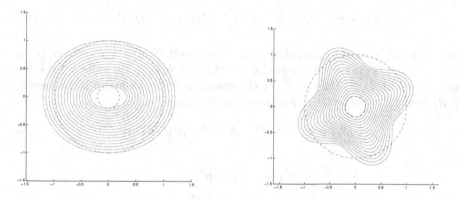

Fig. 2. The pattern of distortions for both experiments

The resulting distortion for size of the internal hole $\rho = 0.2$ at the radius $r = 0.3$ are shown in Fig. 1 (solid line - undeformed, dashed - deformed ring, dotted - deformed ball):

Bubble-like deformation. Now we take $u_r = s_0 \sin 4\theta$. Hence

$$[U_k,\ k \in \mathbb{Z}] = [\ldots, \mu s_0 i, 0, 0, 0, A_0 = 0, 0, 0, 0, -\mu s_0 i, \ldots]. \tag{74}$$

The resulting distortions for $\rho = 0.2$ and $r = 0.3$ shows also Fig. 1, using the same types of lines.

In the second numerical experiment - bubble - only U_{-4} and U_4 were nonzero, which means that the difference between positions of the contour $r = 0.3$ for full circle and the ring should behave like ρ^6. In the first experiment it should be ρ^2, i.e. the influence of boundary condition should vanish quicker. The deformations for $\rho = 0.2$ and several intermediate radii (dashed - undeformed, solid - deformed contours) are visible in Fig. 2 and they confirm this observation.

5   Correction Term for Steklov-Poincaré Operator

The elastic energy contained in the ring has the form

$$2\mathscr{E}(\rho,R) = \int_{C(\rho,R)} \sigma(u_\rho): \varepsilon(u_\rho)\,dx = \int_{\Gamma_R} u_\rho\sigma(u_\rho).n\,ds. \tag{75}$$

Since $u_\rho = u$ on Γ_R,

$$2\mathscr{E}(\rho,R) = \int_{\Gamma_R} u\sigma(u_\rho).n\,ds. \tag{76}$$

Now $\sigma(u_\rho)$ is in fact of the form $\sigma(u_\rho) = \sigma_\rho(u)$, because $u_\rho = u$ on Γ_R, which means that $u_\rho = u_\rho(u)$. If we split σ_ρ into

$$\sigma_\rho(u) = \sigma^0 + \rho^2\sigma^1(u) + O(\rho^4) \tag{77}$$

then

$$2\mathscr{E}(\rho,R) = 2\mathscr{E}(0,R) + \rho^2 \int_{\Gamma_R} u\sigma^1(u).n\,ds + O(\rho^4). \tag{78}$$

Thus finding $\mathscr{A}_1$ reduces to computing $\sigma^1(u)$. From (49), (50) we know that $\sigma_\rho(u)$ is a linear function of infinite vectors $a = [a_k,\ k \in \mathbb{Z}]$, $b = [b_k,\ k \in \mathbb{Z}]$, while $\sigma^0(u)$ is the same function of a^0, b^0. Here a^0, b^0 are computed for $B(R)$, while a, b correspond to $C(\rho,R)$. In order to obtain $\sigma^1(u)$ it is enough to express a, b as

$$a = a^0 + \rho^2 a^1 + O(\rho^4), \quad b = b^0 + \rho^2 b^1 + O(\rho^4) \tag{79}$$

because then

$$\sigma^1(u) = \sigma^1(a^1, b^1). \tag{80}$$

In addition, the only nonzero terms in a^1, b^1 are $a_3^1, a_1^1, a_{-1}^1, b_{-1}^1, b_1^1$.

Taking into account that $A = 0$ in (50) for our problem,

$$\phi = \phi^0 + \rho^2\phi^1 + O(\rho^4), \quad \psi = \psi^0 + \rho^2\psi^1 + O(\rho^4) \tag{81}$$

where

$$\phi^1 = a_{-1}^1\frac{1}{z} + a_1^1 z + a_3^1 z^3, \quad \psi^1 = b_{-1}^1\frac{1}{z} + b_1^1 z. \tag{82}$$

Using formulas derived in preceding section, we may explicitly compute the coefficients appearing in (82).

$$a_{-1}^1 = -\bar{b}_1^0, \quad a_3^1 = \frac{1}{\kappa R^4}b_1^0, \quad b_1^1 = \frac{3+\kappa^2}{\kappa R^2}b_1^0,$$

$$a_1^1 = -\frac{2}{(\kappa-1)R^2}\Re a_1^0, \quad b_{-1}^1 = -2\Re a_1^0. \tag{83}$$

As is obvious from earlier calculations, only U_0, U_2, U_{-2} will contribute to these corrections, Since

$$U_k = \frac{\mu}{\pi}\int_0^{2\pi}(u_r + iu_\theta)e^{-ik\theta}\,d\theta \tag{84}$$

as well as

$$u_r + iu_\theta = (u_1 + iu_2)e^{-i\theta} \tag{85}$$

then

$$U_0 = \frac{\mu}{\pi} \int_0^{2\pi} (u_1 + iu_2)e^{-i\theta}\,d\theta$$

$$U_2 = \frac{\mu}{\pi} \int_0^{2\pi} (u_1 + iu_2)e^{-3i\theta}\,d\theta \tag{86}$$

$$U_{-2} = \frac{\mu}{\pi} \int_0^{2\pi} (u_1 + iu_2)e^{+i\theta}\,d\theta.$$

After collecting all formulas we obtain the final expression

$$\int_{\Gamma_R} u^\top \sigma^1(u).n\,ds = \frac{1}{R^2}\left[\frac{2(\kappa-2)}{(\kappa-1)^2}(\Re U_0)^2 - (\kappa+1)|U_{-2}|^2\right.$$
$$\left. - \frac{9(\kappa+1)}{\kappa^2}|U_2|^2 - \frac{6(\kappa+1)}{\kappa}\Re(U_2 U_{-2})\right]. \tag{87}$$

From (86) it follows that

$$\Re U_0 = \frac{\mu}{\pi}\int_0^{2\pi}(u_1\cos\theta + u_2\sin\theta)\,d\theta$$

$$U_2 = \frac{\mu}{\pi}\int_0^{2\pi}(u_1\cos 3\theta + u_2\sin 3\theta)\,d\theta + i\frac{\mu}{\pi}\int_0^{2\pi}(u_2\cos 3\theta - u_1\sin 3\theta)\,d\theta \tag{88}$$

$$U_{-2} = \frac{\mu}{\pi}\int_0^{2\pi}(u_1\cos\theta - u_2\sin\theta)\,d\theta + i\frac{\mu}{\pi}\int_0^{2\pi}(u_2\cos\theta + u_1\sin\theta)\,d\theta.$$

Here values of displacements are taken as $u_i(R\cos\theta, R\sin\theta)$. After discretization these integrals constitute weighted sums of values of u_i at certain points on Γ_R. If we assume piecewise linear approximation over triangles, then it is well known that

$$u_i^h(x) = x^\top \begin{bmatrix} x_1^1 & x_2^1 & 1 \\ x_1^2 & x_2^2 & 1 \\ x_1^3 & x_2^3 & 1 \end{bmatrix}^{-1} U_i^h = x^\top M^{-1} U_i^h \tag{89}$$

and

$$x^\top M^{-1} U_i^h = (M^{-\top} x)^\top U_i^h = c^\top U_i^h \tag{90}$$

where $u_i^h(x)$ is a value of the approximation of u_i at a point x inside the triangle defined by vertices x^1, x^2, x^3 and U_i^h is a vector of the values of u_i^h at these vertices. Observe that c is a vector of weights with which nodal values enter into the expression for $u_i^h(x)$.

Let now $U^h = [u_1^{h1}, u_2^{h1}, \ldots, u_1^{hK}, u_2^{hK}]^\top$ be a vector of nodal values of u^h for the global triangulation. Then we may write down the following formulae

$$\frac{\mu}{\pi} \int_0^{2\pi} u_1 \cos\theta \, d\theta = c_{11}^\top U^h \qquad \frac{\mu}{\pi} \int_0^{2\pi} u_2 \sin\theta \, d\theta = s_{21}^\top U^h$$

$$\frac{\mu}{\pi} \int_0^{2\pi} u_1 \cos 3\theta \, d\theta = c_{13}^\top U^h \qquad \frac{\mu}{\pi} \int_0^{2\pi} u_2 \sin 3\theta \, d\theta = s_{23}^\top U^h$$

$$\frac{\mu}{\pi} \int_0^{2\pi} u_1 \sin\theta \, d\theta = s_{11}^\top U^h \qquad \frac{\mu}{\pi} \int_0^{2\pi} u_2 \cos\theta \, d\theta = c_{21}^\top U^h \tag{91}$$

$$\frac{\mu}{\pi} \int_0^{2\pi} u_1 \sin 3\theta \, d\theta = s_{13}^\top U^h \qquad \frac{\mu}{\pi} \int_0^{2\pi} u_2 \cos 3\theta \, d\theta = c_{23}^\top U^h.$$

Here s_{ij}, c_{ij} are sparse vectors of weights with which nodal values of u enter into appropriate integrals. In this notation

$$(\Re U_0)^2 = \|(c_{11} + s_{21})^\top U^h\|^2$$

$$|U_2|^2 = \|(c_{13} + s_{23})^\top U^h\|^2 + \|(c_{23} - s_{13})^\top U^h\|^2$$

$$|U_{-2}|^2 = \|(c_{11} - s_{21})^\top U^h\|^2 + \|(c_{21} + s_{11})^\top U^h\|^2 \tag{92}$$

$$\Re(U_2 U_{-2}) = (U^h)^\top (c_{13} + s_{23})(c_{11} - s_{21}) U^h$$

$$- (U^h)^\top (c_{23} - s_{13})(c_{21} + s_{11}) U^h.$$

Taking into account (87) we may conclude that the first term in the correction of energy is a well defined quadratic form. Similar, only more complicated expressions may be obtained for further asymptotics corresponding to ρ^4.

References

1. Ammari, H.: Polarization and Moment Tensors: with Applications to Inverse Problems and Effective Medium Theory. Applied Mathematical Sciences Series, vol. 162. Springer, New York (2007)
2. Eschenauer, H.A., Kobelev, V.V., Schumacher, A.: Bubble method for topology and shape optimization of structures. Struct. Optimiz. 8, 42–51 (1994)
3. Garreau, S., Guillaume, Ph., Masmoudi, M.: The topological asymptotic for PDE systems: the elasticity case. SIAM Journal on Control and Optimization 39, 1756–1778 (2001)
4. Gross, W.A.: The second fundamental problem of elasticity applied to a plane circular ring. Zeitschrift für Angewandte Mathematik und Physik 8, 71–73 (1957)
5. Hlaváček, I., Novotny, A.A., Sokołowski, J., Żochowski, A.: Energy change in elastic solids due to a spherical or circular cavity, considering uncertain input data. JOTA (2009) (in press)
6. Il'in, A.M.: Matching of Asymptotic Expansions of Solutions of Boundary Value Problems. Translations of Mathematical Monographs, vol. 102. AMS (1992)
7. Kachanov, M., Shafiro, B., Tsukrov, I.: Handbook of Elasticity Solutions. Kluwer Academic Publishers, Dordrecht (2003)
8. Lewinski, T., Sokołowski, J.: Energy change due to the appearance of cavities in elastic solids. Int. J. Solids Struct. 40, 1765–1803 (2003)
9. Masmoudi, M.: The toplogical asymptotic. In: Glowinski, R., Kawarada, H., Periaux, J. (eds.) Computational Methods for Control Applications, Gakuto (2002)

10. Mazja, W.G., Nazarov, S.A., Plamenevskii, B.A.: Asymptotic theory of elliptic boundary value problems in singularly perturbed domains, vol. 1, 2. Birkhäuser Verlag, Basel (2000)

11. Muskhelishvili, N.I.: Some Basic Problems on the Mathematical Theory of Elasticity. Noordhoff (1952)

12. Nazarov, S.A., Sokołowski, J.: Asymptotic analysis of shape functionals. Journal de Mathématiques pures et appliquées 82, 125–196 (2003)

13. Nečas, J., Hlaváček, I.: Mathematical Theory of Elastic and Elasto-plastic Bodies: An Introduction. Elsevier, Amsterdam (1981)

14. Novotny, A.A., Feijóo, R.A., Padra, C., Taroco, E.A.: Topological sensitivity analysis. Computer Methods in Applied Mechanics and Engineering 192, 803–829 (2003)

15. Novotny, A.A., Feijóo, R.A., Padra, C., Taroco, E.A.: Topological sensitivity analysis for three-dimensional linear elasticity problems. Computer Methods in Applied Mechanics and Engineering (to appear)

16. Rao, M., Sokołowski, J.: Tangent sets in Banach spaces and applications to variational inequalities. Tech. Rep. Les prépublications de l'Institut Élie Cartan 42/2000 (2000)

17. Sokołowski, J., Zolésio, J.-P.: Introduction to Shape Optimization. Shape Sensitivity Analysis. Springer, Heidelberg (1992)

18. Sokołowski, J., Żochowski, A.: On topological derivative in shape optimization. SIAM Journal on Control and Optimization 37(4), 1251–1272 (1999)

19. Sokołowski, J., Żochowski, A.: Topological derivatives of shape functionals for elasticity systems. Mechanics of Structures and Machines 29, 333–351 (2001)

20. Sokołowski, J., Żochowski, A.: Optimality conditions for simultaneous topology and shape optimization. SIAM Journal on Control and Optimization 42, 1198–1221 (2003)

21. Sokołowski, J., Żochowski, A.: Modelling of topological derivatives for contact problems. Numerische Mathematik 102(1), 145–179 (2005)

Model Order Reduction for Nonlinear IC Models

Arie Verhoeven[1], Jan ter Maten[2], Michael Striebel[3], and Robert Mattheij[4]

[1] Eindhoven University of Technology (CASA), Den Dolech 2,
5600 MB Eindhoven, The Netherlands
Arie.Verhoeven@na-net.ornl.gov
[2] NXP Semiconductors
jan.ter.maten@nxp.com
[3] Technische Universität Chemnitz
michael.striebel@mathematik.tu-chemnitz.de
[4] Dept of Mathematics and Computer Science, TU Eindhoven, PO Box 513,
5600 MB Eindhoven The Netherlands
r.m.m.mattheij@tue.nl

Abstract. Model order reduction is a mathematical technique to transform nonlinear dynamical models into smaller ones, that are easier to analyze. In this paper we demonstrate how model order reduction can be applied to nonlinear electronic circuits. First we give an introduction to this important topic. For linear time-invariant systems there exist already some well-known techniques, like Truncated Balanced Realization. Afterwards we deal with some typical problems for model order reduction of electronic circuits. Because electronic circuits are highly nonlinear, it is impossible to use the methods for linear systems directly. Three reduction methods, which are suitable for nonlinear differential algebraic equation systems are summarized, the Trajectory piecewise Linear approach, Empirical Balanced Truncation, and the Proper Orthogonal Decomposition. The last two methods have the Galerkin projection in common. Because Galerkin projection does not decrease the evaluation costs of a reduced model, some interpolation techniques are discussed (Missing Point Estimation, and Adapted POD). Finally we show an application of model order reduction to a nonlinear academic model of a diode chain.

1 Introduction

The dynamics of electrical circuits at time t can be generally described by the nonlinear, first order, differential-algebraic equation (DAE) system of the form:

$$\begin{cases} \frac{d}{dt}\left[\mathbf{q}(\mathbf{x})\right]+\mathbf{j}(\mathbf{x})+\mathbf{Bu}=\mathbf{0}, & \mathbf{x}(0)=\mathbf{x}_0, \\ \mathbf{y}=\mathbf{h}(\mathbf{x}), \end{cases} \tag{1}$$

where $\mathbf{x}: \mathbb{R} \rightarrow \mathbb{R}^d$ represents the unknown vector of circuit variables in time t, the vector-valued functions $\mathbf{q}, \mathbf{j}: \mathbb{R} \times \mathbb{R}^d \rightarrow \mathbb{R}^d$ represent the contributions of, respectively, reactive elements (such as capacitors and inductors) and of nonreactive elements (such as resistors) and $\mathbf{B} \in \mathbb{R}^{d \times m}$ is the distribution matrix for the excitation vector $\mathbf{u}: \mathbb{R} \rightarrow \mathbb{R}^m$ that controls the output response $\mathbf{y}: \mathbb{R} \rightarrow \mathbb{R}^p$. We assume that $d >> m, p$. There are

A. Korytowski et al. (Eds.): System Modeling and Optimization, IFIP AICT 312, pp. 476–491, 2009.

several established methods, such as sparse-tableau, modified nodal analysis, etc. which generate the system (1) from the netlist description of electrical circuit. The dimension d of the unknown vector $\mathbf{x}$ is of the order of the number of elements in the circuit, which means that it can be extremely large, as today's VLSI (Very Large Scale Integrated) circuits have hundreds of millions of elements.

Mathematical model order reduction (MOR) aims to replace the original model (1) by a system of much smaller dimension, which can be solved by suitable DAE solvers within acceptable time. Because we are only interested in the relationship between $\mathbf{u}$ and $\mathbf{y}$ in the time-domain, the model can be replaced by a low-order model for $\mathbf{z} : \mathbb{R} \to \mathbb{R}^r$, like

$$\begin{cases} \frac{d}{dt}[\tilde{\mathbf{q}}(\mathbf{z})] + \tilde{\mathbf{j}}(\mathbf{z}) + \tilde{\mathbf{B}}\mathbf{u} = \mathbf{0}, \quad \mathbf{z}(0) = \mathbf{z}_0, \\ \qquad\qquad \mathbf{y} = \tilde{\mathbf{h}}(\mathbf{z}). \end{cases} \tag{2}$$

At present, however, only linear MOR techniques are well-enough developed and properly understood to be employed [1]. To that end, we either linearize the system (1) or decouple it into nonlinear and linear subcircuits (interconnect macromodeling of parasitic subcircuits [9]). For dynamical systems the observability and controllability functions [1] are defined by

$$L_c(\mathbf{x}_0) = \min\{\tfrac{1}{2}\int_{-\infty}^{0}\|\mathbf{u}(t)\|^2 dt : \mathbf{u} \in L_2(-\infty,0), \, \mathbf{x}(-\infty) = 0, \, \mathbf{x}(0) = \mathbf{x}_0\}, \tag{3}$$

$$L_o(\mathbf{x}_0) = \tfrac{1}{2}\int_{0}^{\infty}\|\mathbf{y}(t)\|^2 dt, \; \forall_{\tau\in[0,\infty)}\mathbf{u}(\tau) = \mathbf{0}, \; \mathbf{x}(0) = \mathbf{x}_0. \tag{4}$$

They represent the minimum amount of input energy to reach state $\mathbf{x}_0$ and the output energy that comes free when starting at state $\mathbf{x}_0$ (compare kinetic and potential energy in mechanical systems). The system is in balanced form at basis $\mathbf{V}$ if the (energy) ratio $\frac{L_o(\mathbf{Vz})}{L_c(\mathbf{Vz})}$ is balanced. For linear time-invariant (LTI) systems as

$$\begin{cases} \dot{\mathbf{x}} = \mathbf{A}\mathbf{x} + \mathbf{B}\mathbf{u}, \quad \mathbf{x}(0) = \mathbf{x}_0, \\ \qquad \mathbf{y} = \mathbf{C}\mathbf{x}, \end{cases} \tag{5}$$

we have $L_c(\mathbf{x}_0) = \tfrac{1}{2}\mathbf{x}_0^T\mathbf{W}^{-1}\mathbf{x}_0$ and $L_o(\mathbf{x}_0) = \tfrac{1}{2}\mathbf{x}_0^T\mathbf{M}\mathbf{x}_0$, where $\mathbf{W}, \mathbf{M} \in \mathbb{R}^{d\times d}$ are the controllability and observability Gramians, which are symmetric positive definite matrices. They satisfy the well-known Lyapunov equations

$$\mathbf{A}\mathbf{W} + \mathbf{W}\mathbf{A}^T = -\mathbf{B}\mathbf{B}^T, \tag{6}$$

$$\mathbf{A}^T\mathbf{M} + \mathbf{M}\mathbf{A} = -\mathbf{C}^T\mathbf{C}. \tag{7}$$

An LTI system is balanced w.r.t. basis $\mathbf{V}$ if $\mathbf{W} = \mathbf{V}\Sigma\mathbf{V}^T$ and $\mathbf{M} = \mathbf{V}^{-T}\Sigma\mathbf{V}^{-1}$ are simultaneously diagonalized, such that

$$\frac{L_o(\mathbf{x})}{L_c(\mathbf{x})} = \frac{\mathbf{x}^T\mathbf{M}\mathbf{x}}{\mathbf{x}^T\mathbf{W}^{-1}\mathbf{x}} = \frac{\mathbf{x}^T\mathbf{V}^{-T}\Sigma\mathbf{V}^{-1}\mathbf{x}}{\mathbf{x}^T\mathbf{V}^{-1}\Sigma^{-1}\mathbf{V}^{-T}\mathbf{x}} = \frac{\mathbf{z}^T\Sigma^2\mathbf{z}}{\mathbf{z}^T\mathbf{z}}. \tag{8}$$

For redundant systems the singular values of Σ converge rapidly to zero. This allows to obtain an accurate reduced model by Truncated Balanced Realization (TBR). There also exist many other much cheaper MOR techniques for LTI systems, like PRIMA, PVL, PMTBR, SPRIM, etc. For the special case $\mathbf{A} = \mathbf{A}^T, \mathbf{B} = \mathbf{C}^T$ it follows from (6),

(7) that $\mathbf{W} = \mathbf{M}$. Then it is possible to find an orthogonal $\mathbf{V}$ such that $\mathbf{W} = \mathbf{M} = \mathbf{V}\Sigma\mathbf{V}^T$ are balanced.

For nonlinear systems as (1) it is no longer possible to apply these linear MOR techniques. Then we try to exploit the (piecewise) linear structure as well as possible. The reduced model can be constructed for a benchmark simulation, such that it is accurate if the solution is in the neighborhood of the benchmark solution. In this paper we present the application of some promising nonlinear reduction methods on some electronic circuit models. These are the Trajectory Piecewise Linear approach (TPWL) [13] and the Proper Orthogonal Decomposition (POD) [2] supported by the Missing Point Estimation (MPE) technique [5]. This paper does not include an error analysis but interested readers could look at [7,10,13].

2   Model Order Reduction for Subcircuits

A continuously increasing number of functions is combined in each single integrated circuit. Therefore, complex devices are designed in a modular manner. Functional units like e.g., decoders, mixers, and operational amplifiers, are developed by different experts and stored in device libraries. Other circuit designers then choose these models according to their requirements and instantiate them in higher level circuits. To enable the combination of different blocks, each model is equipped with its own number of junctions, or *pins*, by which a communication with the outside world is possible.

In the first instance, numerical simulations are run to verify a design. Hence, it is desirable to have, besides the exact circuit schematic, a suitable description of the individual model that enables fast simulations, i.e., a library of reduced subcircuit models.

In circuits that are developed to act as a subcircuit in higher hierarchies a subset of its nodes are terminals. To these nodes both known inner elements as well as elements whose nature may change with different instantiations of the model are connected. Due to Kirchhoff's current law, the sum of all currents flowing into each single node is zero at each timepoint. In terms of the network equations (1) the contribution of currents from inner elements at the terminals is covered by $\frac{d}{dt}\mathbf{q}(\mathbf{x})$ and $\mathbf{j}(\mathbf{x})$, respectively. As the nature, i.e. reactive or nonreactive, of the adjacent elements in the final circuit is not known, when the cell is designed, additional unknowns $\mathbf{j_{pin}}$, i.e. *pin currents*, are introduced on the subcircuit level. We assume that the cell under consideration has d_e nodes and $d_{pin} < d_e$ of them are terminals. Then we can extend (1) to

$$\frac{d}{dt}[\mathbf{q}(\mathbf{x})] + \mathbf{j}(\mathbf{x}) + \mathbf{B}\mathbf{u}(t) + \mathbf{A_{pin}}\mathbf{j_{pin}} = \mathbf{0}, \tag{9}$$

with $\mathbf{j_{pin}} \in \mathbb{R}^{d_{pin}}$ and where $\mathbf{A_{pin}} \in \{0,1\}^{d \times d_{pin}}$ with d_{pin} columns containing exactly one non-zero element is an incidence matrix describing the topological distribution of the pin.

The pin currents $\mathbf{j_{pin}}$ can be determined when there is an external circuitry available, completing the network equations. During the process of developing the single cell a suitable test bench that emulates the typical environment the subcircuit will operate in later has to be defined by the designer.

Communication amongst electrical devices is done in terms of time varying voltages and currents. Regarding the cell (9) we can either inject the currents $\mathbf{j_{pin}}$ and get the voltage response at the terminals or supply the voltages at the pins and receive the according currents. The state $\mathbf{x}$ comprises the node voltages and the currents through inductors and voltage sources. With the pin currents' incidence matrix $\mathbf{A_{pin}}$ we can access the node voltages at the terminals by

$$\mathbf{x_{pin}} = \mathbf{A}_{\mathbf{pin}}^T \cdot \mathbf{x}. \tag{10}$$

Now, *current injection* means regarding $\mathbf{j_{pin}}$ as inputs returning $\mathbf{x_{pin}}$ as the output. Therefore, we can write

$$0 = \tfrac{d}{dt}\,[\mathbf{q}(\mathbf{x})] + \mathbf{j}(\mathbf{x}) + (\mathbf{B}\mathbf{A_{pin}}) \begin{pmatrix} \mathbf{u}(t) \\ \mathbf{j_{pin}}(t) \end{pmatrix}, \tag{11}$$

$$\mathbf{y} = \mathbf{A}_{\mathbf{pin}}^T \mathbf{x}. \tag{12}$$

Voltage injection on the other hand implies that the node voltages at the terminals are prescribed and corresponding pin currents are additional unknowns, i.e., they are added to the state vector:

$$0 = \tfrac{d}{dt} \begin{pmatrix} \mathbf{q}(\mathbf{x}) \\ 0 \end{pmatrix} + \begin{pmatrix} \mathbf{j}(\mathbf{x}) + \mathbf{A_{pin}}\mathbf{j_{pin}} \\ -\mathbf{A}_{\mathbf{pin}}^T\mathbf{x} \end{pmatrix} + \begin{pmatrix} \mathbf{B} \\ & \mathbf{I} \end{pmatrix} \begin{pmatrix} \mathbf{u}(t) \\ \mathbf{x_{pin}}(t) \end{pmatrix} \tag{13}$$

$$\mathbf{y} = (\mathbf{0} \ \ \mathbf{I}) \begin{pmatrix} \mathbf{x} \\ \mathbf{j_{pin}} \end{pmatrix} \tag{14}$$

Finally, the common structure of both approaches is

$$0 = \tfrac{d}{dt}\,[\bar{\mathbf{q}}_\lambda(\bar{\mathbf{x}}_\lambda)] + \bar{\mathbf{j}}_\lambda(\bar{\mathbf{x}}_\lambda) + (\bar{\mathbf{B}}_\lambda \ \ \mathbf{C}_\lambda) \begin{pmatrix} \mathbf{u}_\lambda(t) \\ \mathbf{u}_{\mathbf{pin},\lambda} \end{pmatrix}, \tag{15a}$$

$$\mathbf{y}_\lambda = \mathbf{C}_\lambda^T \bar{\mathbf{x}}_\lambda \in \mathbb{R}^{d_{\mathrm{pin},\lambda}}, \tag{15b}$$

where we introduce λ as an identifier for the cell, taken from some set $\mathscr{I}$ of indices. Viewed from the outside, the cell (15) appears just in terms of its input-output behavior, i.e., given $\mathbf{u}_{\mathbf{pin},\lambda} \in \mathbb{R}^{d_{\mathrm{pin},\lambda}}$ it returns $\mathbf{y}_\lambda$.

Now, we turn our attention to the circuitry a cell might be embedded in. We assume that the state space of this circuit level has dimension d, i.e., it is described by the states $x \in \mathbb{R}^d$. Furthermore, we let this level consist of $r \in \mathbb{N}$ instantiations of cells, i.e., $\mathscr{I} = \{1,2,\dots,r\}$, only. After due consideration we see that this electrical system is described by

$$0 = \sum_{\lambda \in \mathscr{I}} \mathbf{A}_\lambda^T \mathbf{y}_\lambda, \quad \text{with } \mathbf{A}_\lambda \in \{0,1\}^{d_{\mathrm{pin},\lambda} \times d} \tag{16}$$

where $\mathbf{y}_\lambda$ is determined by (15) with $\mathbf{u}_{\mathbf{pin},\lambda} = \mathbf{A}_\lambda \mathbf{x}$ for all $\lambda \in \mathscr{I}$.

As all the instantiated cells appear just in terms of their input-output behavior, we are free to reduce the order of single models (15) and use them again on level (16). Furthermore, as subcircuits are regarded as special elements, we can also include other elements on this level. Hence we can write in general form again

$$\frac{d}{dt}\,[\mathbf{q}(\hat{\mathbf{x}})] + \mathbf{j}(\hat{\mathbf{x}}) + \hat{\mathbf{B}}\hat{\mathbf{u}} = 0, \quad \text{with } \hat{\mathbf{x}} = \left(\mathbf{x}^T, \mathbf{y}_1^T, \dots, \mathbf{y_r}^T\right)^T, \tag{17}$$

which can be seen as a subcircuit on another level again. In this way, a hierarchical model order reduction would be possible.

3   Trajectory Piecewise Linear Model Order Reduction

The idea behind the Trajectory Piecewise Linear (TPWL) method is to linearize (1) several times along a given trajectory $\tilde{\mathbf{x}}(t)$ (corresponding to some typical input $\tilde{\mathbf{u}}(t)$) that satisfies

$$\frac{d}{dt}[\mathbf{q}(\tilde{\mathbf{x}})] + \mathbf{j}(\tilde{\mathbf{x}}) + \mathbf{B}\tilde{\mathbf{u}} = \mathbf{0}. \tag{18}$$

Note that in [16] an alternative version of TPWL is described where the nonlinear functions $\mathbf{q}(t,\mathbf{x}), \mathbf{j}(t,\mathbf{x})$ are linearized around the Linearization Tuples $(t_i, \tilde{\mathbf{x}}(t_i))$. Below the nonlinear system itself is linearized around the complete trajectory $\tilde{\mathbf{x}}(t)$. Furthermore we can use just Linearization Points (LPs) $\tilde{\mathbf{x}}(t_i)$ instead of Linearization Tuples because the system in (1) does not depend explicitly on t and behaves linearly with respect to $\mathbf{u}$. Define $\mathbf{y}(t) = \mathbf{x}(t) - \tilde{\mathbf{x}}(t)$ and $\bar{\mathbf{u}}(t) = \mathbf{u}(t) - \tilde{\mathbf{u}}(t)$. Linearizing the nonlinear equation (1) gives us

$$\frac{d}{dt}[\mathbf{q}(\tilde{\mathbf{x}})] + \mathbf{j}(\tilde{\mathbf{x}}) + \mathbf{B}\tilde{\mathbf{u}} + \frac{d}{dt}[\mathbf{C}(\tilde{\mathbf{x}})\mathbf{y}] + \mathbf{G}(\tilde{\mathbf{x}})\mathbf{y} + \mathbf{B}\bar{\mathbf{u}} = \mathbf{0}. \tag{19}$$

Because the trajectory $\tilde{\mathbf{x}}(t)$ satisfies (18) we obtain the following time-varying linear system for $\mathbf{y}$

$$\frac{d}{dt}[\mathbf{C}(\tilde{\mathbf{x}}(t))\mathbf{y}(t)] + \mathbf{G}(\tilde{\mathbf{x}}(t))\mathbf{y}(t) + \mathbf{B}\bar{\mathbf{u}}(t) = \mathbf{0}. \tag{20}$$

The main idea of TPWL is to approximate the time-varying Jacobian matrices $\mathbf{C}(\tilde{\mathbf{x}}(t)), \mathbf{G}(\tilde{\mathbf{x}}(t))$ by a weighted combination of piecewise constant matrices. Then a (finite) sequence of linearized local systems is used to create a globally reduced subspace. The final TPWL model is constructed as a weighted sum of all locally linearized reduced systems. The disadvantage of standard linearization methods is that they only deliver good results in the neighborhood of the chosen linearization point (LP) $\mathbf{x}(t_i)$ (see Fig. 1). To overcome this, several linearized models are created in TPWL. The LPs can be computed simultaneously with the numerical time-integration of (18) for the trajectory $\tilde{\mathbf{x}}(t)$. This procedure can be described by the following steps:

1. Set an absolute accuracy factor $\varepsilon > 0$, set $i = 1$.
2. Linearize the system around $\tilde{\mathbf{x}}_i = \tilde{\mathbf{x}}(t_i)$. This implies:

$$\mathbf{C}_i\dot{\mathbf{y}} + \mathbf{G}_i\mathbf{y} + \mathbf{B}\bar{\mathbf{u}}(t) = \mathbf{0}, \tag{21}$$

with $\mathbf{C}_i = \frac{\partial}{\partial \mathbf{x}}\mathbf{q}(\tilde{\mathbf{x}})\big|_{\tilde{\mathbf{x}}_i}$ and $\mathbf{G}_i = \frac{\partial}{\partial \mathbf{x}}\mathbf{j}(\tilde{\mathbf{x}})\big|_{\tilde{\mathbf{x}}_i}$, where $\tilde{\mathbf{x}}_i$ stays for $\tilde{\mathbf{x}}(t_i)$. Save $\mathbf{C}_i$, and $\mathbf{G}_i$.

3. Reduce the linearized system to dimension $r \ll d$ by an appropriate linear MOR method, like "Poor Man's TBR" [12] or by Krylov-subspace methods [11]. This implies

$$\mathbf{C}_i^r\dot{\mathbf{z}} + \mathbf{G}_i^r\mathbf{z} + \mathbf{B}_i^r\bar{\mathbf{u}}(t) = \mathbf{0}, \tag{22}$$

where $\mathbf{C}_i^r = \mathbf{V}_i^T\mathbf{C}_i\mathbf{V}, \mathbf{G}_i^r = \mathbf{V}_i^T\mathbf{G}_i\mathbf{V}_i, \mathbf{B}_i^r = \mathbf{V}_i^T\mathbf{B}$ with $\mathbf{V}_i \in \mathbb{R}^{d \times r_i}, \mathbf{z} \in \mathbb{R}^{r_i}$ and $\mathbf{y} \approx \mathbf{V}_i\mathbf{z}$. Save the local projection matrix $\mathbf{V}_i$. To preserve sparsity it could be preferable to diagonalize the reduced systems afterwards, although this destroys the orthogonality.

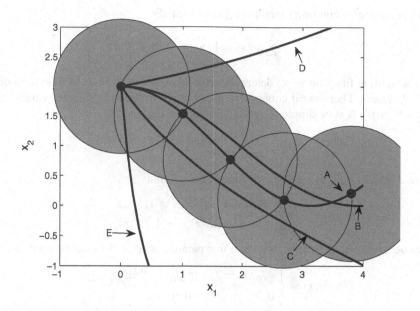

Fig. 1. The Linearization Points of this TPWL model are derived from the trajectory A. Because solutions B and C are in the neighborhood of the surrounding balls, they can be efficiently simulated using a TPWL model. But this is not the case for the solutions D and E.

4. Integrate the original system (1) until $\frac{\|\tilde{\mathbf{x}}(t_k) - \tilde{\mathbf{x}}_i\|}{\|\tilde{\mathbf{x}}_i\|} > \varepsilon$. Then we choose $\tilde{\mathbf{x}}(t_k)$ as $(i + 1)$-th LP. Set $i = i + 1$ and go to step 2.

Steps 2 to 4 are repeated until the end of the given trajectory. In this way, a finite number of locally reduced subspaces with bases $\mathbf{V}_1, \ldots, \mathbf{V}_s$ are created corresponding to the LPs $\{\tilde{\mathbf{x}}(t_1), \ldots, \tilde{\mathbf{x}}(t_s)\}$. All locally reduced subspaces are merged into a globally reduced subspace and each locally linearized system (21) is now projected onto this global subspace. The procedure can be described by the following steps:

1. Define $\tilde{\mathbf{V}} = [\mathbf{V}_1, \ldots, \mathbf{V}_s] \in \mathbb{R}^{d \times (r_1 + \ldots + r_s)}$.
2. Calculate the SVD of $\tilde{\mathbf{V}}$: $\tilde{\mathbf{V}} = \mathbf{U} \boldsymbol{\Sigma} \mathbf{W}^T$ with $\mathbf{U} = [\mathbf{u}_1, \ldots, \mathbf{u}_d] \in \mathbb{R}^{d \times d}, \boldsymbol{\Sigma} \in \mathbb{R}^{d \times \bar{r}s}$ and $\mathbf{W} \in \mathbb{R}^{\bar{r}s \times \bar{r}s}$, where $\bar{r} = (r_1 + \ldots + r_s)/s$.
3. Define the new global projection matrix $\mathbf{V} \in \mathbb{R}^{d \times r}$ as $[\mathbf{u}_1, \ldots, \mathbf{u}_r]$.
4. Project each local linearized system (21) onto $\mathbf{V}$.

Because of the construction of the global projection matrix $\mathbf{V}$ it is approximately true that $\mathscr{R}(\mathbf{V}_i) \subset \mathscr{R}(\mathbf{V})$ for $i = 1, \ldots, s$. All locally reduced linearized reduced systems are combined in a weighted sum to build the global TPWL model. Note that the TPWL model in [16] directly approximates $\mathbf{x}$ instead of $\mathbf{y} = \mathbf{x} - \tilde{\mathbf{x}}$ by $\mathbf{Vz}$. Then it is necessary to add the defect of the trajectory $\tilde{\mathbf{x}}$ to the new input vector. But if the original state $\mathbf{x} = \tilde{\mathbf{x}} + \mathbf{y}$ is approximated by $\tilde{\mathbf{x}} + \mathbf{Vz}$ the reduced state $\mathbf{z} \in \mathbb{R}^r$ satisfies

$$\sum_{i=1}^{s} w_i(\mathbf{z}) \left[\mathbf{V}^T \mathbf{C}_i \mathbf{V} \dot{\mathbf{z}} + \mathbf{V}^T \mathbf{G}_i \mathbf{V} \mathbf{z} + \mathbf{V}^T \mathbf{B} \bar{\mathbf{u}}(t) \right] = \mathbf{0}. \tag{23}$$

In (23) we need weighting functions $w_i(\mathbf{z})$ that satisfy

$$\sum_{i=1}^{s} w_i(\mathbf{z}) = 1, \; w_i(\mathbf{z}) \in [0,1]. \tag{24}$$

The weighting function $w_i(\mathbf{z})$ determines the influence of the i-th local system on the global system. Therefore it equals zero if $\mathbf{z}$ is far from the i-th projected Linearization Point $\mathbf{V}^T \mathbf{x}(t_i)$. A very simple weighting function is defined by

$$w_i(\mathbf{z}) = \begin{cases} 1 \text{ if } i = \min\{j \mid d_j(\mathbf{z}) = d_{\min}(\mathbf{z})\}, \\ 0 \text{ otherwise.} \end{cases} \tag{25}$$

Here $d_i(\mathbf{z})$ and $d_{\min}(\mathbf{z})$ are distance functions such that

$$d_i(\mathbf{z}) = \|\mathbf{z} - \mathbf{V}^T \mathbf{x}(t_i)\|, \; i = 1, \ldots, s, \tag{26}$$

$$d_{\min}(\mathbf{z}) = \min\{d_i(\mathbf{z}), \; i = 1, \ldots, s\}. \tag{27}$$

A more advanced alternative, with two free parameters $\alpha, \varepsilon > 0$, can be used like

$$\bar{w}_i(\mathbf{z}) = \begin{cases} \exp(-\frac{\alpha d_i(\mathbf{z})}{d_{\min}(\mathbf{z})}) \text{ if } \exp(-\frac{\alpha d_i(\mathbf{z})}{d_{\min}(\mathbf{z})}) > \varepsilon, \\ 0 \qquad\qquad \text{otherwise.} \end{cases} \tag{28}$$

$$w_i(\mathbf{z}) = [\textstyle\sum_{k=1}^{s} \bar{w}_k(\mathbf{z})]^{-1} \bar{w}_i(\mathbf{z}). \tag{29}$$

The TPWL method delivers reduced models that are cheap to simulate because the reduced model (23) does not need any evaluations of the original functions $\mathbf{q}, \mathbf{j}$ and Jacobian matrices $\mathbf{C}, \mathbf{G}$, because all matrices $\mathbf{V}^T \mathbf{C}_i \mathbf{V}, \mathbf{V}^T \mathbf{G}_i \mathbf{V}$ and $\mathbf{V}^T \mathbf{B}$ can be computed before the simulation. The reduction error of a TPWL method consists of a linearization and a truncation part. This error can be controlled by use of the Linearization Points [16,17]. Clearly the accuracy becomes higher for a large number of them. For strongly nonlinear systems the price is that a large number of Linearization Points is required to keep the linearization error sufficiently small. If the weighting functions $w_i(\mathbf{z})$ are not updated within the Newton method this will imply additional stepsize restrictions.

In the next three sections we will show how nonlinear systems can be reduced without linearization. Then the reduced models are obtained by Galerkin projection of the original model.

4   Empirical Balanced Truncation

For LTI systems the controllability and observability Gramians also satisfy

$$\mathbf{W} = \int_0^\infty e^{\mathbf{A}t} \mathbf{B} \mathbf{B}^T e^{\mathbf{A}^T t} dt, \; \mathbf{M} = \int_0^\infty e^{\mathbf{A}^T t} \mathbf{C}^T \mathbf{C} e^{\mathbf{A}t} dt. \tag{30}$$

Consider $\mathbf{X}(t) = [\mathbf{x}_1, \ldots, \mathbf{x}_m] = e^{\mathbf{A}t} \mathbf{B}$ and $\mathbf{Y}(t) = [\mathbf{y}_1, \ldots, \mathbf{y}_n] = \mathbf{C} e^{\mathbf{A}t}$. Let $\delta(t)$ be Dirac's delta function, then $\mathbf{x}_i$ and $\mathbf{y}_j$ satisfy

$$\frac{d}{dt}[\mathbf{q}(\mathbf{x}_i)] + \mathbf{j}(\mathbf{x}_i) = \mathbf{b}_i \delta(t), \; \mathbf{x}_i(0) = \mathbf{0}, \quad i = 1, \ldots, m, \tag{31}$$

$$\begin{cases} \frac{d}{dt}[\mathbf{q}(\mathbf{x}_j)] + \mathbf{j}(\mathbf{x}_j) = \mathbf{0}, \; \mathbf{x}_j(0) = \mathbf{e}_j, \\ \mathbf{y}_j = \mathbf{h}(\mathbf{x}_j), \end{cases} \quad j = 1, \ldots, n. \tag{32}$$

Then it follows for LTI systems that the Gramians can be expressed in terms of the correlations of the states and outputs

$$\mathbf{W} = \int_0^\infty e^{\mathbf{A}t}\mathbf{B}\mathbf{B}^T e^{\mathbf{A}^T t} dt = \int_0^\infty \mathbf{X}(t)\mathbf{X}(t)^T dt = \sum_{i=1}^m \int_0^\infty \mathbf{x}_i(t)\mathbf{x}_i(t)^T dt, \quad (33)$$

and

$$\mathbf{M} = \int_0^\infty e^{\mathbf{A}^T t}\mathbf{C}^T\mathbf{C}e^{\mathbf{A}t} dt = \int_0^\infty \mathbf{Y}(t)^T\mathbf{Y}(t) dt = \sum_{i=1}^n \int_0^\infty \mathbf{y}_i(t)^T\mathbf{y}_i(t) dt. \quad (34)$$

If the states $[\mathbf{x}_1, \ldots, \mathbf{x}_m]$ and $[\mathbf{y}_1, \ldots, \mathbf{y}_n]$ are available, these Gramians $\mathbf{W}, \mathbf{M}$ can be numerically integrated as follows

$$\mathbf{W} \approx \hat{\mathbf{W}} = \sum_{i=1}^m \frac{1}{N} \sum_{k=1}^N \mathbf{x}_i(t_k)\mathbf{x}_i(t_k)^T, \quad \mathbf{M} \approx \hat{\mathbf{M}} = \sum_{i=1}^n \frac{1}{N} \sum_{k=1}^N \mathbf{y}_i(t_k)^T\mathbf{y}_i(t_k). \quad (35)$$

For LTI systems we have that $\hat{\mathbf{W}} \to \mathbf{W}$, $\hat{\mathbf{M}} \to \mathbf{M}$ if $N \to \infty$. Empirical balanced truncation (EBT) applies these formulae for $\hat{\mathbf{W}}, \hat{\mathbf{M}}$ to nonlinear systems with a larger set of inputs and initial values to include also the nonlinear properties. It is a powerful method because it really approximates the relationship between the input and output and neglects all other phenomena but also needs a lot of experiments. Then TBR or another linear MOR technique is used to balance $\hat{\mathbf{W}}, \hat{\mathbf{M}}$ by solving a system of Lyapunov equations. Thus a basis $\mathbf{V}$ can be constructed by truncation. The reduced model for $\mathbf{z} \in \mathbb{R}^r$ is constructed by Galerkin projection.

5 Proper Orthogonal Decomposition

The Proper Orthogonal Decomposition (POD), also known as the Principal Component Analysis (PCA) and the Karhunen-Loéve expansion, is a special case of Empirical Balanced Truncation. It approximates the controllability Gramian $\hat{\mathbf{W}}$ by using only one trajectory

$$\hat{\mathbf{W}} = \frac{1}{N} \sum_{k=1}^N \mathbf{x}_1(t_k)\mathbf{x}_1(t_k)^T = \mathbf{V}\mathbf{\Sigma}\mathbf{V}^T. \quad (36)$$

Because the two Gramians are assumed to be equal, the POD basis can be found from the singular value decomposition

$$\hat{\mathbf{W}} = \mathbf{V}\mathbf{\Sigma}\mathbf{V}^T, \quad (37)$$

where $\mathbf{V} \in \mathbb{R}^{d \times d}$ is an orthogonal matrix and $\mathbf{\Sigma}$ a positive real diagonal matrix.

Thus the POD basis $\mathbf{V}_r = (\mathbf{v}_1 \ldots \mathbf{v}_r)$ is an orthonormal basis and derived from the collected state evolutions (snapshots)

$$\mathbf{X} = (\mathbf{x}(t_1) \ldots \mathbf{x}(t_N)). \quad (38)$$

The POD method is particularly popular for systems governed by nonlinear partial differential equations describing computational fluid dynamics. Analytical solutions do not exist for such systems and the collected data may serve as the only adequate description of the system dynamics. The POD basis is found by minimizing the time-averaged approximation error given by

$$av\left(\| \mathbf{x}(t_k) - \mathbf{x}_n(t_k) \|_2\right). \tag{39}$$

The averaging operator $av(\cdot)$ is defined as:

$$av(f) := \frac{1}{N} \sum_{k=1}^{N} f(t_k). \tag{40}$$

Solving the minimization problem of (39) is equivalent to computing the eigenvalue decomposition of $\frac{1}{N}\mathbf{XX}^T$. Because $\frac{1}{N}\mathbf{XX}^T$ is a symmetric positive definite matrix there exists an orthogonal matrix $\mathbf{V}_r \in \mathbb{R}^{d \times r}$ and a positive real diagonal matrix $\Lambda_r \in \mathbb{R}^{r \times r}$ such that

$$\frac{1}{N}\mathbf{XX}^T\mathbf{V}_r = \mathbf{V}_r\Lambda_r. \tag{41}$$

The term $\frac{1}{N}\mathbf{XX}^T$ equals the state covariance matrix. The POD basis is a subset of the eigenvectors of this covariance matrix and is stored by the matrix $\mathbf{V}_r$. The most important POD basis function is the eigenvector corresponding to the first eigenvalue. The truncation degree is determined from the eigenvalue distribution in $\Lambda_r = \text{diag}(\lambda_1, \ldots, \lambda_r)$. Based on the commonly adopted ad-hoc criterion, the truncation degree r should at least capture 99% of the total energy. The POD basis minimizes, in Least Squares sense, (39) over all possible bases. Error estimates for the solutions obtained from the reduced model are available in [10].

6   Galerkin Projection

For each t let the state $\mathbf{x}(t) \in \mathbb{R}^d$ belong to a separable Hilbert space $\mathscr{X}$, equipped with the Euclidian inner product. Then for all t the state $\mathbf{x}$ can be expanded in a basis $\mathbf{V} = \begin{pmatrix} \mathbf{v}_1 & \ldots & \mathbf{v}_d \end{pmatrix}$

$$\mathbf{x}(t) = \sum_{i=1}^{d} z_i(t)\mathbf{v}_i. \tag{42}$$

The basis is derived from various criteria based on the approximation quality of the original state $\mathbf{x}$ by its truncated expansion $\mathbf{x}_r$ as defined in (43)

$$\mathbf{x}(t) \approx \mathbf{x}_r(t) = \sum_{i=1}^{r} z_i(t)\mathbf{v}_i. \tag{43}$$

The order r of the truncated expansion is lower than the order d of the original expansion. Different reduction methods yield different bases.

The reduced order model is the model that describes the dynamics of the basis coefficients or the reduced state $\mathbf{z} = \{z_1, \ldots, z_r\}$. In many methods the reduced order model is derived by replacing the original state $\mathbf{x}$ by its truncated expansion $\mathbf{x}_r$ and projecting the original equations onto a truncated basis

$$\mathbf{W}_r = \begin{pmatrix} \mathbf{w}_1 \ \ldots \ \mathbf{w}_r \end{pmatrix}. \tag{44}$$

Galerkin projection of (1) onto $\mathbf{V}_r$ along $\mathbf{W}_r$ results in the reduced DAE model

$$\begin{cases} \frac{d}{dt} \left[\mathbf{W}_r^T \mathbf{q}(\mathbf{V}_r \mathbf{z}) \right] + \mathbf{W}_r^T \mathbf{j}(\mathbf{V}_r \mathbf{z}) + \mathbf{W}_r^T \mathbf{B}\mathbf{u} = \mathbf{0}, \quad \mathbf{z}(0) = \mathbf{z}_0, \\ \qquad\qquad\qquad \mathbf{y} = \mathbf{h}(\mathbf{V}_r \mathbf{z}). \end{cases} \tag{45}$$

The original d-dimensional DAE model is reduced to an r-dimensional DAE reduced order model by means of the Galerkin projection. Unfortunately, the resulting reduced order model (45) for $\mathbf{z} \in \mathbb{R}^r$ is not always solvable for any arbitrary truncation degree r. Furthermore, in contrast to TPWL this reduced model still needs evaluations of the original model, because the functions $\mathbf{V}_r^T \mathbf{q}(t, \mathbf{V}_r \mathbf{z})$ and $\mathbf{V}_r^T \mathbf{j}(t, \mathbf{V}_r \mathbf{z})$ cannot be expanded before the simulation.

For circuit models the snapshots can be collected from a transient simulation with fixed parameters and sources. The reduced model can also be used to approximate the model for different parameters or sources as long as the solution still approximately lies in the projected space. For circuit models with a lot of redundancy the reduced model can have a much smaller dimension. Unfortunately, direct application of POD to circuit models does not work well in practice. Firstly, for Differential Algebraic Equations the Galerkin projection scheme may yield an unsolvable reduced order model. This problem has been studied in more detail in [5,14]. Secondly, the computational effort required to solve the reduced order model and the original model is about the same in nonlinear cases. This is due to the fact that the evaluation costs of the reduced model (45) are not reduced at all because $\mathbf{V}_r$ will be a dense matrix in general.

7   Missing Point Estimation (MPE)

As mentioned before, many MOR techniques for nonlinear systems as (1) use Galerkin projection to obtain a reduced model of the following type

$$\frac{d}{dt} \left[\mathbf{W}^T \mathbf{q}(\mathbf{V}\mathbf{z}) \right] + \mathbf{W}^T \mathbf{j}(\mathbf{V}\mathbf{z}) + \mathbf{W}^T \mathbf{B}\mathbf{u} = \mathbf{0}. \tag{46}$$

The original state can be obtained by $\mathbf{x} = \mathbf{V}\mathbf{z}$. Thus indeed it is assumed that $\mathbf{x} \in \mathcal{R}(\mathbf{V})$. If $\mathbf{x} \in \mathbb{R}^d$ and $\mathbf{V} \in \mathbb{R}^{d \times r}$ where $r \ll d$ it is clear that the reduced model (46) is of much smaller size than the original model (1). For LTI systems with $\mathbf{q}(t, \mathbf{x}) = \mathbf{C}\mathbf{x}$ and $\mathbf{j}(t, \mathbf{x}) = \mathbf{G}\mathbf{x} - \mathbf{s}(t)$ it is really possible to reduce the simulation time for small r. In particular if the reduced model is diagonalized, we certainly get a model that is very cheap to solve. For the general case it is much worse because then the evaluation costs are not reduced at all. But if the linear algebra part is dominant, we still can expect a speed-up. Despite the resulting low dimensional model, the computational effort required to solve the reduced order model and the original model is relatively the same in nonlinear cases. It may even occur that the original model is cheaper to evaluate than the reduced order model. The low dimensionality is obtained by means of projection, either by the Galerkin projection method or the least square method. In the projection schemes, the original numerical model must be projected onto the projection space. It implies that the original model must be re-evaluated when the original numerical model is time-varying, which is the general case for nonlinear systems. A consequence is that the evaluation costs for the reduced model are not reduced at all.

Missing Point Estimation (MPE) is a well-known technique that modifies the matrix **V** such that only a part of the equations of the original model have to be evaluated. This makes POD applicable for model order reduction of nonlinear DAEs. The Missing Point Estimation (MPE) was proposed in [2] as a method to reduce the computational cost of reduced order, nonlinear, time-varying models. The method is inspired by the Gappy-POD approach that was introduced by Everson and Sirovich in [8]. More details can be found in [5,15].

7.1 Adapted POD Method

Assume that we have a benchmark solution $\tilde{\mathbf{x}}(t)$ of the DAE (1). Consider the snapshot matrix $\mathbf{X} \in \mathbb{R}^{d \times N}$. Consider the singular value decomposition of $\mathbf{X}$:

$$\mathbf{X} = \mathbf{U}\Sigma\mathbf{V}^T, \tag{47}$$

where $\mathbf{U} \in \mathbb{R}^{d \times d}, \mathbf{V} \in \mathbb{R}^{N \times N}$ are orthogonal matrices and $\Sigma \in \mathbb{R}^{d \times N}$. Thus the correlation matrix satisfies $\mathbf{W} = \frac{1}{N}\mathbf{X}\mathbf{X}^T = \frac{1}{N}\mathbf{U}\Sigma\Sigma^T\mathbf{U}^T$. Because $\Sigma\Sigma^T \in \mathbb{R}^{d \times d}$ is a positive real diagonal matrix we can write $\Sigma\Sigma^T = \Gamma^2$, where $\Gamma \in \mathbb{R}^{d \times d}$ is another positive real diagonal matrix.

In contrast to POD we introduce the matrix $\mathbf{L} = \mathbf{U}\Gamma \in \mathbb{R}^{d \times d}$, such that also $\mathbf{W} = \frac{1}{N}\mathbf{L}\mathbf{L}^T$. Note that the columns of $\mathbf{L}$ are still an orthogonal basis but not orthonormal. Then we transform the original system (1) by writing $\mathbf{x} = \mathbf{L}\mathbf{y}$ and using orthogonal Galerkin projection as follows

$$\frac{d}{dt}\left[\mathbf{L}^T\mathbf{q}(\mathbf{L}\mathbf{y})\right] + \mathbf{L}^T\mathbf{j}(\mathbf{L}\mathbf{y}) + \mathbf{L}^T\mathbf{B}\mathbf{u} = \mathbf{0}, \quad \mathbf{x} = \mathbf{L}\mathbf{y}. \tag{48}$$

Note that we are able to compute the matrix $\mathbf{L}^T\mathbf{B}$ before the simulation in contrast to the nonlinear functions $\mathbf{L}^T\mathbf{q}(\mathbf{L}\mathbf{y})$ and $\mathbf{L}^T\mathbf{j}(\mathbf{L}\mathbf{y})$. Therefore we are going to approximate $\mathbf{L}^T$ and $\mathbf{L}$ such that $\mathbf{L}^T\mathbf{q}(\mathbf{L}\mathbf{y})$ and $\mathbf{L}^T\mathbf{j}(\mathbf{L}\mathbf{y})$ become cheaper to evaluate. Note that we will use different approximations for $\mathbf{L}$ and $\mathbf{L}^T$. Because $\mathbf{L} = \mathbf{U}\Gamma$ we can approximate it by $\mathbf{U}_r\Gamma_r\mathbf{P}_r = \mathbf{L}\mathbf{P}_r^T\mathbf{P}_r$ where $\mathbf{U}_r \in \mathbb{R}^{d \times r}$ and $\Gamma_r \in \mathbb{R}^{r \times r}$ consists of the r most dominant singular values of Γ and $\mathbf{P}_r \in \{0,1\}^{r \times d}$ is a selection matrix. The matrices $\mathbf{U}_r, \Gamma_r$ and $\mathbf{P}_r$ easily follow from the singular value decomposition. But if we use this approximation we still have the problem that for each function $\mathbf{f}$ the projected function $\mathbf{L}^T\mathbf{f} \approx \mathbf{P}_r^T\Gamma_r\mathbf{U}_r^T\mathbf{f}$ needs all elements of $\mathbf{f}$. Therefore we use here also another approximation

$$\mathbf{L}^T \approx \mathbf{T}_g\mathbf{P}_g = \mathbf{L}^T\mathbf{P}_g^T\mathbf{P}_g, \tag{49}$$

where $\mathbf{P}_g \in \{0,1\}^{g \times d}$ is another selection matrix and $\mathbf{T}_g \in \mathbb{R}^{d \times g}$ contains the g columns of $\mathbf{L}^T$ with largest norm. If the singular values of Γ decrease rapidly we often need just a small number g of columns. This means that the aliasing error $\|\mathbf{T}_g\mathbf{P}_g - \mathbf{L}^T\|$ also converges rapidly to zero. Now we can approximate the transformed DAE (48) by

$$\frac{d}{dt}\left[\mathbf{L}^T\mathbf{P}_g^T\mathbf{P}_g\mathbf{q}(\mathbf{L}\mathbf{P}_r^T\mathbf{P}_r\mathbf{y})\right] + \mathbf{L}^T\mathbf{P}_g^T\mathbf{P}_g\mathbf{j}(\mathbf{L}\mathbf{P}_r^T\mathbf{P}_r\mathbf{y}) + \mathbf{L}^T\mathbf{B}\mathbf{u} = \mathbf{0}, \quad \mathbf{x} = \mathbf{L}\mathbf{y}. \tag{50}$$

Because $\mathbf{L} \approx \mathbf{L}\mathbf{P}_r^T\mathbf{P}_r$ and $\mathbf{L}^T \approx \mathbf{L}^T\mathbf{P}_g^T\mathbf{P}_g$ it also follows that

$$\mathbf{L}^T \approx \mathbf{P}_r^T\mathbf{P}_r\mathbf{L}^T\mathbf{P}_g^T\mathbf{P}_g. \tag{51}$$

Writing $\mathbf{a} = \mathbf{P}_r\mathbf{y} \in \mathbb{R}^r$ we get the following truncated system of r equations

$$\frac{d}{dt}\left[\mathbf{P}_r\mathbf{L}^T\mathbf{P}_g^T\mathbf{P}_g\mathbf{q}(\mathbf{L}\mathbf{P}_r^T\mathbf{a})\right] + \mathbf{P}_r\mathbf{L}^T\mathbf{P}_g^T\mathbf{P}_g\mathbf{j}(\mathbf{L}\mathbf{P}_r^T\mathbf{a}) + \mathbf{P}_r\mathbf{L}^T\mathbf{B}\mathbf{u} = \mathbf{0}, \quad \mathbf{x} = \mathbf{L}\mathbf{P}_r^T\mathbf{a}. \tag{52}$$

Because $\mathbf{L} = \mathbf{U}\Gamma$ and $\mathbf{L}^T = \Gamma\mathbf{U}^T$ we can also write this system as

$$\frac{d}{dt}\left[\Gamma_r\mathbf{U}_r^T\mathbf{P}_g^T\mathbf{P}_g\mathbf{q}(\mathbf{U}_r\Gamma_r\mathbf{a})\right] + \Gamma_r\mathbf{U}_r^T\mathbf{P}_g^T\mathbf{P}_g\mathbf{j}(\mathbf{U}_r\Gamma_r\mathbf{a}) + \Gamma_r\mathbf{U}_r^T\mathbf{B}\mathbf{u} = \mathbf{0}, \quad \mathbf{x} = \mathbf{U}_r\Gamma_r\mathbf{a}. \tag{53}$$

This system is still badly scaled. Therefore we have to multiply all equations by Γ_r^{-1} and write $\mathbf{z} = \Gamma_r\mathbf{a}$, such that we get

$$\frac{d}{dt}\left[\mathbf{U}_r^T\mathbf{P}_g^T\mathbf{P}_g\mathbf{q}(\mathbf{U}_r\mathbf{z})\right] + \mathbf{U}_r^T\mathbf{P}_g^T\mathbf{P}_g\mathbf{j}(\mathbf{U}_r\mathbf{z}) + \mathbf{U}_r^T\mathbf{B}\mathbf{u} = \mathbf{0}, \quad \mathbf{x} = \mathbf{U}_r\mathbf{z}. \tag{54}$$

We need just g elements of the functions $\mathbf{q},\mathbf{j}$ in this case. Define $\bar{\mathbf{q}} = \mathbf{P}_g\mathbf{q}, \bar{\mathbf{j}} = \mathbf{P}_g\mathbf{j}$ and the matrices $\mathbf{W}_{r,g} = \mathbf{P}_r\mathbf{U}^T\mathbf{P}_g^T = \mathbf{U}_r^T\mathbf{P}_g^T \in \mathbb{R}^{r \times g}$, $\bar{\mathbf{B}}_r = \mathbf{U}_r^T\mathbf{B}$. Then we get indeed

$$\frac{d}{dt}\left[\mathbf{W}_{r,g}\bar{\mathbf{q}}(\mathbf{U}_r\mathbf{z})\right] + \mathbf{W}_{r,g}\bar{\mathbf{j}}(\mathbf{U}_r\mathbf{z}) + \bar{\mathbf{B}}_r\mathbf{u} = \mathbf{0}, \quad \mathbf{x} = \mathbf{U}_r\mathbf{z}. \tag{55}$$

Because the g selected elements $\bar{\mathbf{q}},\bar{\mathbf{j}}$ of $\mathbf{q},\mathbf{j}$ only need a small subset of the elements of $\mathbf{U}_r\mathbf{z}$, it is possible to replace the dense matrix $\mathbf{U}_r$ by a sparse matrix $\mathbf{P}_h^T\mathbf{P}_h\mathbf{U}_r$ such that all unused rows of $\mathbf{U}_r$ are replaced by zero rows. The selection matrix $\mathbf{P}_h$ can easily be found from the average absolute values of the Jacobian matrices $\mathbf{C},\mathbf{G}$ along the benchmark solution. For many applications, e.g. circuit models, the required number h of rows is just slightly larger than g. Thus the matrix $\bar{\mathbf{U}}_{r,h} = \mathbf{P}_h\mathbf{U}_r \in \mathbb{R}^{h \times r}$ is often of a relatively small size. In this manner we finally get the following reduced model for $\mathbf{z} \in \mathbb{R}^r$

$$\frac{d}{dt}\left[\mathbf{W}_{r,g}\bar{\mathbf{q}}(\mathbf{P}_h^T\bar{\mathbf{U}}_{r,h}\mathbf{z})\right] + \mathbf{W}_{r,g}\bar{\mathbf{j}}(\mathbf{P}_h^T\bar{\mathbf{U}}_{r,h}\mathbf{z}) + \bar{\mathbf{B}}_r\mathbf{u} = \mathbf{0}, \quad \mathbf{x} = \mathbf{U}_r\mathbf{z}. \tag{56}$$

This reduced model can be simulated very efficiently because it does not need expensive function evaluations.

8 Applications

We consider the academic diode chain model shown in Fig. 2, that is described by the following equations

$$V_1 - U_{in}(t) = 0,$$
$$i_E - g(V_1, V_2) = 0,$$
$$g(V_1, V_2) - g(V_2, V_3) - C\frac{d}{dt}V_2 - \frac{1}{R}V_2 = 0,$$
$$\vdots$$
$$g(V_{N-1}, V_N) - g(V_N, V_{N+1}) - C\frac{d}{dt}V_N - \frac{1}{R}V_N = 0,$$
$$g(V_N, V_{N+1}) - C\frac{d}{dt}V_{N+1} - \frac{1}{R}V_{N+1} = 0,$$

$$g(V_a, V_b) = \begin{cases} I_s(e^{\frac{V_a - V_b}{V_T}} - 1) & \text{if } V_a - V_b > 0.5 \text{ V}, \\ 0 & \text{otherwise}, \end{cases}$$

$$U_{in}(t) = \begin{cases} 20 & \text{if } t \leq 10 \text{ ns}, \\ 170 - 15t & \text{if } 10 \text{ ns} < t \leq 11 \text{ ns}, \\ 5 & \text{if } t > 11 \text{ ns}. \end{cases}$$

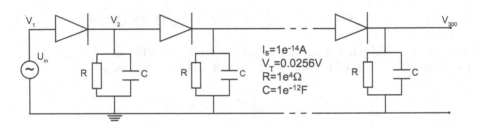

Fig. 2. Structure of the test circuit

Fig. 3 shows the numerical solution (nodal voltage in each node) of the original model at $[0, 70 \text{ ns}]$, computed in MATLAB by the Euler Backward method with fixed step sizes of 0.1 ns.

Fig. 4 indicates the redundancy of the model, as most of the eigenvalues of the correlation matrix $\frac{1}{N}\mathbf{XX}^T$ can be neglected (left) and also the aliasing error of $\mathbf{L}^T$ rapidly converges to zero (right). Fig. 5 shows the relative errors over all nodes in the time interval $[0, 70 \text{ ns}]$, defined as $\frac{\|\mathbf{Vz-x}\|}{\|\mathbf{x}\|}$, for the reduced models of different orders constructed by TPWL (left) and POD (right). For TPWL the relative error is most of the time lower then the chosen error bound $\varepsilon = 0.025$. Furthermore, for higher order reduced models, a smaller number of LPs has been used than for the reduced models with lower order, as the local systems with higher orders are more accurate. E.g. for a reduced model of order 100 we have used 42 LPs and for smaller reduced models 60 LPs. The POD models are, as expected, more accurate, but much slower to simulate than the TPWL models (see the corresponding extraction and simulation times in Table 1). However, a

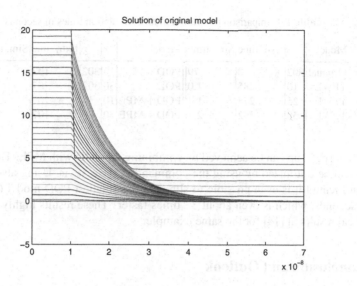

Fig. 3. Numerical solution of the full-scale nonlinear diode chain model

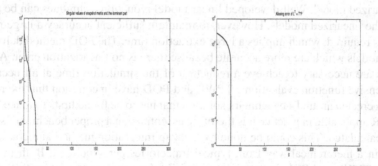

Fig. 4. The eigenvalues of the correlation matrix $\frac{1}{N}XX^T$ (left) and the aliasing error of $\mathbf{L}^T$ (right)

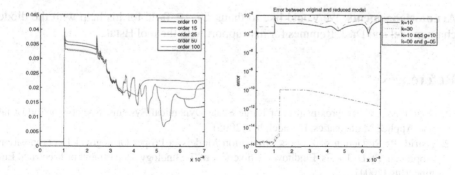

Fig. 5. Relative errors over all nodes for the reduced models created by TPWL (left) and by POD (right)

Table 1. Comparison of extraction and simulation times in seconds

Model	r	Extr. time	Sim. time	Model	r	g	Extr. time	Sim. time
Original	302	0	79	POD	10	302	107	72
TPWL	10	285	1.0	POD	30	302	107	74
TPWL	25	278	1.4	POD + MPE	10	10	107	3.9
TPWL	50	202	2.4	POD + MPE	30	35	107	10.2

significant speed up can be achieved by combining the POD with MPE. The MATLAB scripts can be optimized by using the command pcode *.m. Using also a modified Newton method it is even possible to simulate the smallest POD model ($k = g = 10$) in 2.4 seconds, which is even about 33 times faster! These results highly improve the numerical results in [14] for the same example.

9   Conclusion and Outlook

In this paper we studied how nonlinear IC models can be reduced by TPWL and POD. The first method has the advantage that it really approximates the system behavior of the linearized model. Well-developed linear model reduction techniques can be used to reduce the linearized models. However, to maintain sufficient accuracy a large number of LPs is required, which implies a large extraction time. The POD method delivers reduced models which are more accurate because there is no linearization error. Adapted versions are necessary to achieve a reduction of the simulation time at all because of the expensive function evaluations. TPWL and POD have in common that the reduced model is created around a benchmark solution that has to be found first. To make nonlinear MOR applicable in practice it is therefore essential that a proper benchmark solution can be calculated. This could be done by a cheap integration method at a coarse time-grid or in a hierarchical way from typical trajectories per subcircuit. Both the MOR methods TPWL and POD seems to be promising for reducing the simulation time for nonlinear DAE systems. They offer a good starting point for further research on MOR of non-linear dynamical systems.

Acknowledgements. We would like to thank Dr. B. Tasić for his help with the diode chain model and Dr. J. Rommes for his support with the tool Hstar.

References

1. Antoulas, A.C.: Approximation of Large-Scale Dynamical Systems. Society for Industrial and Applied Mathematics, Philadelphia (2005)
2. Astrid, P.: Reduction of Process Simulation Models: a Proper Orthogonal Decomposition Approach: Ph.D. Thesis, Eindhoven University of Technology, Department of Electrical Engineering (2004)
3. Astrid, P., Weiland, S.: On the construction of POD models from partial observations. In: Proceedings of the 44th IEEE Conf. on Decision and Control, Spain, pp. 2272–2277 (2005)

4. Astrid, P., Weiland, S., Willcox, K., Backx, A.: Missing Point Estimation in models described by Proper Orthogonal Decomposition. In: Proceedings of the 43th IEEE Conf. on Decision and Control, Bahamas, vol. 2, pp. 1767–1772 (2004)
5. Astrid, P., Verhoeven, A.: Application of Least Squares MPE technique in the reduced order modeling of electrical circuits. In: Yamamoto, Y., Sugie, T., Ohta, Y. (eds.) Proceedings of 17th Int. Symposium on Mathematical Theory of Networks and Systems (CDROM), Japan, pp. 1980–1986 (2006)
6. Benner, P., Mehrmann, V., Sorensen, D.C.: Dimension Reduction of Large-Scale Systems. Springer, Heidelberg (2006)
7. d'Elia, M.: Reduced Basis Method and Model Order Reduction for Initial Value Problems of Differential Algebraic Equations: Master's Thesis, Politecnico di Milano, Facoltà di Ingegneria dei Sistemi, Milano Leonardo (2007)
8. Everson, R., Sirovich, L.: The Karhunen-Luève procedure for gappy data. Journal Opt. Soc. Am. 12, 1657–1664 (1995)
9. Freund, R.W.: Krylov-subspace methods for reduced order modeling in circuit simulation. Journal of Computational and Applied Mathematics 123, 395–421 (2000)
10. Homescu, C., Petzold, L.R., Serban, R.: R Serban: Error estimation for reduced-order models of dynamical systems. SIAM Journal on Numerical Analysis 43, 1693–1714 (2005)
11. Odabasioglu, A., Celik, M., Pileggi, L.T.: PRIMA: Passive reduced-order interconnect macromodeling algorithm. IEEE Transactions on Computer-aided Design of Integrated Circuits and Systems 17(8), 645–654 (1998)
12. Phillips, J., Silvera, L.M.: Poor Man's TBR: A simple model reduction scheme. IEEE Transactions on Computer-aided Design of Integrated Circuits and Systems 14(1) (2005)
13. Rewienski, M., White, J.: A trajectory piecewise-linear approach to model order reduction and fast simulation of nonlinear circuits and micromachined devices. In: Proc. of the Int. Conf. on CAD, pp. 252–257 (2001)
14. Verhoeven, A.: Redundancy Reduction of IC Models by Multirate Time-Integration and Model Order Reduction: Ph.D. thesis, Eindhoven University of Technology, Department of Mathematics and Computer Science, Eindhoven (2008)
15. Verhoeven, A., Voss, T., Astrid, P., ter Maten, E.J.W., Bechtold, T.: Model order reduction for nonlinear problems in circuit simulation. Presented at the ICIAM Conf., Zürich (2007)
16. Voss, T.: Model Reduction for Nonlinear Differential Algebraic Equations: Master's Thesis, University of Wuppertal (2005)
17. Voss, T., Pulch, R., ter Maten, E., El Guennouni, A.: Trajectory piecewise linear approach for nonlinear differential-algebraic equations in circuit simulation. In: Ciuprina, G., Ioan, D. (eds.) Scientific Computing in Electrical Engineering, Sinaia, Romania, pp. 167–173. Springer, Heidelberg (2007)
18. Voss, T., Verhoeven, A., Bechtold, T., ter Maten, J.: Model order reduction for nonlinear differential-algebraic equations in circuit simulation. In: Bonilla, L.L., Moscoso, M., Vega, J.M. (eds.) Progress in Industrial Mathematics at ECMI 2006, Madrid, Spain, pp. 518–523. Springer, Heidelberg (2006)
19. Willcox, K.: Unsteady flow sensing and estimation via the gappy proper orthogonal decomposition. Computers and Fluids 35, 208–226 (2006)

Boundary Smoothness of the Solution of Maxwell's Equations of Electromagnetics

Jean-Paul Zolésio[1] and Michel C. Delfour[2]

[1] CNRS and INRIA, INRIA, 2004 route des Lucioles, BP 93,
06902 Sophia Antipolis Cedex, France
`jean-paul.zolesio@sophia.inria.fr`
[2] Centre de recherches mathématiques et Département de mathématiques et de statistique,
Université de Montréal, C.P. 6128, succ. Centre-ville,
Montréal (Qc), Canada H3C 3J7
`delfour@CRM.UMontreal.CA`

Abstract. We address the regularity of the solution to the time dependent Maxwell equations of electromagnetics in the case of metallic boundary condition under minimal regularity of the data. We extend the so-called extractor technique that we introduced in 1995 for wave equation in several cases (including the non-cylindrical case of moving domains for which the sharp-hidden regularity [10] was still an open problem). Concerning the electrical vector field we consider its normal component e at the boundary and, using a specific version of the so called pseudo-differential extractor (that we recently introduced in a different context), we obtain new sharp regularity results that are quantified in terms of curvature through the *oriented distance function* and all the intrinsic geometry we developed in the book [6].

1 Introduction

This paper deals with the regularity of the solution at the boundary of the 3D time-dependent solution E, H of Maxwell's equations of Electromagnetics. We show a *hidden regularity* result at the boundary for the electric field on a metallic obstacle. We consider a domain Ω with boundary Γ on which the boundary condition $E_\Gamma = 0$ is applied[1]. Assuming divergence-free initial data $E_i \in H^i(\Omega, \mathbb{R}^N)$, $i = 0, 1$, and divergence-free current $J \in L^2(0, \tau; L^2(\Omega, \mathbb{R}^N))$ we show that, at the boundary, the magnetic field verifies $H \in H^{1/2}(0, \tau; L^2(\Gamma, \mathbb{R}^3))$ while $\text{curl} E \in H^{-1/2}(0, \tau; L^2(\Gamma, \mathbb{R}^3))$. The proof makes use of the *Extractor technique* introduced at ICIAM 1995 [5] and in several papers ([1,3,2]); we first prove that $(DE.n)_\Gamma \in L^2(]0, \tau[\times \Gamma, \mathbb{R}^3)$, $E.n \in H^{1/2}(0, \tau; L^2(\Gamma))$ and $\nabla_\Gamma E.n \in H^{-1/2}(0, \tau; L^2(\Gamma))$. The proof of this last regularity follows a pseudo-differential extractor technique which is developed in a forthcoming paper [7].

2 Divergence-Free Solutions of Maxwell and Wave Equations

As E is the electrical field, we deal with vector functions, say $E \in C^0([0, \tau], H^1(\Omega, \mathbb{R}^N))$, where Ω is a bounded smooth domain with boundary Γ and $I =]0, \tau[$ is the

---

[1] For $N = 3$ this condition can be written $E \times n = 0$.

A. Korytowski et al. (Eds.): System Modeling and Optimization, IFIP AICT 312, pp. 492–506, 2009.

time interval. Throughout this paper we shall be concerned with divergence-free initial conditions E_0, E_1 and right-hand side F for the classical wave equation formulated in the cylindrical evolution domain $Q = I \times \Omega$. We shall discuss the boundary conditions on the lateral boundary $\Sigma = I \times \Gamma$.

2.1 Wave Deriving from Maxwell Equation

Assuming perfect media ($\varepsilon = \mu = 1$) the Ampère law is

$$\operatorname{curl} \mathbf{H} = \frac{\partial}{\partial t} E + J, \tag{1}$$

where J is the electric current density. The Faraday's law is

$$\operatorname{curl} E = -\frac{\partial}{\partial t} \mathbf{H}. \tag{2}$$

The conservation laws are

$$\operatorname{div} E = \rho, \quad \operatorname{div} \mathbf{H} = 0, \tag{3}$$

where ρ is the volume charge density. From (1) and (2), as $\operatorname{div} \operatorname{curl} = 0$, we obtain

$$\operatorname{div} J = -\operatorname{div}(E_t) = -\rho_t. \tag{4}$$

We assume that $\rho = 0$, which implies that $\operatorname{div} J = 0$. Under this assumption any E solving (1) is divergence-free as soon as the initial condition E_0 is. We shall also assume $\operatorname{div} E_0 = 0$ so that (6) will be a consequence of (1).

With $F = -J_t$, we similarly get $\operatorname{div} F = 0$ and E is solution of the usual Maxwell equation:

$$E_{tt} + \operatorname{curl} \operatorname{curl} E = F, \quad E(0) = E_0, \quad E_t(0) = E_1. \tag{5}$$

Lemma 1. *Assume that* $\operatorname{div} F = \operatorname{div} E_0 = \operatorname{div} E_1 = 0$. *Then any solution E to Maxwell equation (5) verifies the conservation condition (3) (with* $\rho = 0$*):*

$$\operatorname{div} E = 0. \tag{6}$$

We have the classical identity

$$\operatorname{curl} \operatorname{curl} E = -\Delta E + \nabla(\operatorname{div} E) \tag{7}$$

so that E is also solution of the following wave equation problem

$$E_{tt} - \Delta E = F, \quad E(0) = E_0, \quad E_t(0) = E_1. \tag{8}$$

2.2 Boundary Conditions

The physical boundary condition for metallic boundary is $E \times n = 0$ which can be written as the homogeneous Dirichlet condition on the tangential component of the field E:

$$E_\Gamma = 0 \text{ on } \Gamma. \tag{9}$$

We introduce the following Fourier-like boundary condition involving the mean curvature $\Delta b_\Omega = \lambda_1 + \lambda_2$ of the surface Γ

$$\Delta b_\Omega E.n + \langle DE.n, n \rangle = 0 \text{ on } \Gamma. \tag{10}$$

In flat pieces of the boundary this condition reduces to the usual Neumann condition.

Proposition 1. *Let E be a smooth element ($E \in \mathscr{H}^2$, see below) and the three divergence-free elements $(E_0, E_1, F) \in H^2(\Omega, \mathbb{R}^3) \times H^1(\Omega, \mathbb{R}^3) \times L^2(0, \tau; H^1(\Omega, \mathbb{R}^3))$. Then we have the following conclusions.*

  i) *Let E be solution to Maxwell-metallic system (5), (9). Then E solves the mixed wave problem (8), (9), (10) and, from Lemma 1, E solves also the free divergence condition (6).*
 ii) *Let E be solution to the wave equation (8) with "metallic" b.c. (9). Then E verifies the Fourier-like condition (10) if and only if E verifies the free divergence condition (6).*
iii) *Let E be a divergence-free solution to the "metallic" wave problem (8), (6), (9), then E solves the Maxwell problem (5), (9), (10).*

Proof. We consider $e = \operatorname{div} E$; if E is solution to Maxwell problem (5) then e solves the scalar wave equation with initial conditions $e_i = \operatorname{div} E_i = 0$, $i = 0, 1$ and right hand side $f = \operatorname{div} F = 0$. If E solves (10) then we get $e = 0$, as from the following result we get $e = 0$ on the boundary:

Lemma 2. *Let $E \in H^2(\Omega)$ solving the tangential Dirichlet condition (9), then we have the following expression for the trace of $\operatorname{div} E$:*

$$\operatorname{div} E|_\Gamma = \Delta b_\Omega \langle E, n \rangle + \langle DE.n, n \rangle \text{ on } \Gamma. \tag{11}$$

Proof. The divergence successively decomposes as follows at the boundary (see [13,12]):

$$
\begin{aligned}
\operatorname{div} E|_\Gamma &= \operatorname{div}_\Gamma(E) + \langle DE.n, n \rangle = \operatorname{div}_\Gamma(E.n\mathbf{n}) + \operatorname{div}_\Gamma(E_\Gamma) + \langle DE.n, n \rangle \\
&= \langle \nabla_\Gamma(E.n), n \rangle + E.n \operatorname{div}_\Gamma(\mathbf{n}) + \operatorname{div}_\Gamma(E_\Gamma) + \langle DE.n, n \rangle.
\end{aligned}
\tag{12}
$$

Obviously $\langle \nabla_\Gamma(E.n), n \rangle = 0$, the mean curvature of the surface Γ is $\Delta b_\Omega = \operatorname{div}_\Gamma(\mathbf{n})$ and if the field E satisfies the tangential Dirichlet condition (9) we get the following simple expression for the restriction to the boundary of the divergence:

$$\operatorname{div}(E)|_\Gamma = \Delta b_\Omega \langle E, n \rangle + \langle DE.n, n \rangle. \tag{13}$$

$\square$

Then if E satisfies the extra "Fourier-like" condition (10) we get $e = 0$ on Γ, so that $e = 0$. $\square$

2.3   The Wave-Maxwell Mixed Problem

From previous considerations, it follows that under the divergence-free assumption for the three data E_0, E_1, F, the following three problems are equivalent (in the sense that any smooth solution of one of them is solution to the two others): Maxwell problem (5), (9), Free-Wave problem (8), (6), (9), and Mixed-Wave problem (8), (9), (10). We *emphasize* that any solution to Maxwell problem satisfies the divergence-free condition (6) and the Fourier-like condition (10). Any solution to the Mixed-Wave problem satisfies (for free) the divergence-free condition (6). Any solution to the Free-Wave problem satisfies (for free) the Fourier-like condition (10). The object of this paper is to develop the proof of the following regularity result.

Theorem 1. *Let* (E_0, E_1, J) *be divergence-free vector fields in*

$$H^1(\Omega, \mathbb{R}^3)^2 \times L^2(\Omega, \mathbb{R}^3) \times H^1(I, L^2(\Omega, \mathbb{R}^3)) \tag{14}$$

with zero tangential components: $(E_0)_\Gamma = 0$. *Assume also* $\operatorname{curl} E_1 = 0$. *The Maxwell problem (5), (9) has a unique solution*

$$E \in C^0(\bar{I}, H^1(\Omega, \mathbb{R}^3)) \cap C^1(\bar{I}, L^2(\Omega, \mathbb{R}^3)) \tag{15}$$

verifying the boundary regularity:

$$\operatorname{curl} E|_\Gamma \in H^{-1/2}(I \times \Gamma, \mathbb{R}^3) \tag{16}$$

so that the magnetic field **H** *at the boundary verifies*

$$\mathbf{H}|_\Gamma \in H^{1/2}(I, L^2(\Gamma, \mathbb{R}^3)). \tag{17}$$

Moreover, we have

$$E|_\Gamma \in H^{1/2}(I, L^2(\Gamma, \mathbb{R}^3)). \tag{18}$$

Furthermore, if $J|_\Gamma \in L^2(I, L^2(\Gamma))$, *from Ampère law (1) we obtain*

$$\operatorname{curl} \mathbf{H}|_\Gamma \in H^{-1/2}(I, L^2(\Gamma, \mathbb{R}^3)). \tag{19}$$

2.3.1   Tangential Decomposition

For any vector field $G \in H^1(\Omega, \mathbb{R}^N)$ denote by G_Γ the tangential part $G_\Gamma = G|_\Gamma - <G, n> \mathbf{n}$ and (see [12,8,6,11]) consider its tangential Jacobian matrix $D_\Gamma G = D(Gop_\Gamma)|_\Gamma$ and its transpose D_Γ^*. To derive the regularity result we shall be concerned with the following three terms at the boundary:

$$(DE.n)_\Gamma, \quad \nabla_\Gamma(E.n), \quad E_t. \tag{20}$$

Lemma 3. *For all* $E \in H^2(\Omega, \mathbb{R}^N)$, *we have by direct computation:*

$$DE|_\Gamma = DE.n \otimes n + D_\Gamma E. \tag{21}$$

Obviously, as $E = E_\Gamma + \langle E, n \rangle n$, we have:

$$D_\Gamma E = D_\Gamma E_\Gamma + D_\Gamma (E.nn) \tag{22}$$

so that

$$E_\Gamma = 0 \quad \Rightarrow \quad D_\Gamma E = D_\Gamma (\langle E, n \rangle n). \tag{23}$$

Now as $D_\Gamma(\langle E, n \rangle n) = \langle E, n \rangle D_\Gamma(n) + n \otimes \nabla_\Gamma(\langle E, n \rangle)$ and as $D_\Gamma(n) = D^2 b_\Omega$, we get the following result.

Lemma 4. *Assume that $E_\Gamma = 0$. Then we have*

$$D_\Gamma E = \langle E, n \rangle D^2 b_\Omega |_\Gamma + n \otimes \nabla_\Gamma(\langle E, n \rangle). \tag{24}$$

Moreover as

$$\operatorname{div}_\Gamma E := \operatorname{div} E|_\Gamma - \langle DE.n, n \rangle \tag{25}$$

when $\operatorname{div} E = 0$, we get $\langle DE.n, n \rangle = -\operatorname{div}_\Gamma E$, and if, in addition, $E_\Gamma = 0$, we have $\langle DE.n, n \rangle = -\operatorname{div}_\Gamma(\langle E, n \rangle n)$, that is the following result.

Lemma 5. *Denote by $H = \Delta b_\Omega$ the mean curvature. Then*

$$E_\Gamma = 0 \text{ and } \operatorname{div} E = 0 \tag{26}$$

implies the following identities

i)
$$\langle DE.n, n \rangle = -H E.n, \tag{27}$$

ii)
$$DE.n = \langle DE.n, n \rangle n + (DE.n)_\Gamma = -H E.nn + (DE.n)_\Gamma \tag{28}$$

and

$$|DE.n|^2 = H^2 |E.n|^2 + |(DE.n)_\Gamma|^2 \tag{29}$$

iii)
$$DE = -H E.nn \otimes n + (DE.n)_\Gamma \otimes n + E.n D^2 b + n \otimes \nabla_\Gamma(E.n), \tag{30}$$

iv)
$$DE..DE = H^2 |E.n|^2 + |(DE.n)_\Gamma|^2 + |E.n|^2 D^2 b..D^2 b + |\nabla_\Gamma(E.n)|^2. \tag{31}$$

Proposition 2. *Let $E \in H^2(\Omega, \mathbb{R}^N)$, $\operatorname{div} E = 0$, $E_\Gamma = 0$, then:*

$$DE..DE|_\Gamma = (H^2 + D^2 b..D^2 b)|E.n|^2 + |(DE.n)_\Gamma|^2 + |\nabla_\Gamma(E.n)|^2, \tag{32}$$

that is

$$DE..DE|_\Gamma = |DE.n|^2 + |E.n|^2 D^2 b..D^2 b + |\nabla_\Gamma(E.n)|^2. \tag{33}$$

2.4   Boundary Estimate of DE

Define $2\varepsilon \overset{\text{def}}{=} DE + D^*E$ and $2\sigma \overset{\text{def}}{=} DE - D^*E$ so that

$$DE = \varepsilon(E) + \sigma(E) \tag{34}$$

and

$$\|\operatorname{curl} E\|_{L^2(\Gamma,\mathbb{R}^3)}^2 \leq 4\,\|DE\|_{L^2(\Gamma,\mathbb{R}^{N^2})}^2. \tag{35}$$

From the decomposition (21) we have:

$$\|DE\|_{L^2(I,L^2(\Gamma,\mathbb{R}^3))} \leq \|DE.n \otimes n\| + \|D^2bE.n\|, \tag{36}$$

but

$$\|DE.n \otimes n\|^2 = \int_0^\tau \int_\Gamma (DE.n \otimes n)..(DE.n \otimes n)\,dt\,d\Gamma. \tag{37}$$

That is

$$\begin{aligned}
\|DE.n \otimes n\|_{L^2(I,L^2(\Gamma,\mathbb{R}^3))}^2 &\leq \int_0^\tau \int_\Gamma |DE.n|^2\,dt\,d\Gamma \\
&= \int_0^\tau \int_\Gamma \{|(DE.n)_\Gamma|^2 + |<DE.n,n>|^2\}\,dt\,d\Gamma,
\end{aligned} \tag{38}$$

but, as $\langle DE.n,n \rangle = -\langle E,n \rangle D^2b_\Omega$, we get the estimate (35).

2.5   Extractor Identity

Let $I = \,]0,\tau[$ be the time interval and for any integer $k \geq 1$ define the spaces

$$H^k \overset{\text{def}}{=} C^0\left(\bar{I}, H^k(\Omega,\mathbb{R}^3)\right) \cap C^1\left(\bar{I}, H^{k-1}(\Omega,\mathbb{R}^3)\right), \tag{39}$$

$$\mathscr{H}^k \overset{\text{def}}{=} \left\{E \in H^k : \operatorname{div} E = 0, \ E_\Gamma = 0 \text{ on } \Gamma\right\}. \tag{40}$$

Let $F \in L^2(I,L^2(\Omega,\mathbb{R}^3))$, $E_0 \in H^1(\Omega,\mathbb{R}^3)$, $E_1 \in L^2(\Omega,\mathbb{R}^3)$ with $\operatorname{div} E_0 = \operatorname{div} E_1 = 0$. Consider $E \in \mathscr{H}^1$ solution of the equations

$$A.E := E_{tt} - \Delta E = F \in L^2(I,L^2(\Omega,\mathbb{R}^3)), \tag{41}$$

$$E(0) = E_0, \quad E_t(0) = E_1. \tag{42}$$

2.5.1   The Extractor $e(V)$

Let $E \in \mathscr{H}^2$, and $V \in C^0([0,\tau[,C^2(D,\mathbb{R}^N))$, $\langle V(t,.),n \rangle = 0$ on ∂D. Consider its flow mapping $T_s = T_s(V)$ and the derivative:

$$e(V) \overset{\text{def}}{=} \frac{\partial}{\partial s}\mathscr{E}(V,s)\Big|_{s=0}, \tag{43}$$

where

$$\mathscr{E}(V,s) \overset{\text{def}}{=} \int_0^1 \int_{\Omega_s} \left(|E_t \circ T_s^{-1}|^2 - D(E \circ T_s^{-1})..D(E \circ T_s^{-1})\right) dx\,dt. \tag{44}$$

By change of variable

$$D(E \circ T_s^{-1}) \circ T_s = DE.DT_s^{-1} \tag{45}$$

we get the second expression

$$\mathscr{E}(V,s) = \int_0^1 \int_\Omega (|E_t|^2 - (DE.DT_s^{-1})..(DE.DT_s^{-1}))J(t)\,dx\,dt. \tag{46}$$

We have two expressions (44) and (46) for the same term $\mathscr{E}(V,s)$. The first one is an integral on a mobile domain $\Omega_s(V)$ while the second one is an integral over the fixed domain Ω. So taking the derivative with respect to the parameter s we shall obtain two different expressions for $\mathbf{e}$ that we shall respectively denote by $\mathbf{e_1}$ and $\mathbf{e_2}$.

2.5.2   Expression for $\mathbf{e_1}$

As the element E is smooth, $E \in \mathscr{H}^2$, we can directly apply the classical results from [12]. For simplicity, assume that $\operatorname{div} V = 0$ so that $J(t) = 1$. In this specific case we get

$$\mathbf{e} = \left.\frac{\partial}{\partial s}\mathscr{E}\right|_{s=0}, \tag{47}$$

and

$$\begin{aligned}
\mathbf{e_1} = 2\int_0^1 \int_\Omega \{E_t.(-DE_t.V) - DE..D(-DE.V)\}\,dx\,dt \\
+ \int_0^1 \int_\Gamma \{|E_t|^2 - DE..DE\}v\,d\Gamma\,dt.
\end{aligned} \tag{48}$$

2.5.3   Green-Stokes Theorem

Using integration by parts:

$$\begin{aligned}
\int_0^1 \int_\Omega \{DE..D(DE.V)\}\,dx\,dt = \int_0^1 \int_\Omega \langle -\Delta E, DE.V\rangle \,dx\,dt \\
+ \int_0^1 \int_\Gamma \langle DE.n, DE.V\rangle \,d\Gamma(x)\,dt.
\end{aligned} \tag{49}$$

2.5.4   Time Integration by Parts

Then

$$\begin{aligned}
\int_0^1 \int_\Omega E_t.(DE_t.V)\,dx\,dt = \int_0^1 \int_\Omega (-E_{tt}.(DE.V) + E_t.(DE.W))\,dx\,dt \\
- \int_\Omega E_t(0).(DE(0).W)\,dx.
\end{aligned} \tag{50}$$

Furthermore, assuming that the initial condition is of the form

$$E_0 \in H^1(\Omega, \mathbb{R}^3), \quad E_1 \in L^2(\Omega, \mathbb{R}^3), \tag{51}$$

we get

$$
\mathbf{e}_1 = 2 \int_0^1 \int_\Omega \{ E_{tt}.(DE.V) - E_t.(DE.W) - \langle \Delta E, DE.V \rangle \} \, dx \, dt
$$

$$
+ 2 \int_\Omega E_1.(DE_0.W) \, dx \tag{52}
$$

$$
+ \int_0^1 \int_\Gamma \{ (|E_t|^2 - DE..DE) \langle V, n \rangle + 2 \langle DE.n, DE.V \rangle \} \, d\Gamma(x) \, dt.
$$

The discussion is now on the last boundary integral.

2.5.5   Specific Choice for V at the Boundary

As the boundary $\Gamma = \partial\Omega \in C^2$ we can apply all intrinsic geometry material introduced in [6]. Denoting by $p = p_\Gamma$ the projection mapping onto the manifold Γ (which is smoothly defined in a tubular neighborhood of Γ) we consider the oriented distance function $b = b_\Omega = d_{\Omega^c} - d_\Omega$ where $\Omega^c = \mathbb{R}^N \setminus \Omega$, and its "localized version" defined as follows (see [4]): let $h > 0$ be "a small" positive number and $\rho_h(.) \geq 0$ be a cutting scalar smooth function such that $\rho_h(z) = 0$ when $|z| > h$ and $\rho(z) = 1$ when $|z| < h/2$. Then set

$$
b_\Omega^h \stackrel{\text{def}}{=} \rho_h \circ b_\Omega \tag{53}
$$

and define the associate localized projection mapping

$$
p_h \stackrel{\text{def}}{=} I_d - b_\Omega^h \nabla b_\Omega^h \tag{54}
$$

smoothly defined in the tubular neighborhood

$$
\mathcal{U}_h(\Gamma) \stackrel{\text{def}}{=} \{ x \in D : |b_\Omega(x)| < h \}. \tag{55}
$$

Let any smooth element $v \in C^0(\Gamma)$ be given and consider the vector field V of the following form

$$
V(t,x) \stackrel{\text{def}}{=} W(x)(1-t), \quad W(x) \stackrel{\text{def}}{=} v \circ p_h \nabla b_\Omega^h. \tag{56}
$$

Then the last term (boundary integral) in (52) takes the following form:

$$
\int_0^1 \int_\Gamma \{ (|E_t|^2 - DE..DE) + 2 \langle DE.n, DE.n \rangle \} v(1-t) \, d\Gamma(x) \, dt. \tag{57}
$$

We get:

$$
\mathbf{e}_1 = \int_0^1 \int_\Gamma (|E_t|^2 - DE..DE + 2|DE.n|^2) v \, d\Gamma \, dt
$$

$$
+ 2 \int_Q (E_{tt}.DE.V - \langle \Delta E, DE.V \rangle) \, dx \, dt - \int_\Omega \langle E_t(0), DE(0).W \rangle \, dx. \tag{58}
$$

As from (33) we have

$$
DE..DE = |DE.n|^2 + D^2 b_\Omega..D^2 b_\Omega |E.n|^2 + |\nabla_\Gamma E.n|^2 \tag{59}
$$

and as

$$
|DE.n|^2 = |(DE.n)_\Gamma|^2 + (\Delta b_\Omega)^2 |E.n|^2 \tag{60}
$$

we obtain the following result.

Proposition 3

$$\mathbf{e}_1 = \int_0^1 \int_\Gamma (\tau - t)\{|E_t|^2 + |(DE.n)_\Gamma|^2 - |\nabla_\Gamma(E.n)|^2$$
$$+ |E.n|^2(H^2 - D^2b..D^2b)\}v\,d\Gamma dt \qquad (61)$$
$$+ 2\int_Q \langle A.E, DE.V \rangle dQ - 2\int_\Omega \langle E_1, D(E_0).W \rangle dx.$$

2.5.6 Second Expression for e

From (46) we obtain the s derivative as a distributed integral term as follows

$$\mathbf{e}_2 = \int_Q \{(|E_t|^2 - DE..DE)\operatorname{div}V(0) - 2DE..(-DE.DV)\}\,dx dt. \qquad (62)$$

2.5.7 Extractor Identity

As $\mathbf{e} = \mathbf{e}_1 = \mathbf{e}_2$ we get

$$\int_\Sigma (\tau - t)\{(|E_t|^2 - |\nabla_\Gamma(E.n)|^2 + |(DE.n)_\Gamma|^2 + |E.n|^2(H^2 - D^2b..D^2b))\}v\,d\Sigma$$
$$= \int_Q \{(|E_t|^2 - DE..DE)\operatorname{div}V - 2DE..(-DE.DV)\}\,dx dt \qquad (63)$$
$$- \int_Q 2(E_{tt} - \Delta E).DE.V\,dQ + \int_\Omega 2\langle E_1, DE_0.W \rangle\,dx.$$

That is

$$\int_\Sigma (\tau - t)\{(|E_t|^2 - |\nabla_\Gamma(E.n)|^2 + (DE.n)_\Gamma|^2)\}v\,d\Sigma$$
$$= \int_Q \{(|E_t|^2 - DE..DE)\operatorname{div}V - 2DE..(-DE.DV)\}\,dx dt$$
$$- 2\int_Q 2\langle A.E, DE.V \rangle dQ + \int_\Omega 2\langle E_1, DE_0.W \rangle\,dx \qquad (64)$$
$$+ \int_\Sigma (1 - t)|E.n|^2(D^2b..D^2b - H^2)v\,d\Sigma.$$

Notice that the curvature terms

$$D^2b..D^2b - H^2 = \lambda_1^2 + \lambda_2^2 - (\lambda_1 + \lambda_2)^2 = -2\kappa, \qquad (65)$$

where $\kappa = \lambda_1\lambda_2$ is the *Gauss curvature* of the boundary Γ.

3 Regularity at the Boundary

We apply twice this last identity.

3.1 Tangential Field E^τ

In a first step consider the "tangential vector field" obtained as $E^\tau \overset{\text{def}}{=} E - E.\nabla b_\Omega^h \nabla b_\Omega^h$. We get

$$E_{tt}^\tau - \Delta E^\tau = (E_{tt} - \Delta E) - (E_{tt} - \Delta E).\nabla b_\Omega^h \nabla b_\Omega^h + C. \tag{66}$$

That is $A.E^\tau = (A.E)^\tau + C$, where the commutator $C \in L^2(0,T,L^2(\Omega,\mathbb{R}^3))$ is given by

$$C \overset{\text{def}}{=} -E.\Delta b_\Omega^h \nabla b_\Omega^h - 2D^2 b_\Omega^h.\nabla b_\Omega^h \nabla b_\Omega^h - E.\nabla b_\Omega^h d^2 b_\Omega^h - 2D^2 b_\Omega^h.\nabla(E.b_\Omega^h). \tag{67}$$

The conclusion formally derives as follows: as $E^\tau \in L^2(I,H^1(\Omega,\mathbb{R}^3))$ we get the traces terms

$$E^\tau.n = E_t^\tau = 0 \in L^2(I,H^{1/2}(\Gamma)). \tag{68}$$

Since $e_1 = e_2$, we conclude by choosing the vector field of the form

$$V(t,x) = (\tau - t)\nabla b_\Omega^h = (\tau - t)\rho_h' \circ b_\Omega \nabla b_\Omega. \tag{69}$$

That is $v = 1$ and as for $0 < t \le \tau/2$ we have $\tau/2 \le \tau - t$, we get:

$$\begin{aligned}
\tau/2 \int_0^{\tau/2} \int_\Gamma |(DE^\tau.n)_\Gamma|^2 \, d\Gamma \, dt &\le \int_0^{\tau/2} (\tau - t) \int_\Gamma |(DE^\tau.n)_\Gamma|^2 \, d\Gamma \, dt \\
&\le \int_0^\tau (\tau - t) \int_\Gamma |(DE^\tau.n)_\Gamma|^2 \, d\Gamma \, dt \\
&= \int_Q (\tau - t)\left\{ (|E_t^\tau|^2 - DE^\tau..DE^\tau)\,\mathrm{div}\,(\nabla b_\Omega^h) - 2DE^\tau..(-DE^\tau.D(\nabla b_\Omega^h)) \right\} dQ \\
&\quad - 2\int_Q (\tau - t)\left\langle A.E^\tau, DE^\tau.(\nabla b_\Omega^h) \right\rangle dQ + \int_\Omega 2\left\langle E_1^\tau, DE_0^\tau.(\nabla b_\Omega^h) \right\rangle dx.
\end{aligned} \tag{70}$$

As for $0 < t < \tau$ we have $2/\tau(\tau - t) \le 2$ we get, with $T = \tau/2$

$$\begin{aligned}
\int_0^T \int_\Gamma |(DE^\tau.n)_\Gamma|^2 \, d\Gamma \, dt &\\
\le 2\int_0^{2T} \int_\Omega &\left\{ (|E_t^\tau|^2 - DE^\tau..DE^\tau)\,\mathrm{div}\,(\nabla b_\Omega^h) - 2DE^\tau..(-DE^\tau.D(\nabla b_\Omega^h)) \right\} dx\,dt \\
- 4\int_0^{2T} \int_\Omega &\left\langle A.E^\tau, DE^\tau.(\nabla b_\Omega^h) \right\rangle dx\,dt + 4/T \int_\Omega \left\langle E_1^\tau, DE_0^\tau.(\nabla b_\Omega^h) \right\rangle dx,
\end{aligned} \tag{71}$$

there exists a constant $M > 0$ such that

$$\begin{aligned}
\int_0^T \int_\Gamma &\{|E_t^\tau|^2 + |(DE^\tau.n)_\Gamma|^2\} \, d\Gamma \, dt \\
\le M \|\nabla b_\Omega^h\|_{W^{1,\infty}(\Omega,\mathbb{R}^N)} \cdots &\\
\cdot \Big\{ \|E^\tau\|_{\mathscr{H}^1(0,2T)}^2 &+ |A.E^\tau|_{L^2([0,2T]\times\Omega,\mathbb{R}^3)} + 1/T(|E_0^\tau|_{H^1(\Omega,\mathbb{R}^3)}^2 + |E_1^\tau|_{L^2(\Omega,\mathbb{R}^3)}^2) \Big\}.
\end{aligned} \tag{72}$$

Notice that

$$\nabla b_\Omega^h = \rho_h' \circ b_\Omega \nabla b_\Omega, \tag{73}$$

so that

$$\|\nabla b_\Omega^h\|_{L^\infty(\mathbb{R}^N,\mathbb{R}^N)} \leq \text{Max}_{0\leq s\leq h}|\rho_h'(s)|. \tag{74}$$

Moreover

$$D^2 b_\Omega^h = D(\rho_h' \circ b_\Omega \nabla b_\Omega) = \rho_h'' \circ b_\Omega \nabla b_\Omega \times \nabla b_\Omega + \rho_h' \circ b_\Omega D^2 b_\Omega \tag{75}$$

so that

$$\|D^2 b_\Omega^h\|_{L^\infty(\mathbb{R}^N,\mathbb{R}^{N^2})} \leq \tag{76}$$
$$\text{Max}_{0\leq s\leq h}|\rho_h''(s)| + \text{Max}_{0\leq s\leq h}|\rho_h'(s)| \, \|D^2 b_\Omega\|_{L^\infty(\mathscr{U}_h(\Gamma),\mathbb{R}^{N^2})}.$$

By choice of ρ_h in the form $\rho_h(s) = f(2s/h - 1)$ when $h/2 < s < h$ and $F(x) = 2x^3 - 3x^2 + 1$, we obtain

$$\|\rho_h\|_{C^2([0,h])} \leq \frac{8}{h^2}. \tag{77}$$

So the previous estimate is in the form

$$\|D^2 b_\Omega^h\|_{L^\infty(\mathbb{R}^N,\mathbb{R}^{N^2})} \leq C_0 \frac{1}{h^2} \|D^2 b_\Omega\|_{L^\infty(\mathscr{U}_h(\Gamma),\mathbb{R}^{N^2})} \tag{78}$$

for the *larger* h such that the following condition holds

$$D^2 b_\Omega \in L^\infty(\mathscr{U}_h(\Gamma),\mathbb{R}^{N^2}). \tag{79}$$

3.1.1   Regularity Result for E^τ

Proposition 4. *Let Ω be a bounded domain in $\mathbb{R}^3$ with boundary Γ being a C^2 manifold. Let h verify condition (79).*
   There exists a constant $M > 0$ such that for any data $(E_0, E_1, F) \in L^2(\Omega,\mathbb{R}^3) \times H^1(\Omega,\mathbb{R}^3) \times L^2(\Omega,\mathbb{R}^3)$, the vector

$$E^\tau \in \mathscr{H}^1(0,2\tau) \stackrel{\text{def}}{=} C^0\left([0,2\tau],H^1(\Omega,\mathbb{R}^3)\right) \cap C^1\left([0,2\tau],L^2(\Omega,\mathbb{R}^3)\right) \tag{80}$$

verifies

$$(DE^\tau.n)_\Gamma \in L^2(0,\tau;L^2(\Gamma,\mathbb{R}^3)) \tag{81}$$

and

$$\int_0^T \int_\Gamma \{|(DE^\tau.n)_\Gamma|^2\} d\Gamma \, dt$$
$$\leq M \|\nabla b_\Omega^h\|_{W^{1,\infty}(\Omega,\mathbb{R}^N)} \cdots \tag{82}$$
$$\cdot \{\|E\|_{\mathscr{H}^1(0,2T)}^2 + |F|_{L^2([0,2T]\times\Omega,\mathbb{R}^3)}^2 + 1/T|E_0^\tau|_{H^1(\Omega,\mathbb{R}^3)}^2 + 1/T|E_1^\tau|_{L^2(\Omega,\mathbb{R}^3)}^2\}$$

It can be verified that

$$D(E^\tau).n = (DE.n)_\Gamma. \tag{83}$$

3.2   The Normal Vector Field e

Set

$$e \stackrel{\text{def}}{=} E.\nabla b_\Omega^h. \tag{84}$$

Lemma 6

$$e_{tt} - \Delta e = (E_{tt} - \Delta E).\nabla b_\Omega^h + \theta, \tag{85}$$

where

$$\theta = D^2 b_\Omega^h..DE + \text{div}\,(D^2 b_\Omega^h.E) \quad and \quad \frac{\partial}{\partial n}e = \langle DE.n,n \rangle = -\Delta b_\Omega e \text{ on } \Gamma. \tag{86}$$

Then e is solution of the wave problem:

$$e_{tt} - \Delta e = \Theta, \tag{87}$$

where

$$\Theta = F.\nabla b_\Omega^h + D^2 b_\Omega^h..DE + \text{div}\,(D^2 b_\Omega^h.E). \tag{88}$$

3.3   Extension to $\mathbb{R}$

Let

$$\rho \in C^2(\mathbb{R}), \quad \rho \geq 0, \quad \text{supp}\,\rho \subset [-2\tau, +2\tau], \quad \rho = 1 \text{ on } [-\tau, +\tau]. \tag{89}$$

Define

$$\tilde{e} \stackrel{\text{def}}{=} \rho(t)e(t), \quad t \geq 0, \quad \tilde{e} = \rho(t)(e_0 + te_1), \quad t < 0, \tag{90}$$

which turns to be solution on $\mathbb{R}$ to the wave problem

$$\tilde{e}_{tt} - \Delta\tilde{e} = H \quad and \quad \frac{\partial}{\partial n}\tilde{e} = g, \tag{91}$$

where

$$g \stackrel{\text{def}}{=} \begin{cases} -\Delta b_\Omega \tilde{e} & \text{for } t > 0 \\ \rho(t)(\dfrac{\partial}{\partial n}e_0 + t\dfrac{\partial}{\partial n}e_1) & \text{for } t < 0, \end{cases} \quad \text{on } \Gamma \tag{92}$$

and $H \in L^2(\mathbb{R}, L^2(\Omega))$ verifies

$$H \stackrel{\text{def}}{=} \begin{cases} \rho(t)\Theta + \rho''e + 2\rho'\dfrac{\partial}{\partial t}e & \text{for } t > 0 \\ \rho''(e_0 + te_1) + 2\rho'e_1 - \rho(\Delta e_0 + t\Delta e_1) & \text{for } t < 0. \end{cases} \tag{93}$$

4 Fourier Transform

Define

$$z(\zeta)(x) \stackrel{\text{def}}{=} \int_{-\infty}^{+\infty} exp(-i\zeta t)\,\tilde{e}(t,x)\,dt, \tag{94}$$

which turns to be solution to

$$\frac{\partial}{\partial n}z = \mathscr{F}.g \text{ on } \Gamma. \tag{95}$$

Consider the perturbed domain $\Omega_s = T_s(V)(\Omega)$ with boundary $\Gamma_s = T_s(V)(\Gamma)$, the following integral and derivative

$$\mathscr{E}(s,V) \stackrel{\text{def}}{=} \int_{-\infty}^{+\infty} d\zeta \int_{\Omega_s(V)} \left(|\zeta||z \circ T_s(V)^{-1}|^2 + \frac{1}{1+|\zeta|}|\nabla(z \circ T_s(V)^{-1})|^2 \right) dx, \tag{96}$$

$$e \stackrel{\text{def}}{=} \frac{d}{ds}\mathscr{E}(s,V)\Big|_{s=0}, \tag{97}$$

and compute the derivative in the two different ways.

4.1 By Moving Domain Derivative

Let

$$e_1 \stackrel{\text{def}}{=} \int_{-\infty}^{+\infty} d\zeta \int_{\Omega} \left(|\zeta|2\mathscr{R}e\{\langle z, \nabla\bar{z}.(-V)\rangle\} + \frac{1}{1+|\zeta|}2\mathscr{R}e\{\langle \nabla z, \nabla(\nabla\bar{z}(-V))\rangle\} \right) dx$$

$$+ \int_{-\infty}^{+\infty} d\zeta \left(\int_{\Gamma} \left\{ |\zeta||z|^2 + \frac{1}{1+|\zeta|}|\nabla z|^2 \right\} \langle V,n\rangle\,d\Gamma(x) \right). \tag{98}$$

By Stokes theorem we get,

$$\int_{-\infty}^{+\infty} d\zeta \int_{\Omega} \frac{1}{1+|\zeta|}2\mathscr{R}e\{\nabla(z).\nabla(\nabla\bar{z}(-V))\}\,dx$$

$$= \int_{-\infty}^{+\infty} d\zeta \int_{\Omega} \frac{1}{1+|\zeta|}2\mathscr{R}e\{\Delta(z),(\nabla\bar{z}.V)\}\,dx \tag{99}$$

$$- \int_{-\infty}^{+\infty} \int_{\Gamma} \frac{1}{1+|\zeta|}2\mathscr{R}e\{\langle \nabla z.n, \nabla\bar{z}.V\rangle\}\,d\Gamma\,dt.$$

As $V = vn$ on Γ, we get for the last term:

$$- \int_{-\infty}^{+\infty} \int_{\Gamma} \frac{1}{1+|\zeta|}2\mathscr{R}e\{\langle \nabla z.n, \nabla\bar{z}.n\rangle\}v\,d\Gamma\,dt, \tag{100}$$

but on Γ we have

$$\langle \nabla z.n, \nabla\bar{z}.n\rangle = |\mathscr{F}.g|^2. \tag{101}$$

Finally, we get

$$
e_1 = \int_{-\infty}^{+\infty} \int_{\Gamma} \{ |\zeta| |z|^2 + \frac{1}{1+|\zeta|} \mathscr{R}e\{\langle \nabla z, \nabla \bar{z} \rangle\} - 2|\mathscr{F}.g|^2 \} v \, d\Gamma \, dt
$$
$$
+ \int_{-\infty}^{+\infty} d\zeta \int_{\Omega} (|\zeta| 2\mathscr{R}e\{\langle z, \nabla \bar{z}.(-V) \rangle\} + \frac{1}{1+|\zeta|} 2\mathscr{R}e\{\Delta z, (\nabla \bar{z}.V)\}) \, dx. \tag{102}
$$

Then

$$
\int_{-\infty}^{+\infty} \int_{\Gamma} \{ |\zeta| |z|^2 + \frac{1}{1+|\zeta|} |\nabla_\Gamma z|^2 \} v
$$
$$
= \int_{-\infty}^{+\infty} \int_{\Gamma} \frac{1}{1+|\zeta|} |\mathscr{F}.g|^2 \, d\Gamma \, dt
$$
$$
- \int_{-\infty}^{+\infty} d\zeta \int_{\Omega} \frac{1}{1+|\zeta|} 2\mathscr{R}e\{ |\zeta|^2 z + \mathscr{F}.H(\nabla \bar{z}.V) \} \, dx \tag{103}
$$
$$
+ \int_{-\infty}^{+\infty} d\zeta \int_{\Omega} |\zeta| 2\mathscr{R}e\{ \langle z, \nabla \bar{z}.V \rangle \} \, dx + e_2.
$$

Hence there exists $M > 0$ such that

$$
\int_{-\infty}^{+\infty} \int_{\Gamma} \{ |\zeta| |z|^2 + \frac{1}{1+|\zeta|} |\nabla_\Gamma z|^2 \} v \leq M \left\{ \|z\|_{L^2(\mathbb{R}, H^1(\Omega))}^2 + \|z\|_{L^2(\mathbb{R}, L^2(\Gamma))}^2 \right\}. \tag{104}
$$

We have $\sqrt{|\zeta|} z \in L^2(\mathbb{R}_\zeta, L^2(\Gamma))$ and $\frac{1}{\sqrt{|\zeta|}} \nabla_\Gamma z \in L^2(\mathbb{R}_\zeta, L^2(\Gamma, \mathbb{R}^N))$. By a density argument, we conclude that

$$
E.n \in H^{1/2}(I, L^2(\Gamma)) \cap L^2(I, H^{1/2}(\Gamma)) \tag{105}
$$
$$
\nabla_\Gamma(E.n) \in H^{-1/2}(I, L^2(\Gamma, \mathbb{R}^N)). \tag{106}
$$

References

1. Cagnol, J., Zolésio, J.P.: Hidden boundary smoothness for some classes of differential equations on submanifolds. In: Optimization Methods in Partial Differential Equations (South Hadley, MA, 1996). Contemp. Math., vol. 209, pp. 59–73. Amer. Math. Soc., Providence (1997)
2. Cagnol, J., Zolésio, J.P.: Hidden shape derivative in the wave equation. In: Systems Modelling and Optimization, Detroit, MI, 1997. Chapman & Hall/CRC Res. Notes Math., vol. 396, pp. 42–52. Chapman & Hall/CRC, Boca Raton (1999)
3. Cagnol, J., Zolésio, J.P.: Shape control in hyperbolic problems. In: Optimal Control of Partial Differential Equations, Chemnitz, 1998. Internat. Ser. Numer. Math., vol. 133, pp. 77–88. Birkhäuser, Basel (1999)
4. Cagnol, J., Zolésio, J.P.: Evolution equation for the oriented distance function. SIAM J. Optim. (2003)
5. Delfour, M.C., Zolésio, J.P.: Curvatures and skeletons in shape optimization. Z. Angew. Math. Mech. 76(3), 198–203 (1996)
6. Delfour, M.C., Zolésio, J.P.: Shapes and Geometries. Analysis, Differential Calculus, and Optimization. Advances in Design and Control. Society for Industrial and Applied Mathematics (SIAM), Philadelphia (2001)

7. Delfour, M.C., Zolésio, J.P.: Pseudodifferential extractor and hidden regularity (to appear)
8. Kawohl, B., Pironneau, O., Tartar, L., Zolésio, J.P.: Optimal Shape Design. Lecture Notes in Mathematics, vol. 1740. Springer, Berlin (2000)
9. Lagnese, J.E.: Exact boundary controllability of Maxwell's equations in a general region. SIAM J. Control and Optimiz. 27(2), 374–388 (1989)
10. Lasiecka, I., Triggiani, R.: Control Theory for Partial Differential Equations: Continuous and Approximation Theories. II. Abstract Hyperbolic-like Systems over a Finite Time Horizon. In: Encyclopedia of Mathematics and its Applications, vol. 75. Cambridge University Press, Cambridge (2000)
11. Moubachir, M., Zolésio, J.P.: Moving Shape Analysis and Control, Application to Fluid Structure Interaction. Monographs and Textbooks in Pure and Applied Mathematics, vol. 277. Chapman and Hall/ CRC, Taylor and Francis Group, Boca Raton (2006)
12. Sokołowski, J., Zolésio, J.P.: Introduction to Shape Optimization. Shape Sensitivity Analysis. Springer Series in Computational Mathematics, vol. 16. Springer, Berlin (1992)
13. Zolésio, J.P.: Material derivative in shape analysis. In: Optimization of Distributed Parameter Structures, Iowa City, Iowa, 1980. NATO Adv. Study Inst. Ser. E: Appl. Sci., vol. 50, pp. 1457–1473. Nijhoff, The Hague (1981)

Author Index